T Cell Receptors

T Cell Receptors

Edited by

JOHN I. BELL
Nuffield Professor of Clinical Medicine
Nuffield Department of Medicine
John Radcliffe Hospital, Oxford

MICHAEL J. OWEN
Imperial Cancer Research Fund
Lincoln's Inn Fields, London

and

ELIZABETH SIMPSON
Transplantation Biology Group
MRC Clinical Sciences Centre
Royal Postgraduate Medical School
Hammersmith Hospital, London

Oxford New York Tokyo
OXFORD UNIVERSITY PRESS
1995

Oxford University Press, Walton Street, Oxford OX2 6DP
Oxford New York
Athens Auckland Bangkok Bombay
Calcutta Cape Town Dar es Salaam Delhi
Florence Hong Kong Istanbul Karachi
Kuala Lumpur Madras Madrid Melbourne
Mexico City Nairobi Paris Singapore
Taipei Tokyo Toronto
and associated companies in
Berlin Ibadan

Oxford is a trade mark of Oxford University Press

Published in the United States
by Oxford University Press Inc., New York

A catalogue record for this book is available from the British Library

Library of Congress Cataloging in Publication Data
T cell receptors / edited by John I. Bell, Michael J. Owen, and Elizabeth Simpson
Includes bibliographical references and index.
1. T cells–Receptors. 2. Immunogenetics. I. Bell, John I.
II. Owen, M. J. III. Simpson, Elizabeth, Dr.
[DNLM: 1. Receptors, Antigen, T-Cell. 2. Signal Transduction.
QW 573 T249 1995]
QR185.8.T2T336 1995 616.07'9–dc20 95–10484

ISBN 0 19 262418 0 (Hbk)
ISBN 0 19 262419 9 (Pbk)

Typeset by Footnote Graphics, Warminster, Wiltshire
Printed in Great Britain by
Bookcraft (Bath) Ltd, Midsomer Norton, Avon

Foreword

MARK M. DAVIS

In reading over the contents of this excellent volume, it is hard to believe that a little over ten years ago the area of T cell recognition by its antigen receptor was such a source of confusion and anguish that it had been described by Hugh McDevitt as 'an embarassment to the field of immunology.' Paradigms shot up, blossomed under the admiring smiles of the multitude and then withered away for lack of confirming data. Those were very frustrating times for T cell biologists, but as indicated in works described here, there has been such an explosion of activity since those first reports that this must be at least partial compensation for those difficulties.

In this period of time, as recorded in these very complete chapters, we have acquired extensive information on the TCR gene repertoire of both helper and cytotoxic T cells and found a completely new type of TCR bearing cell ($\gamma\delta$ T cells) whose function in the immune system remains a mystery. Differentiation and selection in the thymus is also understood to a much greater degree; the whole reason for this organ and the cells that pass through is to develop and test a TCR repertoire which represents a best 'fit' for an individual's MHC molecules. We have also progressed to an increasingly sophisticated biochemical understanding of TCR-mediated recognition with direct demonstration and quantitation of TCR binding to peptide–MHC and superantigen–MHC complexes as well as the wealth of information on MHC structures. There has also been impressive progress in understanding the intracellular signalling process in T cells which starts with TCR engagement. These are but a few of the many achievements of these past few years that provide an excellent foundation for work in the next decade and beyond on this very interesting set of molecules. As the authors put forth very clearly in their respective chapters, there are still many questions to be answered and more excitement to come. The authors and the editors are to be congratulated in compiling such a comprehensive treatise on this subject.

Contents

3 The function of αβ T cells and the role of the co-receptor molecules, CD4 and CD8 46

ROSE ZAMOYSKA AND PAUL TRAVERS

4 γδ T cell specificity and function 70

ADRIAN C. HAYDAY

Part III T cell genes

Part IV T cell proteins

Contributors

Hans Acha-Orbea Ludwig Institute for Cancer Research, Lausanne Branch, Ch. des Boveresses 155, CH-1066 Epalinges, Switzerland.

Bernhard Arden Paul-Ehrlich-Institute, Federal Agency for Sera and Vaccines, Paul-Ehrlich-Str. 51–59, 63225 Langen, Germany.

John I. Bell University of Oxford, Nuffield Department of Clinical Medicine, John Radcliffe Hospital, Headington, Oxford OX3 9DU, UK.

Christophe Benoist Laboratoire de Génétique Moléculaire des Eucaryotes du CNRS et Unité 184, de Biologie Moléculaire de L'INSERM, Institut de Chimie Biologique, 11, rue Humann, 67085 Strasbourg Cédex, France.

Stephen P. Clark AmGen Center, Thousand Oaks, California 91320, USA.

P. Concannon Virginia Mason Research Center, 1000 Seneca Street, Seattle, WA 98101, USA.

Doreen A. Cantrell Lymphocyte Activation Laboratory, Imperial Cancer Research Fund, London WC2A 3PX, UK.

I. Nicholas Crispe Section of Immunobiology, Yale University School of Medicine, New Haven, CT 06510, USA.

M. Davis Howard Hughes Medical Institute, Stanford University, Beckman Building, Stanford, CA 94305, USA.

Klaus Eichmann Max-Planck Institute for Immunobiology, Stubeweg 51, W7800, Freiburg, Germany.

Mark Exley ImmuLogic Pharmaceutical Company, 610, Linoln Street, Waltham, MA 02154, USA.

John D. Fraser Department of Molecular Medicine, School of Medicine, The University of Auckland, Private Bag, Auckland, New Zealand.

Nicholas R. J. Gascoigne Department of Immunology, The Scripps Research Institute, 10666 North Torrey Pines Road, La Jolla, CA 92037, USA.

Susan Gilfillan Laboratoire de Génétique Moléculaire des Eucaryotes du CNRS et Unité 184, de Biologie Moléculaire de L'INSERM, Institut de Chimie Biologique, 11, rue Humann, 67085 Strasbourg Cédex, France.

Craig Hall Division of Immunology, Beth Israel Hospital, Harvard Medical School, Boston, MA 02115, USA.

Adrian C. Hayday Department of Biology and Section of Immunobiology, Yale University, New Haven, CT 06511, USA.

Katherine L. Hilyard Department of Biology, Roche Products Limited, Roche Research Centre, 40, Broadwater Road, Welwyn Garden City, Hertfordshire AL7 3AY, UK.

Georg Holländer Division of Paediatric Oncology, Dana-Farber Cancer Institute, Harvard Medical School, Boston, MA 02115, USA.

Manley Huang GenPharm International, 297, North Bernardo Avenue, Mountain View, CA 94043, USA.

Keith R. Hudson Department of Molecular Medicine, School of Medicine, The University of Auckland, Private Bag, Auckland, New Zealand.

Ottmar Janssen Division of Tumor Immunology, Dana-Farber Cancer Institute and Department of Pathology, Harvard Medical School, Boston, MA 02115, USA.

Dieter Kabelitz Paul-Ehrlich-Institute, Federal Agency for Sera and Vaccines, Paul-Ehrlich-Str. 51–59, 63225 Langen, Germany.

Christian Levelt Max-Planck Institute for Immunobiology, Stubeweg 51, W7800 Freiburg, Germany.

Tak W. Mak The Ontario Cancer Institute, 500 Sherbourne Street, Toronto, Ontario, Canada.

Bernard Malissen Centre d'Immunologie, INSERM-CNRS de Marseille-Luminy, case 906, 13288 Marseille Cedex 9, France.

Marie Malissen Centre d'Immunologie, INSERM-CNRS de Marseille-Luminy, case 906, 13288 Marseille Cedex 9, France.

Diane Mathis Laboratoire de Génétique Moléculaire des Eucaryotes du CNRS et Unité 184, de Biologie Moléculaire de L'INSERM, Institut de Chimie Biologique, 11, rue Humann, 67085 Strasbourg Cédex, France.

Andrew L. Mellor National Institute for Medical Research, The Ridgeway, Mill Hill, London NW7 1AA, UK.

Jacques F. A. P. Miller The Walter and Eliza Hall, Institute of Medical Research, Post Office Royal Melbourne Hospital, Victoria 3050, Australia.

Paul A. H. Moss University of Oxford, Institute of Molecular Medicine, John Radcliffe Hospital, Headington, Oxford OX3 9DU, UK.

Pamela S. Ohashi Ontario Cancer Institute, Department of Medical Biophysics, University of Toronto, 500 Sherbourne Street, Toronto, Ontario M4X IK9, Canada.

Michael J. Owen Imperial Cancer Research Fund, London WC2A 3PX, UK.

Manolo Izquierdo Pastor Lymphocyte Activation Laboratory, Imperial Cancer Research Fund, London WC2A 3PX, UK.

Hidde Ploegh Department of Biology, Center for Cancer Research, Massachusetts Institute of Technology, 77, Massachusetts Avenue, Cambridge, MA 02139–4307, USA.

K. V. S. Prasad Division of Tumor Immunology, Dana-Farber Cancer Institute and Department of Pathology, Harvard Medical School, Boston, MA 02115, USA.

Haydn M. Prosser Howard Hughes Medical Institute, Center for Cancer Research and Department of Biology, Massachusetts Institute of Technology, Cambridge, Massachusetts 02139, USA.

Monika Raab Division of Tumor Immunology, Dana-Farber Cancer Institute and Department of Pathology, Harvard Medical School, Boston, MA 02115, USA.

Karen Reif Lymphocyte Activation Laboratory, Imperial Cancer Research Fund, London WC2A 3PX, UK.

M. A. Robinson Laboratory of Immunogenetics, Twinbrook II Facility, National Institute of Allergy and Infectious Diseases, 12441, Parklawn Drive, Rockville, MD 20852, USA.

Christopher E. Rudd Division of Tumor Immunology, Dana-Farber Cancer Institute and Department of Pathology, Harvard Medical School, Boston, MA 02115, USA.

David G. Schatz Section of Immunobiology, Yale University School of Medicine, New Haven, CT 06510, USA.

Eric Sebza Ontario Cancer Institute, Department of Medical Biophysics, University of Toronto, 500 Sherbourne Street, Toronto, Ontario M4X IK9, Canada.

Jian She Division of Immunology, Beth Israel Hospital, Harvard Medical School, Boston, MA 02115, USA.

Antonio da Silva Division of Tumor Immunology, Dana-Farber Cancer Institute and Department of Pathology, Harvard Medical School, Boston, MA 02115, USA.

Elizabeth Simpson Transplantation Biology Group, MRC Clinical Sciences Centre, Royal Postgraduate Medical School, Hammersmith Hospital, London, UK.

Steve Simpson Division of Immunology, Beth Israel Hospital, Harvard Medical School, Boston, MA 02115, USA.

Jack L. Strominger Department of Biochemistry and Molecular Biology, Harvard University, 7, Divinity Avenue, Cambridge, Massachusetts 02138, USA.

Janice C. Telfer Division of Tumor Immunology, Dana-Farber Cancer Institute and Department of Pathology, Harvard Medical School, Boston, MA 02115, USA.

Cox Terhorst Division of Immunology, Beth Israel Hospital, Harvard Medical School, Boston, MA 02115, USA.

Susumu Tonegawa Howard Hughes Medical Institute, Center for Cancer Research and Department of Biology, Massachusetts Institute of Technology, Cambridge, Massachusetts 02139, USA.

Paul Travers Department of Crystallography, Birkbeck College, Mallet Street, London WC1 E7HX, UK.

BaoPing Wang Division of Immunology, Beth Israel Hospital, Harvard Medical School, Boston, MA 02115, USA.

Tom Wileman Division of Immunology, Beth Israel Hospital, Harvard Medical School, Boston, MA 02115, USA.

Alan F. Williams MRC Cellular Immunology Unit, Sir William Dunn School of Pathology, University of Oxford, South Parks Road, Oxford OX1 3RE, UK.

Melissa Woodrow Lymphocyte Activation Laboratory, Imperial Cancer Research Fund, London WC2A 3PX, UK.

Rose Zamoyska Department of Molecular Immunology, NIMR, The Ridgeway, Mill Hill, London NW7 1AA, UK.

1 Introduction: historical overview

JACQUES F. A. P. MILLER

To my mind one of the most important advances in immunology that we may look forward to in the next 5 or 10 years is an explanation of the mechanisms of the 'cellular' immunities.

Medawar (1958)

The outstanding feature of the development of immunology in the last 10 years has been the recognition of the function of the lymphocyte and of the importance of the thymus in the immune process.

Burnet (1966)

In the last decade, hardly an issue of a journal reporting immunological investigations has appeared without some reference to T lymphocytes. Nowadays, as a result of the widespread coverage in the lay press of the disease, acquired immunodeficiency syndrome or AIDS, the term T cell has become a household name. In marked contrast, before 1960, neither the thymus nor its cells were considered to play any role in immunity. In the 1950s, for example, some immunologists concluded from their work that they had obtained 'evidence that the thymus gland does *not* participate in the control of the immune response' (MacLean *et al.* 1957).

The immunological competence of the small lymphocyte had been established unequivocally in the late 50s and early 60s by Gowans and his collaborators (Gowans *et al.* 1962). Yet, even though the thymus was known to be a lymphocyte-producing organ, immunologists were most reluctant to consider an immunological function for the thymus or its lymphocytes. Medawar (1963) even suggested that 'we shall come to regard the presence of lymphocytes in the thymus as an evolutionary accident of no very great significance'. What then was responsible for turning the tide?

1 Early work on thymectomy

A totally unforeseen consequence of an experiment designed to understand how a particular leukaemogenic virus induced leukaemia in low-leukaemic strain mice was that neonatally thymectomized mice died prematurely from causes unrelated to leukaemia induction. This suggested 'that the thymus at birth may be essential to life' (Miller 1961*a*). Further investigations showed clearly that mice thymectomized at one day of age, but not later than a few days, were highly susceptible to infections, had a marked deficiency of lymphocytes in the circulation and in lymphoid tissues, and were unable to reject foreign skin grafts. These results led to the hypothesis that 'during embryogenesis the thymus would produce the originators of immunologically competent cells many of which would have migrated to other sites at about the time of birth. This would suggest that lymphocytes leaving the thymus are specially selected cells' (Miller 1961*b*). In adult mice, thymectomy had for long been known *not* to have any untoward effects, and it was partly this fact that had led many to discredit the thymus as having an immunological role. If, however, adult thymectomized mice were exposed to total body irradiation, the recovery of lymphoid and immune functions was shown to be thymus-dependent (Miller 1962*a*). As expected, implanting thymus tissue into neonatally thymectomized or adult thymectomized and irradiated mice allowed a normal immune system to develop. When the thymus tissue was obtained from a foreign strain, the neonatally thymectomized recipients became specifically immunologically tolerant of the histocompatibility antigens of the donor. This led to the suggestion that 'when one is inducing a state of immunological tolerance in a newly born animal', for example by the classical technique of injecting allogeneic bone marrow cells at birth, 'one is in effect performing a selective or immunological thymectomy' (Miller 1962*b*). In other words, lymphocytes developing in the thymus in the presence of foreign cells would be deleted, implying that the thymus should be the site where self tolerance is imposed.

Initially, these unorthodox conclusions raised eyebrows, particularly at various meetings where they were presented and disputed (e.g. Harris 1962). Nevertheless they were soon confirmed by groups working independently (Arnason *et al.* 1962; Martinez *et al.* 1962; Parrott 1962) and fully ratified when athymic *nunu* mice became available (Rygaard 1973). These mice were immunodeficient but became competent following thymus implantation. Moreover, allogeneic thymus grafts established tolerance of thymus-donor tissues (Owen *et al.* 1986). We of course now know in detail the pathway of lymphocyte differentiation in the thymus and the various positive and negative selection procedures which indeed show that the thymus is the seat where self tolerance is achieved and that the mature T lymphocytes are specially selected cells. In the present book, we are given lucid accounts of some of the details relating to thymus cell differentiation, positive and negative selection (Chapters 2, 7, and 10), repertoire selection (Chapters 5 and 6), and tolerance induction (Chapters 10 and 11).

2 Communications between lymphocytes

In the late 50s and early 60s, there was no reason to believe that mammalian small lymphocytes could belong to entirely distinct subsets. In birds, on the other hand, it was known that the bursa of Fabricius, an organ analogous to the thymus and situated near the cloaca, was involved in antibody production

(Glick *et al.* 1956). Szenberg and Warner (1962) were the first to show a division of labour among avian lymphocytes, early bursectomy being associated with defects in antibody formation and early thymectomy with defects in cellular immune responses. Since mice do not have a bursa and since neonatal thymectomy in that species prevented both cellular immune responses and normal antibody production (Miller 1962*b*, 1963) it was widely held that the mammalian thymus fulfilled the functions of both the avian thymus and bursa. Nevertheless an explanation had to be found for the fact that neonatally thymectomized mice had a deficiency of lymphocytes limited to areas of lymph nodes and spleen known to be associated with changes induced by cellular immune responses and not in areas where antibody-producing cells appeared (Parrott *et al.* 1966). A clue came from the experiments of Claman *et al.* (1966): they showed that irradiated mice receiving a mixed population of marrow and thymus cells produced far more antibody than when given either cell source alone. These investigators were, however, unable to determine the origin of the antibody-forming cells in their model for lack of genetic markers.

Davies and his collaborators (1967) attempted to unravel the situation by using adult thymectomized irradiated mice given bone marrow and thymus grafts from donors which differed slightly immunogenetically. When the spleens from these mice were transferred soon after challenge with sheep erythrocytes into irradiated recipients pre-sensitized against either the thymus or the marrow donor, those capable of rejecting cells of thymus-donor type were able to produce antibody. Those immunized against marrow donors produced much less. The transfer experiments were, however, performed 30 days after irradiation and thymus grafting. At this time it was already known that the lymphoid cell population of the thymus graft had been entirely replaced by cells of marrow origin (Dukor *et al.* 1965). Thus haemolysins detected in the irradiated recipients pre-sensitized against the thymus donor might well have been produced by marrow-derived cells that had repopulated and migrated from the thymus graft. Hence the cells producing antibody could have had the immunogenetic characteristics of the marrow donor and yet be thymus-derived. In the words of Davies himself, 'it may be that thymus-derived cells can produce antibody, but only in the presence of cells of bone marrow origin. Equally cells of bone marrow origin may be the cells whose immunological potential is enhanced by association with cells of thymic origin. These are not problems which the present analysis can resolve' (Davies *et al.* 1968).

In independent studies designed to determine the ability of various cell types to restore immune functions to thymectomized mice, the introduction of genetically marked cells into neonatally thymectomized or thymectomized irradiated hosts clearly established beyond doubt and for the first time that antibody-forming cell precursors (subsequently known as B cells) were derived from bone marrow, and that thymus-derived cells (now called T cells) were essential to allow B cells to respond to antigen by producing antibody (Miller and Mitchell 1967, 1968; Mitchell and Miller 1968).

The existence of two distinct lymphocyte subsets was first regarded with surprise and skepticism. Gowans (1969), who had clearly shown that the recirculating small lymphocyte could initiate *both* cellular and humoral immune responses (Gowans *et al.* 1962), stated: 'Had it not been for Dr Miller's experiments I would have assumed that a single variety of small lymphocyte was involved in each of our experiments'. Good (1969) was 'concerned at separating thymus-derived from marrow-derived cells' since the former 'are in fact, marrow-derived cells'. Nevertheless, as we now know, the existence of T and B cells was not only confirmed but led to a re-investigation of numerous immunological phenomena including the carrier effect (Mitchison 1971; Rajewsky 1971), memory (Miller and Sprent 1971), tolerance (Miller and Mitchell 1970; Basten 1989), autoimmunity (Sinha *et al.* 1990), and genetically determined unresponsive states (Mitchell *et al.* 1972). T cells were clearly responsible for the 'cellular' immunities, the mechanism of which Medawar (1958) wished to elucidate, and T cells were themselves soon subdivided into subsets based on function, cell surface markers, and secreted products or interleukins (reviewed by Sprent and Webb 1987).

3 Antigen recognition

Burnet's clonal selection theory (Burnet 1957, 1959*a*) clearly predicted heterogeneity of lymphocytes in terms of their antigen recognition potential. Evidence in support of the theory was provided by Nossal and Lederberg (1958) when they showed that a given antibody-forming cell from multiply immunized animals never produced more than one antibody specificity. Nevertheless antigen could have committed a multipotential antibody-forming cell precursor to a single specificity, as had been suggested as early as 1900 by Paul Ehrlich. That B lymphocytes were already committed to a single specificity was formally demonstrated by Naor and Sulitzeanu (1967) and Ada and Byrt (1969) who found that a small proportion of *naïve* lymphocytes could specifically bind labelled antigen and that this binding could be blocked by anti-immunoglobulin antibody. Yet T cells could never be shown to bind antigen even though their function could be specifically inactivated by radioactively labelled antigen (Basten *et al.* 1970). Concern with the notion of single cellular commitment for T cells had previously been raised by Simonsen (1967) as a result of his studies of graft-versus-host (GVH) reactions in chickens: cells reacting to a single histocompatibility antigen were present in a random sample of as few as 50 donor lymphocytes. 'One *had* to wonder if there would be enough monospecific clones in the immune system to cope with more than the minute portion of the antigenic universe that could be assumed to be filled by a single H-locus' (Simonsen 1990). This assumption was of course erroneous but resolution of the dilemma posed by the GVH studies had to await elucidation of the structure and function of the products of the major histocompatibility complex (MHC).

The discovery of MHC restriction of T cell reactivities by Zinkernagel and Doherty (1974) was soon confirmed by other investigators studying different T cell responses (Bevan 1975; Gordon *et al.* 1975; Miller *et al.* 1975). MHC restriction together with the finding that immune responsiveness (Ir) genes were MHC linked (Benacerraf and Katz 1975; McDevitt *et al.* 1976) provided a satisfactory explanation of how T cells recognized antigen: it had first to be processed and presented as peptides by 'professional' antigen-presenting cells (APC). The high frequency of alloreactive cells was thus no longer a mystery: as Simonsen neatly puts it, 'for the T-cell population of an individual, MHC is not just a portion, but rather the (antigenic) universe itself' (Simonsen 1990). Thanks to the work of several investigators, including Bjorkman *et al.* (1987), we now have an understanding of the molecular structure of MHC gene products and of the mechanisms involved in MHC restriction and antigen processing, as given in detail in Chapters 2, 5, 11, 18–20 of this book.

In the 70s, great controversy raged over the nature of the receptor on T cells (TCR) that enabled them to recognize antigen. Some considered the TCR to be some form of immunoglobulin and called it IgT (Marchalonis 1977), but others strongly disputed this view and produced experimental evidence against it (Kemp *et al.* 1982; Miller *et al.* 1983). The use of monoclonal antibodies and of T cell clones gave some insight into the true nature of the TCR (Haskins *et al.* 1983). The breakthrough came when Davis and his collaborators used the refined technology of subtractive cDNA/mRNA hybridization to B and T lymphocytes to isolate a number of cDNA clones encoding the α, β (Hedrick *et al.* 1984), and δ chains (Chien *et al.* 1987) of the murine TCR. The sequences of these cDNAs and the immunoglobulin-like nature of the rearranging genes which gave rise to them were the crucial factors in settling the very contentious and long-standing issue of what types of molecules governed T cell recognition. In independent investigations, the genes encoding the human TCR (Yanagi *et al.* 1984) and the γ chain of the murine TCR were also cloned (Saito *et al.* 1984).

Identification and characterization of the different genetic loci coding for TCR enabled direct studies to be made on T cell differentiation and selection and allowed T cells to be separated into two subsets—a major one utilizing the αβ TCR and a minor one the γδ TCR. The lineage relationships and the functions of both of these are discussed in this book by Crispe and Schatz, Zamoyska and Travers, and by Hayday. Furthermore, Chapters 12, 14, 15, and 16 give fascinating accounts of the genes involved in coding for the TCR and of the enzymes responsible for generating recombination events.

4 T cell activation and signalling

Several years ago, Bretscher and Cohn (1970) and Lafferty and Woolnough (1977) suggested that T cells required two signals in order to be activated: the first signal was antigen and the second some costimulatory factor. Much evidence, primarily from transplantation studies, supported the view that professional APC were required for the activation of T cells *in vivo*. For example, islet tissue could be transplanted across a major histocompatibility barrier without rejection provided passenger leucocytes (professional APC) were removed by culturing the islets *in vitro* under high oxygen tension. If antigen was subsequently supplied by injecting leucocytes expressing graft antigens, the islet cells were rejected. Hence, before leucocyte priming the islet tissue expressed antigen but could not deliver the appropriate costimulator signals to initiate a T cell response. The second signal must thus have been provided by the injected leucocytes (Lafferty *et al.* 1983). This set the scene for the study of T cell signalling.

The critical event that allows T cells to respond is signalling through the binding of specific MHC–peptide complexes by the heterodimeric TCR–CD3 complex. This, of course, is responsible for selectivity of the response. Ligand binding leads to TCR clustering and activates receptor-associated protein kinases, as discussed by Cantrell *et al.*, Telfer *et al.*, and Terhorst *et al.* in this book (Chapters 8, 9, and 17). TCR aggregation *per se* may, however, not be sufficient to allow signalling: for this to occur ligand binding may, in addition to inducing TCR aggregation, have to produce a conformational change in the TCR (Janeway *et al.* 1989). The CD4 and CD8 co-receptors, as well as numerous other T cell surface molecules, such as CD45, also contribute to transmembrane signalling (Janeway 1992).

The affinity of the interaction between TCR and MHC–peptides is remarkably low ($K_D = 10^{-4}$–10^{-5} M) (Matsui *et al.* 1991; Weber *et al.* 1992). This suggests that T cells can recognize a particular peptide–MHC complex only if the cells to be surveyed have the correct ligands for T cell adhesion molecules. Furthermore, in order for T cells to proliferate and differentiate to effector cells, they must produce interleukin-2 (IL-2). Recent studies have shown this to occur when the B7/BB1 molecules on the APC interact with the T cell surface receptors, CD28 (Linsley *et al.* 1991*a*) and CTLA-4 (Linsley *et al.* 1991*b*). Lack of costimulation is associated with inadequate IL-2 production and a block both in IL-2 gene transcription and in responsiveness to IL-4 (Schwartz 1992).

5 Immunoregulation, tolerance, and autoimmunity

An inverse relationship has often been noted between the humoral and cell-mediated arms of the immune response to foreign proteins or infectious organisms (Parish 1972). This is likely to reflect a dynamic state in which different T cell subsets with distinct lymphokine profiles are mutually antagonistic. Evidence for this notion is supported by the work of Mosmann and Coffman (1989) who identified two subsets of helper T cells, Th1 secreting interleukin-2 (IL-2) and interferon-γ, and Th2 secreting IL-4, IL-5, and

IL-10. Since the Th2 lymphokines, IL-4 and IL-10, inhibit the induction of Th1 responses, a powerful Th2 response may well prevent cell-mediated immunity. The mechanisms which control the activation of these Th subsets are unclear, although it is likely that MHC class II genes, types of APC involved, and TCR ligand density all play a decisive role. Differential lymphokine gene expression is undoubtedly a major factor in the regulation of T cell-dependent immune responses and is likely to be the focus of attention during the next decade.

Burnet's clonal selection theory predicted that 'forbidden clones' should be deleted for self-tolerance to be established (Burnet 1959*b*). Working with B lymphocytes, Nossal (1983) extended the concept to include other mechanisms such as clonal abortion and clonal anergy. Proof that a clonal deletional mechanism exists for T cells was, however, obtained some 30 years after Burnet's prediction when techniques became available to detect individually specific T lymphocytes, as discussed by Mellor and Acha-Orbea in Chapters 10 and 11 of this book. The dominant toleragenic mechanism for T cells indeed appears to be deletion of clones of self-reactive cells occurring during intrathymic development (Kappler *et al.* 1987; Kisielow *et al.* 1988; MacDonald *et al.* 1988). In the case of extrathymic self-molecules which are potential targets of autoimmune T cell aggression, it is not clear whether tolerance is imposed by peripheral deletion or silencing (anergy or suppression), or whether such T cells normally ignore these autoantigens (Miller 1992). Cells lacking B7/BB1 or correct ligands for T cell adhesion molecules are likely to be ignored by mature peripheral T cells, unless the latter are activated by cross-reactive antigens presented on professional APC (Miller and Heath 1993). On the other hand, evidence is mounting for the existence of some T cell-dependent suppression of potentially autoaggressive T cells (Fowell and Mason 1993) or of T cells involved in transplantation immunity (Qin *et al.* 1993). The idea of suppressor T cells was first suggested by Gershon (1974), but the failure to isolate a distinct subset of suppressor T cells led many to question their existence (e.g. Möller 1988). Nevertheless, the well-documented evidence for two types of helper T cells, Th1 and Th2 with their distinct lymphokine profiles (referred to above), has resurrected the concept and strongly suggests that T cell-dependent immunoregulation of lymphocyte responses is a reality that needs further exploration at both the cellular and molecular levels.

The selectivity of peptide binding to the groove of MHC molecules of defined haplotypes and the polymorphism of the antigen processing machinery provide us with a molecular basis for the pathogenesis of autoimmune disease (Fairchild and Wraith 1992). Moreover, genetic linkage studies have implicated TCR variable region genes in the development of such diseases (Moss *et al.* 1992). An increased knowledge of TCR genes and their ligands, as detailed in this book, may hopefully allow the design of new and more specific therapeutic strategies aimed at preventing or reversing autoimmune processes.

References

Ada, G. L. and Byrt, P. (1969). Specific inactivation of antigen-reactive cells with ^{125}I-labelled antigen. *Nature*, **222**, 1291–2.

Arnason, B. G., Jankovic, B. D., and Waksman, B. H. (1962). Effect of thymectomy on 'delayed' hypersensitive reactions. *Nature*, **194**, 99–100.

Basten, A. (1989). Self-tolerance: the key to autoimmunity. *Proc. Roy. Soc. London Ser. B*, **238**, 1–23.

Basten, A., Miller, J. F. A. P., Warner, N. L., and Pye, J. (1970). Specific inactivation of thymus-derived (T) and non-thymus-derived (B) lymphocytes by ^{125}I-labelled antigen. *Nature New Biol.*, **231**, 104–6.

Benacerraf, B. and Katz, D. H. (1975). The histocompatibility-linked immune response genes. *Adv. Cancer Res.*, **21**, 121–73.

Bevan, M. J. (1975). The major histocompatibility complex determines susceptibility to cytotoxic T cells directed against minor histocompatibility antigens. *J. Exp. Med.*, **142**, 1349–64.

Bjorkman, P. J., Saper, M. A., Samraoui, B., Bennett, W. S., Strominger, J. L., and Wiley, D. C. (1987). Structure of the human class I histocompatibility antigen, HLA-2. *Nature*, **329**, 506–11.

Bretscher, P. A. and Cohn, M. (1970). A theory of self-nonself discrimination: paralysis and induction involve the recognition of one and two determinants on an antigen, respectively. *Science*, **169**, 1042–9.

Burnet, F. M. (1957). A modification of Jerne's theory of antibody production using the concept of clonal selection. *Austr. J. Sci.*, **20**, 67–9.

Burnet, F. M. (1959*a*). *The clonal selection theory of acquired immunity*. Cambridge University Press.

Burnet, F. M. (1959*b*). Auto-immune disease. II. Pathology of the immune response. *Br. Med. J.*, **2**, 720–5.

Burnet, F. M. (1966). Chairman's opening remarks. In *The thymus: experimental and clinical studies* (ed. G. E. W. Wolstenholme and R. Porter), pp. 1–2. *Ciba Found. Symp.*, Churchill, London.

Chien, Y., Iwashima, M., Kaplan, K., Elliott, J., and Davis, M. M. (1987). A new T cell receptor gene located within the α locus and expressed early in T cell differentiation. *Nature*, **327**, 677–82.

Claman, H. N., Chaperon, E. A., and Triplett, R. F. (1966). Thymus-marrow cell combinations—synergism in antibody production. *Proc. Soc. Exp. Biol. Med.*, **122**, 1167–71.

Davies, A. J. S., Leuchars, E., Wallis, V., Marchant, R., and Elliott, E. V. (1967). The failure of thymus-derived cells to produce antibody. *Transplantation*, **5**, 222–31.

Davies, A. J. S., Leuchars, E., Wallis, V., Sinclair, N. R. St. C., and Elliott, E. V. (1968). The selective transfer test. An analysis of the primary response to sheep red cells. In *Advance in transplantation* (ed. J. Dausset, J. Hamburger, and G. Mathé), pp. 97–100. Munksgaard, Copenhagen.

Dukor, P., Miller, J. F. A. P., House, W., and Allman, V. (1965). Regeneration of thymus grafts. I. Histological and cytological aspects. *Transplantation*, **3**, 639–68.

Ehrlich, P. (1900). The Croonian lecture. On immunity with special reference to cell life. *Proc. Roy. Soc. London Ser. B*, **66**, 424–48.

Fairchild, P. J. and Wraith, D. C. (1992). Peptide–MHC interaction in autoimmunity. *Curr. Opin. Immunol.*, **4**, 748–53.

Fowell, D. and Mason, D. (1993). Evidence that the T cell repertoire of normal rats contains cells with the potential to cause diabetes. Characterization of the $CD4^+$ T cell subset that inhibits this autoimmune potential. *J. Exp. Med.*, **177**, 627–36.
Gershon, R. K. (1974). T-cell control of antibody production. *Contemp. Top. Immunobiol.*, **3**, 1–40.
Glick, B., Chang, T. S., and Japp, R. G. (1956). The bursa of Fabricius and antibody production. *Poultry Sci.*, **35**, 224–5.
Good, R. A. (1969). Discussion. In *Immunological tolerance. A reassessment of mechanisms of the immune response* (ed. M. Landy and W. Braun), p. 136. Academic Press, New York.
Gordon, R. D., Simpson, E., and Samelson, L. E. (1975). *In vitro* cell-mediated immune responses to the male-specific (H-Y) antigen in mice. *J. Exp. Med.*, **142**, 1108–20.
Gowans, J. L. (1969). Discussion. In *Immunological tolerance. A reassessment of mechanisms of the immune response* (ed. M. Landy and W. Braun), p. 169. Academic Press, New York.
Gowans, J. L., McGregor, D. D., Cowen, D. M., and Ford, C. E. (1962). Initiation of immune responses by small lymphocytes. *Nature*, **196**, 651–5.
Harris, R. J. C. (1962). Discussion after Miller, J. F. A. P. Role of the thymus in virus-induced leukaemia. In *Tumour viruses of murine origin* (ed. G. E. W. Wolstenholme and M. O'Connor), p. 283. *Ciba Found. Symp.*, Churchill, London.
Haskins, K., Kubo, R., White, J., Pigeon, M., Kappler, J., and Marrack, P. (1983). The major histocompatibility complex restricted antigen receptor on T cells. I. Isolation with a monoclonal antibody. *J. Exp. Med.*, **157**, 1149–69.
Hedrick, S. M., Cohen, D. I., Nielsen, E. A., and Davis, M. M. (1984). Isolation of cDNA clones encoding T cell-specific membrane-associated proteins. *Nature*, **308**, 149–53.
Janeway, C. A. Jr. (1992). The T cell receptor as a multicomponent signalling machine: CD4/CD8 coreceptors and CD45 in T cell activation. *Annu. Rev. Immunol.*, **10**, 645–74.
Janeway, C. A. Jr., Dianzani, U., Portoles, P., Rath, S., Reich, E.-P., Rojo, J., *et al.* (1989). Cross-linking and conformational change in T-cell receptors: role in activation and in repertoire selection. *Cold Spring Harbor Symp. Quant. Biol.*, **54**, 657–66.
Kappler, J. W., Roehm, M., and Marrack, P. (1987). T cell tolerance by clonal elimination in the thymus. *Cell*, **49**, 273–80.
Kemp, D. J., Adams, J. M., Mottram, P. L., Thomas, W. R., Walker, I. D., and Miller, J. F. A. P. (1982). A search for messenger RNA molecules bearing immunoglobulin V_H nucleotide sequences in T cells. *J. Exp. Med.*, **156**, 1848–53.
Kisielow, P., Blüthmann, H., Staerz, U. D., Steinmetz, M., and von Boehmer, H. (1988). Tolerance in T-cell-receptor transgenic mice involves deletion of nonmature $CD4^+8^+$ thymocytes. *Nature*, **333**, 742–6.
Lafferty, K. J. and Woolnough, J. (1977). The origin and mechanism of the allograft reaction. *Immunol. Rev.*, **35**, 231–62.
Lafferty, K. J., Prowse, S. J., and Simeonovic, C. J. (1983). Immunobiology of tissue transplantation: a return to the passenger leukocytc concept. *Annu. Rev. Immunol.*, **1**, 143–73.
Linsley, P. S., Brady, W., Grosmarie, L., Aruffo, A., Damie, N. K., and Ledbetter, J. A. (1991*a*). Binding of the B cell activation antigen B7 to CD28 costimulates T cell proliferation and interleukin-2 mRNA accumulation. *J. Exp. Med.*, **173**, 721–30.

Linsley, P. S., Brady, W., Urnes, M., Grosmarie, L., Damie, N. K., and Ledbetter, J. A. (1991*b*). CTLA-4 is a second receptor for the B cell activation antigen B7. *J. Exp. Med.*, **174**, 561–70.
MacDonald, H. R., Sneider, R., Lees, R. K., Howe, R. C., Acha-Orbea, H., Fetenstein, H., *et al.* (1988). T cell receptor V use predicts reactivity and tolerance to Mls^a-encoded antigens. *Nature*, **332**, 40–5.
MacLean, L. D., Zak, S. J., Varco, R. L., and Good, R. A. (1957). The role of the thymus in antibody production: an experimental study of the immune response in thymectomized rabbits. *Transplant Bull.*, **41**, 21–2.
Marchalonis, J. J. (1977). The lymphocyte plasma membrane: isolation and properties of biologically relevant proteins. In *The lymphocyte structure and function* (ed. J. J. Marchalonis), pp. 373–432. Dekker, New York.
Martinez, C., Kersey, J., Papermaster, B. W., and Good, R. A. (1962). Skin homograft survival in thymectomized mice. *Proc. Soc. Exp. Biol. Med.*, **109**, 193–6.
Matsui, K., Boniface, J. J., Reay, P. A., Schild, H., Fazekas de St. Groth, B., and Davis, M. M. (1991). Low affinity interaction of peptide–MHC complexes with T cell receptors. *Science*, **254**, 1788–91.
McDevitt, H. O., Delovitch, T. L., Press, J. L., and Murphy, D. B. (1976). Genetic and functional analysis of the Ia antigens: their possible role in regulating the immune response. *Transplant. Rev.*, **30**, 197–235.
Medawar, P. B. (1958). The Croonian lecture. The homograft reaction. *Proc. Roy. Soc. London Ser. B*, **149**, 145–68.
Medawar, P. B. (1963). Discussion after Miller, J. F. A. P. and Osoba, D. The role of the thymus in the origin of immunological competence. In *The immunologically competent cell* (ed. G. E. W. Wolstenholme and J. Knight), p. 70. *Ciba Found. Study Group*, Churchill, London.
Miller, J. F. A. P. (1961*a*). Analysis of the thymus influence in leukaemogenesis. *Nature*, **191**, 248–9.
Miller, J. F. A. P. (1961*b*). Immunological function of the thymus. *Lancet*, **2**, 748–9.
Miller, J. F. A. P. (1962*a*). Immunological significance of the thymus of the adult mouse. *Nature*, **195**, 1318–19.
Miller, J. F. A. P. (1962*b*). Effect of neonatal thymectomy on the immunological responsiveness of the mouse. *Proc. Roy. Soc. London Ser. B*, **156**, 410–28.
Miller, J. F. A. P. (1963). Tolerance in the thymectomized animal. In *La tolérance acquise et la tolérance naturelle à l'égard de substances antigéniques définies. Colloque du C.N.R.S.* pp. 47–75. Paris.
Miller, J. F. A. P. (1992). The Croonian lecture. The key role of the thymus in the body's defence strategies. *Phil. Trans. R. Soc. B*, **337**, 105–24.
Miller, J. F. A. P. and Heath, W. R. (1993). Self-ignorance in the peripheral T cell pool. *Immunol. Rev.*, **133**, 131–50.
Miller, J. F. A. P. and Mitchell, G. F. (1967). The thymus and the precursors of antigen-reactive cells. *Nature*, **216**, 659–63.
Miller, J. F. A. P. and Mitchell, G. F. (1968). Cell to cell interaction in the immune response. I. Hemolysin-forming cells in neonatally thymectomized mice reconstituted with thymus or thoracic duct lymphocytes. *J. Exp. Med.*, **128**, 801–20.
Miller, J. F. A. P. and Mitchell, G. F. (1970). Cell to cell interaction in the immune response. V. Target cells for tolerance induction. *J. Exp. Med.*, **131**, 675–99.
Miller, J. F. A. P. and Sprent, J. (1971). Cell to cell interaction in the immune response. VI. Contribution of thymus-derived and antibody-forming cell precursors to immunological memory. *J. Exp. Med.*, **134**, 66–82.
Miller, J. F. A. P., Vadas, M. A., Whitelaw, A., and Gamble, J. (1975). H-2 gene

complex restricts transfer of delayed-type hypersensitivity in mice. *Proc. Natl Acad. Sci. USA*, **72**, 5095–8.

Miller, J. F. A. P., Morahan, G., and Walker, I. D. (1983). T cell antigen receptors: fact and artefact. *Immunol. Today*, **4**, 141–3.

Mitchell, G. F. and Miller, J. F. A. P. (1968). Cell to cell interaction in the immune response. II. The source of hemolysin-forming cells in irradiated mice given bone marrow and thymus or thoracic duct lymphocytes. *J. Exp. Med.*, **128**, 821–37.

Mitchell, G. F., Grumet, F. C., and McDevitt, H. O. (1972). Genetic control of the immune response. The effect of thymectomy on the primary and secondary response of mice to poly-L-(Tyr,Glu)-poly-D,L-Ala-poly-L-Lys. *J. Exp. Med.*, **135**, 126–35.

Mitchison, N. A. (1971). The carrier effect in the secondary response to hapten protein conjugates. II. Cellullar cooperation. *Eur. J. Immunol.*, **1**, 18–27.

Möller, G. (1988). Do suppressor T cells exist? *Scand. J. Immunol.*, **27**, 247–50.

Mosmann, T. R. and Coffman, R. L. (1989). TH1 and TH2 cells: different patterns of lymphokine secretion lead to different functional properties. *Annu. Rev. Immunol.*, **7**, 145–73.

Moss, P. A. H., Rosenberg, W. M. C., and Bell, J. I. (1992). The human T cell receptor in health and disease. *Annu. Rev. Immunol.*, **10**, 71–96.

Naor, D. and Sulitzeanu, D. (1967). Binding of radioiodinated serum albumin to mouse spleen cells. *Nature*, **214**, 687–8.

Nossal, G. J. V. (1983). Cellular mechanisms of immunologic tolerance. *Annu. Rev. Immunol.*, **1**, 33–62.

Nossal, G. J. V. and Lederberg, J. (1958). Antibody production by single cells. *Nature*, **181**, 1419–20.

Owen, J. J. T., Jenkinson, E. J., and Kingston, R. (1986). Thymic stem cells: their interaction with the thymic stroma and tolerance induction. *Curr. Top. Microbiol. Immunol.*, **126**, 35–41.

Parish, C. R. (1972). The relationship between humoral and cell-mediated immunity. *Transplant. Rev.*, **13**, 35–66.

Parrott, D. M. V. (1962). Strain variation in mortality and runt disease in mice thymectomized at birth. *Transplant. Bull.*, **29**, 102–4.

Parrott, D. M. V., de Sousa, M. A. B., and East, J. (1966). Thymus-dependent areas in the lymphoid organs of neonatally thymectomized mice. *J. Exp. Med.*, **123**, 191–204.

Qin, S., Cobbold, S. P., Pope, H., Elliott, J., Kioussis, D., Davies, J., *et al.* (1993). 'Infectious' transplantation tolerance. *Science*, **259**, 974–7.

Rajewsky, K. (1971). The carrier effect and cellular cooperation in the induction of antibodies. *Proc. Roy. Soc. London Ser. B*, **176**, 385–92.

Rygaard, J. (1973). *Thymus and self: immunobiology of the mouse mutant nude*. Wiley, London.

Saito, H., Kranz, D. M., Takagaki, Y., Hayday, A. C., Eisen, H. N., and Tonegawa, S. (1984). A third rearranged and expressed gene in a clone of cytotoxic T lymphocytes. *Nature*, **312**, 36–40.

Schwartz, R. H. (1992). Costimulation of T lymphocytes: the role of CD28, CTLA-4 and B7/BB1 in interleukin-2 production and immunotherapy. *Cell*, **71**, 1065–8.

Simonsen, M. (1967). The clonal selection hypothesis evaluated by grafted cells reacting against their host. *Cold Spring Harbor Symp. Quant. Biol.*, **32**, 517–23.

Simonsen, M. (1990). Alloreactive T cells. *Scand. J. Immunol.*, **32**, 565–75.

Sinha, A. A., Lopez, M. T., and McDevitt, H. O. (1990). Autoimmune diseases: the failure of self tolerance. *Science*, **248**, 1380–8.

Sprent, J. and Webb, S. R. (1987). Function and specificity of T cell subsets in the mouse. *Adv. Immunol.*, **41**, 39–133.

Szenberg, A. and Warner, N. L. (1962). Dissociation of immunological responsiveness in fowls with a hormonally arrested development of lymphoid tissue. *Nature*, **194**, 146–7.

Weber, S., Traunecker, A., Oliveri, F., Gerhard, W., and Karjalainen, K. (1992). Specific low-affinity recognition of major histocompatibility complex plus peptide by soluble T-cell receptor. *Nature*, **356**, 793–6.

Yanagi, Y., Yoshikai, Y., Leggett, K., Clark, S. P., Aleksander, I., and Mak, T. W. (1984). A human T cell-specific cDNA clone encodes a protein having extensive homology to immunoglobulin chains. *Nature*, **308**, 145–9.

Zinkernagel, R. M. and Doherty, D. C. (1974). Immunological surveillance against self components by sensitized T lymphocytes in lymphocytic choriomeningitis. *Nature*, **251**, 547–8.

Part I

T cell populations

2 Development of T cells in the thymus

I. NICHOLAS CRISPE AND
DAVID G. SCHATZ

1 Introduction

Ten years ago the thymus was often represented as a 'black box', into which disappeared a small number of progenitor cells, and out of which emerged self-tolerant, self-MHC restricted, fully functional T cells. The lineage relationships between cell populations inside the thymus were a mystery, and the majority of thymocytes were cells that appeared to be destined only for death. The last decade has seen an explosive growth in understanding of the thymus. It is now possible to trace the lineage relationships between most of the known subsets of thymocytes, there is a general understanding of the timing of T cell receptor gene rearrangement processes, and tools are available to study both positive and negative selection of the T cell repertoire in molecular terms. In this review we will summarize the progress that has been made, and explore some of the issues which remain in thymocyte development. This review deals mainly with thymocyte development in the postnatal mouse thymus. The ontogeny of mouse thymocytes in the fetus appears to be broadly similar, but some processes clearly differ between the fetal and the adult thymus; in particular, the developmental relationships of αβ and γδ T cells.

While there is a consensus view on thymocyte lineage relationships, areas of uncertainty remain. The precise relationships between several minor, potentially important subpopulations of thymocytes are unknown. The commitment of progenitor cells to the T cell lineage, the induction of rearrangement, and the control mechanisms that govern sequential rearrangement of the T cell receptor (TCR) β and then α genes are not understood. The timing and mechanism of the lineage divergence between αβ and γδ T cells is not known. And the process of positive selection, which is so clearly illustrated in TCR transgenic mice, has proven to be more complex than anyone imagined. Rather than proposing a premature grand synthesis, we will use this review to draw attention to current areas of debate.

2 Subsets and lineage relationships

2.1 Haematopoietic stem cells and early thymocytes

The differentiation potential of different subpopulations of thymocytes has been defined in repopulation experiments, in which highly purified cell subsets isolated by cell sorting are assayed by adoptive transfer into recipients that express different allelic variants of the cell surface markers Thy-1 or Ly-5 (Goldschneider *et al.* 1986; Crispe *et al.* 1987). In these experiments, cells that result from the differentiation of the transferred progenitors are identified and characterized by multicolour flow cytometry. The time-course of the appearance of more mature subsets of thymocytes, such as $CD4^+CD8^+$ (double-positive, DP) cells or TCR αβ high, $CD4^+CD8^-$ or $CD4^-CD8^+$ (single-positive, SP) cells, serves as an indication of the degree of maturity of the donor cell population (Shimonkevitz *et al.* 1987; Scollay *et al.* 1988). Using these techniques, a very rare population of thymocytes has been identified that seems to include very early progenitors. These cells express the early thymocyte markers HSA, CD44 (Pgp-1), and high levels of MHC class I, but lack CD3 or any other TCR components. They are CD8 negative, but express easily detectable levels of CD4 (Wu *et al.* 1991). In adoptive transfer experiments, these cells give rise to B cells as well as T cells, but not to myeloid cells.

Multipotential haematopoietic stem cells are found in the adult bone marrow, and are defined as self-renewing progenitor cells that are able to give rise to all haematopoietic lineages. Such cells have been phenotypically defined and very extensively purified (Spangrude *et al.* 1988). These cells are negative for a panel of lineage-specific markers, but express a low level of Thy-1, a high level of MHC class I, and Ly-6A/E (Sca-1). Within the multipotential stem cell pool there are resting cells with a low level of metabolic activity, and activated cells that nevertheless retain the potential to give rise to all haematopoietic cell lineages (Li and Johnson 1992). The expression of CD4 on stem cells remains controversial; but other haematopoietic progenitor populations in both mouse and human are $CD4^+$ (Frederickson and Basch 1989; Basch *et al.* 1990). Thus in all external particulars, there are no clear differences between stem cells and the earliest thymocytes. Adoptive transfer of purified stem cells directly into the thymus of recipient mice leads to the production of T cells, but only after a long delay. In irradiated recipients, a burst of myeloid differentiation of donor origin precedes the development of a detectable number of thymocytes (Spangrude and Scollay 1990); and the burst size of each donor-derived cell clone is very much larger when stem cells, rather than $CD4^+$ very early thymocytes, are used as the donor population (Ezine *et al.* 1984; Scollay *et al.* 1988).

These data allow two possibilities for the relationship between stem cells and early thymocytes. Either there is a minute population of true stem cells in the thymus, not yet identified; or some differentiation from stem cells to lymphoid restricted progenitors occurs before cells traffic to the thymus. Since stem cells and the earliest thymocytes are so similar in surface phenotype, one would expect that true intrathymic stem cells would be co-purified with the $CD4^+$ very early thymocytes. Yet this population is devoid of non-lymphoid progenitor activity; and in unseparated thymocytes, CD8-depleted thymocytes and DN thymocytes the frequency of erythromyeloid progenitors (day 12 CFU-S) is extremely low: less than one cell per thymus (I. N. Crispe, unpublished data). These data make the most sense if thymus-seeding cells have already differentiated beyond the multipotential stem cell stage. Yet while very early B cell progenitors devoid of T cell progenitor activity have been described (Muller-Sieburg *et al.* 1986), neither pre-thymic T lineage nor common lymphoid precursor cells have been identified. Perhaps the bone marrow has been the wrong place to look for them, and they will turn up somewhere else.

2.2 Double- and triple-negative early thymocytes

While the earliest thymocytes express CD4, their differentiation proceeds through a series of $CD4^-CD8^-$ (double-negative, DN) stages. These DN cells were the first population of thymocytes identified as progenitors, through their ability to home to and repopulate the thymus of an irradiated recipient (Fowlkes *et al.* 1985). They were soon found to include minor populations of both TCR $\alpha\beta$ and TCR $\gamma\delta$ positive cells, without progenitor activity, and a major population of $TCR^-CD4^-CD8^-$ (triple-negative, TN) cells, which were able to repopulate the thymus. The TN cells were further subdivided by surface markers into a group of $CD44^+$, HSA intermediate cells, a group of $CD44^-$, $CD25^+$, HSA high cells, and a population with trace levels of CD4 and/or CD8 plus very high HSA (see Figure 2.1). Assay of each of these subsets in adoptive transfers suggested that they form three sequential stages in the early thymocyte developmental pathway (Pearse *et al.* 1989). 'Early TN' are cells that express the CD44 marker, which is a receptor for hyaluronic acid (Aruffo *et al.* 1990); do not yet express CD25, the low affinity p55 chain of the IL-2 receptor (Shimonkevitz *et al.* 1987); and retain the ability to home to the thymus following intravenous injection (Lesley *et al.* 1988). 'Intermediate TN' are cells that express a higher level of HSA, have lost expression of CD44, and no longer home to the thymus. These cells express CD25 but, although they bind to IL-2, the receptor–ligand complex is not internalized and the cells do not proliferate (Malek *et al.* 1985). The role of the CD25 molecule in T cell development is unknown, but anti-CD25 monoclonal antibodies inhibit thymocyte development (Zuniga-Pflucker and Kruisbeek 1990). 'Late TN' are one of several populations of intermediates between TN and DP thymocytes. They co-purify with other DN cell types in antibody and complement depletion protocols, but in fact express very low levels of either CD4, CD8, or CD3 (Nikolic-Zugic and Bevan 1988). These cells differentiate rapidly *in vivo*, and spontaneously *in vitro* into DP cells (Petric *et al.* 1990*a,b*). All of the thymocyte cell types described here, and their lineage relationships, are illustrated in Figure 2.1.

Two other intermediates between TN and DP cells have been described. Immature $CD4^-CD8^+$ cells appear in ontogeny ahead of DP cells, in the mouse by one day (Husmann *et al.* 1988). Such cells exist in the adult thymus, where they express a high level of HSA, lack TCR, and are devoid of T cell

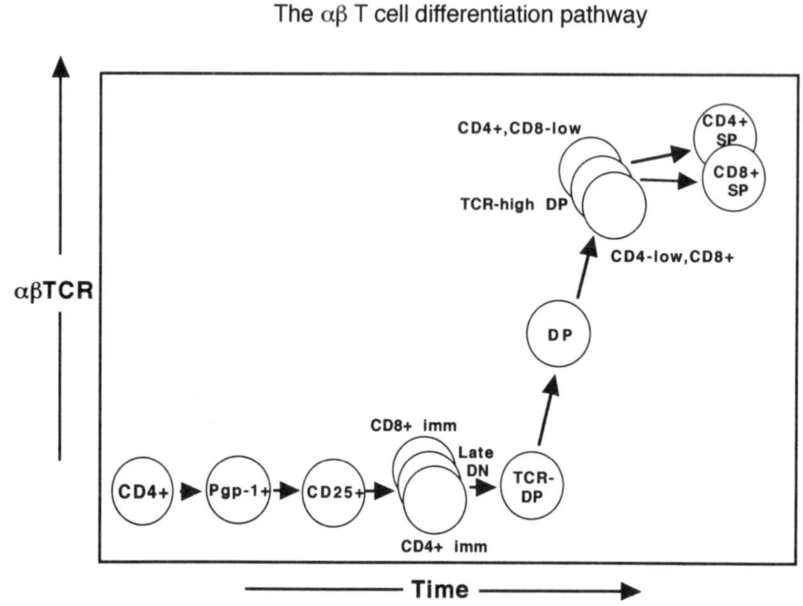

Fig. 2.1 The pathway of intrathymic T cell development. Individual cell populations are represented on axes indicating the time resident in the thymus, versus the level of TCR αβ on the cell surface. Subpopulations of cells shown overlapping are believed to be functionally equivalent.

function (Crispe and Bevan 1987; Bluestone *et al.* 1987; Shortman *et al.* 1988). An equivalent population of $CD4^{+}CD8^{-}TCR^{-}$ cells has been described, although they are much rarer in most mouse strains (Hugo *et al.* 1991*b*). There is no reason to think that late TN and TCR negative SP populations are functionally distinct; instead they probably reflect variation in the order in which CD4 and CD8 genes are activated in individual thymocytes.

The expression of TCR components by cells in transition from TN to DP is controversial. Late TN cells have been reported to express very low levels of CD3, at the limit of resolution of the FACS (Petrie *et al.* 1990*a*). Yet a substantial proportion of DP cells are TCR negative (Richie *et al.* 1988), and the rearrangement of TCR α genes is continuing in this population. These discrepancies may be resolved if normal thymocytes at this stage of development express an immature TCR that consists of TCR β chain dimers associated with some CD3 components. Receptors of this type have been described in cell lines and on the thymocytes of TCR transgenic mice (Kishi *et al.* 1991), as well as in TCR α deficient mutant mice created by gene inactivation in embryonic stem cells (Mombaerts *et al.* 1992*a*). This TCR ββ receptor has been proposed to participate in a signalling event which terminates TCR β locus rearrangement, and/or activates TCR α locus rearrangement (Borgulya *et al.* 1992), but the latter possibility seems unlikely since TCR α locus rearrangement proceeds apace in TCR β deficient mutant mice (Mombaerts *et al.* 1992*b*). In addition to a role in TCR β gene allelic exclusion, a receptor that signals the existence of an in-frame TCR β rearrangement capable of cell surface expression may provide the trigger for the proliferation of DP cells. But the expression and functional importance of TCR ββ receptors on normal thymocytes has yet to be demonstrated clearly.

2.3 Subpopulations of DP thymocytes

The vast majority of cells in the thymic cortex, and about 70–80% of all thymocytes, are DP. These cells mostly express TCR αβ at a low level, but around 30% of the cells have no detectable CD3 or TCR αβ (Richie *et al.* 1988), while another 5% express a high level of TCR αβ similar to that on SP thymocytes and peripheral T cells (Hugo *et al.* 1991*a*). Both the TCR negative and the TCR low cells contain blast populations, and overall around 20% of DP cells are in cycle. The DP cells undergo several rounds of division, but once a TCR is expressed, the cells divide once or at most twice more. The TCR positive cells become susceptible to selection on the basis of their specificity. Within the TCR low subset of DP cells, there seems to be heterogeneity with respect to the consequences of signalling through different components of the TCR. Treatment of thymocytes with anti-Cβ antibody leads to apoptosis of only a subset of the TCR αβ–$CD3^+$ cells (Finkel *et al.* 1989). This is an indication that functional maturation of DP cells involves a change in the coupling of the TCR α and β chains to the CD3 γδε signalling unit.

2.4 Transition to the TCR high SP phenotype

The small subset of DP thymocytes with a high level of TCR αβ is expanded in TCR transgenic mice in the presence of positively selecting MHC molecules, but absent when the correct MHC is missing. These cells must therefore be the immediate products of positive selection. In adoptive transfer experiments *in vivo*, they rapidly differentiate into SP cells. But it is not clear whether all thymocytes must transit through a TCR high DP subset. The situation may be analogous to the various intermediates between TN and DP cells, where a set of phenotypic changes occur more or less at the same time, but not in a tightly regulated sequence. Thus, some cells may become TCR high as DP, but others may become SP and then up-regulate their TCR.

The absence of a defined series of intermediates between the DP and the mature SP thymocytes leads to a problem in defining the exact developmental stage at which the repertoire is modified. One study examined the consequences of positive and negative selection in cells phenotypically intermediate between DP and SP. In a three-colour FACS analysis, thymocytes were classified as either DP, or $CD4^+$ CD8 intermediate, or $CD4^+$ CD8 low, or $CD4^+$ $CD8^-$. Their expression of T cell receptor Vβ elements was examined, in both deleting and positively-selecting environments. In this study, positive selection effects on the T cell repertoire were seen at an earlier stage than negative selection effects, which was used to argue that these processes occur sequentially in T cell development (Guidos *et al.* 1990). However, it is also possible that the order of these changes in the TCR Vβ repertoire is determined by the nature of the antigen-presenting cells, or the location of the self-antigens causing deletion. Thus the order in which positive and negative selection occur is still open to interpretation.

In parallel with these changes in TCR density, co-receptor expression and repertoire, the cells are translocated from the thymic cortex to the medulla. This change is associated with modification of the glycocalyx, as carbohydrate structures that bind the lectin peanut agglutinin are obscured by the addition of sialic acid residues. The heavily sialylated glycocalyx may be important for mature T cell recirculation (Reisner *et al.* 1976; Gillespie *et al.* 1993). The cells also acquire functional competence; $CD8^+$ SP thymocytes reproduce the functions of $CD8^+$ peripheral T cells, and $CD4^+$ SP cells function like $CD4^+$ T cells (Ceredig *et al.* 1982, 1983). Most SP thymocytes are thought to leave the thymus, but *in vivo* labelling experiments suggest that a subset are long-term thymic residents (Scollay 1983), the biological significance of which is not known.

3 Which TCR?

Both αβ and γδ T cells develop in the thymus. While γδ cells will be discussed in more detail elsewhere in this volume (see Chapter 4), the question of the common origin of αβ and γδ cells is discussed here. These two cell types both express CD3-associated membrane receptors, and neither appears to function by secreting antigen-specific molecules; they therefore appear to be more closely related than either are to B cells. The CD4 intermediate very early thymocytes give rise to γδ cells as well as αβ cells and B cells (Wu *et al.* 1991), which comes as no surprise. But the point at which thymocytes become committed to the αβ T cell lineage and lose their γδ cell differentiation potential is hard to determine. One indication that this separation is unexpectedly late in the developmental sequence comes from experiments in which sorted TN cells were allowed to differentiate *in vivo* or *in vitro*. After adoptive transfer, these cells give rise mainly to αβ TCR^+ DP cells, but γδ cells are also detectable in the progeny. In tissue culture, late stage TN will spontaneously express a low level of αβ TCR. Under the influence of a cytokine cocktail including IL-1α, IL-2, IL-7, and SCF (stem cell factor, Steel factor) the same cell population will give rise to a majority of γδ cells (Petrie *et al.* 1992). Two interpretations of these experiments are possible: either the lineage divergence between αβ and γδ cells occurs at this stage of development; or the αβ and γδ lineages diverge at an earlier stage but the cells undergo parallel developmental pathways with identical surface phenotypes.

The *in vivo* repopulation and *in viro* differentiation experiments with DN subsets deal with the developmental potential of heterogeneous cell populations and as yet have not yielded a clear answer on the timing of lineage divergence. Another approach to this issue is to study the molecular events, particularly TCR gene rearrangements, that occur in developing αβ and γδ thymocytes. A priori, one might have hoped that the gene rearrangements would be diagnostic of lineage commitment: that is, that thymocytes committed to the αβ lineage would contain α and/or β gene rearrangements, but not γ and δ gene rearrangements, and that the opposite would be true of γδ committed thymocytes. This is not the case, however, since D to Jβ, γ, and perhaps also δ gene rearrangements are promiscuous, occurring in cells of both lineages. Considerable uncertainty has therefore existed concerning the relationship of the two lineages, and several models have arisen to try to accommodate the data. More recently, the analysis of mice containing TCR transgenes has clarified the situation somewhat, and led to a growing (though certainly not unanimous) consensus that the two lineages develop independently of one another.

3.1 Models for the αβ–γδ lineage split

Experimental data will be considered in the context of the three models presented below and diagrammed in Figure 2.2.

1. The sequential rearrangement model (Allison and Lanier 1987; Pardoll *et al.* 1987): This model states that γ and δ gene rearrangements are attempted first, and that if productive (in-frame) rearrangements of both genes are obtained, the cell becomes a γδ T cell. However, if a productive γ or a productive δ rearrangement is not obtained, then the cell goes on to rearrange its β and α genes and to become an αβ T cell.
2. The competitive rearrangement model: According to this model, γ, δ, and β gene rearrangements occur simultaneously in uncommitted thymocytes. If γ and δ genes are productively rearranged first, the cell becomes a γδ T cell, while if a productive β gene is assembled first, the cell commits to the αβ lineage.
3. The different lineages model: This model holds that the αβ and γδ lineages are independent. This model is distinguished from the other two because it states that lineage commitment is *independent of the outcome of any and all gene rearrangement events*. In Figure 2.2, lineage commitment is shown as occurring before any gene rearrangements happen, but the model need not specify when lineage commitment occurs in relation to the onset of gene rearrangement. It merely states that the decision is not influenced by the results of gene rearrangement.

Each of the models makes predictions concerning the status of gene rearrangements in αβ and γδ T cells, the perturbations on thymocyte development expected to be caused by TCR transgenes, and the effects of disrupting

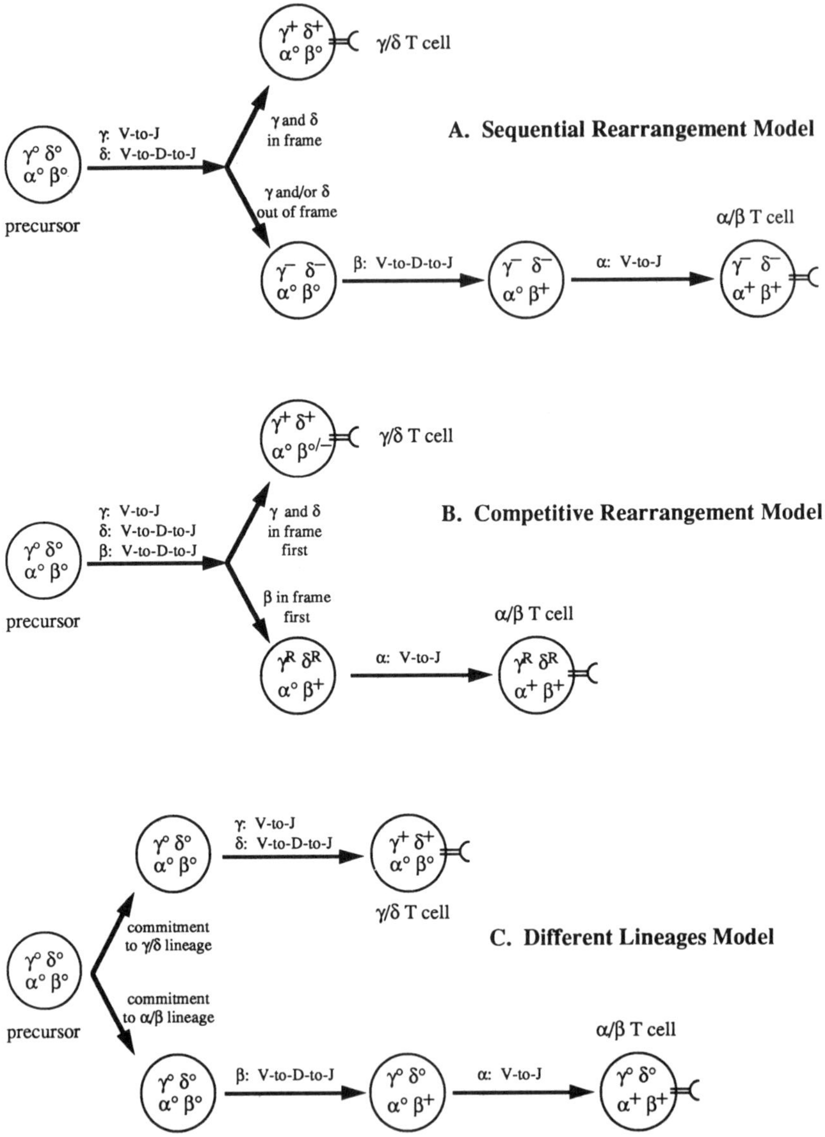

individual TCR loci by homologous recombination. As discussed below, the different lineages model is able best to accommodate the current data.

3.2 Patterns of TCR gene rearrangements in αβ and γδ T cells

Early support for the sequential rearrangement model came from studies of the pattern of TCR gene rearrangement and expression during fetal development

Fig. 2.2 Three models for the lineage relationship between αβ and γδ T cells. The models are described in detail in the text. Beginning with a thymic precursor cell, a series of rearrangement events take place, as indicated above the horizontal arrows. In the sequential rearrangement model (A), the lineage chosen by the cell depends on whether γ and δ gene rearrangements are in-frame, whereas in the competitive rearrangement model (B), lineage choice depends on whether productive β or γδ gene rearrangements are made first. In the sequential rearrangement model (C), lineage commitment occurs independently of and before gene rearrangement events. Lineage commitment is indicated by branching arrows. The rearrangement status of the different TCR loci at each stage of development is indicated as follows: °, unrearranged; $^{+}$, productively rearranged; $^{-}$, non-productively rearranged; $^{o/-}$, unrearranged or non-productively rearranged; ^{R}, some but not all alleles rearranged. The models are simplifications in that after lineage commitment, only productive rearrangements are depicted. Cells that fail to make the productive rearrangements necessary to express surface TCR die.

(summarized in Raulet 1989). γ, δ, and D to Jβ gene rearrangements begin simultaneously in the thymus at day 13 of murine fetal development, with $\gamma\delta^{+}$ cells detectable as early as day 14. V to DJβ rearrangements are not found until day 16, and surface expression of the αβ TCR is not seen until day 17, coincident with the first detection of α gene rearrangements. Taken together, this sequence of events suggested the possibility of a precursor–product relationship between cells rearranging γδ and those rearranging αβ (Allison and Lanier 1987; Pardoll *et al.* 1987).

Both the sequential and competitive rearrangement models predict that extensive δ gene rearrangement should occur in αβ T cell precursors. Rearrangements of the δ locus cannot be studied directly in mature αβ T cells because any α rearrangement deletes the δ locus, and most αβ T cells have α rearrangements on both alleles (Hue *et al.* 1990). The by-product of a deletional V(D)J recombination event, however, is a circular DNA molecule containing the deleted DNA, and such circles can be isolated from thymocytes. This has allowed determination of the structure of the δ locus in cells that are actively rearranging their α loci (and are therefore committed to the αβ lineage). An extensive analysis of such circular products found that the δ locus was unrearranged in all cases (Winoto and Baltimore 1989*a*), suggesting that the δ locus does not rearrange in αβ T cell precursors and therefore that δ rearrangement does not begin until after lineage commitment has occurred. Two similar studies, however, obtained the opposite result, finding that the δ locus was often rearranged on circles isolated from αβ T cells (Okazaki and Sakano 1988; Takeshita *et al.* 1989). It is difficult to reconcile the discrepancies between the data, but it is worth noting that if the δ locus is rearranged on excised DNA circles, it is not possible to rule out that δ rearrangements happened on the circles *after* α rearrangement had occurred. It is therefore not possible to determine from these data the actual order of rearrangement events during thymocyte development.

In general, two-thirds of all genes assembled by V(D)J recombination will be out-of-frame unless there is some selection for or against productive rearrangements. The sequential rearrangement model predicts that αβ T cells should contain extensive γ and δ gene rearrangements, both in-frame and out-of-frame, but an individual cell should not contain both in-frame γ and δ rearrangements. Therefore there should be some modest amount of selection against in-frame γ rearrangements in αβ T cells, resulting in αβ T cells containing somewhat more than two-thirds of out-of-frame γ rearrangements. Indeed, early studies found predominantly out-of-frame γ rearrangements in αβ T cell clones (Heilig and Tonegawa 1986; Reilly *et al.* 1986; Rupp *et al.* 1986; Moisan *et al.* 1989). Other studies, however, identified a variety of productive γ rearrangements in αβ T cell lines (Traunecker *et al.* 1986; Heilig and Tonegawa 1987). The conflict in the data appears to arise from the fact that while almost one-third of Vγ1.2–Jγ2 rearrangements are in-frame in αβ T cell lines, Vγ2–Jγ1 rearrangements are essentially always out-of-frame (Raulet *et al.* 1991). It is possible therefore that productive Vγ2 rearrangements are selected against in αβ T cells, but because the γ locus is expressed poorly in αβ T cells (Garman *et al.* 1986; Raulet *et al.* 1991; and see below), it is perhaps more reasonable to think that the rearrangement status of the γ locus is largely irrelevant to αβ T cell development (as would be predicted by the different lineages model). Thus, γ rearrangements are prevalent in αβ T cells and both productive and non-productive rearrangements can be found. These data do not help in deciding between the three models.

The sequential rearrangement model predicts that no β rearrangements should be found in γδ T cells, while the competitive rearrangement model predicts that VDJβ rearrangements should be common in γδ T cells, but that all such rearrangements should be out-of-frame. Neither of these simple predictions is correct: D to Jβ rearrangements are very common in γδ T cells, but complete V to DJβ rearrangements (either in-frame or out-of-frame) are extremely rare (Lew *et al.* 1986; Asarnow *et al.* 1988; Haas and Tonegawa 1992). The data argue strongly against the competitive rearrangement model, but are not a compelling argument against the sequential rearrangement model since the D to Jβ rearrangements seen in γδ cells could have occurred after lineage commitment, or could be considered irrelevant since they do not yield a functional β gene.

All three models predict that α gene rearrangement should occur after the lineage split, and only in αβ T cell precursors. Indeed, α gene rearrangement would be detrimental in γδ committed cells since such rearrangements would delete the δ locus from the chromosome. Therefore, it was not surprising to find that while the α locus is heavily rearranged in αβ T cells, it is almost never rearranged in γδ T cells (Asarnow *et al.* 1988; Korman *et al.* 1988; Takagaki *et al.* 1989).

What molecular mechanism acts to prevent α gene rearrangement while its close neighbour, the δ locus, is undergoing rearrangement? While the molecular mechanisms that regulate rearrangement of immunoglobulin (Ig) and TCR genes are not well understood, there is a strong correlation between transcription of a gene segment and its rearrangement (see Chapter 15). In particular, there is growing evidence that the enhancer elements that flank Ig and TCR genes exert positive effects on both transcription and gene rearrangement. Therefore, elements that suppress transcription might also suppress rearrangement. It was therefore of particular interest when a negative regulatory element was discovered that flanks the α enhancer and prevents enhancer-mediated transcription in γδ but not in αβ T cells (Winoto and Baltimore 1989*b*). This 'α-silencer' element was proposed to play a role in αβ versus γδ lineage determination by blocking α rearrangement. No direct support for this hypothesis has yet been obtained.

In summary, the data on TCR gene rearrangements in αβ and γδ T cells do not strongly support any of the models. The lack of VDJβ rearrangements in γδ T cells, however, argues against the competitive rearrangement model.

3.3 Perturbations of thymocyte development caused by TCR transgenes

Mice containing a transgenic γδ TCR should provide an excellent test of the three models: both the sequential and competitive rearrangement models predict that αβ T cell development should be blocked in such mice, while the different lineages model predicts that αβ T cells should develop normally. The data support the different lineages model, but not in a straightforward manner.

In a provocative series of experiments with γδ transgenic mice, Tonegawa's group found that the effect on αβ T cell development depended on the γ transgene used. When a large (40 kb) portion of the γ locus was used, αβ T cell development was essentially normal, but if a smaller (15 kb) γ transgene was used, αβ development was severely disrupted (Bonneville *et al.* 1990; Ishida *et al.* 1990). The large γ transgene was not expressed in αβ T cells, but when mice were made with a small γ transgene alone (in the absence of a δ transgene), αβ T cells expressed significant levels of γ transgene transcripts. This led to the hypothesis that a γ transcriptional 'silencer' element, present in the large transgene but absent from the small one, was responsible for the difference in expression. This in turn suggested that the critical event in the αβ versus γδ lineage decision is the activation (or repression) of the γ silencer: αβ T cells can develop normally in the presence of functionally rearranged γ and δ transgenes if the γ silencer prevents γ gene expression. Therefore the lineage decision is made independently of the status of TCR gene rearrangements, as required by the different lineages model.

A separate series of experiments by Hedrick and his colleagues with γδ transgenic mice yielded somewhat different results but additional support for the different lineages model (Dent *et al.* 1990). The γ transgene used in these experiments was a short version apparently lacking the silencer element, and the transgenes encoded an alloreactive γδ TCR which recognized a class I MHC antigen in H-2^b but not H-2^d mice. Two phenotypes were observed in transgenic H-2^d (alloantigen negative) mice: type 1 mice had large numbers of transgene positive cells and no αβ T cells, while type 2 mice also had significant numbers of transgene positive cells but a virtually normal number and distribution of αβ T cells. It was also noted that mature αβ T cell clones from type 2 mice did not express both γ and δ transgene transcripts (Hedrick and Dent 1991), in part because one of the transgenes has been deleted from the chromosome (Dent 1991). Similar to the Tonegawa laboratory's results, this implies that αβ T cell precursors must avoid expressing the products of the transgenes in order to develop normally. Taken together, the results in H-2^d transgenic mice support the different lineages model by demonstrating that αβ T cells can develop normally despite containing γδ transgenes (type 2 mice), but this only happens if αβ T cell precursors avoid transgene expression. This conclusion is further strengthened by results with H-$2^{b/d}$ transgenic mice, discussed below.

In transgenic mice in which αβ T cell development is blocked (the Tonegawa mice with the small γ transgene, and the Hedrick type 1 H-2^d mice), are additional γδ T cells produced to compensate for the missing αβ T cells? In both studies, the answer was no: the αβ T cell deficient transgenic mice show at least a tenfold reduction in thymus cell number when compared to the non-transgenic controls. Thus even when αβ T cell development is disrupted, the number of γδ T cells does not increase significantly, as if a fixed number of thymocytes can commit to the γδ lineage regardless of the status of the αβ thymocyte development. Again, this supports the different lineages model.

In H-$2^{b/d}$ (alloantigen positive) transgenic mice, quite different results were obtained: αβ T cell development and total thymocyte number were normal, and cells with surface expression of the transgenic γδ TCR were absent from the thymus. Further, peripheral T cells expressed none or significantly reduced levels of the transgene. This demonstrates that self-reactive γδ thymocytes can be eliminated by a process that appears analogous to negative selection of αβ thymocytes. In addition, because αβ T cell development was normal in the face of efficient deletion of transgene positive cells, it was concluded that thymocytes in the αβ lineage need not pass through a stage in which γ and δ are expressed. This conclusion is not compatible with either the sequential or competitive rearrangement models because both of these models require that uncommitted thymocytes transcribe rearranged γ and δ genes in order to determine whether the rearrangements are in-frame.

If developing αβ T cells never express the γδ transgenes, how can one explain the dramatic effect of γδ transgenes on αβ T cell development in type 1 H-2^d mice and in the Tonegawa mice containing the short γ transgene? The most likely explanation is that early, high level expression of the transgenic γδ TCR *on γδ lineage cells* interferes with αβ T cell development by some as yet unknown mechanism. If this is correct, then when the γ transgene contains the silencer, expression must occur later and/or at lower levels, allowing normal αβ T cell development. Therefore, while direct γδ gene expression in αβ lineage cells is likely to be detrimental to their development, the primary effect of γδ transgenes on αβ T cell development is probably indirect, being mediated by transgene expression on γδ lineage cells.

Studies with γδ transgenic mice suggest several conclusions: first, that there is an interaction between abnormal expression of γδ genes and αβ T cell development; second, that αβ T cells can develop normally even when all thymocytes contain correctly assembled γδ genes; third, that αβ lineage thymocytes need not pass through a γδ expressing stage of development; fourth, that the effect of a γδ transgene depends on its levels and/or timing of expression; and fifth, that the γδ lineage does not generate additional T cells to compensate for the absence of αβ T cells. The second, third, and fifth of these conclusions together provide strong support for the different lineages model.

Few data are available concerning the effects of αβ transgenes on γδ T cell development. One study found that a β transgene dramatically inhibited certain γ rearrangements and the development of adult γδ T cells (von Boehmer *et al.* 1988). These results were taken as support for the competitive rearrangement model. It is not clear, however, by what mechanism the β transgene had its effect: was it due to β expression in γδ precursors, or a more indirect effect? Further, the biological relevance of the observation is uncertain since as discussed above, complete V to DJβ rearrangements (in-frame or out-of-frame) are not found in γδ T cells. Therefore, in normal thymocyte development there is never an opportunity for complete β rearrangements and a β polypeptide to interfere with γδ T cell development. These results with αβ transgenic mice do not rescue the competitive rearrangement model.

3.4 The effects of TCR gene knock-outs on thymocyte development

Homologous recombination in embryonic stem cells has been used to create mice unable to express TCR α, β, or δ. The results are unambiguous and consistent with the different lineages model, but do not in fact directly test any of the models. Disruption of TCR α (Mombaerts *et al.* 1992*a*; Philpott *et al.* 1992) or TCR β (Mombaerts *et al.* 1992*a*) eliminates all mature αβ T cells but γδ T cell development proceeds normally and yields normal numbers of $\gamma\delta^+$ thymocytes and $\gamma\delta^+$ peripheral T cells. Reciprocally, mice unable to express δ lack γδ T cells but have a normal complement of αβ T cells (Itohara *et al.* 1993). The results demonstrate that each lineage can develop normally in the complete absence of mature cells of the other lineage, consistent with the two lineages being independent. The results also emphasize the inability of the γδ T cell lineage to expand to compensate for the lack of $\alpha\beta^+$ T cells or thymocytes.

3.5 A model of the lineage relationship of αβ and γδ T cells

The data summarized thus far are consistent with the different lineages model but conflict with the other two models in a variety of ways, most significantly in the finding that αβ T cells can develop normally under some circumstances in γδ transgenic mice. Thus it appears that the αβ versus γδ lineage commitment decision itself is not determined by what TCR gene rearrangements have occurred in the developing thymocyte. On what event or events does the lineage decision hinge? This question has yet to be answered.

A second and more approachable question raised by the independence of the two lineages is: At what developmental stage do they diverge? What events (particularly TCR gene rearrangements) occur in a thymic progenitor cell before lineage commitment occurs? While a definitive answer cannot be given, some developmental boundaries within which lineage commitment occurs can be defined. There is no evidence as to whether lineage commitment by progenitor cells occurs before or after their entry into the thymus. But if commitment occurs before entry, early αβ and γδ cells must undergo a similar, if not identical, series of phenotypic changes before TCR expression. For this reason, it is reasonable to suppose that commitment occurs in the thymus. In addition, both V to DJβ and α rearrangements appear to be specific for αβ lineage cells; therefore, commitment occurs before the onset of these rearrangements. Since the onset of α gene rearrangement occurs as thymocytes transit from the DN to the DP stage of development, lineage commitment must occur in DN thymocytes (or earlier). Rearrangements of the γ, D to Jβ, and perhaps δ loci are found in both lineages; therefore these loci may rearrange before lineage divergence. Alternatively, lineage divergence may occur before any gene rearrangements take place. The current evidence does not allow one to specify the timing of lineage commitment relative to γ, D to Jβ and δ gene rearrangements.

Recent experiments have shed further light on the question of when the two lineages diverge. Early work showed that when a transgene containing unrearranged Vβ, Dβ, and Jβ gene segments was introduced into mice, no rearrangement of the transgene occurs unless an Ig heavy chain enhancer was introduced into the construct (Ferrier *et al.* 1990). More recently, similar transgenes containing either the TCR β enhancer or the 'core' TCR α enhancer (lacking the α silencer) in place of the Ig heavy chain enhancer have been introduced into mice. With the TCR β enhancer, the transgene began rearranging in thymocytes at day 14 of fetal development and was extensively rearranged in adult γδ T cells; thus the β enhancer conferred on the transgene the rearrangement timing and potential seen of the endogenous DβJβ locus. With the core TCR α enhancer, rearrangement began at day 17 of fetal development and no rearrangement was found in adult γδ T cells (P. Ferrier, personal communication). These results strengthen the conclusion that γδ T cells diverge before the α enhancer becomes active (and before any α gene rearrangement takes place), and that the β enhancer is active in γδ precursors.

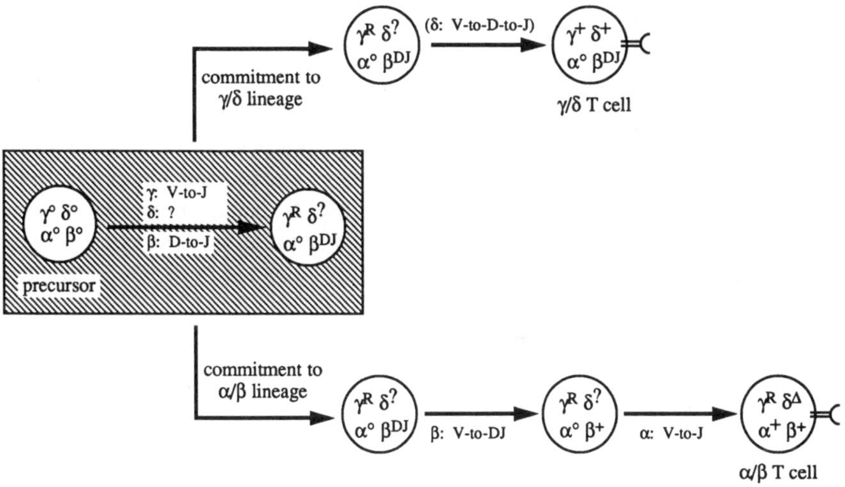

Fig. 2.3 A model for the lineage relationship between αβ and γδ T cells. The αβ and γδ T cell lineages develop independently from a common precursor cell, with lineage commitment occurring before any gene rearrangements take place or at some point during the process of γ, D to Jβ, and perhaps δ gene rearrangements (represented as cells within the shaded box). γ, D to Jβ, and perhaps δ gene rearrangements are promiscuous, occurring in either lineage. Therefore, if lineage commitment occurs before gene rearrangement begins, these rearrangements occur in the committed cells. For simplicity, the cells that result from lineage commitment are depicted as already containing these promiscuous rearrangements. In cells committed to the γδ lineage, D to Jβ rearrangements are irrelevant since V to DJβ rearrangements do not occur, while in αβ committed cells, γ gene rearrangements are irrelevant because the rearranged genes are not transcribed. V to DJβ and α gene rearrangements occur only after lineage commitment. D to Jβ and V to DJβ rearrangements precede α rearrangements and occur in DN thymocytes; α rearrangements occur in DP thymocytes. As in Figure 2.2, only productive rearrangements are indicated after lineage divergence. ^{DJ} represents a D to J rearranged locus; $^{?}$ represents a locus whose rearrangement status is uncertain; $^{\Delta}$ represents a locus that has been deleted from the chromosome; other notation as described in the legend to Figure 2.2.

They also call into question the importance of the α silencer in preventing α rearrangements in γδ committed thymocytes.

A model for the lineage relationship of αβ and γδ T cells, incorporating the data discussed above, is presented in Figure 2.3. Lineage commitment is depicted as occurring at the same time as γ, D to Jβ and, perhaps δ rearrangements (inside the box in the Figure 2.3). Both γ and D to Jβ rearrangements are promiscuous in that they can occur in either lineage; it is unclear if the same is true of δ rearrangements. Presumably it is not vital to suppress γ rearrangements in αβ precursors because the γ silencer prevents transcription of the assembled γ genes, and similarly, D to Jβ rearrangements are tolerated in γδ precursors because they can not encode a relevant protein in the absence of V to DJβ rearrangements. It may be much more important to prevent promiscuous α and δ gene rearrangements because of the physical linkage between the two loci, but the mechanisms that operate to achieve this may not be relevant to the lineage commitment decision. The critical features of the model are that lineage commitment occurs independently of the nature of the rearrangements that actually take place, that γ, D to Jβ and perhaps δ rearrangements can take place before or after lineage commitment, and that V to DJβ and α gene rearrangements occur only after lineage commitment. If the model is roughly correct, then one focus of future experiments must be the molecular 'switch(es)' that trigger commitment to either the αβ or the γδ T cell lineages.

4 Blocks in αβ T cell development

A set of natural and artificial mutants in key genes involved in thymocyte maturation result in developmental arrests at three points during T cell development (Figure 2.4). The fact that arrest at each of these points is

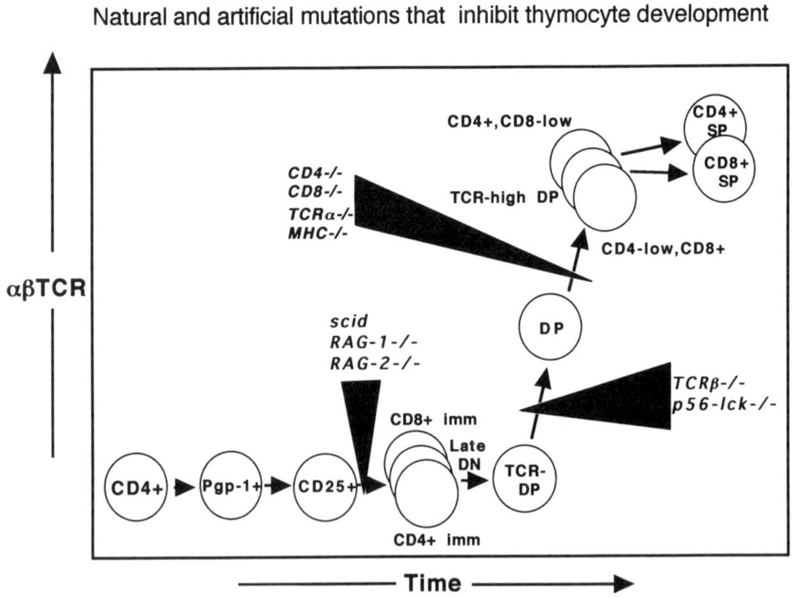

Fig. 2.4 One natural and a series of artificial mutations that arrest thymocyte development at distinct stages. In *scid* and recombinase deficient mice, there is a thymus of several million cells, and development is arrested at the $CD25^+$ DN stage. In both TCR β deficient, and $p56^{lck}$ deficient mutants, the thymus consists of around ten million cells, half of which are TCR negative DP. In mice lacking the CD4 or CD8 co-receptors or their MHC ligands, or in TCR α deficient mice, a normal number of cortical-type thymocytes is produced (around 100 million), but differentiation progresses no further.

caused by several different mutations suggests that these points occur at developmental stages where physiological controls act on T cell development. Following the usage of Shortman (1992), we term these 'control points'.

4.1 Intermediate TN

In mice that are unable to recombine TCR genes, there is a small thymus containing 1–2% of the normal number of thymocytes. In the natural mutant *scid*, and in the RAG-1 and RAG-2 deficient mice produced by gene knock-out techniques, there is a developmental arrest at the HSA high, $CD25^+$ 'intermediate TN' stage (Carroll and Bosma 1991; Shinkai *et al.* 1992; Mombaerts *et al.* 1992*b*). This suggests that some product of recombination is necessary for the cell to make a transition into the late TN stage, from which cells spontaneously express CD4 and CD8. An obvious candidate is TCR β, but thymocytes devoid of TCR β protein can progress beyond this stage and are in fact arrested at the next developmental control point (Mombaerts *et al.* 1992*a*). It is possible that the cells progress beyond this point only when a partial rearrangement of one of the TCR loci has been achieved. The intermediate TN in normal mice express an excess of the 1.0 kb TCR β mRNA that results from the transcription of DJ rearrangements (Crispe *et al.* 1987; Pearse *et al.* 1988), and this mRNA could act as a signal, possibly by its translation into a truncated polypeptide that acts intracellularly.

The interpretation of the developmental arrest in *scid* mice is complicated by experiments in which *scid* bone marrow was mixed with normal marrow, and the mixture used to reconstitute irradiated host mice. In such chimeras, cells of *scid* genotype progressed to the DP stage although they did not express TCRs. Thus the developmental arrest in *scid* thymocytes may result from an abnormality of the thymic stroma, which in turn is due to the lack of TCR^+ cells. The same could apply to the RAG-1 and RAG-2 knock-out mice. In this case, one could argue that the progression of thymocytes beyond the intermediate DN stage requires a signal from the thymic stroma. The nature of this signal is unknown, but the CD25 molecule that is diagnostic of this developmental stage is known to be connected to the $p56^{lck}$ tyrosine kinase, and is an obvious candidate for the signal receptor.

4.2 Expansion of the DP population

Culture *in vitro* of late TN or TCR negative SP thymocytes results in the production of some DP cells, but not in the expansion that is observed *in vivo*. Developmentally arrested thymocyte phenotypes that suggest a lack of DP thymocyte expansion are also seen in two types of gene knock-out mice, those that lack the TCR β constant region and those that lack $p56^{lck}$ (Mombaerts *et al.* 1992*a*; Molina *et al.* 1992). The thymus of a young adult normal mouse contains $100–200 \times 10^6$ thymocytes, mostly DP cells. In both of these mutants, the thymus consists of around 10×10^6 cells, half of which are TN and half of which are DP. In *scid* mice, the introduction of a TCR β transgene is sufficient to restore DP thymocyte expansion and reproduce the phenotype of TCR α deficient mutant mice (Kishi *et al.* 1991). Thus it is reasonable to postulate that TCR β chains plus a p56^{lck}-mediated signal are essential for expansion of the DP subset. In terms of the economics of the thymus, this makes complete sense. Since TCR β rearrangement occurs first, appears to be allelically excluded, and requires the generation of a D–J joint followed by an in-frame V–DJ joint, the a priori probability of obtaining an in-frame rearrangement on any one chromosome is one-third. The cumulative probability of obtaining a successful VDJ rearrangement on either chromosome is $1/3 + (1/3 \times 2/3) = 5/9$. Thus nearly half of thymocytes would be expected to fail to make an in-frame TCR β rearrangement. A cell that has succeeded becomes a valuable cell, and is then induced to undergo expansion. In this way, the thymus avoids wasting resources by not expanding clones of cells that have failed to rearrange TCR β productively.

4.3 Positive selection and exit from the DP subset

Mutant mice deficient in the TCR α chain, and TCR transgenic *scid* mice in which the selecting MHC molecule is absent, generate normal numbers of DP thymocytes but do not produce SP cells (Scott *et al.* 1989; Mombaerts *et al.* 1992*a*; Philpott *et al.* 1992). Similarly, mice deficient in the expression of MHC class II produce very few CD4$^+$SP cells, while mice that express very low levels of MHC class I due to a deficiency of β2-microglobulin contain very few CD8$^+$SP cells (Zijlstra *et al.* 1990; Cosgrove *et al.* 1991). Thus it appears that T cells that contain an in-frame TCR β rearrangement undergo full expansion as DP cells, but cannot undergo further maturation to SP cells unless all of the components of the TCR–MHC interaction are present. There is limited evidence concerning the other requirements for this interaction. The role of the co-receptor-associated tyrosine kinase, p56^{lck}, cannot be evaluated because p56^{lck} deficient mice are developmentally arrested at an earlier stage.

5 The complexity of positive selection

5.1 The control of T cell restriction

Antigen-specific T cells recognize antigens in association with individual allelic variants of MHC molecules (MHC restriction), and this was shown to be imposed by the thymus in thymus grafting experiments (Fink and Bevan 1978; Zinkernagel *et al.* 1978). The bone marrow-derived components of the thymus, both T cells and antigen-presenting cells, have little if any influence on positive selection, while fetal thymus grafts depleted of haematopoietic cells (which consist only of thymus epithelial cells) are fully competent to impose MHC restriction specificity on T cells which develop in the graft (Lo and Sprent 1986; Blackman *et al.* 1989). Thus, thymocytes with the potential to recognize exogenous antigens in the context of self-MHC molecules expressed on thymus epithelium are selected to complete their maturation. This is positive selection. The other known selective influence on the T cell repertoire, negative selection, will be dealt with at length elsewhere in this volume (see Chapters 7, 10, and 11). Negative selection appears to be primarily a result of the interaction of thymocytes with bone marrow-derived cells. Specialized antigen-presenting cells such as dendritic cells are highly effective in inducing negative selection (Matzinger and Guerder 1989), but thymocytes are able to negatively select one another, at least for MHC class I-related specificities (Shimonkevitz and Bevan 1988).

5.2 TCR Vβ expression and TCR transgenic mice

The phenomenology of positive selection, defined by thymus grafting experiments, gave rise to the expectation that T cells carrying receptors with the potential for self-restricted recognition would survive thymocyte maturation, while the same T cells in a non-permissive MHC environment would fail to mature. In unmanipulated mice, individual clonal specificities are so rare that there is no way to follow them through maturation in different MHC environments. Fortunately, subpopulations of T cells defined by their usage of TCR Vβ chains are subject to positive selection *en masse*. The best-defined example is the control of Vβ17a expression in the repertoire of SWR and C57L mice.

SWR mice are H-2^q, and express Vβ17a on around 17% of their $CD4^+$ T cells. C57L mice are H-2^b, and only 3% of their $CD4^+$ cells are $V\beta17a^+$. In (SWR × C57L)F1 mice, the high expression of Vβ17a in $CD4^+$ cells is dominant, albeit incompletely (Table 2.1). The same dominance was seen in F1 mice, hemizygous for Vβ17a expression but matched for their non-MHC genetic background. Thus, expression of H-2^q causes positive selection of $V\beta17a^+$ DP cells into the $CD4^+$ T cell subset. Using mice made congenic for the Vβa genotype, which expresses Vβ17a, the positive selection effect was also shown by fetal thymus lobe grafting to be controlled by thymus epithelium (Blackman *et al.* 1989).

In TCR transgenic mice, a majority of thymocytes express the transgenic receptor, although there is some rearrangement of the endogenous TCR α loci, which may generate TCRs of different specificity (see Chapter 16). While individual lines of TCR transgenic mice differ in detail, lines expressing both class I and class II restricted TCRs follow the rules of positive selection. In the presence of the MHC element recognized by the transgenic TCR, a majority of thymocytes express both chains of the transgenic TCR, and there are a large number of transgene positive SP cells which are either $CD4^+$ or $CD8^+$, depending on the restriction specificity of the TCR (Kisielow *et al.* 1988; Sha *et al.* 1988; Berg *et al.* 1989). When the MHC recognition element is absent, differentiation of transgene expressing T cells stops at the DP stage,

Table 2.1 Positive selection of CD4$^+$ T cells with Vβ17a TCRs

Strain	Vβ17 genotype	H-2 genotype	Per cent Vβa17$^+$		Per cent Vβ6$^+$	
			CD4$^+$SP	CD8$^+$SP	CD4$^+$SP	CD8$^+$SP
SWR	a/a	q/q	15.2	1.8	5.7	5.0
C57L	a/a	b/b	2.6	4.1	9.4	17.8
S × L	a/a	b/q	11.1	4.4	2.4	6.3
L × B10	a/b	b/b	1.3	2.1	11.8	13.1
L × B10.G	a/b	b/q	4.5	1.2	10.7	13.2
S × B10	a/b	b/q	6.3	1.0	8.8	10.3
S × B10.G	a/b	q/q	6.4	1.5	9.9	15.6

The data are percentages of single-positive (SP) thymocytes that express either Vβ17a or Vβ6, in mice either homozygous (a/a) or hemizygous (a/b) for Vβ17a. Results are the means of groups of five to ten mice; the standard deviation was always less than 2.0%. In the F1 hybrids, SWR is designated S and C57L is designated L. The table shows that high expression of Vβ17a is dominant in F1 hybrids between strains that express high (SWR) and low (C57L) Vβ17a levels, as would be predicted if the differences in the percentage of Vβ17a$^+$ cells are due to positive selection. In F1 mice with either B10 (H-2^b) or B10.G (H-2^q), expression of H-2^q was essential for high Hβ17a expression in CD4$^+$SP thymocytes.

and any SP thymocytes result from cells that express endogenous TCR α chains, usually in combination with the transgenic TCR β chain. When the transgenic TCRs are backcrossed on to the *scid* mutation, endogenous TCR gene rearrangement does not occur and the MHC environment determines absolutely the thymocyte and peripheral T cell phenotype (Scott *et al.* 1989).

These experiments support a model of the control of T cell restriction in which TCRs are generated by the recombination of TCR β and then TCR α genes, and the cells expressing these TCRs absolutely require an interaction with an MHC molecule to achieve maturation. Individual TCRs appear to be potentially either MHC class I or MHC class II restricted, and their MHC class specificity determines the final CD4/CD8 phenotype of those DP thymocytes that mature. Maturation of DP cells results in: an extended life-span, up-regulation of the surface density of TCR molecules, loss of either CD4 or CD8 expression, the acquisition of functional competence, and the ability to leave the thymus.

Various aspects of this synthesis have been questioned. One problem is the mechanism by which DP cells are driven into either the CD4$^+$SP or the CD8$^+$SP subset. If DP cells are in fact 'instructed' to become either CD4$^+$ or CD8$^+$ depending on the class of the MHC molecules with which they interact, a problem arises as to the mechanism of their MHC class consciousness. If a DP thymocyte's TCR is engaged by an MHC class I molecule, it is likely that the CD8 co-receptor is involved in the interaction; conversely the TCR of a DP cell recognizing class II molecules will be juxtaposed to the CD4 co-receptor. The problem arises because the signalling mechanism employed by the CD4 and CD8 co-receptors appears to be the same: activation through phosphorylation of the tyrosine kinase p56^{lck} (Veillette *et al.* 1989;

Zamoyska *et al.* 1989). There are quantitative differences between the extent of $p56^{lck}$ association with the two co-receptors, so one possible mechanism is that a weak $p56^{lck}$ signal leads to suppression of CD4 expression with preservation of CD8, while a stronger $p56^{lck}$ signal leads to the converse. Alternatively, either CD4 or CD8 could transmit their signals through another molecule, in addition to $p56^{lck}$. There is direct evidence for differential signalling by CD4 and CD8 from an experiment in which a hybrid co-receptor was expressed in transgenic mice. The extracellular domain of CD8 α was attached to the transmembrane and signalling domains of CD4, and this construct was co-expressed with an MHC class I restricted transgenic TCR (Seong *et al.* 1992). These mice developed $CD4^+SP$, transgenic TCR expressing T cells, arguing that the extracellular region of the CD8 α-CD4 construct had been engaged by MHC class I in the thymus, but that the signals transmitted through the CD4 cytoplasmic domain has instructed the cell to assume a $CD4^+SP$ phenotype.

An alternative model of positive selection has been proposed, which is independent of the mechanism of co-receptor signalling. In this 'stochastic/selective' model, DP thymocytes tend to become either $CD4^+CD8^-$ or $CD4^-CD8^+$ at random. Those that carry a TCR/co-receptor combination that is able to recognize MHC molecules are allowed to complete their maturation, while those that express either a non-selectable TCR or the inappropriate co-receptor are eliminated (Robey and Axel 1990). This model has the advantage of economy of mechanism: the only signalling event that need be postulated is one that leads to cell survival. But it has one major problem: in TCR transgenic *scid* mice there should be a subset of thymocytes with an inappropriate CD4/CD8 phenotype for the transgenic TCR, which do not mature and are not represented in the peripheral T cell pool. Such cells, if they exist, must either be extremely short-lived or fall well within the usual limits of the DP population on FACS analysis, because they are extremely difficult to observe.

Nevertheless, such cells may exist and experiments have been performed to test the predictions of the stochastic/selective model. Mice that express a CD4 transgene, but a normal repertoire of TCRs, would be expected to contain $CD4^+CD8^-$ and $CD4^+CD8^+$ peripheral T cells, and indeed they do. The model predicts that within the $CD4^+CD8^+$ T cell population there should be both cells with MHC class I restricted TCRs, which had been positively selected with the involvement of the CD8 molecule; and cells with MHC class II restricted TCRs, that were rescued by positive selection involving the transgenic CD4 molecule. The $CD4^+CD8^+$ T cells were tested for their competence to react to allogeneic MHC class I or class II, and were found to be responsive to both MHC classes (Robey *et al.* 1991*a*; Teh *et al.* 1991). But since there is no information on relative contributions of the TCR and the co-receptors to alloreactivity versus MHC restricted recognition, this doesn't really argue that the $CD4^+CD8^+$ MHC class II alloreactive cells resulted from the rescue of potentially MHC class II restricted $CD8^+SP$ immature thymocytes.

More direct experiments were performed in mice transgenic for both an MHC class I restricted TCR, and CD8. In this case, the experiments were designed to test for the rescue of cells expressing the transgenic TCR, that would have become $CD4^+$SP but for the expression of the CD8 transgene. The results from two research groups are broadly in agreement; enhanced maturation of the $CD8^+$SP cells was observed, but there was no rescue of $CD4^+CD8^+$SP cells expressing the transgenic TCR (Robey *et al.* 1991*b*; Borgulya *et al.* 1991). In this case, there was no evidence for a stochastic/selective mechanism. But new experiments continue to emerge. A current study of a different MHC class I restricted TCR co-expressed with a CD8 transgene has shown an increase in $CD4^+$, endogenous $CD8^-$ thymocytes, compatible with the rescue of MHC class I restricted $CD4^+$SP cells by the transgenic CD8 (E. Robey, personal communication). This supports the stochastic/selective model.

Any model that can be described as 'selective' or 'Darwinian' has an undoubted aesthetic and intellectual appeal. But until there is some clarification of the extent to which positive selection of the T cell repertoire is either 'instructive' or 'stochastic/selective', this appeal will need to be resisted.

5.3 Positive selection and TCR gene recombination

The allelic exclusion of immunoglobulin heavy chain genes in individual B cells is best explained by a model in which the successful rearrangement of a heavy chain gene on either chromosome results in the cessation of heavy chain gene rearrangement and the induction of light chain gene rearrangement. While the mechanisms involved are not clear, this model works well for IgH and TCR β loci. But T cell clones have been described in which there are two in-frame TCR α gene rearrangements (Malissen *et al.* 1988), suggesting that TCR α is not governed by these rules. Similarly, in TCR αβ transgenic mice, rearrangement of the endogenous TCR β loci is suppressed by the transgene but endogenous TCR α locus rearrangement occurs, and in conditions of selection against the transgenic TCR an endogenous TCR α repertoire may be produced (Bluthmann *et al.* 1988).

The ability of developing T cells to generate a second in-frame TCR α rearrangement when they already have an expressible TCR α chain raises the question of what mechanism actually controls TCR α locus rearrangement. A clue comes from the expression of the RAG-1 and RAG-2 genes in thymocytes of TCR transgenic mice. Here, lack of positive selection correlates with high RAG-1/2 mRNA expression, while positively selected cells become RAG-1/2 mRNA negative (Borgulya *et al.* 1992). We have observed the same phenomenon in a comparison of RAG-1 and RAG-2 levels in thymocyte subsets of normal mice (H. T. Petrie, F. Livak, D. G. Schatz, and I. N. Crispe, unpublished data). If RAG-1 and RAG-2 are in fact determinants of the level of TCR α locus recombination, this argues that positive selection shuts off TCR α locus recombination, and stabilizes the expression of a particular TCR as well as leading to cell survival and commitment to a SP subset. This model is attractive because it allows each cell multiple attempts to generate a selectable TCR, with less wastage of thymocytes. But the formal proof will come only from a direct analysis of the extent of endogenous TCR α locus recombination in TCR transgenic mice of selecting and non-selecting MHC backgrounds.

If DP thymocytes are engaged in a cycle of rearrangement and attempts at positive selection, a number of mechanisms could account for their limited life-span *in vivo*. Do they simply run out of TCR α germline elements to rearrange? Is their limited life-span governed by an internal clock, such as the accumulation of apoptosis-promoting molecules such as Fas, or the depletion of apoptosis-inhibiting molecules such as Bcl-2? Or are they simply ejected from the protected environment of the thymus after three or four days, to undergo breakdown elsewhere?

5.4 The death of useless thymocytes

In *in vivo* thymidine labelling experiments, a young adult mouse with 200×10^6 thymocytes generates $20–40 \times 10^6$ new cells per day by cell division (Shortman and Jackson 1974). Yet estimates of the number of cells that exit the thymus and seed the peripheral T cells pool are very much smaller (Scollay *et al.* 1980), leading to the conclusion that a majority of thymocytes fail to mature. Thymocytes might die for three reasons: failure to generate a TCR by gene recombination, expression of a TCR that fails to undergo positive selection, or expression of a self-reactive TCR leading to death by negative selection. The relative contribution of these three sources of unwanted thymocytes is not known, but the fact remains that most DP thymocytes are destined for destruction.

Given this argument, one might expect to see histological evidence of cell death in the thymus. Yet thymus sections do not reveal significant numbers of pyknotic nuclei, and unmanipulated thymocytes do not contain the characteristic DNA fragments corresponding to monomers and oligomers of the length of DNA associated with single nucleosomes. Such DNA fragments are found in tissues undergoing apoptosis. These incongruities, plus the similar turnover kinetics of thymidine and iododeoxyuridine labels in thymocytes despite the fact that one is potentially re-utilized while the other is not (discussed by Rothenberg 1990), suggest that unwanted thymocytes may leave the thymus before undergoing apoptosis. Unselected cells might be expected to be DP, but a large pool of DP cells undergoing apoptosis has not been reported at any anatomical site. Nevertheless the cells may be changing their phenotype as they leave the thymus; or they may be rapidly expelled from the body, for instance through the epithelium of the gut. In the final section of this chapter, a speculative argument is advanced that these cells may rapidly change from a DP to a DN phenotype, and are likely to undergo apoptosis in the liver.

5.5 Unselected thymocytes, DN T cells, and *lpr* disease

While the fate of thymocytes that die as a result of repertoire selection is not known, some educated guesses may be made. A clue comes from the repertoire of the rare subset of αβ T cells that express neither CD4 nor CD8, and that are expanded in some T cell receptor transgenic mice and in *lpr* mutants. These DN T cells are found both in the thymus and in the periphery. Their repertoire has been examined for negative and positive selection of TCR Vβ regions, and has been found to be unselected (Huang and Crispe 1992). A phenotypically identical population of cells is enormously expanded in mice homozygous for the *lpr* mutation (Davidson *et al.* 1986), a defect in the *fas* gene which encodes a cell surface molecule, Fas (Watanabe-Fukunaga *et al.* 1992*a*). Cross-linking of Fas molecules on the surface of human tumour cells by anti-Fas antibody leads to apoptosis (Yonehara and Yonehara 1989), and it has been argued that Fas is a physiological apoptosis trigger. If this line of reasoning is sound, the normal immune system must be generating large numbers of DN T cells. However, these cells are normally eliminated by apoptosis induced by signalling through Fas.

Messenger RNA for Fas is expressed in the thymus, in activated peripheral T cells, and in several non-lymphoid tissues including the ovary (Watanabe-Fukunaga *et al.* 1992*b*). Within the thymus, *fas* expression increases as cells mature from the DN to the DP subset (D. P. M. Hughes and I. N. Crispe, unpublished observations). This is compatible with a role for Fas in the process of the destruction of thymocytes that failed repertoire selection in the thymus. To a first approximation, *lpr* mutant mice exhibit normal clonal deletion of self-reactive T cells (Kotzin *et al.* 1988), although this may vary with the self-antigen under study and the genetic background (Mountz *et al.* 1990; Smyth *et al.* 1992). Thus if the accumulating DN cells of *lpr* mutant mice originate from cells that were unselected in the thymus, the mechanism that produces them is likely to be failure of positive selection.

The proposition that unselected thymocytes become DN before they die by apoptosis is difficult to reconcile with the scarcity of these cells in thymocyte suspensions. But since cell death is not occurring in the thymus at a detectable rate, thymocytes destined for apoptosis must be ejected, to die somewhere else. One candidate for the site of death is the liver. A population of T cells, most of which are DN, are found in the normal liver (Seki *et al.* 1991); and an expanded population is present in *lpr* mutant mice (Ohteki *et al.* 1990). If thymocyte-derived, TCR αβ expressing DN cells are undergoing apoptosis in the normal liver, but failing to die in the *lpr* liver, how is this process regulated? It may be that Fas expression by itself is insufficient to allow the induction of apoptosis, perhaps because the natural ligand(s) of Fas are limited in their anatomical distribution. A more detailed analysis of the distribution of Fas in the thymus, peripheral lymphoid tissue, liver, and other possible sources of T cell elimination such as the intestinal epithelium will clarify the role of this molecule, which may be central in controlling life and death during T cell development.

Acknowledgements

I. N. C. is supported by an Investigator award from the Cancer Research Institute, and D. G. S. is supported by the Howard Hughes Medical Institute.

References

Allison, J. P. and Lanier, L. L. (1987). The T-cell antigen receptor gamma gene: rearrangement and cell lineages. *Immunol. Today*, **8**, 293–6.

Aruffo, A., Stamenkovic, I., Melnick, M., Underhill, C. B., and Seed, B. (1990). CD44 is the principal cell surface receptor for hyaluronidate. *Cell*, **61**, 1303–13.

Asarnow, D. M., Kuziel, W. A., Bonyhadi, M., Tigelaar, R. E., and Tucker, P. W. Allison, J. P. (1988). Limited diversity of γ/δ antigen receptor genes of Thy-1^+ dendritic epidermal cells. *Cell*, **55**, 837–47.

Basch, R. S., Kouri, Y. H., and Karpatkin, S. (1990). Expression of CD4 by human megakaryocytes. *Proc. Natl. Acad. Sci. USA*, **87**, 8085–9.

Berg, L. J., Pullen, A. M., Fazekas de St. Groth, B., Mathis, D., Benoist, C., and Davis, M. M. (1989). Antigen/MHC specific T cells are preferentially exported from the thymus in the presence of their ligand. *Cell*, **58**, 1035–46.

Blackman, M. A., Marrack, P., and Kappler, J. (1989). Influence of the Major Histocompatibility Complex on positive selection of V beta 17a T cells. *Science*, **244**, 214–16.

Bluestone, J. A., Pardoll, D., Sharrow, S. O., and Fowlkes, B. J. (1987). Characterization of murine thymocytes with CD3 associated T cell receptor structures. *Nature*, **326**, 82–4.

Bluthmann, H., Kisielow, P., Uematsu, Y., Malissen, M., Krimpenfort, P., Berns, A., *et al.* (1988). T cell specific deletion of T cell receptor transgenes allows functional rearrangements of endogenous alpha and beta genes. *Nature*, **334**, 156–9.

Bonneville, M., Itohara, S., Krecko, E. G., Mombaerts, P., Ishida, I., Katsuki, M., *et al.* (1990). Transgenic mice demonstrate that epithelial homing of gamma/delta T cells is determined by cell lineages independent of T cell receptor specificity. *J. Exp. Med.*, **171**, 1015–26.

Borgulya, P., Kishi, H., Muller, U., Kirberg, J., and von Boehmer, H. (1991). Development of the CD4 and CD8 lineages of T cells: instruction versus selection. *EMBO J.*, **10**, 913–18.

Borgulya, P., Kishi, H., Uematsu, Y., and von Boehmer, H. (1992). Exclusion and inclusion of alpha and beta T cell receptor alleles. *Cell*, **69**, 529–38.

Carroll, A. M. and Bosma, M. J. (1991). T-lymphocyte development in *scid* mice is arrested shortly after the initiation of T-cell receptor δ gene recombination. *Genes Dev.*, **5**, 1357–66.

Ceredig, R., Glasebrook, A. L., and MacDonald, H. R. (1982). Phenotypic and functional properties of murine thymocytes.1. Precursors of cytolytic T lymphocytes and interleukin-2 producing cells are all contained within a population of mature thymocytes as analyzed by monoclonal antibodies and flow microfluorimetry. *J. Exp. Med.*, **155**, 358–582.

Ceredig, R., Dialynas, D. P., Fitch, F. W., and MacDonald, H. R. (1983). Precursors of T cell growth factor producing cells in the thymus: ontogeny, frequency and quantitative recovery in a subpopulation of phenotypically mature thymocytes defined by monoclonal antibody GK1.5. *J. Exp. Med.*, **158**, 1654–71.

Cosgrove, D., Gray, D., Derich, A., Kaufman, J., Lemeur, J., Benoist, C., *et al.* (1991). Mice lacking MHC class II molecules. *Cell*, **66**, 1051–66.

Crispe, I. N. and Bevan, M. J. (1987). Expression and functional significance of the J11d marker on mouse thymocytes. *J. Immunol.*, **138**, 2013–18.

Crispe, I. N., Moore, M. W., Husmann, L. A., Smith, L., Bevan, M. J., and Shimonkevitz, R. P. (1987). Differentiation potential of subsets of CD4−, CD8− thymocytes. *Nature*, **329**, 336–9.

Davidson, W. F., Dumont, F. J., Bedigian, H. G., Fowlkes, B. J., and Morse, H. C. III. (1986). Phenotypic, functional, and molecular genetic comparisons of the abnormal lymphoid cells of C3H-lpr and C3H-gld/gld mice. *J. Immunol.*, **136**, 4075–84.

Dent, A. L. (1991). Studies on T lymphocyte development using γ/δ antigen receptor transgenic mice. Ph.D. dissertation, University of California, San Diego.

Dent, A. L., Matis, L. A., Hooshmand, F., Widacki, S. M., Bluestone, J. A., and Hedrick, S. M. (1990). Self-reactive γ/δ cells are eliminated in the thymus. *Nature*, **343**, 714–19.

Ezine, S., Weissman, I. L., and Rouse, R. V. (1984). Bone marrow cells give rise to distinct cell clones within the thymus. *Nature*, **309**, 629–31.

Ferrier, P., Krippl, B., Blackwell, T. K., Furley, A. J., Suh, H., Winoto, A., *et al.* (1990). Separate elements control DJ and VDJ rearrangement in a transgenic recombination substrate. *EMBO J.*, **9**, 117–25.

Fink, P. J. and Bevan, M. J. (1978). H-2 antigens of the thymus determine lymphocyte specificity. *J. Exp. Med.*, **148**, 766–72.

Finkel, T. H., Cambier, J. C., Kubo, R. T., Born, W. K., Marrack, P., and Kappler, J. W. (1989). The thymus has two functionally distinct populations of alpha-beta T cells: one population is deleted by ligation of the alpha-beta TCR. *Cell*, **58**, 1047–54.

Fowlkes, B. J., Edison, L., Mathieson, B. J., and Chused, T. M. (1985). Early T lymphocytes: differentiation *in vivo* of adult intrathymic precursor cells. *J. Exp. Med.*, **162**, 802–22.

Frederickson, G. G. and Basch, R. S. (1989). L3T4 antigen expression on hemopoietic precursor cells. *J. Exp. Med.*, **169**, 1473–8.

Garman, R. D., Doherty, P. J., and Raulet, D. H. (1986). Diversity, rearrangement and expression of murine T cell gamma genes. *Cell*, **45**, 733–42.

Gillespie, W., Paulson, J. C., Kelm, S., Pang, M., and Baum, L. G. (1993). Regulation of alpha-2,3-sialyltransferase with conversion of Peanut Agglutinin $(PNA)^+$ to PNA^- phenotype in developing thymocytes. *J. Biol. Chem.*, **268**, 3801–4.

Goldschneider, I., Komschlies, K. L., and Greiner, D. L. (1986). Studies of thymocytopoiesis in rats and mice. I. Kinetics of appearance of thymocytes using a direct intrathymic adoptive transfer assay for thymocyte precursors. *J. Exp. Med.*, **163**, 1–17.

Guidos, C. J., Danska, J. S., Fathman, C. G., and Weissman, I. L. (1990). T cell receptor-mediated negative selection of autoreactive T lymphocyte precursors occurs after commitment to the CD4 or CD8 lineages. *J. Exp. Med.*, **172**, 835–45.

Haas, W. and Tonegawa, S. (1992). Development and selection of γ/ δ T cells. *Curr. Opin. Immunol.*, **4**, 147–55.

Hedrick, S. M. and Dent, A. (1991). A model for γ/δ T-cell development: rearranged γ-and δ-chain genes incorporated into the germline of mice. *Res. Immunol.*, **141**, 588–92.

Heilig, J. and Tonegawa, S. (1986). Diversity of murine gamma genes and expression in fetal and adult T lymphocytes. *Nature*, **322**, 836–40.

Heilig, J. S. and Tonegawa, S. (1987). T-cell gamma gene is allelically but not isotypically excluded and is not required in known functional T-cell subsets. *Proc. Natl. Acad. Sci. USA*, **84**, 8070–4.

Huang, L. and Crispe, I. N. (1992). Distinctive selection mechanisms govern the T cell receptor repertoire of peripheral CD4−CD8− α/β T cells. *J. Exp. Med.*, **176**, 699–706.

Hue, I., Trucy, J., McCoy, C., Couez, D., Mallissen, B., and Malissen, M. (1990). A novel type of aberrant T cell receptor alpha-chain gene rearrangement. Implications for allelic exclusion and the V-J recombination process. *J. Immunol.*, **144**, 4410–19.

Hugo, P., Boyd, R. L., Waanders, G. A., Petrie, H. T., and Scollay, R. (1991*a*). Timing of deletion of autoreactive V beta 6^+ cells and down-modulation of either CD4 or CD8 on phenotypically distinct $CD4^+8^+$ subsets of thymocytes expressing intermediate or high levels of T cell receptor. *Int. Immunol.*, **3**, 265–72.

Hugo, P., Waanders, G. A., Scollay, R., Petrie, H. T., and Boyd, R. L. (1991*b*). Characterisation of immature $CD4^+$CD8−CD3− thymocytes. *Eur. J. Immunol.*, **21**, 835–8.

Husmann, L. A., Shimonkevitz, R. P., Crispe, I. N., and Bevan, M. J. (1988). Thymocyte subpopulations during early fetal development in the BALB/c mouse. *J. Immunol.*, **141**, 736–40.

Ishida, I., Verbeek, S., Bonneville, M., Itohara, S., and Berns, A. (1990). T-cell receptor γ/δ and γ transgenic mice suggest a role of a γ gene silencer in the generation of α/β T cells. *Proc. Natl. Acad. Sci. USA*, **87**, 3067–71.

Itohara, S., Mombaerts, P., Lafaille, J., Iacomini, J., Nelson, A., Clarke, A. R., *et al.* (1993). T cell receptor δ gene mutant mice independent generation of α/β T cells and programmed rearrangements of γ/δ TCR genes. *Cell*, **72**, 337–48.

Kishi, H., Borgulya, P., Scott, B., Karjalainen, K., Traunecker, A., Kaufman, J., *et al.* (1991). Surface expression of the beta T cell receptor (TCR) chain in the absence of other TCR or CD3 proteins on immature T cells. *EMBO J.*, **10**, 93–100.

Kisielow, P., Teh, H.-S., Bluthmann, H., and von Boehmer, H. (1988). Positive selection of antigen-specific T cells in the thymus by restricting MHC molecules. *Nature*, **335**, 730–3.

Korman, A., Marusic-Galesic, S., Spencer, D., Kruisbeek, A., and Raulet, D. H. (1988). Predominant variable region gene usage by γ/δ T cell receptor-bearing cells in the adult thymus. *J. Exp. Med.*, **168**, 1021–40.

Kotzin, B. L., Babcock, S. K., and Herron, L. R. (1988). Deletion of potentially self-reactive T cell receptor specificities in L3T4−, Lyt-2− T cells of lpr mice. *J. Exp. Med.*, **168**, 2221–9.

Lesley, J., Schulte, R., and Hyman, R. (1988). Kinetics of thymus repopulation by intrathymic progenitors after intravenous injection: evidence for successive repopulation by an IL-$2R^+$, Pgp-1− and by an IL-2R−, Pgp-1^+ progenitor. *Cell. Immunol.*, **117**, 378–88.

Lew, A. M., Pardoll, D. M., Maloy, W. L., Fowlkes, B. J., Kruisbeek, A., Cheng, S. F., *et al.* (1986). Characterization of T cell receptor gamma chain expression in a subset of murine thymocytes. *Science*, **234**, 1401–5.

Li, C. L. and Johnson, G. R. (1992). Rhodamine123 reveals heterogeneity within murine Lin−, Sca-1^+ hematopoietic stem cells. *J. Exp. Med.*, **175**, 1443–7.

Lo, D. and Sprent, J. (1986). Identity of cells that imprint H-2 restricted T cell specificity in the thymus. *Nature*, **319**, 672–4.

Malek, T. R., Schmidt, J. A., and Shevach, E. M. (1985). The murine IL-2 receptor. III. Cellular requirements for the induction of IL-2 receptor expression on T cell subpopulations. *J. Immunol.*, **134**, 2405–13.

Malissen, M., Trucy, J., Letourneur, F., Rebal, N., Dunn, D. E., Fitch, F. W., *et al.* (1988). A T cell clone expresses two T cell receptor alpha genes but uses one alpha beta heterodimer for allorecognition and self MHC-restricted antigen recognition. *Cell*, **55**, 49–59.

Matzinger, P. and Guerder, S. (1989). Does T-cell tolerance require a dedicated antigen-presenting cell? *Nature*, **338**, 74–6.

Moisan, J. P., Bonneville, M., Bouyge, I., Moreau, J. F., Soulillou, J. P., and Lefranc, M. P. (1989). Characterization of T-cell-receptor gamma (TRG) gene rearrangements in alloreactive T-cell clones. *Hum. Immunol.*, **24**, 95–110.

Molina, T. J., Kishihara, K., Siderovski, D. P., Van Ewijk, W., Narendran, A., Timms, E., *et al.* (1992). Profound block in thymocyte development in mice lacking p56-lck. *Nature*, **357**, 161–4.

Mombaerts, P., Clark, A. R., Rudnicki, M. A., Iacomini, J., Itohara, S., Lafaille, J., *et al.* (1992*a*). Mutations in T cell receptor genes alpha and beta block thymocyte development at different stages. *Nature*, **360**, 225–31.

Mombaerts, P., Iacomini, J., Johnson, R. S., Herrup, K., Tonegawa, S., and Papaiannou, V. E. (1992*b*). RAG-1 deficient mice have no mature B and T lymphocytes. *Cell*, **68**, 869–77.

Mountz, J. D., Smith, T. M., and Toth, K. S. (1990). Altered expression of self-reactive T cell receptor Vβ regions in autoimmune mice. *J. Immunol.*, **144**, 2159–66.

Muller-Sieburg, C. E., Whitlock, C. A., and Weissman, I. L. (1986). Isolation of two early B lymphocyte progenitors from mouse marrow: a committed pre-pre-B cell and a clonogenic Thy-1-low haematopoietic stem cell. *Cell*, **44**, 653–62.

Nikolic-Zugic, J. and Bevan, M. J. (1988). Thymocytes expressing CD8 differentiate into $CD4^+$ cells following intrathymic injection. *Proc. Natl. Acad. Sci. USA*, **85**, 8633–7.

Ohteki, T., Seki, S., Abo, T., and Kumagai, K. (1990). Liver is a possible site for the proliferation of abnormal $CD3^+4-8-$ double-negative lymphocytes in autoimmune MRL-lpr/lpr mice. *J. Exp. Med.*, **172**, 7–12.

Okazaki, K. and Sakano, H. (1988). Thymocyte circular DNA excised from T cell receptor alpha-delta gene complex. *EMBO J.*, **7**, 1669–74.

Pardoll, D. M., Fowlkes, B. J., Bluestone, J. A., Kruisbeek, A., Maloy, W. L., Coligan, J. E., *et al.* (1987). Differential expression of two distinct T-cell receptors during thymocyte development. *Nature*, **326**, 79–81.

Pearse, M., Gallagher, P., Wilson, A., Wu, L., Fisicaro, N., Miller, J. F. A. P., *et al.* (1988). Molecular characterization of T-cell antigen receptor expression by subsets of CD4− CD8− murine thymocytes. *Proc. Natl. Acad. Sci. USA*, **85**, 6082–6.

Pearse, M., Li, W., Egerton, M., Wilson, A., Shortman, K., and Scollay, R. A. (1989). Murine early thymocyte developmental sequence is marked by transient expression of the interleukin-2 receptor. *Proc. Natl. Acad. Sci. USA*, **86**, 1614–18.

Petrie, H. T., Pearse, M., Scollay, R., and Shortman, K. (1990*a*). Development of immature thymocytes: Initiation of CD3, CD4 and CD8 acquisition parallels down-regulation of the interleukin-2 receptor alpha chain. *Eur. J. Immunol.*, **20**, 2810–13.

Petrie, H. T., Scollay, R., and Shortman, K. (1990*b*). Lineage relationships and developmental kinetics of imature thymocytes: CD3, CD4 and CD8 acquisition *in vivo* and *in vitro*. *J. Exp. Med.*, **172**, 1580–3.

Petrie, H. T., Scollay, R., and Shortman, K. (1992). Commitment to the TCR alpha-beta or gamma-delta lineages can occur just prior to the onset of CD4 and CD8 expression amongst immature thymocytes. *Eur. J. Immunol.*, **22**, 2185–8.

Philpott, K. L., Viney, J. L., Kay, G., Rastan, S., Gardiner, E. M., Chae, S., *et al.* (1992). Lymphoid development in mice congenitally lacking T-cell receptor alpha/beta expressing cells. *Science*, **256**, 1448–52.

Raulet, D. H. (1989). The structure, function, and molecular genetics of the γ/δ T cell receptor. *Annu. Rev. Immunol.*, **7**, 175–207.

Raulet, D. H., Spencer, D. M., Hsiang, Y. H., Goldman, J. P., Bix, M., Liao, N. S., *et al.* (1991). Control of γ/δ T-cell development. *Immunol. Rev.*, **120**, 185–204.

Reilly, E. B., Krenz, D. M., Tonegawa, S., and Eisen, H. N. (1986). A functional γ gene formed from known γ-gene segments is not necessary for antigen-specific responses of murine cytotoxic T lymphocytes. *Nature*, **321**, 878–90.

Reisner, Y., Linker-Israeli, M., and Sharon, N. (1976). Separation of mouse thymocytes into two subpopulations by the use of peanut agglutinin. *Cell. Immunol.*, **25**, 129–34.

Richie, E. R., McEntire, B., Crispe, N., Kimura, J., Lanier, L. L., and Allison, J. P. (1988). Alpha-beta T cell receptor gene and protein expression occurs at early stages of thymocyte differentiation. *Proc. Natl. Acad. Sci. USA*, **85**, 1174–8.

Robey, E. and Axel, R. (1990). CD4: Collaborator in immune recognition and HIV infection. *Cell*, **60**, 697–700.

Robey, E., Ramsdell, F., Elliott, J., Raulet, D., Kioussis, D., Axel, R., *et al.* (1991*a*). Expression of CD4 in transgenic mice alters the specificity of CD8 T cells for allogeneic major histocompatibility complex. *Proc. Natl. Acad. Sci. USA*, **88**, 608–12.

Robey, E., Fowlkes, B. J., Gordon, J. W., Kioussis, D., von Boehmer, H., Ramsdell, F., *et al.* (1991*b*). Thymic selection in CD8 transgenic mice supports an instructive model for commitment to a CD4 or CD8 lineage. *Cell*, **64**, 99–107.

Rothenberg, E. (1990). Death and transfiguration of cortical thymocytes: a reconsideration. *Immunol. Today*, **11**, 116–19.

Rupp, R., French, G., Hengartner, H., Zinkernagel, R. M., and Joho, R. (1986). No functional γ-chain transcripts detected in an alloreactive cytotoxic T-cell clone. *Nature*, **321**, 876–78.

Scollay, R. (1983). The long-lived medullary thymocyte re-visited: precise quantitation of very small subset. *J. Immunol.*, **132**, 1085–8.

Scollay, R., Butcher, E., and Weissman, I. L. (1980). Thymus cell migration. Quantitative aspects of cellular traffic from the thymus to the periphery in mice. *Eur. J. Immunol.*, **10**, 210–18.

Scollay, R., Wilson, A., D'Amico, A., Kelly, K., Egerton, M., Pearse, M., *et al.* (1988). Developmental status and reconstitution potential of subpopulations of murine thymocytes. *Immunol. Rev.*, **104**, 81–120.

Scott, B., Bluthmann, H., Teh, H. S., and von Boehmer, H. (1989). The generation of mature T cells requires interaction of the alpha beta T cell receptor with major histocompatibility antigens. *Nature*, **338**, 591–3.

Seki, S., Abo, T., Ohteki, T., Sugiura, K., and Kumagai, K. (1991). Unusual α/β-T cells expanded in autoimmune lpr mice are probably a counterpart of normal T cells in the liver. *J. Immunol.*, **147**, 1214–21.

Seong, R. K., Chamberlain, J. W., and Parnes, J. R. (1992). Signal for T cell differentiation to a CD4 cell lineage is delivered by CD4 transmembrane region and/or cytoplasmic tail. *Nature*, **356**, 718–20.

Sha, W. C., Nelson, C. A., Newberry, R. D., Kranz, D. M., Russell, J. H., and Loh, D. Y. (1988). Positive and negative selection of an antigen receptor on T cells in transgenic mice. *Nature*, **336**, 73–6.

Shimonkevitz, R. P. and Bevan, M. J. (1987). Split tolerance induced by the intrathymic adoptive transfer of thymocyte stem cells. *J. Exp. Med.*, **168**, 143–56.

Shimonkevitz, R. P., Husmann, L. A., Bevan, M. J., and Crispe, I. N. (1987). Transient expression of IL-2 receptor precedes the differentiation of immature thymocytes. *Nature*, **329**, 157–9.

Shinkai, Y., Rathbun, G., Lam, K.-P., Oltz, E. M., Stewart, V., Mendelsohn, M., *et al.* (1992). RAG-2-deficient mice lack mature lymphocytes owing to inability to initiate V(D)J rearrangement. *Cell*, **68**, 855–67.

Shortman, K. (1992). Cellular aspects of early T cell development. *Curr. Opin. Immunol.*, **4**, 140–6.

Shortman, K. and Jackson, H. (1974). The differentiation of T lymphocytes. I. Proliferation kinetics and interrelationships of subpopulations of mouse thymus cells. *Cell. Immunol.*, **12**, 230–46.

Shortman, K., Wilson, A., Egerton, M., Pearse, M., and Scollay, R. (1988). Immature $CD4^-$, $CD8^+$ murine thymocytes. *Cell. Immunol.*, **113**, 462–79.

Smyth, L., Howell, M., and Crispe, I. N. (1992). Abnormal $CD4^+$ T cell populations and defective clonal deletion vary independently in MRL-Mp-$^+/^+$ and MRL-Mp-lpr/lpr mice. *Dev. Immunol.*, **3**, 309–18.

Spangrude, G. J. and Scollay, R. (1990). Differentiation of hematopoietic stem cells in irradiated mouse thymic lobes: kinetics and phenotype of progeny. *J. Immunol.*, **145**, 3661–8.

Spangrude, G. J., Heimfeld, S., and Weissman, I. L. (1988). Purification and characterization of mouse hematopoietic stem cells. *Science*, **241**, 58–62.

Takagaki, Y., Nakanishi, N., Ishida, I., Kanagawa, O., and Tonegawa, S. (1989). T cell receptor-gamma and -delta genes preferentially utilized by adult thymocytes for the surface expression. *J. Immunol.*, **142**, 2112–21.

Takeshita, S., Toda, M., and Yamagishi, H. (1989). Excision products of the T cell receptor gene support a progressive rearrangement model of the α/δ locus. *EMBO J.*, **8**, 3261–70.

Traunecker, A., Oliveri, F., Allen, N., and Karjalainen, K. (1986). Normal T cell development is possible without 'functional' γ chain genes. *EMBO J.*, **5**, 1589–93.

Veillette, A., Bookman, M. A., and Horak, E. M. (1989). Signal transduction through the CD4 receptor involves the activation of the internal membrane tyrosine-protein kinase p56lck. *Nature*, **338**, 257–8.

von Boehmer, H., Bonneville, M., Ishida, I., Ryser, S., Lincoln, G., Smith, R. T., *et al.* (1988). Early expression of a T-cell receptor beta-chain transgene suppresses rearrangement of the V gamma 4 gene segment. *Proc. Natl. Acad. Sci. USA*, **85**, 9729–32.

Watanabe-Fukunaga, R., Brannan, C. I., Copeland, N. G., Jenkins, N. A., and Nagata, S. (1992*a*). Lymphoproliferation disorder in mice explained by defects in Fas antigen that mediates apoptosis. *Nature*, **356**, 314–56.

Watanabe-Fukunaga, R., Brannan, C. I., Itoh, M., Yonehara, S., Copeland, N. G., Jenkins, N., *et al.* (1992*b*). The cDNA structure, expression, and chromosomal assignment of the mouse Fas antigen. *J. Immunol.*, **148**, 1274–9.

Winoto, A. and Baltimore, D. (1989*a*). Separate lineages of T cells expressing the α/β and γ/δ receptors. *Nature*, **338**, 430–2.

Winoto, A. and Baltimore, D. (1989*b*). α/β lineage-specific expression of the α T cell receptor gene by nearby silencers. *Cell*, **59**, 649–55.

Wu, L., Scollay, R., Egerton, M., Pearse, M., Spangrude, G. J., and Shortman, K. (1991). CD4 expressed on earliest T-lineage cells in the adult murine thymus. *Nature*, **349**, 71–4.

Yonehara, S. and Yonehara, M. (1989). A cell-killing monoclonal antibody (anti-Fas) to a cell surface antigen co-downregulated with the receptor of tumor necrosis factor. *J. Exp. Med.*, **169**, 1747–56.

Zamoyska, R., Derham, P., Gorman, S. D., von Hoegen, P., Bolen, J. B., Veillette, A., *et al.* (1989). Inability of CD8 alpha prime polypeptides to associate with p56lck correlates with impaired function *in vitro* and lack of expression *in vivo*. *Nature*, **342**, 278–81.

Zijlstra, M., Bix, M., Simister, N., Loring, J. M., Raulet, D. H., and Jaenisch, R. (1990). Beta-2 microglobulin deficient mice lack CD4–8^+ cytotoxic T cells. *Nature*, **344**, 742–6.

Zinkernagel, R. M., Callahan, G. N., Althage, A., Cooper, S., Klein, P. A., and Klein, J. (1978). On the thymus in the differentiation of H-2 self-recognition by T cells: Evidence for dual recognition. *J. Exp. Med.*, **147**, 882–96.

Zuniga-Pflucker, J. C. and Kruisbeek, A. M. (1990). Intrathymic radioresistant stem cells follow an IL-2/IL-2R pathway during thymic regeneration after sublethal irradiation. *J. Immunol.*, **144**, 3736–40.

3 The function of αβ T cells and the role of the co-receptor molecules, CD4 and CD8

ROSE ZAMOYSKA AND PAUL TRAVERS

1 Introduction

The primary function of T cells bearing αβ T cell receptors is to recognize the presence of pathogens within the body and to activate their disposal, either directly or by recruiting other immune cells. However, the nature and location of the various pathogens with which the immune system has to deal imposes constraints on both the recognition of the pathogen and the mechanisms that must be invoked to dispose of it. In response to these constraints the immune system has diversified both the means used to recognize the presence of pathogens and the effector functions of the responding cells.

The immune system must respond to pathogens that belong to two classes; those that exist outside of the cell, and here we consider that endosomal vesicles are outside of the cell, and those that exist inside the cell. For

Table 3.1 Summary of the major effector responses of αβ T cells to pathogens localized in different cellular compartments

	Cell-mediated immunity		Humoral immunity
Typical pathogens	Vaccinia virus Influenza virus *Listeria* *Trypanosoma cruzi*	*Mycobacterium tuberculosis* *Mycobacterium leprae* *Leishmania donovari*	*Clostridium tetani* *Staphylococcus aureus* *Streptococcus penumoniae* Polio virus
Location	Intracellular cytosol	Macrophage vesicles	Extracellular fluid
Effector T cell	Cytolytic $CD8^+$ T cell	Inflammatory $CD4^+$ T cell (T_h1)	Helper $CD4^+$ T cell (T_h2)
Antigen presentation	Peptide–MHC class I on infected cell	Peptide–MHC class II on infected macrophage	Peptide–MHC class II on specific B cell
Lymphokine/ cytokines produced	Perforin Granzymes Cytolysin IFNγ	IFNγ GM-CSF IL-3 IL-2	IL-4, IL-5, IL-6 IL-3 GM-CSF IL-10, TGFβ
Effector response	Death of infected cell	Activation of infected macrophage	Activation of specific B cell to make antibody

A few examples of typical pathogens which localize to intracellular or extracellular sites are indicated together with the predominant T cell responses they initiate and the main lymphokines or cytokines produced by the different T cell subsets.

example, see Table 3.1, eukaryotic parasites, toxins, and some bacteria exist either in the blood or in the tissue spaces. Many bacteria, for example *Leishmania*, are adapted to live within endocytic vesicles of macrophages and resist degradation by endosomal enzymes. Finally, all viruses and some bacteria, notably Mycobacteria, cross cell membranes and carry out their life cycle within the cytoplasm of a host cell. As we shall see, there are two mechanisms by which the presence of pathogens is notified to the immune system that correspond to this topological division. In terms of effector function, however, there are three major classes of response that correspond to the locations in which pathogens are found (Table 3.1). Inflammatory T cells (Th1) activate macrophages to destroy engulfed bacteria while helper T cells (Th2) activate B cells to secrete antibody to neutralize, opsonize, and induce lysis of extracellular pathogens. Finally cytotoxic T cells (CTL) kill cells infected with cytoplasmic pathogens. Clearly, the effector functions of T cells are closely linked to the sites in which pathogens are found and thus a T cell must not only recognize the presence of foreign antigen but must also be able to identify the cellular compartment from which the antigen was derived. How then does the T cell carry out these tasks?

2 αβ T cells recognize antigen–MHC complexes

Unlike B cells, whose receptors for antigen, membrane bound immunoglobulin molecules, can recognize foreign antigens directly, the antigen receptors of T cells are constrained to recognize a complex of a self-protein, an MHC molecule, with a short peptide fragment, the nominal antigen. It is generally considered that the requirement for recognition of self-MHC molecules as part of the recognition of foreign antigens was first demonstrated by Zinkernagel and Doherty (1974) who showed that T cells acquired the ability to recognize foreign antigen (in their example, lymphocytic choriomeningitis virus was used) only in the context of MHC molecules. However, in the light of current understanding of the role of polymorphism in the MHC molecules in selecting peptides of different sequences, the Zinkernagel/Doherty experiment has an alternative explanation in that the nominal antigen presented for recognition is different, depending on the MHC alleles expressed in the presenting cell. What then is the evidence that T cell receptors recognize MHC molecules at all? At present the best evidence is derived from experiments in mice transgenic for a T cell receptor of a defined specificity for antigen and MHC (Sha *et al.* 1988; Kisielow *et al.* 1988; Berg *et al.* 1989; Scott *et al.* 1989). In the absence of both the antigen and the appropriate MHC molecule, thymocytes bearing the transgenic receptor fail to develop into mature T cells and instead undergo programmed cell death within the thymus. In the presence of the appropriate MHC allele, however, mature T cells are produced. These results indicate that T cells are positively selected by the correct MHC allele and must recognize it independently of the antigenic peptide.

2.1 Class I and class II MHC molecules present antigens derived from distinct locations

There are two distinct classes of MHC molecule capable of presenting antigens to T cells. Class I MHC molecules (HLA-A, B, and C in man and H-2 K, D, and L in the mouse) are expressed on almost all nucleated cells, although the level of expression does vary between cell types. On the other hand, MHC class II molecules (HLA-DR, DQ, and DP in man, H-2E and A in mouse) are expressed only on a subset of cells with specific immunological functions, namely dendritic cells, macrophages, B cells, and thymic epithelial cells. Expression of class II molecules can be induced on other cell types by the cytokines interferon-γ and TNF, which also increase the expression of the class I molecules.

The source of antigens presented by the MHC class I and class II molecules also differs. MHC class I molecules present antigens derived from cytoplasmic peptides (Townsend and Bodmer 1989); in an uninfected cell these would be self-peptides. Cytoplasmic and nuclear proteins of viral origin and proteins from intracytoplasmic bacteria provide the predominant sources of antigenic peptides presented by MHC class I molecules although recent data suggest that any aggregated protein could provide a source of peptides for MHC class I molecules (Pfieffer *et al.* 1993). The site of protein degradation to provide peptide fragments for class I molecules is not well defined. Clearly, degradation within the cytosol of the cell plays a major role in generating the antigens presented by class I molecules but a role for further proteolytic processing within the lumen of the endoplasmic reticulum has not been ruled out. Whether or not the peptide fragment that binds to the class I molecule is produced in its final form within the cytoplasm of the cell, the antigenic fragment must still be transported from the cytosol into the lumen of the ER. This function is carried out by the TAP (*T*ransporter associated with *A*ntigen *P*resentation) protein dimer, a member of a large family of transport proteins (reviewed in Monaco 1992). Within the lumen of the endoplasmic reticulum, class I molecules that have not bound a peptide antigen are found associated with an ER resident protein, calnexin (Galvin *et al.* 1993). Upon binding of peptide, the class I molecule is released from calnexin and is transported from the ER to the cell surface. Calnexin therefore retains the 'empty' class I molecule within the ER until a peptide is bound. Since the binding of peptides to class I molecules is essentially irreversible, this mechanism ensures that the class I molecule is selective for peptides that have been generated within the cytosol and translocated into the endoplasmic reticulum of the antigen-presenting cell.

MHC class II molecules show a converse pattern, being inhibited from binding peptides within the ER and being specialized to present peptide antigens derived predominantly from extracellular sources, and degraded within intracellular vesicles (endosomes or lysosomes, Brodsky and Guagliardi 1991). Within the ER, class II α and β chains are assembled with the invariant chain, Ii, in a nonameric complex consisting of a trimer of Ii and three class II heterodimers (Roche *et al.* 1991); this complex is transported from the ER and is thought to route the nascent class II molecules to an endosomal compartment, although the precise nature of the endosomal compartment is not yet defined. Association of class II with the Ii inhibits the binding of peptides to the class II molecule, thus preventing binding of peptides within the ER, while limited proteolysis of the invariant chain within the endosomal compartment releases the inhibition and allows peptide binding. Further proteolysis of Ii frees the class II molecule to be transported from the endosomal compartment to the cell surface. Thus the class II molecule is prevented from binding peptides transported into the endoplasmic reticulum and is targeted to a compartment where proteolysis of endocytosed, extracellular proteins occurs.

2.2 T cells with distinct effector functions recognize class I and class II MHC molecules

The specialization of MHC molecules to present antigens from cytoplasmic or extracellular sources is reflected in the specialization of effector functions in the T cells that recognize class I and class II molecules respectively (Table 3.1). Class I molecules present antigen to cytolytic T cells. Naïve CTL (CTLp) become activated upon recognition of their class I+peptide ligand and begin synthesis of a set of effector molecules, perforin and granzymes, that become localized in secretory granules (Griffiths and Mueller 1991). The T cell is now armed and any subsequent recognition by the T cell results in the release of the granule contents on to the surface of the target cell. Perforin, which shows homology to the late complement component C9, forms pores in the target cell membrane (Podack *et al.* 1991; Yagita *et al.* 1992), allowing entry of the granzymes into the cytoplasm of the target cell which, in an as yet undefined mechanism, activate the process of apoptosis in the target cell.

Recognition of antigen in association with class I molecules results in the death of the presenting cell; recognition of complexes of antigenic peptides with MHC class II molecules however, results in the activation of the presenting cell. Class II molecules are recognized by T helper cells which, when activated, can be subdivided into two discrete subclasses defined by their patterns of cytokine secretion and consequently their effector phenotype (reviewed in Hayakawa and Hardy 1991; Mossmann *et al.* 1991; Swain *et al.* 1991). Naïve helper T cells become activated by their first contact with antigen–class II complexes and initially secrete a broad range of cytokines including IL-2, 3, 4, 5, 10, and IFNγ. Later, these cells differentiate either into Th1, or inflammatory T cells, which upon a second encounter with antigen secrete principally IL-2, IFNγ, and TNF, and Th2, or helper T cells, which on a second encounter secrete principally IL-4, 5, 6, and 10. The cytokines secreted by Th1 cells are potent activators of macrophages, increasing their ability to kill phagocytosed micro-organisms and, through the release of inflammatory mediators, to recruit macrophages, lymphocytes, and neutrophils to the site of activation. The cytokines secreted by Th2 cells both activate B cells and drive their differentiation into antibody-secreting cells, while cytokines from both Th cell subsets regulate the isotype switching of B cells. Moreover, activated macrophages up-regulate the level of Fc receptors, particularly FcγRI, and can more efficiently phagocytose immune complexes and antibody-coated pathogens. Thus the cells presenting antigen in the context of class II MHC molecules are activated by Th cells to express effector functions that are important in the elimination of extracellular micro-organisms and toxins, and in the elimination of intravesicular parasites, like *Leishmania*.

The effector functions of the T cells that recognize antigen in the context of class I and class II MHC molecules are therefore quite distinct, and are tailored to eliminate pathogens from the two major compartments (intracytoplasmic versus extracellular and vesicular) in which they occur. It is important for the T cell, therefore, that it be able to recognize not just the antigen but also which class of MHC molecule is presenting the antigen in order for the correct effector mechanisms to be activated.

2.3 αβ T cell recognition of MHC–antigen complexes

The structures of a number of MHC class I molecules have now been determined by X-ray crystallography (Bjorkman *et al.* 1987*a*,*b*) and the insights gained from these has clarified the nature of the antigen–MHC complex that is recognized by T cells. The amino-terminal two domains of the MHC class I molecule form an extended groove in which short peptide fragments bind. In MHC class I molecules the ends of the peptide are tightly bound within the structure of the MHC molecule itself, while the central residues protrude from the cleft and can be recognized by the T cell receptor. In the case of the MHC class II molecule, the ends of the peptide fragment are not tightly bound and can protrude from either end of the cleft. In addition, more of the peptide is accessible at the upper surface of the molecule and available for recognition by the T cell receptor. From the limited number of structures of MHC molecules binding a single peptide, it is difficult to imagine that a T cell receptor could contact the peptide fragment without at the same time contacting the MHC molecule. The nature of peptide binding to the MHC molecule therefore imposes the requirement for co-recognition of the MHC molecule.

At present, the precise site at which the T cell receptor contacts the MHC–peptide complex is not known, nor is the orientation of the receptor with respect to the MHC molecule known. Davis and Bjorkman (1988) have postulated a model for the interaction of the T cell receptor with the MHC–peptide complex that correlates the diversity in the CDR3 loops of the T cell receptor V domains with the diverse nature of the antigenic peptides lying in the centre of the binding groove, to suggest that the CDR1 and CDR2 loops of the T cell receptor contact the MHC molecule while the CDR3 loops contact the bound peptide. Certainly, there is experimental data consistent with this model, where alterations in the amino acid residues of the peptide seen by the T cell receptor are correlated with changes in the CDR3 regions (reviewed in Jorgensen *et al.* 1992). However, one corollary of this model, in its strictest form is that particular TCR Vα and Vβ regions would show a preference for either class I or class II molecules, since the CDR1 and CDR2 loops are encoded within the V region and are not affected by junctional diversity nor, in T cells, somatic hypermutation. While there are some differences in V usage between T cell receptors that recognize class I and class II molecules, most V regions can be used to recognize either class of MHC molecule, and in one notable example, identical Vα and Vβ segments were found to be used by two clones with distinct specificities, one restricted by class I MHC and the second by class II MHC molecules (Rupp *et al.* 1987).

If the sequences of the T cell receptors themselves do not dictate whether they will recognize class I or class II MHC, and as the functions of αβ T cells are intimately related to the class of MHC molecule recognized, how is any individual T cell restricted to recognizing antigen only in the context of class I or class II MHC? In addition to the TCR, T cells express a second molecule which is absolutely committed to recognizing only one class of MHC molecule, and which is involved in T cell differentiation and activation. For class II MHC restricted cells this is the CD4 molecule, and for class I MHC restricted cells it is the CD8 molecule (Swain 1983). Thus there are *de facto* two receptors on T cells which together confer antigen/MHC specificity, the variable αβ TCR and an invariant molecule, CD4 or CD8. Janeway coined the term co-receptors for these molecules, in acknowledgement of the intimate part played by CD4 and CD8 in T cell recognition and activation (Janeway *et al.* 1989). This chapter examines the structure of these co-receptors and reviews our current ideas of their roles in αβ T cell function.

3 Co-receptors, CD4 and CD8

A considerable body of information has been gained concerning the co-receptors CD4 and CD8. The genes encoding CD4 and CD8 have been identified and sequenced (for reviews see Littman 1987; Parnes 1989) and the nature of the control elements is beginning to be defined. The proteins have been well-characterized and recently the external domains were crystallized and the structures solved (Ryu *et al.* 1990; Wang *et al.* 1990; Leahy *et al.* 1992). These molecules have two well-documented interactions, one, between the external domain of the co-receptor and the respective MHC ligand and the second between the cytoplasmic domain and a T cell-specific protein tyrosine kinase, $p56^{lck}$. As these interactions are key to their functions they will be discussed initially in relation to the structure of the molecules before going on to consider the role of the co-receptors in T cell activation.

3.1 Structure of CD4 and CD8

The external structures of CD4 and CD8 are quite distinct from one another although both are members of the Ig superfamily. CD4 is a single polypeptide chain with four external Ig-like domains (D1–D4), a transmembrane region, and a cytoplasmic tail. The external domains interact in pairs, D1 with D2 and D3 with D4, in a novel configuration not previously seen for Ig domains, to form two rigid 'cupped hand'-like structures with a flexible connection between (Ryu *et al.* 1990; Wang *et al.* 1990; see Figure 3.1). The binding site on CD4 for class II MHC has been mapped to either side of the concavity formed by the packing of the D1 and D2 domains (Lamarre *et al.* 1989; Clayton *et al.* 1989). The overall size of the external portion of the CD4 molecule is predicted to be about 130 Å, which is compatible with the idea that CD4 can interact with the same MHC molecule as a T cell receptor, extending alongside the receptor to contact the membrane proximal domain of the MHC molecule (Figure 3.2). In contrast to CD4, CD8 can be expressed in two forms, either as a disulfide linked homodimer of two CD8 α polypeptides or as a disulfide linked heterodimer of an α and β polypeptide. Although the genes encoding CD8 α and β are closely linked (Gorman *et al.* 1988) and the two polypeptides are predicted to have a similar secondary structure, nevertheless they have very little sequence homology with each other (Johnson and Williams 1986). Both CD8 α and β have a single Ig-like external domain followed by a polypeptide linker connecting to the transmembrane and cytoplasmic domains (Figure 3.2). The crystal structure of the N-terminal 114 amino acids of CD8 α has been solved (Leahy *et al.* 1992) and shows that this region of the molecule possesses a fold typical of immunoglobulin variable domains with CDR-like loops oriented upwards and to the side (Figure 3.1). These loop regions of CD8 α have been implicated in binding class I MHC from site-directed mutagenesis studies (Sanders *et al.* 1991). It is not clear, however, whether the binding involves simply the upward facing loops or whether the loops to the side also play a role, in which case the possibility arises that CD8 homodimers may be capable of interacting with more than one MHC molecule at a time.

Fig. 3.1 Structures of the co-receptor molecules CD8 and CD4. A. The structure of the external Ig-like domain of the CD8 α homodimer is shown (Leahy *et al.* 1992), with the β-sheets indicated as broad ribbons with arrows indicating their orientation. Interconnecting loops are the regions which are thought to bind the MHC molecule. As yet it is not known whether the interaction with MHC occurs through the upwards pointing loops only or whether some of the side loops are also involved. B. The structure of the two N-terminal most domains of the CD4 molecule is shown (Ryu *et al.* 1990; Wang *et al.* 1990), illustrating the novel packing adopted by these Ig-related structures. The binding region for class II MHC has been mapped to the loops that lie along the broad face which results from the interaction between the two domains. The relative association of the two domains is maintained by the β-strand which runs from the first to the second domain. (Figures were produced on a Silicon Graphics workstation using the program Setor.)

Although the arrangement of the external domains of CD4 and CD8 is quite different, both bind similar sites on the MHC molecule. CD8 αα homodimers have been shown to bind the α3 domain of class I MHC molecules in a non-polymorphic region containing a loop (residues 222–229) which protrudes from the compact Ig β-barrel structure (Figure 3.2; Salter *et al.* 1990). The class I α3 loop is predominantly negatively charged, while the CDR loops of CD8 are mainly positively charged suggesting electrostatic interactions may be important in binding. Interestingly if the class I MHC

A

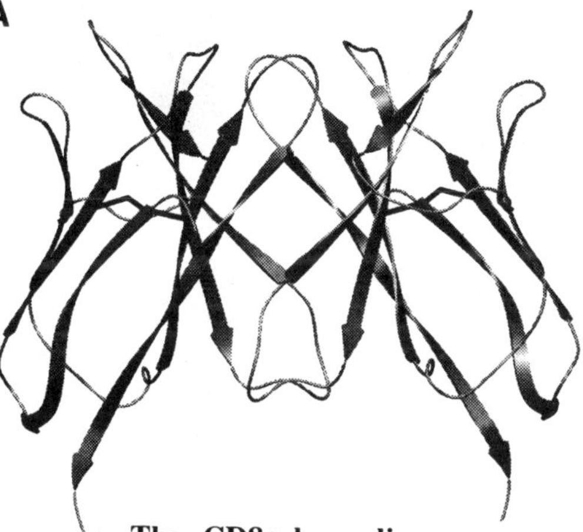

The CD8α homodimer

B

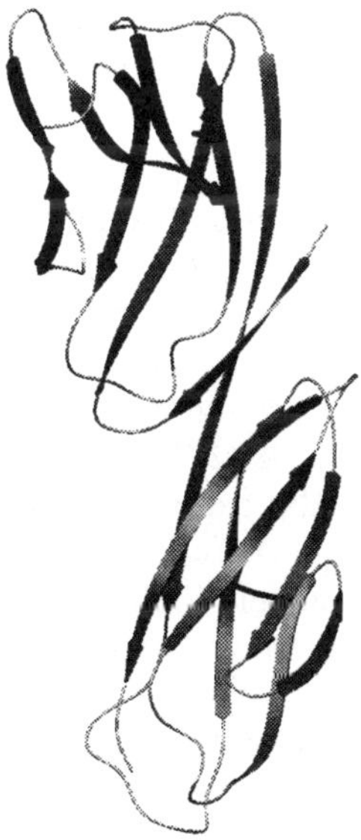

The external domains of CD4 (D1 and D2)

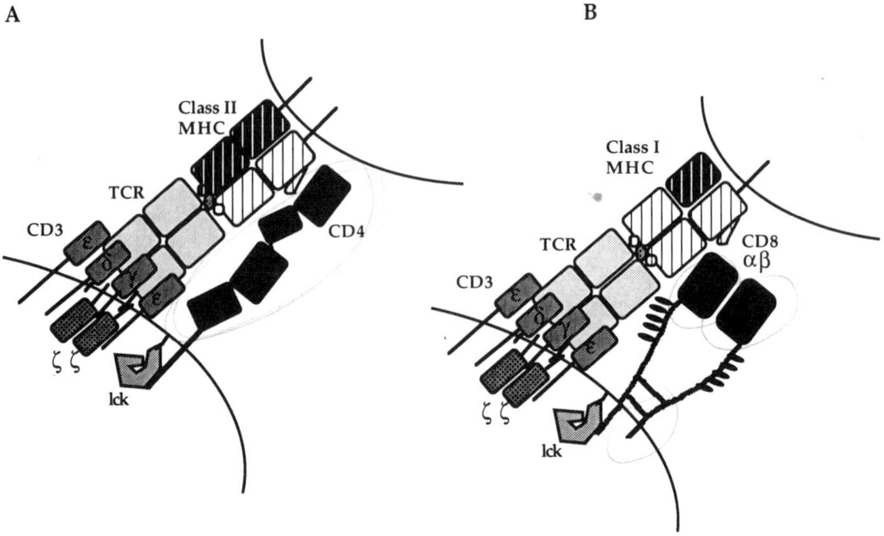

Fig. 3.2 Cartoon of the interaction between the co-receptors, CD4 and CD8 and their MHC ligands. (A) CD4 at approximately 130 Å in length, could comfortably contact the class II MHC molecule in its β2 domain at the same time as the TCR is interacting with the peptide binding domains, as indicated. This interaction would bring the CD4-associated $p56^{lck}$ molecule into proximity of the CD3 chains and any associated kinases to facilitate signal transduction. (B) CD8 heterodimers have only a single external Ig V-like domain (shown as rectangles) which are about 35 Å in length. Thus for CD8 to contact the α3 domain of the class I MHC molecule, the polypeptide region connecting the Ig-like domain to the membrane must be very extended. This region contains a number of *O*-linked sugars (shown as ⬬) which probably protect it from proteolysis. As for CD4, interaction of CD8 with the same MHC molecule that the TCR is contacting, will bring $p56^{lck}$ into proximity of the CD3 molecules.

structure is analysed for mobility of individual residues, the CD8 binding loop is found to be extremely flexible compared to the rest of the molecule (P. Travers unpublished). The region on class II MHC to which CD4 binds has been mapped to the β2 domain in a non-polymorphic region predicted to have similar structure to the CD8 binding loop on the α3 domain of class I MHC (Figure 3.2; König *et al.* 1992; Cammarota *et al.* 1992). However, in contrast to the size of CD4, the external Ig domain of CD8 is only about 30 Å in length. Therefore, for CD8 to interact with the same MHC molecule as the T cell receptor, and presumably achieve similar dimensions to CD4, the connecting peptide region of CD8 (44 to 48 amino acids long, depending on species) must be very extended (Boursier *et al.* 1993; Figure 3.2). In keeping with this, it has been shown that this area contains a number of heavily sialylated *O*-linked sugars (Leahy *et al.* 1992; Classon *et al.* 1992) which may help to keep this region relatively rigid and protected from proteolysis.

The structural and binding studies for CD8 described above have been carried out on CD8 α homodimers. The role of the CD8 β chain has been less well studied, in part because it is only transported to the cell surface when a heterodimer with α, making it difficult to distinguish the contribution of β from that of α. In mouse, both immature and mature $CD8^+$ TCR $\alpha\beta^+$ T cells express more than 95% of their surface CD8 molecules in the form of heterodimers (Walker *et al.* 1984; R. Zamoyska unpublished). However, this may differ between species as there have been reports that some human TCR $\alpha\beta^+$ T cells express CD8 α homodimers in addition to heterodimers (Terry *et al.* 1990; Moebius *et al.* 1991). In both species, cells expressing exclusively CD8 α homodimers have been described, but such cells do not appear to be class I MHC restricted, for example, γδ T cells, whose restriction pattern is unknown (MacDonald *et al.* 1990), NK cells (Moebius *et al.* 1991), and some aberrant class II restricted $CD4^+8^+$ cells (Gallagher *et al.* 1986; Moebius *et al.* 1991). It appears therefore, that cells which see antigen in context of class I MHC favour expression of CD8 αβ heterodimers. Interaction between the CD8 β polypeptide and class I MHC has been implicated in a recent study which demonstrated that hybrid molecules consisting of CD8 β external domains fused to CD8 α transmembrane and cytoplasmic domains could be transported to the cell surface and were able to enhance responses to antigen (Wheeler *et al.* 1992). In addition, some studies have shown that CD8 αβ heterodimers improve the responses of T cells to class I MHC molecules (Karaki *et al.* 1992; Wheeler *et al.* 1992), and yet other studies could not find benefit in expression of CD8 αβ heterodimers rather than CD8 αα homodimers in transfected T cells (Letourneur *et al.* 1990). None of these reports could show any function unique to CD8 β, however we have recently demonstrated that when $CD8^+$ T cells become activated, CD8 β polypeptides alter the sialic acid content of their *O*-linked sugars which are probably situated in the connecting peptide region of the molecule (Casabo *et al.* 1994). Therefore this part of CD8 β which is 10–14 amino acids shorter than the corresponding region of CD8 α acquires a significant increase in negative charge in activated T cells, a modification which does not occur on the CD8 α polypeptide. As yet we do not understand the consequence of these modifications for the function of the CD8 molecule, however they do suggest that β may play a regulatory role, perhaps in directing appropriate interaction with MHC molecules or other structures on the T cell surface.

3.2 CD4 and CD8 co-receptors are critical for the generation and activation of class I and class II MHC restricted T cells

T cells commit to expressing CD4 or CD8 and thus to becoming restricted by class II or class I MHC, during maturation in the thymus. Disruption of expression of CD4 or CD8 molecules in mice by targeted gene knock-out, has convincingly demonstrated the importance of these molecules in T cell development. Animals which can not express CD4 molecules, do not develop

T cells which recognize class II MHC and fulfil the role of Th (Rahemtulla *et al.* 1991). Conversely when the CD8 gene is disrupted, class I restricted T cells are not produced and cytotoxic T cell functions are compromised (Fung-Leung *et al.* 1991). In either instance, development of the other subset of T cells (CD8 in the former and CD4 in the latter) is apparently unaffected. These observations suggest that first, whether a T cell is restricted by class I or class II MHC depends critically on expression of the CD4 and CD8 molecules during development, and secondly, differentiation decisions to develop a helper or cytotoxic phenotype occur co-incident with the decision to express CD4 or CD8.

The precise roles of CD4 and CD8 may well alter during the various developmental stages in the life of a T cell. Thus in addition to their importance during development, CD8 and CD4 molecules are likely to be intimately involved also in the primary encounter of a T cell with antigen. For example, it has been shown *in vivo*, that a profound and long-lasting state of tolerance to antigen can be induced by administration of antibodies to CD4 or CD8 immediately preceding antigen immunization (Cobbold *et al.* 1992). These antibodies do not need to delete the responding T cells, but by binding the co-receptor molecules are able to disrupt normal T cell priming. *In vitro*, Gabert *et al.* (1987) showed that transfection of αβ TCR genes encoding receptors specific for allogeneic class I MHC, could not transfer antigen reactivity to the recipient cell, despite showing the presence of the receptor on the cell surface and its functionality by stimulation with anti-clonotypic antibodies. Transfection of CD8 was required in addition to the TCR genes for the recipient cell to show the same antigenic specificity as the donor.

After multiple exposures to antigen, individual T cells may become less reliant on engagement of their co-receptor molecules for activation or effector function. Such cells are called CD4- or CD8-independent and they are generally insensitive to inhibition with anti-co-receptor antibodies (MacDonald *et al.* 1982). In some cases, CD4- or CD8-independent T cell clones, if fused with thymoma cells to generate T cell hybridomas, will respond to antigen even when selected for loss of expression of the co-receptor (Marrack *et al.* 1983; Kwan Lim *et al.* 1993). Therefore it is clear that of the distinct activities that have been ascribed to CD4 and CD8, at any particular stage of T cell develoment all, some, or none of these activities may be critical for responses to antigen.

3.3 CD4 and CD8 binding of MHC molecules and cell adhesion

The interaction of CD4 with class II MHC and CD8 with class I MHC has been demonstrated in cellular binding assays where transfectants over-expressing the appropriate MHC molecules have been shown to bind specifically cells with high level expression of CD4 (Doyle and Strominger 1987) or CD8 (Norment *et al.* 1988). Such assays have been used to map the contact residues between co-receptor and MHC molecules, by mutating specific amino acids and monitoring the effect of these mutations on binding. However these binding assays only work with unphysiological levels of expression of receptor and ligand suggesting that the affinity between CD4 and CD8 and their ligands is very low and requires extensive multimerization or membrane association to be readily demonstrated.

The affinity of co-receptor and MHC molecule may need to be low in order to prevent unwanted interactions between T cells and cells not expressing foreign antigens but expressing MHC molecules (in the case of class I MHC, most cells of the body). Upon engagement of the TCR the binding affinity of CD8 for class I may increase significantly. O'Rourke *et al.* (1990) found that CTL expressing physiological levels of CD8 showed no detectable binding to purified class I MHC molecules unless their TCR were engaged either with specific antigen or with anti-TCR antibodies, whereupon interaction between CD8 and class I could be detected. Analysis of the association between $CD4^+$ T cells and antigen-presenting cells suggest that the interaction between CD4 and class II MHC becomes apparent only upon activation also. Kupfer and Singer (1989), showed that at the site of interaction between $CD4^+$ T cell and APC couples, CD4 molecules on the T cell would redistribute into the region of contact between the two cells, only if the TCR recognized the peptide–MHC complex. In the absence of specific peptide, cell couples would form but no redistribution of CD4 occurred, despite the presence of the class II MHC molecules on the APC. Thus while CD4 and CD8 molecules can contribute to the adhesion between T cells and their partners, the affinity of this interaction is likely to be very low and dependent both on co-engagement of the TCR and co-aggregation into the area of contact between the two cells.

3.4 Association of co-receptors with $p56^{lck}$

The cytoplasmic domains of CD4 and CD8 α interact with a lymphocyte-specific protein tyrosine kinase, $p56^{lck}$ (Rudd *et al.* 1988; Veillette *et al.* 1988). The association is non-covalent, occurring between a motif which includes two cysteine residues in each of the critical sequences in CD4 and CD8 α cytoplasmic domains and the unique N-terminal region of $p56^{lck}$ (Shaw *et al.* 1989, 1990; Turner *et al.* 1990). In addition, the region immediately N-terminal to this motif in CD4 and CD8 is predominantly basic while that in *lck* is predominantly acidic, suggesting the interaction between the kinase and co-receptor may also be facilitated by electrostatic interactions, while the four cysteine residues (two in the co-receptor and two in *lck*) may interact by co-ordinating a metal ion (Turner *et al.* 1990). Interestingly, CD8 β does not associate with *lck* (Zamoyska *et al.* 1989), therefore while CD8 αα homodimers could potentially interact with two *lck* molecules, CD8 αβ heterodimers, like CD4 molecules, can only associate with a single *lck* molecule.

The proportion of CD4 and CD8 which is associated with *lck* and conversely the proportion of intracellular *lck* which is associated with CD4 and CD8 varies considerably in different estimates (Veillette *et al.* 1989*a*; Luo and Sefton 1990; Julius *et al.* 1993). The degree of association is generally estimated in co-precipitation studies, in which cells are solubilized in detergents, the co-receptors immunoprecipitated, and the amount of associated *lck* assessed by Western blotting and comparison to total cellular *lck*. Thus the experimental procedures may misrepresent the amount of *lck* associated with CD4 or CD8, particularly if, as has been suggested, the CD8 cytoplasmic domain has less stable interaction with *lck*. However, it is generally agreed that less *lck* is associated with CD8 than with CD4. Estimates for associations in normal tissue vary from 30–50% of *lck* associated with CD4 in murine spleen and thymus compared with only 10% associated with CD8 (Veillette *et al.* 1989*a*; Luo and Sefton 1990). The low estimate for numbers of CD8 molecules associated with *lck* in mouse thymus may partly reflect the fact that ~50% of CD8 α molecules expressed in mouse thymus are alternatively spliced products which lack a cytoplasmic domain and are unable to associate with *lck* (Zamoyska *et al.* 1989). The question does arise, however, whether any differences in the level of association of CD4 and CD8 molecules with *lck* has consequences for the functions of the two molecules.

4 CD4 and CD8 contribute to T cell signalling

CD4 and CD8 can contribute to T cell signalling as evidenced by the enhancement of cellular proliferation, particularly for naïve T cells, when the co-receptor is cross-linked with the TCR compared to cross-linking the TCR alone (Owens *et al.* 1987; Anderson *et al.* 1987; Emmrich *et al.* 1988; Eichmann *et al.* 1989). Similarily, antibodies which induce co-association between TCR and CD4 are potent stimulators of proliferation (Rojo *et al.* 1989) and induce phosphorylation of the CD3 ζ chain (Barber *et al.* 1989; Veillette *et al.* 1989*b*; Dianzani *et al.* 1992*a*). While antibody cross-linking is not a particularly physiological signal, a similar phenomenon has been shown in responses to MHC molecules with amino acid changes in the α3 domain which cannot interact with CD8 (Potter *et al.* 1989; Salter *et al.* 1989; Connoly *et al.* 1990). Stimulator cells bearing such variant MHC molecules are not able to activate T cells, despite the potential for interactions between the T cell receptor and the mutated MHC molecule, and between CD8 and wild-type class I MHC molecules co-expressed by the stimulator cell. Such results suggest that the additional affinity contributed by CD8 binding MHC is not sufficient to stimulate the T cell but in addition, CD8 must interact with the same MHC molecule which is being contacted by the TCR.

One candidate for the signalling effects of CD4 and CD8 is the $p56^{lck}$ molecule associated with the cytoplasmic domain. There have been numerous studies which have shown, both for CD4 and CD8, that optimal stimulation is achieved when the co-receptor has a cytoplasmic domain which can interact with *lck*, compared with 'tailless' molecules which should retain the capacity to bind MHC but fail to bind *lck* (Zamoyska *et al.* 1989; Letourneur *et al.* 1990; Miceli *et al.* 1991). Furthermore CD4 and CD8 molecules with specific mutations in the *lck* interaction residues of the cytoplasmic domain, generally behave like tailless mutants (Chalupny *et al.* 1991; Glaichenhaus *et al.* 1991). These data provide strong evidence that *lck* is directly associated with the co-receptor function of these molecules. $p56^{lck}$ kinase is a member of the *src* family of kinases which is critical for T cell development (Molina *et al.* 1992), as well as for normal signal transduction through the TCR (Straus and Weiss 1992). Cross-linking of CD4 or (to a lesser extent) CD8 results in phosphorylation of *lck* on tyrosine residues in the kinase domain, at amino acid 394, and in the regulatory domain, at amino acid 505 (Veillette *et al.* 1989*c*; Luo and Sefton 1990). Monovalent antibody Fab fragments to CD4 or CD8 are not sufficient to induce *lck* phosphorylation, however low levels of phosphorylation occur in the presence of bivalent antibodies, and are enhanced by further cross-linking (Veillette *et al.* 1989*c*; Luo and Sefton 1990). These results suggest that aggregation of co-receptor molecules rather than a conformational change induced by binding antibody, is responsible for *lck* phosphorylation. Furthermore, cross-linking of co-receptors with TCR results in phosphorylation of components of the CD3 complex in mature T cells (Barber *et al.* 1989; Veillette *et al.* 1989*b*) and in thymocytes (Gilliland *et al.* 1991).

It is possible that while the primary interaction of CD4 and CD8 is with *lck*, other kinases may be indirectly associated with these molecules. It has been shown that a 32 kDa GTP binding protein co-precipitates with CD4–*lck* and CD8–*lck* complexes (Telfer and Rudd 1991), suggesting that the co-receptors may interact with part of a wider intracellular signalling cascade.

4.1 Inhibitory effects of antibodies to co-receptors

The relevance of CD4 and CD8 to immune responses was originally suggested from experiments in which it was found that antibodies to the co-receptors had profound inhibitory effects (Nakayama *et al.* 1979; Engelman *et al.* 1981; Wilde *et al.* 1983). As experiments using anti-CD4 and CD8 antibodies to block T cell activation may shed some light on how these molecules function, possible interpretations for this phenomenon will be considered in this section.

One effect of anti-CD4 and CD8 antibodies is likely to be the lowering of affinity of a T cell for its target below the threshold required for stimulation, by preventing the interaction between co-receptors and MHC molecules. The extent to which individual T cells can be blocked by antibody is variable and has been suggested to correlate with the affinity of the T cell receptor for antigen. Thus T cells with 'low affinity' TCR (CD4- or CD8-dependent cells) are easier to block with anti-CD4 or CD8 antibodies than T cells with 'high affinity' TCR (CD4- or CD8-independent cells), which may be completely insensitive to blocking (MacDonald *et al.* 1982). In keeping with this it has been shown that responses to cells with low density of antigen are more susceptible to blocking than those with higher densities of antigen (Shimonkevitz *et al.* 1985; Maryanski *et al.* 1988; Portoles and Janeway 1989; Alexander *et al.* 1991). However, reduction in affinity can not account for all the instances where antibody blocking is observed. For instance, it was found that antibodies could block responses in which CD8 or CD4 could not be contributing to affinity because there were no MHC molecules on the target cells (Hünig 1984; Bank and Chess 1985). Thus for $CD8^+$ T cells which can kill target cells through redirected lysis, that is by engaging their TCRs directly via anti-TCR antibody or lectins on the surface of the target cell, anti-CD8 antibodies can inhibit (Hünig 1984; Van Seventer *et al.* 1986). One interpretation of this inhibition is that the antibodies deliver an 'off' or negative signal to the T cells which prevents response to subsequent TCR stimulation. This is known as the negative signalling hypothesis (Bank and Chess 1985).

The precise nature of the signals suggested by the negative signalling hypothesis have been difficult to determine. There is a correlation between the ability of anti-co-receptor antibodies to block T cell activation and how well they can induce *lck* phosphorylation. Monovalent Fab fragments to CD4, neither block nor induce *lck* phosphorylation; bivalent antibodies to CD4 can block and induce low level phosphorylation; while multivalent IgM antibodies or cross-linking of anti-co-receptor antibodies with a second layer anti-Ig antibody efficiently blocks T cell activation and results in very efficient *lck* phosphorylation (Haque *et al.* 1987; Luo and Sefton 1990; Dianzani *et al.* 1992*b*). One possibility is that in cross-linking co-receptor molecules with antibodies, which does not adequately mimic the clustering of the co-receptor with the T cell receptor during physiological T cell activation and where the natural substrate for the *lck* kinase may be absent, *lck* is autophosphorylated on its regulatory tyrosine, 505 (Veillette *et al.* 1989*c*), and therefore inactivated. Another interpretation for negative signalling is that in the absence of the co-receptor, *lck* is free to diffuse to the site of TCR engagement (this would explain why some T cell hybridomas can respond to antigen even when they no longer express a co-receptor), but expression of the co-receptor tethers the *lck* molecule, restricting its access to the TCR (as suggested by experiments of Haughn *et al.* 1992). Addition of cross-linking antibodies to co-receptor molecules then further restrict the accessibility of *lck* to the TCR and the cell fails to respond. However, inhibition of cells expressing tailless co-receptor molecules which can not associate with *lck*, by anti-co-receptor antibodies, has also been reported indicating that blocking antibodies may have effects other than those directly involving *lck* (Miceli *et al.* 1991; Tanabe *et al.* 1992; Kwan Lim *et al.* 1993).

In a series of experiments designed to examine how anti-co-receptor antibodies may be exerting their inhibitory effects, we generated a number of class I-specific T cell hybridomas, by fusing T cell clones to a thymoma cell (Kwan Lim *et al.* 1993). Such hybridomas turn off the CD8 gene (a property of the fusion partner) but some of the hybridomas nevertheless maintain responsiveness to antigen. We examined whether we could block responses in such cells with antibodies after re-transfection of wild-type and tailless CD8 molecules (the appropriate co-receptor for the TCR) or wild-type and tailless CD4 molecules. We found that antibodies to co-receptors inhibited cells expressing wild-type CD8 molecules very efficiently, and cells expressing tailless CD8 or wild-type CD4 slightly less well, while cells expressing tailless CD4 molecules showed almost no inhibition. As the TCR on these cells has sufficient affinity for its antigen in the absence of any co-receptor, it is difficult to imagine that antibody could reduce the affinity of interaction between T cell and stimulator below the threshold for activation. In addition for the transfectants expressing wild-type CD4 molecules, there was no class II MHC ligand on the stimulator cells to provide increased affinity. For those transfectants expressing wild-type CD8 and CD4, the effects of the blocking antibody could be attributed to cross-linking *lck* via antibodies to the co-receptors, as described above. On the other hand, the tailless CD8 molecules can not associate with *lck*, therefore *lck* induced signalling can not explain the inhibition. We are left with the possibilities that either the co-receptor has as yet unidentified interactions with signalling molecules or that the TCR interacts with the co-receptor in such a way that antibodies to the co-receptor interfere with TCR recognition (see below).

4.2 Co-receptor TCR interactions

There is evidence that TCRs associate with co-receptor molecules independently of antigen. In some instances it has been possible to demonstrate close association of the receptor and co-receptor by fluorescence resonance energy transfer (Mittler *et al.* 1989), and by co-capping and co-modulation, although these effects are normally slight (Rojo *et al.* 1989; Janeway *et al.* 1989). In addition there have been reports that elements of the TCR may co-precipitate with CD8 or CD4 (Gallagher *et al.* 1989; Suzuki *et al.* 1992; Beyers *et al.* 1992). There are other examples of data which though indirect, point also to a physical interaction between TCR and co-receptor. Kanagawa and Maki (1989) found that T cell hybridomas derived from either CD4$^+$ or CD8$^+$ T cell clones which expressed a Vβ8 TCR, could respond to Mls superantigens only when they did not express CD8. Thus a hybridoma from a CD4$^+$ T cell was responsive to Mls unless also transfected with CD8, whereupon it ceased to respond. In contrast, a hybridoma from a CD8$^+$ T cell which in the presence of CD8 failed to respond to Mls, became responsive when CD8 expression was lost. One interpretation of these results is that the CD8 molecule associates with the TCR and blocks the binding of the the Mls–class II complex.

If the TCR can associate with the co-receptor, it raises interesting questions about where such interactions may occur. In the studies of Janeway *et al.* (1989), the associations of CD4 with TCR were observed as a consequence of interaction with anti-TCR antibody, suggesting antibody induced a conformational change in the T cell receptor, presumably in the external domains, which facilitated interaction with CD4. Similarly the interference by CD8 of TCR activation by Mls, described above (Kanagawa and Maki 1989), suggests extracellular association between CD8 and the TCR. Given the difference in structure and sequence between CD4 and CD8 external domains, one might predict that there would be preferences in the selection of individual TCR genes and a particular co-receptor. However, there does not appear to be a clear distinction between the usage of TCR V(D)J or C components and the class of MHC that any particular T cell recognizes (although some bias for particular V region expression by CD4 or CD8 cells has been observed in specific haplotypes, these have been remarkably few and the restriction is not absolute). An alternative possibility is that interactions occur in the transmembrane and or cytoplasmic domains either between the TCR and co-receptor directly or via other, perhaps cytoskeleton-associated, molecules.

4.3 T cell recognition of antigen–MHC and the transition to activation

In this chapter we have focused on T cell surface molecules concerned with recognition of antigen–MHC complexes, namely the variable T cell receptor recognizing the foreign peptide in association with MHC molecules and the co-receptor recognizing a constant portion of the MHC. However the activation of a T cell to proliferate, to differentiate to an effector cell in the case of naïve cells, or to express their effector function when fully differentiated, depends on more than simply recognition of antigen–MHC complexes. T cell activation is a highly regulated event, with only specialized antigen-presenting cells capable of initiating T cell responses. Part of the requirement for specialized antigen presentation is to ensure activation of T cell subsets appropriate for dealing with particular pathogens. This is achieved both through recognition of specific MHC molecules and through the involvement of particular combinations of other cell surface molecules which serves to ensure that the response which is generated is relevant for elimination of the pathogen.

Other molecules on the T cell surface which play a role in activation can be divided into two main types. The first are molecules involved in cell–cell adhesion, these not only regulate migration and homing to lymphoid organs, but also are involved in the establishment of primary contacts and signalling events between T cells and their partners (Pardi *et al.* 1992). The second are molecules which themselves transduce signals. These are numerous and include CD45, a leucocyte-specific protein tyrosine phosphatase which may be involved in moderating the phosphorylation of *lck* amongst other substrates (Trowbridge *et al.* 1992); CD28, a costimulator molecule which upon interaction with its ligand B7, expressed by antigen-presenting cells, transduces distinct signals which modulate T cell lymphokine production (June *et al.* 1990); and CD2, whose precise function is unknown but which may act to amplify signals through the TCR pathway (Bierer *et al.* 1989). However, all of these interactions involve recognition of invariant ligands, and while they play a role in the decision whether or not the T cell becomes activated, they play no role in determining the specificity of the T cell for antigen. For that reason we have not considered these interactions here. The specificity of the T cell is uniquely determined by the T cell receptor, acting in concert with the appropriate co-receptor, CD4 or CD8. It is the recognition mediated by these receptors that is the primary determinant of the activation of T cells.

References

Alexander, M. A., Damico, C. A., Wieties, K. M., Hansen, T. E., and Connolly, J. M. (1991). Correlation between CD8 dependency and determinant density using peptide-induced, L^d-restricted cytotoxic T lymphocytes. *J. Exp. Med.*, **173**, 849–58.

Anderson, P., Blue, M.-L., Morimoto, C., and Schlossman, S. (1987). Cross-linking of T3 (CD3) with T4 (CD4) enhances the proliferation of resting T lymphocytes. *J. Immunol.*, **139**, 678–82.

Bank, I. and Chess, L. (1985). Perturbation of the T4 molecule transmits a negative signal to T cells. *J. Exp. Med.*, **162**, 1294–303.

Barber, E. K., Dasgupta, J., Schlossman, S., Trevillyan, J., and Rudd, C. (1989). The CD4 and CD8 antigens are coupled to a protein-tyrosine kinase ($p56^{lck}$) that phosphorylates the CD3 complex. *Proc. Natl. Acad. Sci. USA*, **86**, 3277–81.

Berg, L. J., Pullen, A., Fazekas de St. Groth, B., Mathis, D., Benoist, C., and Davis, M. (1989). Antigen/MHC-specific T cells are preferentially exported from the thymus in the presence of their MHC ligand. *Cell*, **58**, 1035–46.

Beyers, A. D., Spruyt, L. L., and Williams, A. F. (1992). Molecular associations between the T-lymphocyte antigen receptor complex and the surface antigens CD2, CD4, or CD8 and CD5. *Proc. Natl. Acad. Sci. USA*, **89**, 2945–9.

Bierer, B. E., Sleckman, B. P., Ratnofsky, S. E., and Burakoff, S. J. (1989). The biologic roles of CD2, CD4, and CD8 in T-cell activation. *Annu. Rev. Immunol.*, **7**, 579–99.

Bjorkman, P. J., Saper, M. A., Samraoui, B., Bennet, W. S., Strominger, J. L., and Wiley, D. C. (1987*a*). The foreign antigen binding site and T cell recognition regions of class I histocompatibility antigens. *Nature*, **329**, 512–18.

Bjorkman, P. J., Saper, M. A., Samraoui, B., Bennet, W. S., Strominger, J. L., and Wiley, D. C. (1987*b*). Structure of the human class I histocompatibility antigen, HLA-A2. *Nature*, **329**, 506–12.

Boursier, J., Alcover, A., Herve, F., Laisney, I., and Acuto, O. (1993). Evidence for an extended structure of the T-cell co-receptor CD8α as deduced from the hydrodynamic properties of soluble forms of the extracellular region. *J. Biol. Chem.*, **26**, 2013–20.

Brodsky, F. and Guagliardi, L. (1991). The cell biology of antigen processing and presentation. *Annu. Rev. Immunol.*, **9**, 707–44.

Cammarota, G., Schierle, A., Takacs, B., Doran, D., Knorr, R., Bannworth, W., *et al.* (1992). Identification of a CD4 binding site on the β2 domain of HLA-DR molecules. *Nature*, **356**, 799–801.

Casabó, L., Mamalaki, C., Kioussis, D., and Zamoyska, R. (1994). T cell activation results in physical modification of the mouse CD8β chain. *J. Immunol.*, **152**, 397–404.

Chalupny, N. J., Ledbetter, J. A., and Kavathas, P. (1991). Association of CD8 with $p56^{lck}$ is required for early T cell signalling events. *EMBO J.*, **10**, 1201–7.

Classon, B., Brown, M.H., Garnett, D., Somoza, C., Barclay, A., Willis, A., *et al.* (1992). The hinge region of the CD8 alpha chain: structure, antigenicity, and utility in expression of immunoglobulin superfamily domains. *Int. Immunol.*, **4**, 215–25.

Clayton, L., Sieh, M., Pious, D., and Reinherz, E. (1989). Identification of human CD4 residues affecting class II MHC versus HIV-1 gp120 binding. *Nature*, **339**, 548–51.

Cobbold, S., Qin, S., Leong, L., Martin, G., and Waldmann, H. (1992). Reprogramming the immune system for peripheral tolerance with CD4 and CD8 monoclonal antibodies. *Immunol. Rev.*, **129**, 165–201.

Connoly, J., Hansen, T., Ingold, A., and Potter, T. (1990). Recognition by CD8 on cytotoxic T lymphocytes is ablated by several substitutions in the class I α3 domain: CD8 and the T cell receptor recognise the same class I molecule. *Proc. Natl. Acad. Sci. USA*, **87**, 2137–41.

Davis, M. and Bjorkman, P. (1988). T-cell antigen receptor genes and T-cell recognition. *Nature*, **334**, 395–402.

Dianzani, U., Shaw, A., Al-Ramadi, B. K., Kubo, R. T., and Janeway, C. A. Jr. (1992*a*). Physical association of CD4 with the T cell receptor. *J. Immunol.*, **148**, 658–78.

Dianzani, U., Shaw, A., Fernandez-Cabezudo, M., and Janeway, C. A. Jr. (1992*b*). Extensive CD4 cross-linking inhibits T cell activation by anti-receptor antibody but not by antigen. *Int. Immunol.*, **4**, 995–1001.

Doyle, C. and Strominger, J. (1987). Interaction between CD4 and class II MHC molecules mediates cell adhesion. *Nature*, **330**, 256–9.

Eichmann, K., Boyce, N., Schmidt-Ullrich, R., and Jonsson, J. (1989). Distinct functions of CD8(CD4) are utilised at different stages of T-lymphocyte differentiation. *Immunol. Rev.*, **109**, 39–75.

Emmrich, F., Rieber, P., Kurrie, R., and Eichmann, K. (1988). Selective stimulation of human T lymphocyte subsets by heteroconjugates of antibodies to the T cell receptor and to subset-specific differentiation antigens. *Eur. J. Immunol.*, **18**, 645–8.

Engelman, E., Benike, C., Glickman, E., and Evans, R. (1981). Antibodies to membrane structures that distinguish suppressor/cytotoxic and helper T cell subpopulations block the mixed leukocyte reaction in man. *J. Exp. Med.*, **154**, 193–8.

Fung-Leung, W.-P., Schillham, M., Rahemtulla, A., Kündig, T., Vollenweider, M., Potter, J., *et al.* (1991). CD8 is needed for development of cytotoxic T cells but not helper T cells. *Cell*, **65**, 443–9.

Gabert, J., Langlet, C., Zamoyska, R., Parnes, J. R., Scmitt-Verhulst, A. M., and Malissen, B. (1987). Reconstitution of MHC class I specificity by transfer of the T cell receptor and Lyt-2 genes. *Cell*, **50**, 545–54.

Gallagher, P., Fazekas de St. Groth, B., and Miller, J. (1986). Stable expression of Lyt-2 homodimers on L3T4$^+$ T cell clones. *Eur. J. Immunol.*, **16**, 1413–17.

Gallagher, P., Fazekas de St. Groth, B., and Miller, J. (1989). CD4 and CD8 molecules can physically associate with the same T cell receptor. *Proc. Natl. Acad. Sci. USA*, **86**, 10044–8.

Galvin, K., Krishna, S., Panchel, F., Frohlich, M., Cummings, D., Carlson, R., *et al.* (1993). The major histocompatibility complex class I gene antigen-binding protein p88 is the product of the calnexin gene. *Proc. Natl. Acad. Sci. USA*, **89**, 8452–6.

Gilliland, L. K., Teh, H., Uckun, F., Norris, N., Teh, S.-J., Schieven, G. *et al.* (1991). CD4 and CD8 are positive regulators of T cell receptor signal transduction in early T cell differentiation. *J. Immunol.*, **146**, 1759–65.

Glaichenhaus, N., Shastri, N., Littman, D. R., and Turner, J. M. (1991). Requirement for the association of p56^{lck} with CD4 in antigen-specific signal transduction in T cells. *Cell*, **64**, 511–20.

Gorman, S. D., Sun, Y. H., Zamoyska, R., and Parnes, J. R. (1988). Molecular linkage of the Lyt-3 and Lyt-2 genes: requirement of Lyt-2 for Lyt-3 surface expression. *J. Immunol.*, **140**, 3646–53.

Griffiths, G. and Mueller, C. (1991). Expression of perforin and granzymes *in vivo*: potential diagnostic markers for activated cytotoxic cells. *Immunol. Today*, **12**, 415–19.

Haque, S., Saizawa, K., Rojo, J., and Janeway, C. Jr. (1987). The influence of valance on the functional activities of monoclonal anti-L3T4 antibodies: discrimination of signalling from other effects. *J. Immunol.*, **139**, 3207–12.

Haughn, L., Gratton, S., Caron, L., Sékaly, R.-P., Veillette, A., and Julius, M. (1992). Association of tyrosine kinase p56^{lck} with CD4 inhibits the induction of growth through the αβ T-cell receptor. *Nature*, **358**, 328–31.

Hayakawa, K. and Hardy, R. (1991). Murine CD4^{+} T-cell subsets. *Immunol. Rev.*, **123**, 145–68.

Hünig, T. (1984). Monoclonal anti-Lyt-2.2 antibody blocks lectin-dependent cellular cytotoxicity of H-2-negative target cells. *J. Exp. Med.*, **159**, 551–8.

Janeway, C. J., Rojo, J., Saizawa, K., Dianzani, U., Portoles, P., Tite, J., *et al.* (1989). The co-receptor function of murine CD4. *Immunol. Rev.*, **109**, 77–92.

Johnson, P. and Williams, A. F. (1986). Striking similarities between antigen receptor J pieces and sequence in the second chain of the murine CD8 antigen. *Nature*, **323**, 74–6.

Jorgensen, J., Reay, P., Ehrich, E., and Davis, M. (1992). Molecular components of T cell recognition. *Annu. Rev. Immunol.*, **10**, 835–73.

Julius, M., Maroun, C., and Haughn, L. (1993). Distinct roles for CD4 and CD8 as co-receptors in antigen receptor signalling. *Immunol. Today*, **14**, 177–83.

June, C., Ledbetter, J., Linsley, P., and Thompson, C. (1990). Role of CD28 receptor in T-cell activation. *Immunol. Today*, **11**, 211–16.

Kanagawa, O. and Maki, R. (1989). Inhibition of MHC class II-restricted T cell response by Lyt-2 alloantigen. *J. Exp. Med.*, **170**, 901–12.

Karaki, S., Tanabe, M., Nakauchi, H., and Takiguchi, M. (1992). β-chain broadens range of CD8 recognition for MHC class I molecule. *J. Immunol.*, **149**, 1613–18.

Kisielow, P., Teh, H. S., Bluthmann, H., and von Boehmer, H. (1988). Positive selection of antigen-specific T cells in thymus by restricting MHC molecules. *Nature*, **335**, 730–4.

König, R., Huang, L.-Y., and Germain, R. (1992). MHC class II interaction with CD4 mediated by a region analogous to the MHC class I binding site for CD8. *Nature*, **356**, 796–8.

Kupfer, A. and Singer, S.J. (1989). Cell biology of cytotoxic and helper cell functions: immunofluorescent microsopic studies of single cells and cell couples. *Annu. Rev. Immunol.*, **7**, 309–37.

Kwan Lim, G., Ong, T., Aosai, F., Stauss, H., and Zamoyska, R. (1993). Is CD8 dependence a true reflection of T cell receptor affinity for antigen? *Int. Immunol.*, **5**, 1219–28.

Lamarre, D., Ashkenaze, A., Fleury, S., Smith, D., Sekaly, R., and Capon, D. (1989). The MHC-binding and gp120-binding functions of CD4 are separable. *Science*, **245**, 743–6.

Leahy, D. J., Axel, R., and Hendrickson, W. A. (1992). Crystal structure of a soluble form of the human T cell coreceptor CD8 at 2.6 Å resolution. *Cell*, **68**, 1145–62.

Letourneur, F., Gabert, J., Cosson, P., Blanc, D., Davoust, J., and Malissen, B.

(1990). A signalling role for the cytoplasmic segment of the CD8 alpha chain detected under limiting stimulatory conditions. *Proc. Natl. Acad. Sci. USA*, **87**, 2339–43.

Littman, D. R. (1987). The structure of the CD4 and CD8 genes. *Annu. Rev. Immunol.*, **5**, 561–84.

Luo, K. and Sefton, B. (1990). Cross-linking of T-cell surface molecules CD4 and CD8 stimulates phosphorylation of the lck tyrosine protein kinase at the autophosphorylation site. *Mol. Cell. Biol.*, **10**, 5305–13.

MacDonald, H. R., Glasebrook, A. L., Bron, C., Kelso, A., and Cerottini, J. C. (1982). Clonal heterogeneity in the functional requirement for Lyt-2/3 molecules on CTL: possible implications for the affinity of CTL antigen receptors. *Immunol. Rev.*, **68**, 89–116.

MacDonald, H. R., Schreyer, M., Howe, R., and Bron, C. (1990). Selective expression of CD8 alpha (Ly-2) subunit on activated thymic gamma/delta cells. *Eur. J. Immunol.*, **20**, 927–30.

Marrack, P., Endres, R., Shimonkevitz, R., Zlotnik, A., Dialynas, D., Fitch, F. *et al.* (1983). The major histocompatibility complex-restricted antigen receptor on T cells. II Role of the L3T4 product. *J. Exp. Med.*, **158**, 1077–91.

Maryanski, J. L., Pala, P., Cerottini, J.-C., and MacDonald, H. (1988). Antigen recognition by H-2-restricted cytolytic T lymphocytes: inhibition of cytolysis by anti-CD8 monoclonal antibodies depends upon both concentration and primary sequence of peptide antigen. *Eur. J. Immunol.*, **18**, 1863–6.

Miceli, M., von Hoegen, P., and Parnes, J. (1991). Adhesion versus coreceptor function of CD4 and CD8: Role of the cytoplasmic tail in coreceptor activity. *Proc. Natl. Acad. Sci. USA*, **88**, 2623–7.

Mittler, R. S., Goldman, S. J., Spitalny, G. L., and Burakoff, S. J. (1989). T-cell receptor-CD4 physical association in a murine T-cell hybridoma: Induction by antigen receptor ligation. *Proc. Natl. Acad. Sci. USA*, **86**, 8531–5.

Moebius, U., Kober, G., Griscelli, A., Hercend, T., and Meuer, S. (1991). Expression of different CD8 isoforms on distinct human lymphocyte subpopulations. *Eur. J. Immunol.*, **21**, 1793–800.

Molina, T., Kishihara, K., Siderovski, D., van Ewijk, W., Narendran, A., Timms, E., *et al.* (1992). Profound block in thymocyte development in mice lacking $p56^{lck}$. *Nature*, **357**, 161–3.

Monaco, J. (1992). A molecular model of MHC class-1-restricted antigen processing. *Immunol. Today*, **13**, 173–9.

Mossmann, T., Schumacher, J., Street, N., Budd, R., O'Garra, A., Fong, T., *et al.* (1991). Diversity of cytokine synthesis and function of mouse CD4+ T cells. *Immunol. Rev.*, **123**, 209–29.

Nakayama, E., Shiba, H., Stockert, E., Oettgen, H., and Old, L. (1979). Cytotoxic T cells: Lyt-phenotype and blocking of killing activity by Lyt antisera. *Proc. Natl. Acad. Sci. USA*, **76**, 3486–9.

Norment, A. M., Salter, R. D., Parham, P., Engelhard, V. H., and Littman, D. R. (1988). Cell–cell adhesion mediated by CD8 and MHC class I molecules. *Nature*, **336**, 79–81.

O'Rourke, A., Rogers, J., and Mescher, M. (1990). Activated CD8 binding to class I protein mediated by the T-cell receptor results in signalling. *Nature*, **346**, 187–9.

Owens, T., Fazekas de St Groth, B., and Miller, J. (1987). Coaggregation of the T cell receptor with CD4 and other T cell surface molecules enhances T cell activation. *Proc. Natl. Acad. Sci. USA*, **84**, 9209–12.

Pardi, R., Inverardi, L., and Bender, J. (1992). Regulatory mechanisms in leukocyte adhesion: flexible receptors for sophisticated travelers. *Immunnol. Today*, **13**, 224–30.

Parnes, J. R. (1989). Molecular biology and function of CD4 and CD8. *Adv. Immunol.*, **44**, 265–311.

Pfieffer, J., Wick, M., Roberts, R., Finday, K., Normark, S., and Harding, C. (1993). Phagocytic processing of bacterial antigens for class I MHC presentation to T cells. *Nature*, **361**, 359–62.

Podack, E., Hengartner, H., and Lichtenheld, M. (1991). A central role of perforin in cytolysis. *Annu. Rev. Immunol.*, **9**, 129–57.

Portoles, P. and Janeway, C. J. (1989). Inhibition of the responses of a cloned CD4+ T cell line to different class II major histocompatibility complex ligands by anti-CD4 and by anti-receptor Fab fragments are directly related. *Eur. J. Immunol.*, **19**, 83–7.

Potter, T., Rajan, T., Dick, R., and Bluestone, J. (1989). Substitution at residue 227 of H-2 class I molecules abrogates recognition by CD8-dependent but not CD8-independent, cytotoxic T lymphocytes. *Nature*, **337**, 73–6.

Rahemtulla, A., Fung-Leung, W., Schilham, S., Kundig, T., Sambhara, S., Narendran, A., *et al.* (1991). Normal development and function of CD8+ cells but markedly decreased helper cell activity in mice lacking CD4. *Nature*, **353**, 180–4.

Roche, P. A., Marks, M. S., and Cresswell, P. (1991). Formation of a nine-subunit complex by HLA class II glycoproteins and the invariant chain. *Nature*, **354**, 392–4.

Rojo, J. M., Saizawa, K., and Janeway, C. A. Jr. (1989). Physical association of CD4 and the T cell receptor can be induced by anti-T-cell receptor antibodies. *Proc. Natl. Acad. Sci. USA*, **86**, 3311–15.

Rudd, C., Trevillyan, J., Dasgupta, J., Wong, L., and Schlossman, S. (1988). The CD4 receptor is complexed in detergent lysates to a protein-tyrosine kinase (pp58) from human T lymphocytes. *Proc. Natl. Acad. Sci. USA*, **85**, 5190–4.

Rupp, F., Brecher, J., Giedlin, M., Mosman, T., Zinkernagel, R., Hengartner, H. *et al.* (1987). T-cell antigen receptors with identical variable regions but different diversity and joining region gene segments have distinct specificities but cross-reactive idiotypes. *Proc. Natl. Acad. Sci. USA*, **84**, 219–22.

Ryu, S.-E., Kwong, P., Truneh, A., Porter, T., Arthos, J., Rosenberg, M., *et al.* (1990). Crystal structure of an HIV-binding recombinant fragment of human CD4. *Nature*, **348**, 419–25.

Salter, R. D., Norment, A. M., Chen, B. P., Clayberger, C., Krensky, A. M., Littman, D. R., *et al.* (1989). Polymorphism in the α3 domain of HLA-A molecules affects binding to CD8. *Nature*, **338**, 345–7.

Salter, R. D., Benjamin, R. J., Wesley, P. K., Buxton, S. E., Garrett, T. P. J., Clayberger, C., *et al.* (1990). A binding site for the T-cell co-receptor CD8 on the α3 domain of HLA-A2. *Nature*, **345**, 41–6.

Sanders, S., Fox, R., and Kavathas, P. (1991). Mutations in CD8 that affect interactions with HLA class I and monoclonal anti-CD8 antibodies. *J. Exp. Med.*, **174**, 371–9.

Scott, B., Bluthmann, H., Teh, H. S., and von Boehmer, H. (1989). The generation of mature T cells requires interaction of the alpha beta T-cell receptor with major histocompatibility antigens. *Nature*, **338**, 591–3.

Sha, W. C., Nelson, C., Newberry, R., Kranz, D., Russell, J., and Loh, D. (1988). Positive and negative selection of an antigen receptor on T cells in transgenic mice. *Nature*, **336**, 73–9.

Shaw, A., Amrein, K., Hammond, C., Stern, D., Sefton, B., and Rose, J. (1989). The cytoplasmic domain of CD4 interacts with the tyrosine protein kinase $p56^{lck}$ through its unique amino terminal domain. *Cell*, **59**, 627–36.

Shaw, A., Chalupny, J., Whitney, J., Hammond, C., Amrein, K., Kavathas, P., *et al.* (1990). Short related sequences in the cytoplasmic domains of CD4 and CD8 mediate binding to the amino-terminal domain of the p56^{lck} tyrosine protein kinase. *Mol. Cell. Biol.*, **10**, 1853–62.

Shimonkevitz, R., Luescher, B., Cerottini, J. C., and MacDonald, H. R. (1985). Clonal analysis of cytolytic T lymphocyte-mediated lysis of target cells with inducible antigen expression: correlation between antigen density and requirement for Lyt-2/3 function. *J. Immunol.*, **135**, 892–9.

Straus, D. B. and Weiss, A. (1992). Genetic evidence for the involvement of the lck tyrosine kinase in signal transduction through the T cell antigen receptor. *Cell*, **70**, 585–93.

Suzuki, S., Kupsch, J., Eichmann, K., and Saizawa, M. (1992). Biochemical evidence of the physical association of the majority of CD3δ chains with the accessory/co-receptor molecules CD4 and CD8 on nonactivated T lymphocytes. *Eur. J. Immunol.*, **22**, 2475–9.

Swain, S. (1983). T cell subsets and the recognition of MHC class. *Immunol. Rev.*, **74**, 129–42.

Swain, S., Bradley, L., Croft, M., Tonkonogy, S., Atkins, G., Weinberg, A., *et al.* (1991). Helper T-cell subsets: phenotype, function and the role of lymphokines in regulating their development. *Immunol. Rev.*, **123**, 115–44.

Tanabe, M., Karaki, S., Takiguchi, M., and Nakauchi, H. (1992). Antigen recognition by the T cell receptor is enhanced by CD8 α-chain binding to the α3 domain of MHC class I molecules, not by signalling via the cytoplasmic domain of CD8α. *Int. Immunol.*, **4**, 147–52.

Telfer, J. and Rudd, C. (1991). A 32-kD GTP-binding protein associated with the CD4-p56^{lck} and CD8-p56^{lck} T cell receptor complex. *Science*, **254**, 439–41.

Terry, L., DiSanto, J., Small, T., and Flomberg, N. (1990). Differential expression and regulation of the human CD8α and CD8β chains. *Tissue Antigens*, **35**, 82–91.

Townsend, A. and Bodmer, H. (1989). Antigen recognition by class I-restricted T lymphocytes. *Annu. Rev. Immunol.*, **7**, 601–24.

Trowbridge, I. S., Johnson, P., Ostergaard, H., and Hole, N. (1992). Structure and function of CD45: a leukocyte-specific protein tyrosine phosphatase. *Adv. Exp. Med. Biol.*, **323**, 29–37.

Turner, J., Brodsky, M., Irving, B., Levin, S., Perlmutter, R., and Littman, D. (1990). Interaction of the unique N-terminal region of tyrosine kinase p56^{lck} with cytoplasmic domains of CD4 and CD8 is mediated by cysteine motifs. *Cell*, **60**, 755–65.

Van Seventer, G., Lier, R., Spits, H., Ivanyi, P., and Melief, C. (1986). Evidence for a regulatory role of the T8 (CD8) antigen in antigen-specific and anti-T3-(CD3)-induced lytic activity of allospecific cytotoxic T lymphocyte clones. *Eur. J. Immunol.*, **16**, 1363–71.

Veillette, A., Bookman, M. A., Horak, E. M., and Bolen, J. B. (1988). The CD4 and CD8 T cell surface antigens are associated with the internal membrane tyrosine-protein kinase p56^{lck}. *Cell*, **55**, 301–8.

Veillette, A., Zúñiga-Pflücker, J. C., Bolen, J. B., and Kruisbeek, A. M. (1989*a*). Engagement of CD4 and CD8 expressed on immature thymocytes induces activation of intracellular tyrosine phosphorylation pathways. *J. Exp. Med.*, **170**, 1671.

Veillette, A., Bookman, M., Horak, E., Samelson, L., and Bolen, J. (1989*b*). Signal transduction through the CD4 receptor involves the activation of the internal membrane tyrosine-protein kinase p56^{lck}. *Nature*, **338**, 257–9.

Veillette, A., Bolen, J., and Bookman, M. (1989*c*). Alterations in tyrosine protein phosphorylation induced by antibody-mediated cross-linking of the CD4 receptor of T lymphocytes. *Mol. Cell. Biol.*, **9**, 4441–6.

Walker, I. D., Murray, B. J., Hogarth, P. M., Kelso, A., and McKenzie, I. F. C. (1984). Comparison of thymic and peripheral T cell Ly-2/3 antigens. *Eur. J. Immunol.*, **14**, 906–10.

Wang, J., Yan, Y., Garrett, T., Liu, J., Rodgers, D., Garlick, R., *et al.* (1990). Atomic structure of a fragment of human CD4 containing two immunoglobulin-like domains. *Nature*, **348**, 411–18.

Wheeler, C., von Hoegen, P., and Parnes, J. (1992). An immunological role for the CD8β chain. *Nature*, **357**, 247–9.

Wilde, D., Marrack, P., Kappler, J., Dialynas, D., and Fitch, F. (1983). Evidence implicating L3T4 in class II MHC reactivity; monoclonal antibody GK1.5 (anti-L3T4a) blocks class II MHC antigen-specific proliferation, release of lymphokines and binding by cloned murine helper T lymphocyte lines. *J. Immunol.*, **131**, 2178–83.

Yagita, H., Nakata, M., Kawasaki, A., Shinkai, Y., and Okumura, K. (1992). Role of perforin in lymphocyte-mediated cytolysis. *Adv. Immunol.*, **51**, 215–42.

Zamoyska, R., Derham, P., Gorman, S. D., von Hoegen, P., Bolen, J. B., Veillette, A., *et al.* (1989). Inability of CD8α′ polypeptide to associate with $p56^{lck}$ correlates with impaired function *in vitro* and lack of expression *in vivo*. *Nature*, **342**, 278–81.

Zinkernagel, R. and Doherty, P. (1974). Immunological surveillance against altered self components by sensitised T lymphocytes in lymphocytic choriomeningitis. *Nature*, **251**, 547–8.

4 γδ T cell specificity and function

ADRIAN C. HAYDAY

1 Introduction—how much like who?

γδ T cells were discovered after the accidental cloning in 1984 of a cDNA for a murine γ chain (1). Since no cellular immunology studies had prepared us for the prospect of an entire subset of T cells bearing a distinct receptor from the αβ T cell receptor (TCR), extensive studies were mounted to understand the function of this 'new' set of T cells. Those studies continue today. They are galvanized by the knowledge that γδ T cells appear to be conserved in all vertebrates.

Several γδ T cell lines have been described that have properties very similar to those of αβ T cells (2, 3). However, the failure of γδ T cells to be 'discovered' by the cellular immunogenetic approaches that revealed αβ T cells, suggests that *en masse* γδ T cell populations may have a distinct role in the body; for example, the recognition of generically stressed cells (4, 5), parasitic or bacterial infections (6–8, 102). Possibly, such a distinct role is shared by specialized, poorly understood sets of αβ T cells, such as those that exist within epithelia. Thus, the uniqueness of γδ T cells may reside in the specificities of the receptors encoded by the γ and δ genes. These receptors may recognize processed antigen, free antigen, cell bound native antigen, or some combination of all three. Whatever the means of recognition, γδ T cells may have been retained in vertebrates because their receptors are particularly compatible with commonly encountered, or generic stress antigens. Alternatively, γδ T cells may be essential in neonates, before the αβ T cell repertoire

Table 4.1 γδ Gene nomenclature

	This chapter	Alternatives	Proposed
Mouse	Vγ1	Vγ1.1	Vγ5.1
	Vγ2	Vγ1.2	Vγ5.2
	Vγ3	Vγ1.3	Vγ5.3
	Vγ3	Vγ2	Vγ4
	Vγ5	Vγ3	Vγ1
	Vγ6	Vγ4	Vγ2
	Vγ7	Vγ5	Vγ3
Human	Vγ9	VγII	

In the past, there was no generally agreed upon nomenclature for the mouse and human γ genes. The following shows the commonly used alternatives for the nomenclature used in the chapter. In the third column is a proposed general nomenclature for the murine γ genes.

is fully developed. Happily, a real prospect that the functions of γδ T c can soon be resolved, is provided by the development through gene targeting, of mice in which either most (9) or none (10) of the peripheral T cells are γδ.

2 γδ Localization

γδ T cells are usually less abundant than αβ T cells, comprising in humans, mice, and chickens, only 1–10% of T cells in the systemic circulation, spleen, and lymph nodes (11–13). However, this is not invariably the case: in newborn lambs, γδ T cells comprise ⩾ 80% of peripheral blood T cells, although this decreases to about 30% in sheep (14).

Oftentimes, γδ T cells are distributed quite differently from αβ T cells. For example, splenic γδ T cells in chickens do not localize to conventional T cell areas, but are dispersed (13). Similarly, in 'knock-out' mice that congenitally lack αβ T cells, the thymus lacks the conventional cortical-medullary areas that harbour respectively immature and mature αβ thymocytes, yet thymic γδ T cell maturation and export seems to occur unhindered (9).

In several classes of animals including birds, mice, and some humans, the most conspicuous difference between γδ T cell and αβ T cell distribution is the disproportionate abundance of γδ T cells within intraepithelial T cell repertoires (13, 15–18). This localization has been a major factor in promoting the idea that γδ T cells have a distinct immunological function. It has been extensively studied in mice, in which the skin, uterine, tongue, and intestinal epithelia all contain large numbers of γδ T cells (19). Indeed, in the skin, essentially all intraepithelial lymphocytes (IELs) are γδ (15, 16). Significant populations of γδ T cells have also been reported to occur in the lung and mammary epithelia, and other epithelia remain to be extensively analysed (20).

Attempts have been made to understand γδ T cell function by studies of 'lymphoid' and epithelial γδ T cells. Novel observations have in each case been made, the most striking of which is the capability of γδ T cells to recognize heat shock proteins (hsps) and 'non-classical' major histocompatibility complex (MHC) class I genes (see below). Such properties, in contrast to the highly specific recognition by αβ T cells of foreign peptides presented by classical, highly polymorphic class I and class II MHC antigens, strengthen the view that γδ T cells have a distinct immune function. Moreover, the recognition by γδ T cells of antigens apparently restricted by non-polymorphic MHC or MHC-related molecules would be consistent with the failure to detect γδ T cells using conventional *ex vivo* assays for T cells, such as alloreactivity, and MHC restricted help of antibody production. It is not yet clear, however, that lymphoid and epithelial γδ T cells share a common function (21), and as a result, one should beware the temptation to extrapolate directly from one set to the other. Here each set will be considered separately.

3 Epithelial γδ T cells

3.1 γδ T cells in the murine skin, uterus, and tongue

The localization of sets of lymphocytes within epithelia is provocative, since it would seem to define a first site of contact—*a first line of defence*—between the outside world and the immune system (4). Furthermore, the protection of epithelial integrity is essential to many components of an animal's physiology, including the prevention of dehydration. However, the general importance of epithelial γδ T cells is called into question by the fact that not all animals harbour large numbers of γδ T cells within their intraepithelial T cell repertoires (22), and by the observation that some epithelia in some animals contain virtually no IELs at all. Human skin, for example, contains few IELs.

In response to this, it should be stated that certain epithelia, notably the gut, always seem to contain IELs, and that such epithelia are frequently the site of serious immune pathologies, such as ulcerative colitis and coeliac disease. Therefore, an understanding of IELs would seem to be critical, not only for resolving whether they form a first line of defence against agents crossing the epithelium, but also for resolving whether they are causative to inflammatory disease. As regards γδ IELs in particular, it is possible in cases where they are not the most abundant IEL subtype, that the αβ T cells that take their place are a specialized subset that bears stronger similarity to γδ IELs than to αβ T cells of the systemic circulation. Hence, an understanding of γδ IELs can facilitate our understanding of intraepithelial T cells in general.

Currently, our capacity to understand γδ IEL function is handicapped by the fact that we have essentially no knowledge of either the role or *modus operandi* of intraepithelial lymphocytes. We do not know to which antigen-presenting cells they respond *in vivo*, or even if they recognize native antigen. The possibility is widely considered that IELs recognize their surrounding epithelial cells, and this will be highlighted in this chapter. However, there is as yet no description of an epithelial cell productively presenting a peptide antigen to a T cell in a physiologically meaningful fashion. Moreover, the intraepithelial localization of γδ T cells poses a problem for studying function via adoptive transfer experiments, since the physiologic localization may be difficult to recapitulate with transferred cells.

Assuming that epithelial cells are recognized by intraepithelial T cells, one must consider that since epithelial cells are considerably less mobile than systemic antigen-presenting cells, the intraepithelial T cells would be able to sample far fewer antigens in a set period of time than systemic T cells recognizing macrophages in the lymph nodes. Such a limited set of antigens might include exogenous peptides, such as *N*-formylated peptides of bacteria (23), or a limited set of endogenous autoantigens, such as hsps (5, 7) or non-classical class I MHC antigens (4), that might serve as generic harbingers of infection. If intraepithelial T cells are to be reliably activated by a limited universe of antigens sampled, then they should in return display a complementary, limited TCR repertoire (4).

Studies of murine IELs suggests that this is true for epithelial $\gamma\delta$ T cells. By comparison to TCR α or β chain genes, the numbers of functional Vγ (eight in mice, six in humans) and Jγ (four in mice, five in humans) gene segments are relatively small (24). Likewise, although the δ chain can theoretically utilize any one of the large number of Vα gene segments, studies in the mouse indicate that most δ chains utilize one of only a small number of 'Vδ' gene segments confined to the C-proximal end of the Vα gene family. Hence, the combinatorial diversity of the $\gamma\delta$ TCR is limited by limited gene segment usage.

In murine epithelia, this trend is taken to an extreme by the predominant expression of a single $\gamma\delta$ TCR, characteristic of a particular epithelium. Thus, in murine skin almost all the $\gamma\delta$ T cells express a TCR encoded by Vγ5–Jγ1, and Vδ1–Dδ2–Jδ1 rearrangements (5). Likewise, in the murine tongue and uterus, the same δ rearrangement is paired with a Vγ6–Jγ1 rearrangement (19). In the murine gut, the phenomenon is less striking, although there is frequent pairing of Vγ7 with Vδ4 (25, 26).

Theoretically, $\gamma\delta$ TCRs could display high junctional diversity: Vδ–Jδ junctions can occur with or without single or multiple Dδ rearrangements, and the Dδ segments can be 'read' in all three frames (27). However, in the murine epidermal, oral, and genital epithelia this is not a factor, since the $\gamma\delta$ TCRs characteristic of each of these sites display homogeneous, 'consensus', Vγ–Jγ, Vδ–Dδ, and Dδ–Jδ junctions (5, 24). In short, epidermal $\gamma\delta$ TCRs are about as diverse as the epidermal growth factor receptors in the same tissue.

The occurrence of homogeneous TCRs in a tissue would seem to run counter to the whole concept of using somatic gene rearrangement to imbue antigen receptors with diversity (28, 29). The phenomenon can in part be explained by the relevant developmental biology. $\gamma\delta$ T cells that populate the skin, uterine, and tongue epithelia develop exclusively early in fetal ontogeny (30), when the junctional diversity of TCR gene segment joining is limited by the lack of incorporation of N nucleotides (31, 32). At this time in development, nucleotide homology plays a significant role in selecting the sites on the V and J elements where recombination will occur (33). Thus, even in situations where TCR γ gene rearrangements cannot contribute to productive proteins (for example, in mice transgenic for a mutant Vγ5–Jγ1–Cγ1 minigene (33), or in TCR $\delta^{-/-}$ 'knock-out' mice) (10), the rearrangements that occur are predominantly of the 'consensus' type (10, 33).

In that case, are T cells with consensus joins functionally irrelevant products of unusual developmental processes, operative early in fetal ontogeny? At least three observations suggest the answer to be 'no'. First, the precise homing of $\gamma\delta$ T cells with different TCRs to different tissues suggests that those TCRs have particular compatibility with cells in those tissues. That the TCR is not itself the tissue-specific homing receptor, is evident from mice transgenic for the epidermis-associated Vγ5–Jγ1 gene, since in such animals, cells with the transgene-encoded TCR also populated the gut. However, unlike most of the endogenous, gut-associated $V\gamma7^+V\delta4^+$ IELs, the cells did not acquire the differentiation marker, CD8, suggesting that they were unable to function properly in the gut (34).

Secondly, although consensus rearrangements are common, even when productive TCRs can not be formed (10, 33), the frequency of productive, consensus rearrangements is lower than when they can form part of a productive TCR (33). This suggests that cells expressing such a TCR can participate in interactions (selection) enabling them to proliferate, and subsequently to dominate a particular IEL repertoire. This is supported by the observation that in the fetal liver of mice, where arguably no such interactions will occur, non-consensus Vγ5–Jγ1 and Vδ1–Dδ2–Jδ1 joins are more common than they are in the fetal thymus (35). It is further supported by the observation that αβ epidermal IELs in TCR $\delta^{-/-}$ mice, unlike the epidermal γδ IELs in normal mice, do not adopt a dendritic morphology (10). Such morphology confers on epidermal γδ IELs the term, dendritic epidermal T cell (or 'DEC'), and it allows them to form an anastomotic network, in contact with essentially all the epidermal melanocytes and keratinocytes. This in turn suggests that DEC cells are functionally interacting with these epidermal cells, in a way that the αβ IELs in TCR $\delta^{-/-}$ mice can not.

Thirdly, and possibly most cogent, is the observation that DECs can be stimulated to produce IL-2 *in vitro* by co-culture with autologous keratinocytes (36). That this is a property conferred by the consensus DEC TCR was shown by its capability to render transfected human T cells responsive to mouse keratinocytes, but not to epithelial cells of the reproductive tissue (36). Presumably the converse would be true for the TCR used by uterine γδ IEL. Recognition is enhanced when the keratinocytes are heat shocked (36). These properties of the consensus DEC TCR are consistent with the recognition by epidermal IELs of a single stress antigen expressed on their neighbouring epithelial cells as a result of infection, or cell transformation (4, 5).

Naturally, the nature of the antigen has been the subject of intensive study. The recognition was not inhibited by available anti-hsp antisera, although these data do not exclude recognition of hsp peptides. However, the apparent failure of the DEC TCR to recognize reproductive tissue epithelium suggests that the antigen, unlike well-known hsps (hsp 63, hsp, 70, hsp 90, etc.), is tissue-specific.

Alternatively, tissue specificity may be a property of the antigen-presenting molecule if such are required for γδ T cell recognition. PAM.2.12, the keratinocyte line frequently used to stimulate DECs, is reported to lack expression of classical class I MHC. Moreover, the recognition of keratinocytes by the DEC TCR is not MHC restricted (36). Therefore, if the antigen recognized is an endogenous peptide, it's presumably held on a non-conventional antigen-presenting molecule, supporting the view that IELs in general, and γδ T cells in particular, recognize antigens presented by non-classical class I

MHC (4, 37). Interestingly, whereas classical class I MHC is expressed on most cells (albeit at very different levels), non-classical class I MHC, such as mouse TL, or class I MHC-related molecules, such as human CD1, are expressed in a tissue-specific fashion. Thus, high level expression of the mouse TL genes, T3, T9, and T21 is limited to the gut lining (38, 39). Likewise, the human CD1d gene is expressed in the gut epithelium, while CD1c is expressed in the dermis, and CD1a in the epidermis (40). A candidate molecule on the surface of mouse keratinocytes that can present antigen to DECs is being actively sought. The prediction would be that its expression is epidermis-specific, and that its engineered expression on other epithelia would render them capable of DEC stimulation. Alternatively, γδ T cells may recognize epithelium-associated non-classical MHC as an antigen, not as a presenting molecule.

Since the activation of systemic αβ T cells by macrophages or dendritic cells requires both the engagement of the TCR, and the costimulatory engagement of another ligand, such as CD28 or CTL-A4 (41), it is likely that DEC stimulation likewise requires a second signal from the cells that present antigen to them *in vivo*. The capability for epithelial cells *in vivo* to provide a second signal to T cells is unknown. It might be provided by a cytokine, since in mice, the costimulatory activation of B cells by engagement of surface gp39 can be substituted for by IL-5 (42). Epithelial cells do secrete cytokines, and secretion is increased by cell stress. Thus, IL-6 is released by UV irradiated keratinocytes (43). *In vitro*, DECs show a marked proliferative response to IL-7.

In vivo, murine DECs undergo significant expansion during the first two weeks of life, concomitant with active hair follicle growth. Re-initiation of hair follicle activity in adult animals similarly stimulates local increases in DEC density (44). Likewise, treatment of mice with skin sensitizing agents elicits a local increase in DEC density (44), and DECs also undergo local expansion in response to epidermal infiltration by autoreactive T cells (see below). Taken together, these data strengthen the notion that *in vivo*, the γδ IEL response is to autologous, stress-related ligands on epidermal cells, rather than to ligands expressed by potential pathogens. However, there is as yet little evidence for their response to malignant epithelial cell transformation.

The effector response of DECs to stimulation *in vivo*—whether they kill stressed cells, or whether they recruit lymphoid help—remains unknown (see below). Although their numbers increase under the conditions described, the nature of their effector response may in large part be governed by the nature of the second signal that they receive from the cells that stimulate them. Since either the nature or status of those cells may change, it is difficult to predict whether DECs, and other IELs will show one or multiple effector functions *in vivo*. However, as would be expected from the homogeneity of their TCR, DECs are apparently incapable of protecting an animal in an antigen-specific fashion: TCR $\alpha^{-/-}$ knock-out mice are unable to reject skin grafts congenic at major MHC loci, in spite of there being approximately normal numbers of γδ DECs (45).

3.2 γδ T cells in the murine gut

The murine small intestinal epithelium, a site of non-classical class I MHC expression, becomes populated by clearly detectable numbers of αβ and γδ IELs within one to three weeks of birth (46). The γδ IELs, unlike most γδ cells in the body, are predominantly $CD8^+$ (47). Moreover, both these, and the αβ intestinal IELs express a CD8 αα homodimer that is distinct from the conventional CD8 αβ heterodimer expressed on most $CD8^+$ systemic T cells. This, together with other observations such as the occurrence in the epithelium of an unusual population of TCR αβ CD4, CD8 double-positive T cells, has encouraged speculation that intestinal IELs are extrathymically derived (48, 49). However, despite a clear capability of the gut to foster T cell maturation (50, 51), the site of origin of most intestinal IELs remains unresolved (49). Irrespective of this, their development perinatally results in their TCRs displaying a high degree of junctional diversity, that was absent from the prenatally-derived skin, uterine, and tongue γδ TCRs.

To date, there has been no formal demonstration of an antigen specificity for murine IELs. However, the possession of junctional diversity has raised the issue of whether γδ intestinal TCRs recognize multiple antigens. Since the epithelium of the gastrointestinal tract is the body's first site of contact with myriad pathogens, one possibility is that the IELs recognize predominant pathogen-derived antigens, such as prokaryotic *N*-formylated peptides that have been shown to be presented by non-classical class I MHC (23, 52, 53). Such peptides might be bound to the surface of intestinal epithelial cells after the gastrointestinal digestion of proteins (93). Alternatively, the antigens may be derived or processed from the proteins of intracellular bacteria, such as the *Chlamydobacter* that variably infect intestinal epithelial cells.

Alternatively, or in addition, γδ IELs may 'see' hsp proteins or hsp peptides (5) presented on non-classical class I MHC (4). There have been numerous reports of lymphoid γδ T cells recognizing hsps or peptides derived therefrom (see below). Since such hsps are conserved from prokaryotes, through protozoa, to humans, hsp recognition could enable γδ IELS to respond 'non-specifically' to myriad infections of the epithelium, either through recognition of pathogen-derived hsp, or through recognition of an autologous hsp expressed in response to infection.

Consistent with IEL responses to g.i. tract infection, oral administration to mice of the intestinal, epithelial-tropic, protozoan parasite, *Eimeria vermiformis*, elicits sharp increases in the resident IEL population (54). On the other hand, arguing against a primary role for γδ IELs in anti-microbial responses, γδ IEL cells occur in specific pathogen-free mice in normal numbers, while αβ IELs are depleted, and only restored to normal numbers by exposure of mice to pathogens (55). Moreover, there has been no demonstration that anti-hsp reactivity is a common property of γδ IELs, and to date, the only T cell lines derived that are specific for a *N*-formylated peptide are systemic, αβ T cells (52, 53). No capability of IELs to protect an animal against microbial challenge has as yet been demonstrated, and the γδ IEL repertoire in TCR $\alpha^{-/-}$ knock-out mice does not enable such mice infected with *Eimeria vermiformis* to develop full and rapid immunity against re-infection (59). Thus, there remains a distinct possibility that gut γδ IELs, like their counterparts in the murine skin, recognize one or more autoantigens, possibly induced by stress, and contribute to the host's protection in a distinct manner (see below). It remains to be seen whether antigen recognition is significantly enhanced or altered by the level of TCR junctional diversity that *de facto* occurs in the IEL repertoire.

There has been little success in deriving cloned lines of murine intestinal γδ IELs, and as a result, there has been little clarification of their effector functions. Because they are $CD8^+$, γδ IELs are customarily regarded as cytolytic, and there are reports that epithelial cells from normal mice, but not from β2-microglobulin knock-out mice (that as a result cannot express class I MHC) elicit serine esterase release by γδ IELs (56). It is unclear whether the physiological response of intestinal γδ IELs to activation includes proliferation. Mitotic figures are rare either in sections of infected animals (54), or in biopsies from patients suffering from intestinal inflammation, such as coeliac disease (57) (see below). One can speculate that a clonal expansion of IELs in response to activation, such as occurs for systemic αβ T cells, could have deleterious effects on epithelial integrity. Because IELs can be derived postnatally, it is possible that any expansion *in vivo* is attributable to renewed haematopoiesis, or recruitment of cells from the systemic circulation.

A factor contributing to the difficulties in understanding murine IELs is the paucity of differentiated, intestinal epithelial cells with which to co-culture them. We are almost entirely ignorant of the mechanism, if any, by which epithelial cells present antigen to T cells. However, it is known that gut epithelial cells can be induced to express polymorphic class II MHC (94). The possibility that γδ IELs respond to antigen presented on class II MHC seemed remote, given that most class II-reactive T cells are $CD4^+$, and γδ T cells are not. However, provocative evidence indicates that the proportion of γδ IELs that express Vδ4 varies in mice according to their class II MHC haplotype (58). Whether this reflects a response to polymorphic class II MHC as an antigen-presenting molecule; to class II MHC as a native antigen; to a superantigen bound to class II MHC or to the product of a linked gene, has not been resolved. However, the prospect that γδ IELs respond to class II MHC highlights the fact that we do not know whether the gut is constitutively capable of antigen presentation to IELs, or whether this can only take place after inductive events. Thus, until we resolve the nature of antigen presentation to IELs, we can not judge whether IELs can function as a 'first line of defence', or whether they merely contribute to a response already initiated by other cells (see below).

3.3 γδ T cells in epithelia of other classes of animals—a varied picture

Intestinal γδ IELS have been reported in several other classes of animals, including chickens, and some humans, particularly those suffering from wheat protein induced enteropathies (57). In chickens, as in mice, the γδ IELs are predominantly $CD8^+$, although they are most likely thymic-dependent (60). In most humans, the predominant IEL is TCR αβ. In a provocative study, a set of such αβ IEL cell lines was shown to be specific for CD1 molecules, that are expressed on cells in epithelial sheets, including those in the gut (61). In addition, other studies have indicated that unlike systemic αβ T cells, the IEL αβ TCR repertoire is limited, possibly oligoclonal (62). Although they are rare, human skin-associated IELs are also reported to be $\alpha\beta^+$ with a limited TCR repertoire (63).

Taken together, these observations provide compelling evidence for IEL recognition of epithelial antigens or of peptides presented by unconventional antigen-presenting molecules on relatively sessile, epithelial antigen-presenting cells. However, they stress that these are properties of IELs, $\alpha\beta^+$ or $\gamma\delta^+$, and not simply properties of γδ T cells. Therefore, the retention in several species, of γδ T cells as a significant subpopulation of IELs, probably simply reflects the fact that epithelial γδ T cells perform the 'generic IEL function' particularly efficiently, possibly because the germline γδ repertoire in those species is highly compatible with the limited universe of antigens that will be encountered. Alternatively, γδ T cells may have specialized effector functions.

3.4 Do epithelial γδ T cells prevent breaches of epithelial integrity and regulate systemic inflammation?

Understanding the function of epithelial lymphocytes is plagued by our ignorance of how fundamental immunological processes in epithelia take place. One such process is the development in the gut of oral tolerance to food antigens. According to one model of this, such tolerance is maintained by a T cell population that actively suppresses systemic T cell responses. In this model, γδ IELs have been reported to antagonize the suppression (64). However, if γδ T cells do not show fine antigen specificity, then their immunoregulatory capacity may translate to a particular mix of secreted cytokines. In different contexts, these may promote or suppress immune responses (103).

Thus, in cases of intestinal infection, it is possible that γδ (and αβ) IELs act to suppress local inflammation (54). Acute, local, lymphoid infiltration would seem to pose a grave risk to epithelial integrity, possibly conferring a selective advantage on animals that can down-regulate it. Several observations suggest such down-regulation as a plausible role for IELs. First, the major increases in gut γδ and αβ IEL numbers after *Eimeria* infection occurred after the systemic response was initiated, and approximately coincided with its down-regulation (54). Similar kinetics were reported for the sharp increase in murine lung lavage γδ T cells elicited by intranasal infection with influenza virus (65).

Second are a set of provocative observations made by Shiohara and his colleagues (66). Injection into mice of a T cell line reactive to autologous class II MHC led to severe infiltration of the epidermal layer in response to class II MHC expressed by epidermal Langerhans cells. This infiltration elicited a major local expansion of epidermal IELs. After the infiltration had subsided, through the loss of Langerhans cells, and the mice had recovered, the T cell line was re-injected. This time, the infiltration proceeded only as far as the dermis, the increased numbers of IELs seemingly suppressing epidermal infiltration. However, the difficulty in accepting this as a function of γδ DEC cells is Shiohara's claim that the IEL purely responsible could be reconstituted from a postnatal thymus (67). As was discussed above (in the context of TCR gene rearrangement), DEC cells are formed exclusively in the fetal thymus (68). Thus, the capability to reconstitute the effector function from postnatal thymus suggests that it may be attributable to a non-DEC, γδ IEL.

The putative suppression of inflammation by γδ IELs may be consistent with an emerging view of coeliac disease, and other wheat protein induced enteropathies (57). These are unusual among human diseases in being characterized by distinct increases in the density of γδ, IELs, to the extent that γδ T cells frequently become the major IEL subtype. As a result, γδ, IELs have often been viewed as causal to the pathology. However, the major γδ IEL increases occur in a silent 'infiltrative' stage, that precedes the 'hyperplastic' and 'destructive' stages in which the pathology is manifest (57). Instead, the latter stages are associated with an activation of the lamina propria lymphocytes and macrophages, that acquire activation markers *in vivo*, and which when explanted cause gut epithelial disruption *in vitro*. If lamina propria T cells and other systemic lymphocytes are causal to the pathology, what is the role of the IELs? Possibly, their expansion is promoted (directly or indirectly) by an initial environmental stimulus, such as wheat protein, just as IELs *in vivo* expand in response to *Eimeria* infection. Because the stimulus is chronic, the IELs remain elevated, but, as a result, acute inflammation would be suppressed. Such a metastable state could be maintained for months or years. In summary, it remains highly plausible that γδ T cells maintain tissue integrity but the mechanism(s) is currently unresolved. A greater understanding of epithelial immunity is required.

4 Lymphoid γδ T cells

Lymphoid γδ T cells from several classes of animals have been cloned from the blood, spleens, or lymph nodes, using the paradigms for T cell activation established for circulating γβ T cells. The TCR usage displayed by lymphoid γδ T cells is largely restricted to TCRs not used by resident epithelial γδ

T cells, and so combinatorial diversity is limited. In mice, the predominantly used gene segments are Vγ1 frequently paired with Vδ6, and Vγ4; in humans, the most common lymphoid TCR is encoded by Vγ9–Vδ2 (24). Junctional diversity is apparent, reflective of the peri- and postnatal derivation of many of the cells. Again, it remains to be shown whether the junctional diversity influences reactivity toward a physiologic ligand.

The dominance of the human lymphoid $\gamma\delta^+$ population by a Vγ9–Vδ2 TCR is interesting, since the development of this TCR in the thymus decreases dramatically within three months of birth (69, 70). Instead, the cells seem largely to be derived from extrathymic, postnatal expansion, which, based on twin studies, is at least in part attributable to the environment (71). Thus, human lymphoid Vγ9–Vδ2 cells appear to be responsive to environmental antigens. Indeed, such cells have been shown to respond to Staphylococcal superantigens (72). Moreover, unlike the responses of αβ T cells to superantigens, both chains of the TCR seem to be critical for a full response.

A broad spectrum of specificities has been defined for lymphoid γδ T cells *in vitro*, the extremes of which are indicated in Table 4.2. Cells restricted to conventional polymorphic MHC, and cells restricted to non-classical class I MHC have all been isolated. Whenever receptors are encoded by rearranging genes, a theoretical capability exists to recognize almost anything. However, it seems clear that in terms of physiologic responses, the lymphoid γδ repertoire is not as diverse as the αβ repertoire. When TCR $\alpha^{-/-}$ mice are exposed to *Eimeria* infection, or are skin-painted with sensitizing haptens, lymphoid γδ T cells accumulate, but seem unable to respond specifically to the challenge. The question then is which of the reported reactivities best illustrates the functional capabilities of γδ T cells *in vivo*? This can often be assessed by precursor frequency in unprimed animals.

By contrast to αβ T cells, the precursor frequency of alloreactive γδ T cells is not high. Although examples of γδ clones reactive to polymorphic MHC exist (73), and although the frequency of γδ T cells increases in mixed lymphocyte reactions (74), there is little evidence that γδ T cells *en masse* react to peptides presented by conventional class I or class II MHC. Indeed, recognition of MHC by γδ T cells can be independent of antigen processing (100). This may be related to the fact that unlike most αβ T cells, lymphoid γδ T cells are usually 'double-negative', for the CD4 and CD8 T cell co-receptors. CD4 and CD8 are known to augment αβ TCR recognition of antigen-presenting cells, probably by increasing the adherence of the two cells (75). Studies in the human indicate that double-negative lymphoid αβ T cells may share properties with the small subset of lymphoid γδ T cells that are also double-negative (95), although the generality of this in other species has not been resolved.

It seems that a relatively high precursor frequency toward hsp 63 exists among lymphoid γδ T cells in humans and in mice. Reactivity of γδ T cells to mycobacterial hsp 63 was first noted by O'Brien *et al.* in their study of a large set of γδ T cell hybridomas made from newborn mouse thymus (7).

Table 4.2 Examples of specificities of lymphoid $\gamma\delta$ T cells

Species	Antigen	Reference
Human	Tetanus toxoid restricted by class II MHC	2
	Mycobacterial antigen restricted by CD1b	95
	Protease resistant, low molecular weight mycobacterial antigens, possibly restricted by class II MHC	8
	Human hsp 60/63, non-conventional restriction	81
	Self-tumour immunoglobulins, non-conventional restriction	98
Mouse	Glutamic acid^{50}–tyrosine^{50} restricted by Qa-1	96
	TL27^{b} (or T22^{b})	97
	Peptide of mycobacterial hsp 60/63, non-conventional restriction	76
	MHC antigens in antigen processing-defective cells	100

Moreover, many of the hybrids spontaneously secreted IL-2, which is rare among $\alpha\beta$ hybrids, and which was apparently due to cross-reactivity toward mouse hsp 63, expressed by the hybrids themselves (7). The TCR was implicated, because the reactivity was inhibitable with anti-CD3, and because somatic variants lacking the TCR could not respond (7). A predominant epitope was reported to be the peptide, amino acids 180–196 (76), injection of which into mice can apparently increase the $\gamma\delta$ T cell precursor frequency toward hsp 63 (77).

The nature of the response to the hsp 63 peptide is unclear. It is neither conventionally MHC restricted, nor inhibitable with antibodies against polymorphic MHC molecules, raising once more the possibility of antigen presentation by non-classical MHC antigens, or by novel antigen-presenting molecules. Hsps have themselves been implicated in peptide binding, and might play a role in non-conventional antigen presentation (78).

Most of the $\gamma\delta$ T cells that recognize the peptide express a TCR constructed from the same gene segments, Vγ1–Cγ4, and Vδ6–Cδ (79). Oddly, however, it is claimed that the same peptide is recognized irrespective of junctional diversity (76, 79). Recognition of a ligand by a set of TCRs with the same V gene segments, but with different junctions is normally indicative of a superantigen rather than a peptide response. Likewise, a number of human Vγ9, Vδ2 T cells recognize hsp 63, but apparently as a cell surface protein (80). The existence of cell surface forms of such proteins has been disputed, although the evidence is increasingly compelling. The human cells can react to hsp 63 on Daudi cells, which lack beta-2-microglobulin, indicating that once more, conventional class I MHC is not involved in the recognition (81).

Several observations have suggested that hsp reactive, lymphoid $\gamma\delta$ T cells

Table 4.3 Ten putative properties common among γδ T cell subsets *in vivo*

1.	Reactivity toward cell bound native antigen*
2.	Reactivity toward non-classical class I MHC antigens
3.	Restriction of antigen recognition by non-classical class I MHC
4.	Frequent intraepithelial localization
5.	Derived exclusively from fetal precursors*
6.	Homogeneous receptor repertoires*
7.	High precursor frequency toward hsp 63*
8.	Functional use of two T cell receptors on a single cell*
9.	Protection against primary infection by *Listeria monocytogenes*, *mycobacteria*, plasmodium and other microbial infections
10.	Maturation without MHC linked selection*

*—may be unique to γδ T cells, for details, see text.

may play a part in the host's protection against bacterial infection. Increased numbers of γδ T cells were reported in the lymph nodes of mice infected with *Mycobacterium tuberculosis* (6), and, in humans, there are increased numbers of γδ T cells both in mycobacteria-associated lesions (91), and in sarcoidosis, where an involvement of mycobacteria is suspected (92). Moreover, murine lymphoid γδ T cells which when depleted render mice more susceptible to *Listeria monocytogenes*, react to mycobacterial hsp 63 (82). A high precursor frequency of cells reactive to prokaryote-derived antigens or peptides presented by non-classical class I MHC, may constitute an advantage powerful enough to have selected for retention of lymphoid γδ T cells in all known vertebrates.

The effector functions of γδ T cells may also be particularly suited to defence against certain microbial infections. Unfortunately, those physiologic effector functions have not been fully characterized. Many γδ T cells used to study specificity are hybridomas, from which nothing can be learned about effector function. Studies of several lymphoid γδ T cell lines indicate that they are capable of non-specific, MHC unrestricted cytotoxicity. This, together with the demonstration of anti-self-tumour immunoglobulin specificity (98), has raised the issue of whether γδ T cells recognize and destroy infected and/or transformed target cells (4). However, no course of tumour progression has yet been shown to be naturally influenced by the activity of γδ T cells.

Studies of cell lines also indicate that a variety of cytokines are secreted in response to cell activation. Depending on the context, and the nature of the 'target' cells in the vicinity, these cytokines could effect a variety of functions. For example, recent studies in TCR $\alpha^{-/-}$ knock-out mice indicate that those 'γδ cytokines' may include IL-4, and be sufficient to help B cells make IgG1 antibodies (83). However, the γδ effector functions that are most beneficial to the host have yet to be resolved.

5 Multiple γδ TCR usage and γδ T cell selection

Although murine γδ IELs are usually restricted to the expression of characteristic γδ TCRs, numerous instances have been reported of epithelial expression of the genes for the lymphoid γδ TCR, Vγ1–Jγ1–Cγ4. Such expression showed mouse-to-mouse variation, suggestive of an environmental cause (21). Indeed, in mice adventitiously or deliberately infected with epithelial-tropic agents, the observation was made consistently (21, 54). By sequencing, and by use of a polymerase chain reaction–restriction fragment length polymorphism protocol (84), the genes expressed were productively rearranged, and capable of encoding protein.

Did this represent infiltration into an infected epithelium of lymphoid γδ T cells, or could it mean that epithelial T cells were actively expressing a second TCR? Although the latter would seem at odds with the 'one receptor–clonal selection theory', it was plausible based on the detection of two γ chains on the surface of a murine, skin-derived γδ T cell hybridoma (85), and on the lack of isotypic exclusion among productive γ chains in a further T cell hybridoma (86). The case was most cogently made, however, by the demonstration of two surface TCRs, contributed to by two γ chains, and one δ chain, on the surface of human, peripheral blood γδ T cells (87).

Allelic exclusion operates weakly, if at all, in the TCR α locus (88), and αβ T cells have been reported in which two TCR α chain genes are expressed, but the product of only one is present on the cell surface (89). However, more recently, αβ T cell lines expressing two cell surface α proteins have been described (101), apparently extending the parallels between αβ and γδ T cells. There would, however, remain a major distinction between αβ and γδ T cells if in only the latter both TCRs were functional (Table 4.3). This is possible: it is unlikely that both αβ TCRs on a single cell could be functionally useful, since one 'bad' TCR would be sufficient to effect negative selection of the cell, while it is unlikely that both receptors could 'pass' the rigour of positive selection. Therefore, the essence of the clonal selection theory may not be violated. By contrast, since neither the specificity, nor the means of selection (if any) of γδ T cells is known, it remains possible that both TCRs on a single γδ cell could be functional. Possibly γδ T cells can tolerate this violation of the one receptor–clonal selection rule, because one or more of the following criteria uniquely applies to them.

1. Neither of the TCRs concerned have to be tolerized, because their primary reactivity is toward self-stress antigens.
2. On activation, the cells neither expand, nor initiate an expansive immune response, so limiting the 'spread' of a potentially harmful interaction with local cells.
3. The cells do not circulate much, and/or have a very limited life-span, further restricting potentially harmful interactions with local cells.

The consideration of these issues raises the question of whether γδ T cells are selected during their development, either positively, so that those cells that may interact profitably in the body are enriched for, or negatively, so that potentially autoreactive cells are deleted. The question of selection in the γδ repertoire is still largely unanswered, despite some elegant attempts to resolve it. In mice transgenic for a non-classical class I MHC reactive γδ TCR, the resultant T cells could be negatively selected on the appropriate genetic background, but it is unclear whether physiologically the cells were αβ T cells inappropriately expressing a γδ TCR. Similarly, other transgenic experiments suggest that self-reactive γδ T cells may be anergized in the periphery. What is clear is that if a large number of γδ T cells do not recognize processed antigen, or are restricted to non-classical, non-polymorphic class I MHC, then the paradigms developed for the selection of αβ T cells by polymorphic MHC molecules in the thymus are unlikely to satisfactorily explain γδ T cell maturation. Furthermore, as stated above, in TCR $\alpha^{-/-}$ mice, the lack of thymic cortical-medullary organization does not seem to hinder the development of γδ T cells (Table 4.3).

At the same time, it is also true that γδ T cells are implicated in several autoimmune pathologies, such as human, autoimmune hepatitis, and lupus nephritis (90, 99), raising the possibility that expression of dual TCRs does violate selection processes and can contribute to pathologic autoimmunity.

If the expression of productively rearranged Vγ1 genes in the infected epithelia of mice is attributable to dual receptor expression, it suggests that in response to infection, cells expressing two TCRs are favoured. The nature of their selective advantage can only be known when the nature of the TCR ligands is resolved. However, any capability of γδ T cells to interact with target cells using two TCRs, either synchronously or sequentially would represent a major functional distinction from αβ T cells (Table 4.3).

6 γδ T cell interactions with other lymphocytes—in adults and in neonates

The responses of lymphoid γδ T cells to environmental, antigenic stimulation have been traditionally difficult to study because of the excess of αβ T cells. In TCR $\alpha^{-/-}$ mice, this is not the case. In these mice, environmental antigenic exposure induces expansion of γδ T cells, independent of the absence of αβ T cells. When explanted, the γδ T cells show cytokine secretion, for example IL-4, in response to mycobacterial extracts, such as PPD. This may provide B cell help, raising the question as to how tolerance is maintained in the B cell repertoire. Indeed, in human patients, T cells helping autoreactive, anti-DNA antibody production were shown to be $\gamma\delta^{+}$ (99). Moreover, the expansion *in vivo*, of γδ T cells in TCR $\alpha^{-/-}$ mice appears to be unregulated. As a result, the spleen and lymph nodes frequently swell excessively. This suggests that *de facto* αβ T cells can down-regulate γδ T cell expansion. This may be due to the capability of αβ T cells to rapidly resolve infections, and thereby to remove bulk stimulating antigen. Alternatively, a specific set of immunoregulatory αβ T cells may interact with γδ T cells. Whatever the explanation, the suggestion is that γδ T cell function in the body is effected not in isolation, but in concert with the rest of the immune system. Thus, adults in which there is an unusual accumulation of γδ T cells may reflect a general breakdown in immune regulation, rather than simply an expansion of γδ T cells in response to direct stimulation. This possibility could be taken into account in the assessment of human disease and immune status. By contrast, neonates are naturally deficient in αβ T cell function, and relatively enriched in γ8 T cells. Perhaps the γ8 T cell functions seen in $TCR\alpha^{-/-}$ mice reflect an essential role of γδ T cells in protecting the neonate.

Acknowledgements

I should like to thank my close colleagues for much discussion of these matters, and especially the members of our laboratory for so frequently enlivening the journal club. I acknowledge N.I.H. for grant support—AI27855.

References

1 Saito, H., Kranz, D. M., Takagaki, Y., Hayday, A., Eisen, H. N., and Tonegawa, S. (1984). Complete primary structure of heterodimeric T cell receptor protein deduced from cDNA sequences. *Nature*, **309**, 757–62.

2 Kazbor, D., Trinhieri, G., Monos, D., Isobe, M., Russo, G., Haney, J., *et al.* (1989). Human TCR γδ,CD8+ T lymphocytes recognize tetanus toxoid in an MHC-restricted fashion. *J. Exp. Med.*, **169**, 1847–51.

3 Matis, L. A., Cron, R., and Bluestone, J. A. (1987). MHC-linked specificity of γδ receptor-bearing T lymphocytes. *Nature*, **330**, 262–4.

4 Janeway, C. A. Jr., Jones, B., and Hayday, A. C. (1988). Specificity and function of cells bearing γδ T cell receptors. *Immunol. Today*, **9**, 73–6.

5 Asarnow, D. M., Kuziel, W. A., Bonyhadi, M., Tigelaar, R. E., Tucker, P. W., and Alison, J. P. (1988). Limited diversity of γδ antigen receptor genes of Thy-1(+) dendritic epidermal cells. *Cell*, **55**, 837–47.

6 Janis, E. M., Kaufmann, S. H. E., Schwartz, R. H., and Pardoll, D. M. (1989). Activation of γδ T cells in the primary immune response to mycobacterium tuberculosis. *Science*, **244**, 713–15.

7 O'Brien, R. L., Happ, M. P., Dallas, A., Palmer, E., Kubo, R., and Born, W. K. (1989). Stimulation of a major set of lymphocytes expressing T cell receptor γδ by an antigen derived from *Mycobacterium tuberculosis*. *Cell*, **57**, 667–74.

8 Pfeffer, K., Schoel, B., Gulle, H., Kaufmann, S. H. E., and Wagner, H. (1990). Primary responses of human T cells to mycobacteria: a frequent set of γδ T cells are stimulated by protease resistant ligands. *Eur. J. Immunol.*, **20**, 1175–80.

9 Philpott, K., Viney, J. L., Kay, G., Rastan, S., Gardiner, E. M., Chae, S., *et al.* (1992). Lymphoid development in mice congenitally lacking T cell receptor αβ expressing cells. *Science*, **256**, 1448–52.

10 Itohara, S., Mombaerts, P., Lafaille, J., Iacomini, J., Nelson, A., Clarke, A. R., *et al.* (1993). T cell receptor delta mutant mice: independent generation of αβ T cells, and programmed rearrangements of γδ cell receptor genes. *Cell*, **72**, 337–48.
11 Brenner, M. B., Mclean, J., Dialynas, D., Strominger, J. L., Smith, J. A., Owen, F. L., *et al.* (1986). Identification of a putative second T cell receptor. *Nature*, **322**, 145–49.
12 Itohara, S., Nakanishi, N., Kanagawa, O., Kubo, R., and Tonegawa, S. (1989). Monoclonal antibodies specific to native murine T cell receptor γδ: analysis of γδ T cells during thymic ontogeny and in peripheral lymphoid organs. *Proc. Natl. Acad. Sci. USA*, **86**, 5094–8.
13 Bucy, P., Chen, C. L., Cihak, J., Löaxh, U., and Cooper, M. (1988). Avian T cells expressing γδ receptors localize in the splenic sinusoids and the intestinal epithelium. *J. Immunol.*, **141**, 2200–5.
14 Hein, W. R., Dudler, L., and Morris, B. (1990). Differential peripheral expansion and *in vivo* antigen reactivity of αβ and γδ T cells emigrating from the early lamb thymus. *Eur. J. Immunol.*, **20**, 1805–13.
15 Kuziel, W. A., Takashima, A., Bonyhadi, M., Bergstresser, J., Allison, J. P., Tigelaar, R. E., *et al.* (1987). Regulation of T cell receptor γ chain RNA expression in murine thy-1(+) dendritic epidermal cells. *Nature*, **328**, 263–6.
16 Stingl, G., Koning, F., Yamada, H., Yokoyama, W. M., Tschachler, E., Bluestone, J. A., *et al.* (1987). Thy-1+ dendritic epidermal cells express T3 antigen and the T-cell receptor γ chain. *Proc. Natl. Acad. Sci. USA*, **84**, 4586–9.
17 Goodman, T. and Lefrancois, L. (1988). Expression of the γδ T cell receptor on intestinal CD8(+) intraepithelial lymphocytes. *Nature*, **333**, 855–8.
18 Halstensen, T. S., Scott, H., and Brandtzaeg, P. (1989). Intraepithelial T cells of TCRγδ, CD8− and Vδ1Jδ1 phenotypes and increased in coeliac disease. *Scand. J. Immunol.*, **30**, 665–72.
19 Itohara, S., Farr, A., Lafaille, J. J., Bonneville, M., Takagaki, Y., Haas, W., *et al.* (1990). Homing of a gamma delta thymocyte subset with homogeneous T-cell receptors to mucosal epithelia. *Nature*, **343**, 754–7.
20 Augustin, A., Kubo, R. T., and Sim, G.-K. (1989). Resident pulmonary lymphocytes expressing the γδ T cell receptor. *Nature*, **340**, 239–41.
21 Kyes, S. and Hayday, A. C. (1990). Disparate types of γδ T cells. *Res. Immunol.*, **141**, 582–7.
22 Brandtzaeg, P., Bosnes, V., Halstensen, T. S., Scott, H., Sollid, L. M., and Valnes, K. N. (1989). T lymphocytes in human gut epithelium preferentially express the αβ antigen receptor, and are often CD45/UCHL1(+). *Scand. J. Immunol.*, **30**, 123–8.
23 Shawar, S. M., Vyas, J. M., Rodgers, J. R., Cook, R. G., and Rich, R. R. (1991). Specialised functions of MHC class I molecules. II. Hmt binds N-formylated peptides of mitochondrial and prokaryotic origin. *J. Exp. Med.*, **174**, 941–4.
24 Hayday, A. (1992). T cell receptor γδ. In *Encyclopedia of immunology* (ed. P. Delves and I. Roitt), pp. 1428–33. Academic Press, London.
25 Kyes, S., Carew, E., Carding, S. R., Janeway, C. A. Jr., and Hayday, A. C. (1989). Diversity in T cell receptor γ gene usage in intestinal epithelium. *Proc. Natl. Acad. Sci. USA*, **86**, 5527–31.
26 Takagaki, Y., DeCloux, A., Bonneville, M., and Tonegawa, S. (1989). γδ T cell receptors on murine intraepithelial lymphocytes are highly diverse. *Nature*, **339**, 172–4.
27 Davis, M. M. and Bjorkman, P. (1988). T cell antigen receptor genes and T cell recognition. *Nature*, **334**, 395–402.

28 Dreyer, W. J. and Bennett, J. C. (1965). The molecular basis of antibody formation: a paradox. *Proc. Natl. Acad. Sci. USA*, **54**, 864–9.
29 Tonegawa, S. (1983). Somatic generation of antibody diversity. *Nature*, **302**, 575–81.
30 Havran, W. and Allison, J. P. (1990). Origin of Thy-1+ dendritic epidermal cells of adult mice from fetal thymic precursors. *Nature*, **344**, 68–70.
31 Elliott, J. F., Rock, E. P., Patten, P. A., Davis, M. M., and Chien, Y.-H. (1988). The adult T cell receptor δ chain is diverse and distinct from that of fetal thymocytes. *Nature*, **331**, 627–30.
32 Lafaille, J. J., DeCloux, A., Bonneville, M., Takagaki, Y., and Tonegawa, S. (1989). Junctional sequences of T cell receptor γδ T cell lineages and for a novel intermediate of V-(D)-J joining. *Cell*, **59**, 859–70.
33 Asarnow, D. M., Cado, D., and Raulet, D. H. (1993). Selection is not required to produce invariant T cell receptor γ gene junctional sequences. *Nature*, **362**, 158–60.
34 Bonneville, M., Itohara, S., Krecko, E., Mombaerts, P., Ishida, I., Katsuki, M., *et al.* (1990). Transgenic mice demonstrate that epithelial homing of γδ T cells is determined by cell lineages, independent of T cell receptor specificity. *J. Exp. Med.*, **171**, 1015–26.
35 Kyes, S., Pao, W., and Hayday, A. (1991). Influence of site of expression on the fetal γδ T cell receptor repertoire. *Proc. Natl. Acad. Sci. USA*, **88**, 7830–3.
36 Havran, W., Chien, Y.-H., and Allison, J. P. (1991). Recognition of self antigens by skin-derived T cells with invariant γδ antigen receptors. *Science*, **252**, 1430–2.
37 Murphy, D. B. (1992). Evolutionary junk, or first class tag-along? *Curr. Biol.*, **2**, 529–31.
38 Wu, M., Van Kaer, L., Itohara, S., and Tonegawa, S. (1991). Highly restricted expression of the thymic leukaemia antigens on intestinal epithelial cells. *J. Exp. Med.*, **174**, 213–18.
39 Hershberg, R., Eghtesady, P., Sydora, B., Brorson, K., Cheroutre, H., Modlin, R., *et al.* (1990). Expression of the thymus leukemia antigen in the intestinal epithelium. *Proc. Natl. Acad. Sci. USA*, **87**, 9727–31.
40 Bleicher, P., Balk, S. P., Hagen, S. J., Blumberg, R. S., Flotte, T. J., and Terhorst, C. (1990). Expression of a murine CD1 on gastrointestinal epithelium. *Science*, **250**, 679–82.
41 Schwartz, R. H. (1990). A cell culture model for T lymphocyte clonal anergy. *Science*, **248**, 1349–56.
42 Purkeson, J. and Isakson, P. (1992). A two-signal model for the regulation of immunoglobulin isotype switching. *FASEB J.*, **6**, 3245–52.
43 Luger, T., Schwarz, T., Krutmann, J., Kirnbauer, R., Neuner, P., Kock, A., *et al.* (1989). IL-6 is produced by epidermal cells and plays an important role in the activation of human T lymphocytes and NK cells. *Ann. N.Y. Acad. Sci.*, **557**, 405–14.
44 Tigelaar, R. E. and Lewis, J. Modulation *in vivo* of skin-derived γδ(+) dendritic epidermal T cell lines. Pers. comm.
45 Simpson, E. Analysis of skin-grafting to T cell receptor α-/-mice. Pers. comm.
46 Bandeira, A., Itohara, S., Bonneville, M., Burlen-Defranoux, O., Mota-Santos, T., Coutinho, A., *et al.* (1991). Extrathymic origin of intestinal intraepithelial T cells bearing the γδ T cell receptor. *Proc. Nat. Acad. Sci. USA*, **88**, 43–7.
47 Lefrancois, L. and Goodman, T. (1989). *In vivo* modulation of cytolytic activity and Thy-1 expression in TCRγδ(+) intraepithelial lymphocytes. *Science*, **243**, 1716–18.
48 Lefrancois, L. (1991). Extrathymic differentiation of intraepithelial lymphocytes:

generation of a separate and unequal T cell repertoire. *Immunol. Today*, **12**, 436–8.
49 Hayday, A. C. (1993). Not in the thymus. *Curr. Biol.*, **3**, 525–8.
50 Ferguson, A. and Parrott, D. V. M. (1972). Growth and development of 'antigen-free' grafts of foetal mouse intestine. *J. Pathol.*, **106**, 95–101.
51 Mosley, R. L. and Klein, J. R. (1992). Peripheral engraftment of fetal intestine into athymic mice sponsors T cell development; direct evidence for thymopoietic function of murine small intestine. *J. Exp. Med.*, **176**, 1365–73.
52 Kurlander, R. J., Shawar, S. M., Brown, M. L., and Rich, R. R. (1992). Specialised role for a murine class Ib MHC molecule in prokaryotic host defenses. *Science*, **257**, 678–9.
53 Pamer, E. G., Wang, C. R., Flaherty, L., Fischer-Lindahl, K., and Bevan, M. J. (1990). H-2M3 presents a Listeria Monocytogenes peptide to cytotoxic T lymphocytes. *Cell*, **70**, 215–33.
54 Findly, R. C., Roberts, S. J., and Hayday, A. C. (1993). Dynamic responses of murine gut intraepithelial T cells after infection by the coccidian parasite, *Eimeria. Eur. J. Immunol.*, **23**, 2557–64.
55 Bandeira, A., Mota-Santos, T., Itohara, S., Degermann, S., Heusser, C., Tonegawa, S., *et al.* (1990). Localisation of γδ T cells in the intestinal epithelium is independent of normal microbial colonization. *J. Exp. Med.*, **172**, 239–44.
56 Eghtesady, P. and Kronenberg, M. (1992). Intestinal γδ T lymphocytes and autoreactive for TL-antigens on stressed intestinal epithelial cells. Unpublished communication.
57 Marsh, M. N. (1992). Gluten, Major histocompatibility complex, and the small intestine. *Gastroenterology*, **102**, 330–54.
58 Lefrancois, L., LeCorre, R., Mayo, J., Bluestone, J. A., and Goodman, T. (1990). Extrathymic selection of T cell receptor γδ(+) T cells by class II MHC molecules. *Cell*, **63**, 333–9.
59 Roberts, S. J., Smith, A. L., Wen, L., Viney, J. L., Owen, M. J., Findly, R. C., *et al.* (1995). Mutant mice reveal the critical cells in the resistance of mice to *Eimeria.* In preparation for *J. Exp. Med.*.
60 Dujon, D., Cooper, M. D., and Imhof, B. A. (1993). Thymic origin of embryonic intestinal γδ T cells. *J. Exp. Med.*, **177**, 257–63.
61 Balk, S. P., Ebert, E. C., Blumenthal, R. L., McDermott, F. V., Wucherpfennig, K. W., Landau, S. B., *et al.* (1991). Oligoclonal expansion and CD1 recognition by human intestinal intraepithelial lymphocytes. *Science*, **253**, 1411–15.
62 Van Kerckhove, C., Russell, G. J., Deusch, K., Reich, K., Bhan, A. K., DerSimonian, H., *et al.* (1992). Oligo-clonality of human intestinal intraepithelial T cells. *J. Exp. Med.*, **175**, 57–63.
63 Dunn, D. A., Gadenne, A.-S., Simha, S., Lerner, E. A., Bigby, M., and Bleicher, P. A. (1993). T cell receptor Vβ expression in normal human skin. *Proc. Natl. Acad. Sci. USA*, **90**, 1267–71.
64 Fujihashi, K., Taguchi, T., Aicher, W. K., McGhee, J. R., Bluestone, J. A., Eldridge, J. N., *et al.* (1992). Immunoregulatory functions for murine intraepithelial lymphocytes: γδ T cell receptor positive T cells abrogate oral tolerance, while αβ positive T cells provide B cell help. *J. Exp. Med.*, **175**, 695–707.
65 Carding, S. R., Allen, W., Kyes, S., Hayday, A., Bottomly, K., and Doherty, P. (1990). Late dominance of the inflammatory process in murine influenza by γδ(+) T cells. *J. Exp. Med.*, **172**, 1225–31.
66 Shiohara, T., Moriya, N., Gotoh, C., Hayakawa, H., Nagashima, M., Seizawa, K., *et al.* (1990). Loss of epidermal integrity by T cell mediated attack induces

long term local resistance to subsequent attack. Induction of resistance correlates with increases in Thy1(+) epidermal cell numbers. *J. Exp. Med.*, **171**, 1027–4.
67 Shiohara, T., Moriya, N., Gotoh, C., Hayakawa, J., Ishikawa, H., Siazawa, K., *et al.* (1990). Loss of epidermal integrity by T cell mediated attack induces long term local resistance to subsequent attack. Thymus dependency in the induction of resistance to graft versus host disease. *J. Immunol.*, **145**, 2482–8.
68 Ikuta, K., Kina, T., MacNeil, I., Uchida, N., Peault, B., Chien, Y.-H., *et al.* (1990). A developmental switch in thymic lymphocyte maturation potential occurs at the level of hematopoietic stem cells. *Cell*, **62**, 863–74.
69 Casorati, G., de Libero, G., Lanzavecchia, A., and Mignone, N. (1989). Molecular analysis of human γδ(+) clones from the thymus and peripheral blood. *J. Exp. Med.*, **170**, 1521–35.
70 McVay, L. M., Carding, S. R., Bottomly, K., and Hayday, A. C. (1991). Regulated expression and structure of T cell antigen receptor γδ transcripts in human thymic ontogeny. *EMBO J.*, **10**, 83–91.
71 Parker, C. M., Groh, V., Band, H., Porcelli, S. A., Morita, C., Fabbi, M., *et al.* (1990). Evidence for extrathymic changes in the T cell receptor γδ repertoire. *J. Exp. Med.*, **171**, 1597–612.
72 Rust, C. J. J., Verreck, F., Vietor, H., and Koning, F. (1990). Specific recognition of staphylococcal enterotoxin A by human T cells bearing receptors with the Vγδ region. *Nature*, **346**, 572–4.
73 Bluestone, J. A., Cron, R. Q., Cotterman, M., Houlden, R. A., and Matis, L. A. (1988). Structure and specificity of γδ T cell receptor on major histocompatibility complex antigen-specific CD3+, CD4−, CD8− T lymphocytes. *J. Exp. Med.*, **168**, 1899–916.
74 Jones, B., Carding, S., Kyes, S., Mjolsness, S., Janeway, C., and Hayday, A. (1988). Molecular analyses of TCR gamma gene expression in allo-activated splenic T cells of adult mice. *Eur. J. Immunol.*, **18**, 1907–15.
75 Killeen, N. and Littman, D. R. (1993). Helper T cell development in the absence of CD4-p56lck association. *Nature*, **364**, 729–32.
76 Born, W. K., Hall, L., Dallas, A., Reardon, C., Kubo, R. T., Shinnick, T., *et al.* (1990). Recognition of a peptide antigen by heat shock reactive γδ T lymphocytes. *Science*, **249**, 67–9.
77 Fu, Y.-X., Cranfill, R., Vollmer, M., van der Zee, R., O'Brien, R. L., and Born, W. K. (1993). *In vivo* response of murine γδ T cells to heat shock protein derived peptide. *Proc. Natl. Acad. Sci. USA*, **90**, 322–6.
78 Mann, R., Dudley, E. C., Sano, Y., O'Brien, R. L., Born, W. K., Janeway, C. A., *et al.* (1990). Modulation of murine self antigens by mycobacterial components. *Curr. Top. Microbiol. Immunol.*, **173**, 151–7.
79 Happ, M. P., Kubo, R. T., Palmer, E., Born, W. K., and O'Brien, R. L. (1989). Limited receptor repertoire in a mycobacteria-reactive subset of γδ lymphocytes. *Nature*, **342**, 696–8.
80 Kaur, I., Voss, S. D., Gupta, R. S., Schell, K., Fisch, P., and Sondel, P. M. (1993). Human peripheral γδ T cells recognize hsp60 on Daudi Burkitt's lymphoma cells. *J. Immunol.*, **150**, 2046–55.
81 Fisch, P., Malkovsky, M., Kovats, S., Sturn, E., Braakman, E., Klein, B. S., *et al.* (1990). Recognition by human Vγ9/Vδ2 T cells of a GroEL homolog on Daudi Burkitts lymphoma cells. *Science*, **250**, 1269–73.
82 Hiromatsu, K., Yoshikai, Y., Matsuzaki, G., Ogha, S., Muramori, K., Matsumoto, K., *et al.* (1992). A protective role for γδ T cells in primary infection with *Listeria Monocytogenes* in mice. *J. Exp. Med.*, **175**, 49–56.

83 Wen, L., Roberts, S. J., Viney, J. L., Wong, F. S., Mallick, C. A., Wong, F. S., Findly, R. C., *et al.* (1994). Immunoglobulin synthesis and generalised autoimmunity in mice congenitally deficient in αβ(+) T cells. *Nature*, **369**, 654–9.
84 Mallick, C. A., Dudley, E. C., Viney, J. L., Owen, M. J., and Hayday, A. C. (1993). Rearrangement and diversity of T cell receptor β chain genes in thymocytes: a critical role for the β chain in development. *Cell*, **73**, 513–19.
85 Koning, F., Yokoyama, W. M., Maloy, W. L., Stingl, G., McConnell, T. J., Cohen, D. I., *et al.* (1988). Expression of Cγ4 T cell receptors and lack of isotype exclusion by dendritic epidermal T cell lines. *J. Immunol.*, **141**, 2057–62.
86 Heilig, J. S. and Tonegawa, S. (1987). T-cell γ gene is allelically but not isotypically excluded and is not required in known functional T-cell subsets. *Proc. Natl. Acad. Sci. USA*, **84**, 8070–4.
87 Davodeau, F., Peyrat, M.-A., Houde, I., Mallet, M.-M., de Libero, G., Vie, H., *et al.* (1993). Surface expression of two distinct functional antigen receptors on human γδ T cells. *Science*, **260**, 1800–2.
88 Marolleau, J. P., Fondell, J., Malissen, M., Trucy, T., Barbier, E., Marcu, K. B., *et al.* (1988). The joining of germ-line Vα to Jα genes replaces pre-existing Vα–Jα complexes in a T cell receptor α,β positive T cell line. *Cell*, **55**, 291–300.
89 Malissen, M., Trucy, J., Letourneur, F., Rebai, N., Dunn, D. E., Fitch, F., *et al.* (1988). A T cell clone expresses two T cell receptor α genes but uses only one αβ heterodimer for allorecognition and self MHC restricted antigen recognition. *Cell*, **55**, 49–59.
90 Wen, L., Peakman, M., Vergani, G. M., and Vergani, D. (1992). Elevation of activated γδ T cell receptor bearing T lymphocytes in patients with autoimmune chronic liver disease. *Clin. Exp. Immunol.*, **89**, 78–82.
91 Modlin, R. L., Pirmez, C., Hofman, F. M., Torigian, V., Uyemura, K., Rea, T. H., *et al.* (1989). Lymphocytes bearing antigen specific γδ T cell receptors accumulate in human infectious lesions. *Nature*, **339**, 544–6.
92 Tamura, N., Holroyd, K. J., Banks, T., Kirby, M., Okayama, H., and Crystal, R. (1990). Diversity in junctional sequences associates with the common human Vγ9 and Vδ2 gene segments in normal blood and lung compared with the limited diversity in granulomatous disease. *J. Exp. Med.*, **172**, 169–82.
93 Sperling, A. I. and Bluestone, J. A. (1993). The first line of defence? *Curr. Biol.*, **3**, 294–7.
94 Scott, H., Sollid, L. M., Fausa, O., Brandtzaeg, P., and Thorsby, E. (1987). Expression of major histocompatibility class II subregion products by jejunal epithelium in patients with coeliac disease. *Scand. J. Immunol.*, **26**, 563–771.
95 Porcelli, S., Brenner, M. B., Greenstein, J. L., Bank, S. P., Terhorst, C., and Bleicher, P. (1989). Recognition of cluster of differentiation 1 antigens by human CD4−CD8−cytolytic T lymphocytes. *Nature*, **341**, 447–50.
96 Vidovic, D., Roglic, M., McKune, K., Guerder, S., MacKay, C., and Dembic, Z. (1989). Qa-1 restricted recognition of foreign antigen by a γδ T-cell hybridoma. *Nature*, **340**, 646–50.
97 Ito, K., Van Kaer, L., Bonneville, M., Hsu, S., Murphy, D. B., and Tonegawa, S. (1990). Recognition of the product of a novel MHC TL region gene (27^b) by a mouse γδ T cell receptor. *Cell*, **62**, 549–61.
98 Wright, A. J., Lee, J. E., Link, M. P., Smith, S. D., Carroll, W., Levy, R., *et al.* (1989). Cytotoxic T lymphocytes specific for self tumor immunoglobulins express the TCR δ chain. *J. Exp. Med.*, **169**, 1557–61.
99 Rajagopolan, S., Zordan, T., Tsokos, G. C., and Datta, S. K. (1990). Pathogenic anti DNA autoantibody-inducing T helper cell lines from patients with active lupus

nephritis: isolation of CD4—CD8—T helper lines that express the $\gamma\delta$ T cell antigen receptor. *Proc. Natl. Acad. Sci. USA*, **87**, 7020–4.

100 Schild, H., Mavaddat, N., Litzenberger, C., Ehrich, E.W., Davis, M.M., Bluestone, J.A., *et al.* (1994). The nature of major histocompatibility complex recognition by $\gamma\delta$ T cells. *Cell*, **76**, 29–37.

101 Padovan, E., Casorati, G., Dellabona, P., Meyer, S., Brockhaus, M., and Lanzavecchia, A. (1993). Expression of two T cell receptor α chains: dual receptor T cells. *Science*, **262**, 422–4.

102 Tsuji, M., Mombaerts, P., Lefrancois, L., Nussenzweig, R. S., Zavala, F., and Tonegawa, S. (1994). $\gamma\delta$ T cells contribute to immunity against the liver stages of malaria in $\alpha\beta$ T-cell-deficient mice. *Proc. Natl. Acad. Sci. USA.*, **91**, 345–9.

103 Ferrick, D. A., Schrenzel, M. D., Mulvania, T., Hsieh, B., Ferdin, W. G., and Lepper, H. (1995). Differential production of interferon γ and interleukin 4 in response to Th-1 and Th-2 stimulating pathogens by $\gamma\delta$ T cells *in vivo*. *Nature*, **373**, 255–7.

5 Emergence of the T cell receptor repertoire

SUSAN GILFILLAN, CHRISTOPHE BENOIST, AND DIANE MATHIS

1 Introduction

The emergence of the diverse adult TCR repertoire must be evaluated in the context of thymocyte development in general. Precursors migrate from the murine fetal liver until embryonic day 15 when the bone marrow and spleen take over haematopoiesis (1). T cells expressing $\gamma\delta$ receptors appear in the thymus on embryonic day 13 and constitute the major T cell population until day 18; at this point cells bearing $\alpha\beta$ receptors begin to predominate and the percentage of $\gamma\delta$ cells rapidly declines to the approximate 1% observed in adults (2, 3). As discussed in detail below, $\gamma\delta$ T cells enter the thymus in waves, the first two consisting of cells expressing essentially monoclonal receptors. These are highly specialized cells and eventually make up the vast majority of intraepithelial lymphocytes (IELs) in the skin, female reproductive tract, and the tongue. The ensuing waves of $\gamma\delta$ T cells express a more diverse array of TCRs, as do $\alpha\beta$ cells. The transition from the fetal to the adult repertoire involves increased commitment to the $\alpha\beta$ lineage as well as increased variability within each lineage.

Also critical for understanding the emergence of the adult repertoire is identification of the ligands recognized by the TCRs. $\alpha\beta$ TCRs recognize peptides embedded in the grooves of the major histocompatibility complex (MHC) class I and II molecules, usually aided by the co-receptors CD8 and CD4 respectively. In contrast, $\gamma\delta$ TCRs interact with a variety of structures, including bacterial heat shock proteins, non-classical and (in at least some cases) classical MHC antigens (4, 5). This rather broad panoply of potential ligands has rendered functional assessment of the $\gamma\delta$ TCR repertoire difficult. Although $\gamma\delta$ cells generated early in ontogeny clearly differ from those produced later, comparison is limited to descriptive terms such as the type of receptor expressed and anotomical location. Partially circumventing this problem, recent studies have clarified the roles of programmed gene rearrangement, homology-directed recombination, N region insertion, and selection in generating the $\gamma\delta$ and $\alpha\beta$ repertoires. These data answer many questions regarding how the fetal/adult dichotomy is generated and more sharply focus speculation as to why.

2 Diversity

Like immunoglobulins (Igs), γδ and αβ TCRs are composed of two chains, each encoded by disparate gene segments—variable (V), diversity (D), joining (J), and constant (C) for β and δ; V, J, and C for α and γ. The V, D, and J segments encode the membrane distal, ligand binding domain of the TCR. In immature T cells, V, D, and J segments are joined by site-specific recombination mediated by at least two proteins, the recombination activating genes RAG-1 and RAG-2 (for reviews see references 6 and 7). A primary source of TCR diversity is the multitude of V(D)J fusions created by this process; random association of different chains ($\alpha \times \beta$ or $\gamma \times \delta$) considerably enlarges potential variability. In theory, all V, D, and J segments could be used at equal frequencies throughout life but, in fact, V(D)J rearrangement is not random and different rules seem to apply to each locus. Fortunately, it has become possible to unequivocally distinguish the effects of programmed rearrangement from those of selection at the γδ locus. As discussed below, these experiments have demonstrated that stringently programmed rearrangement at the γδ loci helps ensure that the perinate/adult dichotomy is more marked for γδ than for αβ TCRs.

Additional diversity can be created at the junctions as V, D, or J segments are brought together by:

- heterogeneity at the points of joining
- variable loss of nucleotides from exposed termini
- addition of template-dependent palindromic 'P' nucleotides to intact termini
- addition of template-independent 'N' nucleotides

Although an exonuclease is generally invoked for the loss of nucleotides, both variable nucleotide loss and gain of P nucleotides may be caused by asymmetric scission of closed hairpin termini prior to joining (8). Whatever the mechanism, these two sources of diversity do not differ significantly between perinates and adults. In contrast, template independent addition of N nucleotides is rarely observed in perinatal V(D)J junctions but comprises a major source of junctional diversity in adult TCRs (9–12). Another noteworthy difference between the perinatal and adult repertoires was initially identified in IgH junctions; more than 80% of perinatal Ig junctions contain 1–6 base pairs (bp) that are shared by the flanking V, D, or J regions (13–15). Similar junctions comprised of 2–5 bp of homologous sequence are found at high frequency in fetal γδ V(D)J regions (11, 16). The precision and over-representation of such junctions led to the suggestion they were directed; i.e. that annealing of short stretches of homology present on juxtaposed segments is facilitated during the joining process (13, 15; see also 17, 18). As discussed below, recent experiments have supported this proposition. Thus, the transition from the fetal to the adult repertoire is marked by the gain of N nucleotides and the loss of homology-directed junctions.

The template-independent DNA polymerase, terminal deoxynucleotidyl transferase (TdT), has long been implicated as the enzyme responsible for the addition of N nucleotides. Expressed almost exclusively in immature thymocytes and pro-B cells, TdT is capable of adding N nucleotides to artificial recombination substrates *in vitro* (19–21). In addition, the expression of TdT in cell lines as well as its onset during ontogeny correlate with the presence of N regions in junctional sequences (22–24). Recently, two types of mice were created to directly assess the role of TdT:

1. Mice carrying a mutant TdT gene in the germline were created using homologous recombination in embryonic stem (ES) cells (25).
2. A different mutation was introduced into one ($TdT^{o/+}$) or both TdT ($TdT^{o/o}$) alleles in ES cells, which were injected into blastocysts null for RAG-2 expression (26, 27). Because only the ES cells could develop into mature lymphocytes, the resulting chimeras were equivalent to heterozygote and homozygote 'knock-out' mice in terms of gene rearrangement.

Analysis of hundreds of V(D)J junctions from both types of $TdT^{o/o}$ (TdT°) mice confirmed that TdT does, indeed, catalyse polymerization of the vast majority of N nucleotides. Less than 3% of TdT° junctions contained template-independent insertions and the majority of these 'N region' consisted of a single nucleotide; this was slightly less than has been observed in similar junctions derived from fetal mice.

Somewhat more surprising, analysis of Vγ3–Jγ1 and IgH V(D)J junctions from TdT° mice and age matched controls demonstrated that the presence of TdT almost completely abrogated homology-directed recombination—the frequency of homology-directed junctions was reduced even in sequences lacking N nucleotides derived from adult wild-type (W^{+}) mice. Several mechanisms may account for this:

1. The addition of N nucleotides to segment ends might place stretches of homology too far apart for annealing to occur—in both Ig and TCR γδ genes the homologies residing close to the ends are overwhelmingly favoured.
2. Addition of N nucleotides may create novel homologies that can not be identified in the resulting junctions.
3. the binding of TdT may directly prevent annealing/joining of short homologous stretches of nucleotides.

The data from TdT° mice validate similar conclusions drawn by Gerstein and Lieber who assessed the affects of coding sequence homology by transfecting several extrachromosomal recombination substrates into either pre-B cell lines expressing varying levels of TdT or fibroblast cell lines co-transfected with RAG-1 and RAG-2, with or without a TdT cDNA (28). Homologous sequences directed the formation of repetitive junctions in the absence of

TdT, but few recurrent junctions were observed in the much more diverse rearrangements from TdT^+ transfectants. Interestingly, one or two base pairs of homology was not sufficient to bias junction formation in this experiment, which is at odds with the dominant γδ and Ig junctions that consist of two nucleotide homologues and a few over-represented IgH junctions that include only a single base pair overlap. This underscores the fact that the 'rules' for each homology block may be different. Other factors such as flanking sequence and relative positions of the homologous sequences probably influence junction formation; homology itself may not be the critical factor in generating all repetitive junctions.

Preferential gene rearrangement and differential expression of TdT both mediate the transition from the perinatal to the adult repertoire. As discussed in detail below, recent studies have made it possible to distinguish the relative importance of each in generating the γδ and αβ repertoires.

3 The γδ repertoire

The γδ repertoire is much more limited than the αβ in terms of the number of germline V and J segments available for recombination. Despite this, considerable diversity is possible at the junctions, particularly for the δ chain which, unlike β, can include two D regions interspersed with N nucleotides (29). The murine TCR γ locus consists of four separate V–J–C clusters encoding seven V, four J, and three functional C segments (reviewed in reference 30). Approximately ten Vδ, two Dδ, two Jδ, and a single Cδ segment are located within the α locus—a few of the 3′ Vα regions are occasionally used in δ chains.

During thymic development, rearrangement at the γ locus occurs in distinct waves that have been characterized using both monoclonal antibodies and molecular techniques (2, 3, 31): Vγ3–Jγ1 rearrangements occur first, beginning approximately on fetal day 13, peak around fetal day 15, and then diminish to almost nothing in the adult thymus (nomenclature according to Garman *et al.* 32). Vγ4–Jγ1 rearrangements are less frequent and occur within the same time-frame, probably slightly after Vγ3; like Vγ3 they are extremely rare in the adult thymus. Rearrangement of the more Jγ1 distal Vγ2 segment is evident at fetal day 15, reaches a maximum at fetal day 18, then subsides slightly in adult mice. Vγ5–Jγ1, Vγ1.1–Jγ4, and Vγ1.2–Jγ2 rearrangements are also prominent in adult, not fetal, γδ cells. Rearrangement at the δ locus appears to be similarly controlled during thymic ontogeny (2, 9, 33): Vδ1 (most proximal to Dδ1) rearrangements are dominant in fetal TCRs whereas the more distal V regions, particularly Vδ5, are more frequent in the adult thymus. The δ gene most commonly expressed in fetal thymocytes is Vδ1–D2–Jδ2; Dδ1 segments are extremely rare in these cells. Another significant difference is that rearrangements including both Dδ1 and Dδ2 are common in adult thymocytes but essentially non-existent in fetal thymocytes.

The Vγ3 and Vγ4 cells comprising the first two waves that populate the fetal thymus are unique in that their expressed receptor repertoire is essentially limited to two Vγ–J and a single Vδ–D–J combination. The predominant Vγ3–Jγ1 and Vγ4–Jγ1 junctions both lack N region diversity and contain 2 and 3 bp of V–J homology respectively. They are each expressed with an equally invariant Vδ1–D2–Jδ2 chain which also lacks N region diversity and contains V–D and D–J homology. Although both selection and recombination have been invoked to explain the preponderance of these cells (2, 11, 30, 34–36), recent experiments have allowed unambiguous distinction of these two possibilities.

Using homologous recombination in embryonic stem cells, Itohara *et al.* created a mutation in Cδ which abrogated δ chain cell surface expression (Cδ°) (37). Quantitative PCR analysis of γ and δ genes amplified from fetal and adult thymocyte DNA demonstrated that the fetal/adult patterns of γ and δ rearrangement were not significantly altered in the mutant mice. This provides strong evidence for temporal control of rearrangement at the γ and δ loci. In addition, sequencing of Vγ3–Jγ1, Vγ4–Jγ1, Vδ1–D1–D2–Jδ2, and Vδ1–D1–D2–Jδ1 rearrangements amplified from newborn thymocyte DNA demonstrated that the junctions encoding the invariant Vγ3 and Vγ4 fetal receptors were formed at similar frequencies in both mutant and control mice. In light of this observation, Itohara *et al.* compiled and re-analysed a large body of fetal γδ DNA sequences and found that 12 dominant, 'canonical' junctions represented over 80% of these rearrangements. By considering the possibility of up to five P nucleotides for each segment, they found each of these junctions to contain at least two and up to six nucleotides of V(D)J homology. Two out-of-frame junctions comprised 40% of Vγ3–Jγ1 rearrangements, and another two 50% of Vγ4–Jγ1 junctions, lending further support for directed recombination. Homology-directed recombination may also explain the apparent lack of Dδ1 use in fetal thymocytes: successive homology-directed recombination from Vδ1 to Dδ1 then Dδ2 would effectively mask the contribution of Dδ1. The invariant Vδ1–D2–Jδ2 chain associated with Vγ3 and Vγ4 could be generated in this manner or by direct Vδ1–D2 homology-directed joining.

Using a different approach, Asarnow *et al.* confirmed that the canonical Vγ3 junctions (one in-frame and two out-of-frame) are formed at high frequency in the absence of selection (38). They constructed a transgene containing unrearranged Vγ2, Vγ4, and Vγ3 segments, each modified to include a stop codon within the coding sequences, upstream of unrearranged Jγ1–Cγ1 segments and the Cγ enhancer. Sequencing demonstrated that the three canonical junctions represented 62% of N^- transgene Vγ3–Jγ1 rearrangements amplified from newborn thymus DNA, in agreement with the Cδ° data. Although probable, similar conclusions about Vγ4 rearrangement could not be drawn from the few sequences published in this report.

Extending these observations, sequencing analysis of Vγ3–Jγ1 rearrangements amplified from adult TdT° mice showed that homology-directed recombination is not restricted to fetal thymocytes (25, 27). The Vγ3–Jγ1 canonical junctions represented approximately 65% of sequences derived from both types of adult TdT° mice. In addition, no significant differences were observed in the relative frequencies of these junctions amplified from sorted $CD4^-CD8^-$ and $CD4^+CD8^+$ thymocytes, populations enriched for γδ and αβ cells respectively. Thus, the propensity for homology-directed recombination does not appear to be enhanced in the γδ lineage.

The sequences derived from sorted TdT° thymocytes did, however, hint at a role for selection in exaggerating the survival of Vγ3 cells expressing the in-frame canonical chain. In close agreement with the Cδ° and Asarnow *et al.* data, the TdT° $CD4^+CD8^+$ sequences indicate that, in the absence of selection, the probability of forming the in-frame canonical sequence is approximately 30%. This may not be significantly lower than the 38% consensus figure for fetal thymocytes (37), but it is intriguing that in all data sets derived from fetal thymus DNA this sequence represents 90–99% of in-frame Vγ3 rearrangements. In contrast, it represented 82% of in-frame junctions amplified from both the non-functional transgene and Cδ° rearrangements, and only 38–60% of the adult TdT° $CD4^-CD8^-$ in-frame junctions (S. Gilfillan unpublished data). Although certainly not conclusive, these data suggest that something in the fetal—but not the adult—thymus promotes the accumulation, enhanced survival, or positive selection of γδ cells expressing the Vγ3 canonical junction. More extensive comparison of γ and δ junctions from fetal Cδ° and W^+ mice and/or endogenous and transgenic Vγ3 and Vγ4 junctions from the mice constructed by Asarnow *et al.* might delineate a possible, albeit secondary, role for selection in enhancing the preponderance of these cells in the fetal thymus.

As mentioned above, comparison of Vγ3–Jγ1 junctions from adult TdT° and control mice clearly demonstrated that expression of TdT effectively blocks the formation of canonical junctions. Because TdT is expressed at high levels in the thymus beginning approximately four days after birth, only rearrangements occurring after this contain significant N nucleotides. The stringent temporal control of rearrangement at the γ and δ loci (in the γδ lineage) ensures that the homology-directed Vγ3, Vγ4, and Vδ1 canonical junctions are formed and that N nucleotides are present only in the more diverse receptors that occur later. Interestingly, limited sequencing of Vγ2–Jγ1 and Vγ1.2–Jγ2 junctions derived from adult TdT° thymus RNA revealed a relatively diverse set of junctions, although V–J homology is present in both cases. This suggests that homology-directed recombination may have relatively little influence at these loci regardless of TdT expression (S. Gilfillan unpublished data).

4 The αβ repertoire

The TCR β locus contains 22 variable segments 5′ to two D–J–C clusters, each consisting of one D region, six functional J regions, and a constant region (39, 40). D–Jβ rearrangements are first detected at fetal day 14, and Vβ–D–Jβ at approximately day 16 (41). Sequencing studies have demonstrated that rearrangement at this locus is not random. Most convincing in terms of programmed rearrangement are sequences derived from a non-functional Vβ17b allele and functional Vβ17a sequences from $CD4^+CD8^+$ $CD3^{lo}$ thymocytes (42). This relatively immature population presumably has not undergone the rigorous selection required for differentiation into $CD3^{hi}$ $CD4^+CD8^-$ or $CD4^-CD8^+$ cells (43, 44), hence over-representation of any particular junction is most likely due to the recombination process itself. In these data sets, a definite bias for Vβ17–Jβ2 rearrangements was observed, Jβ2.6 being strongly preferred for the non-functional Vβ17 allele. An additional study indicated that Vβ8 and Vβ5 also rearrange preferentially to Jβ2, but because the sequences were derived from total thymocytes, the effects of selection could not be assessed (12). Although rearrangement at this locus is not random, as yet no data has indicated that rearrangement in neonates differs significantly from that in adults. Comparison of Vβ expression on fetal and adult thymocytes using a panel of anti-Vβ antibodies revealed no differences, nor did sequence analysis of neonatal and adult Vβ17a, 6, 10, 8, 5, and Vα1 V(D)J junctions (10, 12). One exception is a small population of αβ TCR^+ $CD4^-CD8^-$ thymocytes expressing predominately Vβ8 that is first detected approximately three weeks after birth; however, the functional significance of these cells in adult mice is not known (44).

At the TCR α locus, unlike β, there is evidence for progressive gene rearrangement similar to that observed at the IgH locus (45). The α locus consists of approximately 100 V genes, 50 J segments (separated by the δ locus), and a single C segment (46, 47). Comparison of Vα rearrangements in hybridomas made from fetal and adult thymocytes (and splenocytes) indicated that Vα rearrangement in neonates is biased toward 5′ Jα segments (48, 49). Although the effects of selection could not be clearly distinguished from programming in these analyses, rearrangements within the same cell were at similar positions on homologous chromosomes, providing suggestive evidence for programming. Using PCR to amplify Vα–Jα rearrangements from fetal day 18 and adult thymocytes, Roth *et al.* confirmed and clarified these conclusions (50). In general, the V genes most proximal to the Jα region preferentially rearranged to the most 5′ Jα segments; however, a more distal Vα segment rarely rearranged to these same 5′ Jα in adult animals. For at least one Vα, the bias for 5′ Jα segments was more pronounced in fetal mice, and the four most 3′ Jα were not detected in over 200 fetal junctions examined, although they were used in a small percentage of adult rearrangements. As discussed by Roth *et al.*, these data are consistent with successive rearrangement at the α locus; the internal Vs and Js would be deleted during secondary rearrangements resulting in the biases observed. If this is true, the progressive rearrangement of Vα–Jα itself is the same in neonates and adults, but secondary rearrangements are more common in adult thymocytes. Although as yet there is no data directly demonstrating successive rearrangement at the α locus (see discussion in reference 51) this interpretation is appealing in that it would give each αβ cell several chances to make an α chain suitable for positive selection.

Why secondary rearrangements would be more frequent in adult versus fetal thymocytes is not known. One possible reason is that perinatal thymocytes mature more quickly. Indirect evidence for this was obtained by careful correlation of thymic TdT expression and the presence of N region insertion in expressed Vβ6 genes (23). *In situ* hybridization demonstrated that the onset of TdT expression occurred between days three and five after birth. N regions were first observed in Vβ6 junctions from $CD4^+CD8^+$ thymocytes at day six and achieved maximum levels at day eight; significant insertions were observed in $CD4^+CD8^-$ cells at day eight. The intervals represent the time it takes $CD4^-CD8^-$ cells to reach the $CD4^+CD8^+$ stage and the transition from the $CD4^+CD8^+$ to the $CD4^+CD8^-$ stage respectively. That the two days required for maturation from $CD4^+CD8^+$ to $CD4^+CD8^-$ is less than the four to five day estimate for adult thymocytes suggests that neonatal thymocytes mature more quickly (see discussion in reference 52). In addition, like RAG-1 and RAG-2 (53), TdT expression was confined to the cortex. This was confirmed by quantitative PCR analysis of sorted thymocytes and suggests that TdT expression is terminated concomitant with positive selection, which would allow N region diversity in any secondary α rearrangements occurring prior to selection.

Thus, in terms of programmed gene rearrangement, the perinatal αβ repertoire does not appear to differ enormously from that of the adult; the stringent temporal control observed at the γδ loci is clearly absent. Also absent is a dramatic effect of homology-directed recombination. Over-represented junctions containing VDJ homology were not evident in TCR Vβ sequences amplified from perinatal thymocytes or in Vβ8 sequences derived adult TdT° $CD4^+CD8^+CD3^{lo}$ thymocytes (10, 12, 25). Only 22% of the TdT° Vβ8–D–Jβ junctions analysed contained homology of two or more base pairs; this was in sharp contrast to the 70% observed in Vγ3 junctions. More importantly, only one 'repetitive' homology-directed junction was identified (i.e. was present in independent amplifications from two mice), and it was a V–J join, a relatively rare occurrence. Feeney also showed that homology-directed recombination plays a relatively minor role in shaping the fetal TCR β repertoire by re-examination of existing perinatal Vβ sequences and more extensive sequencing of Vβ8 and Vβ5 junctions amplified from fetal thymocytes (54). One reason for this is the dearth of homology near the recombination signal sequences in V, D, and Jβ segments, particularly D and J—the D segments are essentially G nucleotide stretches while few Gs are found in the J segments. Even when homology is present, for instance a terminal GA shared by Vβ8 and Dβ1 and Dβ2, it is used in only 15% of perinatal Vβ8 junctions (54). This, again, underscores the fact that flanking sequences as well as the position and composition of the homologous sequence probably influence the outcome of these reactions. Less extensive data is available for Vα but as yet no prevalent fetal junctions have been noted. In summary, the transition from the perinatal to the adult αβ repertoire is largely marked by increased diversity at the V(D)J junctions—a major difference between the fetal and adult αβ T cell repertertoire is the presence of N nucleotides in adult V(D)J junctions.

5 Control

The γδ fetal/adult dichotomy is pronounced: of the diversity possible, essentially two invariant receptors dominate the early fetal repertoire. As discussed above, this difference is largely programmed—stringently controlled rearrangement directs V and J usage, and the absence of TdT expression allows the formation of homology-directed canonical junctions. The dichotomy in αβ T cells is less pronounced and is largely dependent on regulated expression of TdT. Therefore, how programmed rearrangement and TdT expression are controlled is crucial to understanding generation of the repertoires.

Goldman *et al.* have begun to elucidate the mechanisms involved in temporal programming of rearrangement at the γ locus by demonstrating that the appearance and disappearance of sterile Vγ3, Vγ4, and Vγ2 transcripts correlate with the onset and cessation of the corresponding Vγ–Jγ1 gene rearrangements (31). This is consistent with the 'accessibility' model of gene rearrangement originally proposed for Igs, transcription being an obvious but not required indication that the locus is 'open' for the recombinase machinery (55). The *cis* factors controlling Vγ accessibility have not been identified, but, as Goldman *et al.* suggest, the Vγ promoters themselves are plausible candidates. Of considerable interest is the extent to which this programming and TdT expression are influenced by the environment: at one extreme is the idea that these processes are programmed at the level of the stem cell—fetal and adult stem cells would markedly differ in their capacity for gene rearrangement and TdT expression; at the other extreme is the concept of a single, plastic stem cell completely dependent on environmental signals.

Experiments designed to test these possibilities have produced conflicting results. In a study widely cited as evidence for programming at the stem cell level, Ikuta *et al.* demonstrated that fetal stem cells gave rise to Vγ3 cells in fetal thymus organ culures whereas adult precursors did not (56). Although fetal stem cells isolated from day 15 and 16 fetal liver, spleen, and BM retained this capacity, those isolated from day 12 and 14 fetal liver were more efficient. When injected into adult thymus lobes, the environment appeared to have a dominant effect—neither population generated Vγ3 cells; instead primarily $CD4^+CD8^+$ and $CD4^+CD8^-$ and $CD4^-CD8^+$ (presumably αβ) T cells were evident after three weeks. In slight contrast, Ogimoto *et al.* demonstrated that Vγ3 cells could be detected in adult thymi of irradiated mice 11 days after reconstitution with fetal liver cells, but not after reconstitution with adult BM cells (57). However, $CD4^+CD8^+$ cells were the predominant population in both cases and generation of Vγ3 cells was transient. In a similar study, both day 14 fetal liver and adult BM precursors gave rise to thymocytes with 'adult' Vα and Vγ rearrangement patterns 14 days after transfer into lethally irradiated recipients (58). This analysis was restricted to hybridomas made from thymocytes generated in both types of chimeras and, unexpectedly, the frequency of δ rearrangement was much lower than that observed in fetal or adult thymocytes. Altogether, these data argue for a pronounced effect of the environment and perhaps some degree of intrinsic programming of the stem cell. Both fetal stem cells and an intact fetal environment are apparently required to generate Vγ3-expressing IELs (59).

Notably, Ikuta *et al.* also demonstrated that at least some fetal stem cells are very plastic by limiting dilution fetal thymus organ culture: in two cases a single precursor generated Vγ3, other γδ and αβ cells (56). This experiment also suggested that the fetal liver stem cell population might be heterogeneous in that only one-third of the repopulated lobes contained Vγ3 cells. But, as the authors point out, the proportion of Vγ3 cells generated in the best lobes was much lower than that observed *in vivo*; treatment with 2′-deoxyguanosine and culture conditions undoubtedly disturbed the thymic microenvironment, perhaps retarding development of Vγ3 cells or encouraging differentiation of others. Thus, the main evidence for intrinsic stem cell programming is the inability of adult precursors to generate Vγ3 cells. Interestingly, adult stem cells did generate other $CD3^+$ cells, some expressing Vγ4 transcripts (60), under the same organ culture conditions.

Ample evidence indicates that the environment influences TdT expression. At a superficial level, *in situ* hybridization demonstrated that TdT expression appears rather suddenly in the thymus and is immediately visible in cells throughout the cortex, not localized as might be expected if a wave of precursors expressing TdT synchronously populated the thymus (23). Some time ago, Rothenberg and Triglia showed that removing neonatal thymocytes from the thymic environment was sufficient to allow expression of TdT; thymocytes isolated from fetal day 18 and newborn mice expressed adult levels of TdT protein after overnight culture in suspension (61). Similarly, Larché *et al.* demonstrated that simply placing a day 14 fetal thymus in culture disturbed the thymic microenvironment enough to allow N region insertion—after only two days of organ culture, significant N region insertions were observed in a proportion of Vγ3–Jγ1 junctions (62). This effect was much more pronounced when junctions were amplified from DNA; in data sets amplified from RNA the in-frame canonical junction (lacking N) was predominant. The reasons for this discrepancy are not clear but may be due to over-expression of RNA in those cells expressing the canonical chain—a similar bias has been observed in Vγ3 sequences derived from adult thymus DNA and RNA (63). In an analogous study, a large proportion of Vα2 junctions amplified from fetal thymus lobes contained adult levels of N after only five days of organ culture (64). The Vα-Jα rearrangement patterns were also more typical of those observed in adult thymi and this effect was exaggerated when IL-4 was added to the cultures.

As might be predicted from these observations, the frequency and extent of N region diversity at various junctions is very similar in thymocytes derived from both fetal and adult precursors placed in either fetal thymic organ culture or an adult thymic environment. Ikuta and Weissman demonstrated that day 14 fetal precursors are capable of giving rise to T cells expressing Vγ4 transcripts with N nucleotides after 14 days of organ culture (60). They actually inferred that these early precursors were programmed not to express TdT, but their data set derived from fetal day 14 repopulation was overwhelmed by the in-frame canonical sequence (lacking N). Disregarding the in-frame canonical sequences, there was no difference in the number of sequences with N or the number of N nucleotides per sequence with N among the data sets. Because the sequences were derived from RNA and a single data set was shown for each time point, the predominance of the in-frame canonical junction in sequences derived from fetal precursors is difficult to interpret. In a less ambiguous study, both fetal day 14 and adult BM precursors generated thymocytes expressing Vβ17a transcripts with adult levels of N insertion 14 days after transfer into adult *scid* mice (65).

To date, the data argue that environment plays a more marked role than programming at the level of the stem cell, at least for TdT expression. Further discussion requires consideration of the relationship of γδ and αβ cells. These cells probably share a common precursor, lineage commitment occurring before cell surface expression of the TCR (reviewed in reference 35, see also reference 43). Supporting this, only incomplete TCR D–Jβ rearrangements are observed in γδ cells whereas both γ and δ rearrangements are evident in mature αβ T cells (66); the majority of rearranged γ genes are transcriptionally silent in αβ cells (67). Knock-out experiments have definitively shown that γδ cells develop in the absence of αβ cells and vice versa (37, 68, 69). The molecular basis for the lineage decision is not known, but silencing of the γ and/or α locus via elements flanking Cγ1 and Cα as well as deletion of the δ locus by unique rearrangement sequences (δ Rec/ψJα) may play mechanistic roles (35, 70–72). Why commitment to the γδ and αβ lineage changes so sharply during ontogeny *in vivo* is a crucial issue. The data discussed above are consistent with the idea that the fetal thymus strongly influences the decision to become γδ (α silenced, no rearrangement δ Rec/ψJα?) whereas adult skews toward αβ (γ silenced, rearrangement of δ Rec/ψJα promoted?). Because a variety of rearrangements were seen in fetal thymic organ culture, it is plausible that tightly programmed rearrangement within the γδ lineage is dependent on environmental signals *in vivo* that are, like control of TdT expression, disrupted *in vitro*. The fetal stem cells described to date seem quite plastic and capable of responding to environmental signals from both the fetal and adult thymus. Adult BM stem cells appear less plastic—at least no culture conditions have been defined that suppress TdT expression or promote the differentiation of Vγ3 cells (these phenomena may be related). Varying the oxygen concentrations surrounding fetal thymus lobes cultured submerged in medium can significantly alter the proportion of αβ and γδ cells produced (73). These and similar studies may lead to isolation of the factors influencing what seems to be the most crucial decision made by the stem cell—whether to become γδ or αβ.

6 Selection

As discussed above, selection plays a relatively minor role in generating the quasi-monoclonal Vγ3 and Vγ4 fetal γδ cells. How much it affects other γδ cells is a matter of debate; the factors influencing expansion of particular γδ cells appear to differ markedly from those affecting αβ (35). By these criteria, selection can not be invoked as a significant factor in generating or maintaining the fetal/adult dichotomy for γδ cells.

The situation for αβ T cells is quite different. Peripheral αβ TCR expression is known to be largely influenced by interaction of the TCR with thymic class I and II molecules and their associated peptides, which results in either positive or negative selection (survival or death) of the cell (44). A particularly dramatic effect is the deletion of almost all T cells expressing particular Vβ chains by retrovirus encoded antigens in association with class II molecules (reviewed in reference 74). In adult animals expression of one of these antigens can almost completely abolish peripheral expression of a Vβ, dramatically altering the nature of the peripheral repertoire. Surprisingly, such deletion is inefficient in the fetal thymus (75–79). This defect has no relationship to the presence of N region diversity. Deletion of Vβ6 by the 3′ open reading frame encoded protein of the mouse mammary tumour virus in association with MHC class II E molecules was monitored in parallel with accumulation of N region diversity in Vβ6 chains (23). Marked deletion of $CD4^+CD8^-$ cells occurred in mice six days of age whereas significant addition of N nucleotides was not observed until day eight. Conversely, chains with N region diversity can also escape deletion; a Vβ17a transgene with N region diversity was not deleted in perinatal mice but was effectively deleted in adults (80). Why deletion is defective in neonates is not known (see discussion in reference 80), but at least some of the T cells that escape deletion appear to be self-reactive and give rise to multiorgan autoimmune disease in neonatally thymectomized mice (81–84). Therefore selection is a major factor in generating the αβ fetal/adult dichotomy; Vβ chains forbidden in the periphery of adult mice are allowed to circulate in neonates. In addition, the paucity of N region nucleotides in neonatal V(D)J junctions is in some cases exaggerated by selection (10). Comparison of selected Vβ chains in fetal and adult TdT° mice may help identify temporal changes in the selective environment that further and more subtly shape the αβ repertoire.

At this point it is important to note that several functional and physiological characteristics also distinguish perinatal from adult αβ T cells: Neonatal but not adult thymocytes proliferate and secrete IL-2 when cultured with adult syngeneic spleen cells (84, 85). Anti-CD3 antibody treatment *in vivo* does not cause depletion of $CD4^+CD8^+$ thymocytes in neonates as it does in adults (86–88). Neonatal $CD4^+CD8^-$ thymocytes are a proliferating population of blast cells that respond to IL-2 and IL-7 *in vitro*, whereas the majority of their adult counterparts have a small resting phenotype and do not respond to these interleukins (75, 89). Neonatal $CD4^+CD8^-$ are also unique in their ability to induce IgM production in *scid* mice (90). Both neonatal and adult peripheral T cells secrete IL-2 and proliferate in response to TCR-independent (phorbol ester and calcium ionophore) stimulation, but naïve neonatal T cells resemble primed adult T cells upon stimulation with anti-CD3 antibodies in that they secrete large amounts of IL-4 rather than IL-2 (91). How these differences relate to selection and repertoire is not yet clear, but they may reflect different requirements for tolerance induction in fetal and adult mice.

7 Why?

Although substantial progress has been made in understanding how the perinate/adult TCR dichotomy is generated, why it exists and is maintained remains an enigma. Importantly, the extreme dichotomy in the γδ repertoire is mediated almost entirely by TCR-independent factors whereas the αβ dichotomy is less severe and is considerably influenced by TCR-dependent interactions. These observations raise several specific questions:

Why are the γδ loci organized to ensure the expression of monoclonal receptors in the fetal thymus? One possibility is that these receptors are crucial at this time, either because they offer some sort of necessary protection, perhaps eradicating damaged cells (2, 92), or because they somehow lay the groundwork for the adult immune system and indirectly guide the emergence of the adult repertoire. A similar argument has been put forth for biased V_H usage and over-representation of some homology-directed junctions in the fetal Ig repertoire, particularly the T15 anti-phosphorylcholine antibodies that protect against *Streptococcus pneumonia* (15, 93, 94). For γδ cells, the Cδ° mice weaken this argument considerably in that they do not appear to be seriously immune deficient and have no marked changes in their adult αβ repertoires. Though they may be beneficial, Vγ3 and Vγ4 cells do not seem essential for survival or emergence of the αβ TCR repertoire; further analyses of the Cδ° mice should clarify their role in various pathological situations. It is conceivable that γδ cells are unnecessary remnants of an evolutionarily primitive immune system. Inadequate for providing protection in an adult animal, they have been replaced by the more diverse αβ cells. Production of γδ cells in the fetus may be allowed simply because αβ cells either are not required or are injurious at this time.

How important is N region diversity? There are two (not mutually exclusive) possibilities for the strict regulation of TdT expression—either it is important to generate more TCR diversity in adult animals or it is deleterious to introduce such diversity during the perinatal period. The TdT° mice have allowed direct assessment of the former possibility. To date no serious anomalies have been observed in these mice; the essential lack of N region diversity has no dramatic effect on T or conventional B cell development. The TdT° mice are healthy, breed well, and are capable of mounting both T and B cell responses to the complex antigens KLH and ovalbumin. There is as yet no evidence that cells expressing adult receptors are necessary to control the potentially self-reactive fetal cells that escape deletion in the neonate. At this superficial level, the 'fetal' repertoire is sufficient to maintain the health and well-being of the average laboratory mouse, although it may prove inadequate when challenged with more serious pathogens. More detailed analysis of these mice is clearly required to fully assess the importance of N region diversity.

Why is TdT not expressed during the perinatal period? As suggested above, the increased junctional diversity mediated by TdT may be deleterious during this time. Although eliminating cell surface expression of γδ receptors did not prove terribly traumatic, disrupting homology-directed recombination and introducing N region diversity in the canonical Vγ3 and Vγ4 receptors could be more catastrophic. For instance, if the canonical Vγ3 and Vγ4 receptors do recognize stress induced self-peptides such as the heat shock proteins and, upon recognition, eliminate damaged cells, one can envisage a TdT-mediated diverse repertoire of Vγ3 and Vγ4 receptors capable of recognizing a variety self-peptides with adverse consequences. A similar argument can be made for the perinatal αβ T cells, especially during the period when deletion is inefficient in the fetal thymus. A far more diverse array of self-reactive receptors could be produced and seed the periphery; whether these could be controlled, as the perinatally produced N^- self-reactive receptors appear to be, is an open question. This argument is appealing in that it can be directly tested by creating mice expressing TdT early during thymic development. Such animals and complete characterization of the TdT° mice may provide the answer to Why?

Acknowledgements

We thank Ann Feeney for data prior to publication and Susan Chan for critical reading of the manuscript. Our laboratory is supported by institute funds from the INSERM and the CNRS, and by a grant from the Association pour la Recherche sur le Cancer and the Association Nationale pour la Recherche sur le Sida. S.G. was supported by a fellowship from the American Cancer Society.

References

1 Ikuta, K., Uchida, N., Friedman, J., and Weissman, I. L. (1992). Lymphocyte development from stem cells. *Annu. Rev. Immunol.*, **10**, 759.

2 Allison, J. P. and Havran, W. L. (1991). The immunobiology of T cells with invariant γδ antigen receptors. *Annu. Rev. Immunol.*, **9**, 679.
3 Allison, J. P. (1993). γδ T-cell development. *Curr. Opin. Immunol.*, **5**, 241.
4 O'Brien, R. L. and Born, W. (1991). Heat shock proteins as antigens for γδ cells. *Semin. Immunol.*, **3**, 81.
5 Matis, L. A. and Bluestone, J. A. (1991). Specificity of γδ receptor bearing T cells. *Semin. Immunol.*, **3**, 75.
6 Alt, F. W., Oltz, E. M., Young, F., Gorman, J., Taccioli, G., and Chen, J. (1992). VDJ recombination. *Immunol. Today*, **13**, 306.
7 Schatz, D. G., Oettinger, M. A., and Schlissel, M. S. (1992). V(D)J recombination: molecular biology and regulation. *Annu. Rev. Immunol.*, **10**, 359.
8 Lieber, M. R. (1992). The mechanism of V(D)J recombination: a balance of diversity, specificity, and stability. *Cell*, **70**, 873.
9 Elliot, J. F., Rock, E. P., Patten, P. A., Davis, M. M., and Chien, Y.-h. (1988). The adult T-cell receptor δ-chain is diverse and distinct from that of fetal thymocytes. *Nature*, **331**, 627.
10 Bogue, M., Candéias, S., Benoist, C., and Mathis, D. (1991). A special repertoire of α:β T cells in neonatal mice. *EMBO J.*, **10**, 3647.
11 Lafaille, J. J., DeCloux, A., Bonneville, M., Takagaki, Y., and Tonegawa, S. (1989). Junctional sequences of T cell receptor γδ genes: implications for γδ T cell lineages and for a novel intermediate of V-(D)-J joining. *Cell*, **59**, 859.
12 Feeney, A. J. (1991). Junctional sequences of fetal T cell receptor β chains have few N regions. *J. Exp. Med.*, **174**, 115.
13 Gu, H., Forster, I., and Rajewsky, K. (1990). Sequence homologies, N sequence insertion and J_H gene utilization in V_HDJ_H joining: implications for the joining mechanism and the ontogenetic timing of Ly 1 B cell and B-CLL progenitor generation. *EMBO J.*, **9**, 2133.
14 Feeney, A. J. (1990). Lack of N regions in fetal and neonatal mouse immunoglobulin V-D-J junctional sequences. *J. Exp. Med.*, **172**, 1377.
15 Feeney, A. J. (1992). Predominance of V_H-D-J_H junctions occurring at sites of short sequence homology results in limited junctional diversity in neonatal antibodies. *J. Immunol.*, **149**, 222.
16 Asarnow, D. M., Goodman, T., LeFrancois, L., and Allison, J. P. (1989). Distinct antigen receptor repertoires of two classes of murine epithelium-associated T cells. *Nature*, **341**, 60.
17 Roth, D. B. and Wilson, J. H. (1986). Nonhomologous recombination in mammalian cells: role for short sequence homologies in the joining reaction. *Mol. Cell. Biol.*, **6**, 4295.
18 Alt, F. W. and Baltimore, D. (1982). Joining of immunoglobulin heavy chain gene segments: implications from a chromosome with evidence of three D-J_H fusions. *Proc. Natl. Acad. Sci. USA*, **79**, 4118.
19 Bollum, F. J. (1974). Terminal deoxynucleotidyl Transferase. In *The enzymes* (ed. P. D. Boyer), Vol. 10, p. 145. Academic Press, New York.
20 Landau, N. R., Schatz, D. G., Rosa, M., and Baltimore, D. (1987). Increased frequency of N-region insertion in a murine pre-B-cell line infected with a terminal deoxynucleotidyl transferase retroviral expression vector. *Mol. Cell. Biol.* **7**, 3237.
21 Kallenbach, S., Doyen, N., D'Andon, M. F., and Rougeon, F. (1992). Three lymphoid-specific factors account for all junctional diversity characterisitic of somatic assembly of T-cell receptor and immunoglobulin genes. *Proc. Natl. Acad. Sci. USA* **89**, 2799.
22 Desiderio, S. V., Yancopoulos, G. D., Paskind, M., Thomas, E., Boss, M. A.,

Landau, N., *et al.* (1984). Insertion of N regions into heavy-chain genes is correlated with expression of terminal deoxytransferase in B cells. *Nature*, **311**, 752.

23 Bogue, M., Gilfillan, S., Benoist, C., and Mathis, D. (1992). Regulation of N-region diversity in antigen receptors through thymocyte differentiation and thymus ontogeny. *Proc. Natl. Acad. Sci. USA*, **89**, 11011.

24 Yancopoulos, G. D., Blackwell, T. K., Suh, H., Hood, L., and Alt, F. W. (1986). Introduced T cell receptor variable region gene segments recombine in pre-B cells: evidence that B and T cells use a common recombinase. *Cell*, **44**, 251.

25 Gilfillan, S., Dierich, A., Lemeur, M., Benoist, C., and Mathis, D. (1993). Mice lacking TdT: mature animals with an immature lymphocyte repertoire. *Science*, **261**, 1175.

26 Shinkai, Y., Rathburn, G., Lam, K.-P., Oltz, E. M., Stewart, V., Mendelsohn, M., *et al.* (1992). RAG-2 deficient mice lack mature lymphocytes owing to inability to initiate V(D)J rearrangement. *Cell*, **68**, 855.

27 Komori, T., Okada, A., Stewart, V., and Alt, F. W. (1993). Lack of N regions in antigen receptor variable region genes of TdT-deficient lymphocytes. *Science*, **261**, 1171.

28 Gerstein, R. M. and Lieber, M. R. (1993). Extent to which homology can constrain exon junctional diversity in V(D)J recombination. *Nature*, **363**, 625.

29 Davis, M. M. and Bjorkman, P. J. (1988). T-cell antigen receptor genes and T-cell recognition. *Nature*, **334**, 395.

30 Raulet, D. H. (1989). The structure, function and molecular genetics of the γ/δ T cell receptor. *Annu. Rev. Immunol.*, **7**, 175.

31 Goldman, J. P., Spencer, D. M., and Raulet, D. H. (1993). Ordered rearrangement of variable region genes of the T cell receptor γ locus correlates with transcription of the unrearranged genes. *J. Exp. Med.*, **177**, 729.

32 Garman, R. D., Doherty, P. J., and Raulet, D. H. (1986). Diversity, rearrangement, and expression of murine T cell gamma genes. *Cell*, **45**, 733.

33 Chien, Y.-h., Iwashima, M., Wettstein, D. A., Kaplan, K. B., Elliot, J. F., Born W., *et al.* (1987). T-cell receptor δ gene rearrangements in early thymocytes. *Nature*, **330**, 722.

34 Lafaille, J. J., Haas, W., Coutinho, A., and Tonegawa, S. (1990). Positive selection of γδ T cells. *Immunol. Today*, **11**, 75.

35 Haas, W. and Tonegawa, S. (1992). Development and selection of γδ T cells. *Curr. Biol.*, **4**, 147.

36 Itohara, S. and Tonegawa, S. (1990). Selection of γδ T cells with canonical T-cell antigen receptors in fetal thymus. *Proc. Natl. Acad. Sci. USA*, **97**, 7935.

37 Itohara, S., Mombaerts P., Lafaille, J., Iacomini, J., Nelson, A., Clarke, A. R., *et al.* (1993). T cell receptor δ gene mutant mice: independent generation of αβ T cells and programmed rearrangements of γδ TCR genes. *Cell*, **72**, 337.

38 Asarnow, D. N., Cado, D., and Raulet, D. H. (1993). Selection is not required to produce invariant T-cell receptor γ-gene junctional sequences. *Nature*, **362**, 158.

39 Lindsten, T., Lee, N. E., and Davis, M. M. (1987). Organization of the T-cell antigen-receptor beta-chain locus in mice. *Proc. Natl. Acad. Sci. USA*, **84**, 7639.

40 Chou, H. S., Nelson, C. A., Godambe, S. A., Chaplin, D. D., and Loh, D. Y. (1987). Germline organization of the murine T cell receptor beta-chain genes. *Science*, **238**, 545.

41 Born, W., Yague, J., Palmer, E., Kappler, J., and Marrack, P. (1985). Rearrangement of T-cell receptor β-chain genes during T-cell development. *Proc. Natl. Acad. Sci. USA*, **82**, 2925.

42 Candéias, S., Waltzinger, C., Benoist, C., and Mathis, D. (1991). The $V\beta17^{+}$ T

cell repertoire: skewed Jβ usage after thymic selection; dissimilar CDR3s in $CD4^+$ versus $CD8^+$ cells. *J. Exp. Med.*, **174**, 989.
43 Boyd, R. L. and Hugo, P. (1991). Towards an integrated view of thymopoiesis. *Immunol. Today*, **12**, 71.
44 Nikolic-Zugic, J. (1991). Phenotypic and functional stages in the intrathymic development of αβ T cells. *Immunol. Today*, **12**, 65.
45 Malynn, B. A., Yancopoulos, G. D., Barth, J. E., Bona, C. A., and Alt, F. A. (1990). Biased expression of J_H-proximal V_H genes occurs in the newly generated repertoire of neonatal and adult mice. *J. Exp. Med.*, **171**, 843.
46 Koop, B. F., Wilson, R. K., Wang, K., Vernooij, B., Zaller, D., Kuo, C. L., *et al.* (1992). Organization, structure, and function of 95 kb of DNA spanning the murine T-cell receptor cα/cδ region. *Genomics*, **13**, 1209.
47 Jouvin-Marche, E., Hue, I., Marche, P., Liebe-Gris, C., Marolleau, J.-P., Malissen, B., *et al.* (1990). Genomic organization of the mouse T cell receptor Vα family. *EMBO J.*, **9**, 2141.
48 Hurwitz, J. L., Samaridis, J., and Pelkonen, J. (1989). Immature and advanced patterns of T cell receptor gene rearrangement among lymphocytes in splenic culture. *J. Immunol.*, **142**, 2533.
49 Thompson, S. D., Pelkonen, J., and Hurwitz, J. L. (1990). First T cell receptor α gene rearrangements during T cell ontogeny skew to the 5′ region of the Jα locus. *J. Immunol.*, **145**, 2347.
50 Roth, M. E., Holman, P. O., and Kranz, D. M. (1991). Nonrandom use of Jα gene segments. Influence of Vα and Jα gene location. *J. Immunol.*, **147**, 1075.
51 Malissen, M., Trucy, J., Jouvin-Marche, E., Cazenave, P.-A., Scollay, R., and Malissen, B. (1992). Regulation of TCR α and β gene allelic exclusion during T-cell development. *Immunol. Today*, **13**, 315.
52 Benoist, C. and Mathis, D. (1992). Generation of the αβ T-cell repertoire. *Curr. Opin. Immunol.*, **4**, 156.
53 Turka, L. A., Schatz, D. G., Oettinger, M. A., Chun, J. J. M., Gorka, G., Lee, K., *et al.* (1991). Thymocyte expression of Rag-1 and Rag-2: termination by T cell receptor cross-linking. *Science*, **253**, 773.
54 Feeney, A. J. (1993). Junctional diversity in the absence of N regions: neonatal TCR β chain junctional sequences are more heterogeneous than neonatal γδ or IgH junctions. *J. Immunol.*, **151**, 3094.
55 Alt, F. W., Blackwell, T. K., DePinho, R. A., Reth, M. G., and Yancopoulis, G. D. (1986). Regulation of genome rearrangement events during lymphocyte differentiation. *Immunol. Rev.*, **89**, 5.
56 Ikuta, K., Kina, T., MacNeil, I., Uchida, N., Peault, B., Chien, Y. -h., *et al.* (1990). A developmental switch in thymic lymphocyte maturation potential occurs at the level of hematopoietic stem cells. *Cell*, **62**, 863.
57 Ogimoto, M., Yoshikai, Y., Matsuzaki, G., Matsumoto, K., Kishihara, K., and Nomoto, K. (1990). Expression of T cell receptor Vγ5 in the adult thymus of irradiated mice after transplantation with fetal liver cells. *Eur. J. Immunol.*, **20**, 1965.
58 Larché, M., Manzo, A. R., and Hurwitz, J. L. (1991). Environmental and allele-specific influences on T cell receptor gene rearrangement: skewed α, δ and γ gene rearrangement patterns in chimeric mice. *Eur. J. Immunol.*, **21**, 2943.
59 Havran, W. L., Carbone, A., and Allison, J. P. (1991). Murine T cells with invariant γδ antigen receptors: origin, repertoire, and specificity. *Semin. Immunol.*, **3**, 89.
60 Ikuta, K. and Weissman, I. L. (1991). The junctional modifications of a T cell

receptor γ chain are determined at the level of thymic precursors. *J. Exp. Med.*, **174**, 1279.
61 Rothenberg, E. and Triglia, D. (1983). Clonal proliferation unlinked to terminal deoxynucleotidyl transferase synthesis in thymocytes of young mice. *J. Immunol.*, **130**, 1627,
62 Larché, M. and Hurwitz, J. L. (1993). Fetal thymocyte potential for T cell receptor Vγ3–Jγ1 junctional modification. *Eur. J. Immunol.*, **23**, 1328.
63 Aguilar, L. K. and Belmont, J. W. (1991). Vγ3 T cell receptor rearrangement and expression in the adult thymus. *J. Immunol.*, **146**, 1348.
64 Larché, M., Rencher, S. D. and Hurwitz, J. L. (1992). Environmental influence on T cell receptor α gene rearrangement and expression *in vitro*. *Eur. J. Immunol.*, **22**, 2733.
65 Bogue, M., Mossmann, H., Stauffer, U., Benoist, C., and Mathis, D. (1993). The level of N-region diversity in T cell receptors is not pre-ordained in the stem cell. *Eur. J. Immunol.*, **23**, 1185.
66 Thompson, S. D., Manzo, A. R., Pelkonen, J., Larché, M., and Hurwitz, J. L. (1991). Developmental T cell receptor gene rearrangements: relatedness of the α/β and γ/δ precursor. *Eur. J. Immunol.*, **21**, 1939.
67 Heilig, J. S. and Tonegawa, S. (1986). Diversity of murine gamma genes and expression in fetal and adult T lymphocytes. *Nature*, **322**, 836.
68 Mombaerts, P., Clarke, A. R., Rudnicki, M. A., Iacomini, J., Itohara, S., Lafaille, J. J., *et al.* (1992). Mutations in T-cell antigen receptor genes α and β block thymocyte development at different stages. *Nature*, **360**, 225.
69 Philpott, K. L., Viney, J. L., Kay, G., Rastan, S., Gardiner, E. M., Chae, S., *et al.* (1992). Lymphoid development in mice congenitally lacking T cell receptor αβ-expressing cells. *Science*, **256**, 1448.
70 Winoto, A. (1992). Regulation of the early stages of T-cell development. *Curr. Opin. Immunol.*, **3**, 199.
71 Shimizu, T., Takeshita, S., Muto, M., Kubo, E., Sado, T., and Yamagishi, H. (1993). Mouse germline transcript of TCR α joining region and temporal expression in ontogeny. *Int. Immunol.*, **5**, 155.
72 Takeshita, S., Toda, M., and Yamagishi, H. (1989). Excision products of the T cell receptor gene support a progressive rearrangement model of the α/δ locus. *EMBO J.*, **8**, 3261.
73 Watanabe, Y. and Katsura, Y. (1993). Development of T cell receptor αβ-bearing T cells in the submersion organ culture of murine fetal thymus at high oxygen concentration. *Eur. J. Immunol.*, **23**, 200.
74 Acha-Orbea, H. and Palmer, E. (1991). Mls—a retrovirus exploits the immune system. *Immunol. Today*, **12**, 356.
75 Ceredig, R. (1990). Intrathymic proliferation of perinatal mouse αβ and γδ cell receptor-expressing mature T cells. *Int. Immunol.*, **2**, 859.
76 Jones, L. A., Chin, L. T., Merriam, G. R., Nelson, L. M., and Kruisbeck, A. D. (1990). Failure of clonal deletion in neonatally thymectomized mice: tolerance is preserved through clonal anergy. *J. Exp. Med.*, **172**, 1277.
77 Schneider, R., Lees, R. K., Pedrazzini, T., Zinkernagel, R. M., Hengartner, H., and MacDonald, H. R. (1989). Postnatal disappearance of self-reactive ($V\beta6^+$) cells from the thymus of MLS^a mice. Implications for T cell development and autoimmunity. *J. Exp. Med.*, **169**, 2149.
78 Finkel, T. H., Kappler, J. W., and Marrack, P. C. (1992). Immature thymocytes are protected from deletion early in ontogeny. *Proc. Natl. Acad. Sci. USA*, **89**, 3372.

79 Fisher, A. G., Waltzinger, C., and Ceredig, R. (1992). Selection of murine T cell receptor αβ and γδ cells in organ cultures established from 14-day embryos. *Eur. J. Immunol.*, **22**, 1765.

80 Signorelli, K., Benoist, C., and Mathis, D. (1992). Why is clonal deletion of neonatal thymocytes defective? *Eur. J. Immunol.*, **22**, 2487.

81 Smith, H., Chen, I.-M., Kubo, R., and Tung, K. S. K. (1989). Neonatal thymectomy results in a repertoire enriched in T cells deleted in adult thymus. *Science*, **245**, 749.

82 Andreu-Sanchez, J. L., de Alboran, I. M., Marcos, M. A. R., Sanchez-Movilla, A., Martinez-A, C., and Kroemer, G. (1991). Interleukin 2 abrogates the non-responsive state of T cells expressing a forbidden T cell receptor repertoire and induces autoimmune disease in neonatally thymectomized mice. *J. Exp. Med.*, **173**, 1323.

83 Sakaguchi, S. and Sakaguchi, N. (1990). Thymus and autoimmunity: capacity of the normal thymus to produce pathogenic self-reactive T cells and conditions required for their induction of autoimmune disease. *J. Exp. Med.*, **172**, 537.

84 Howe, M. L., Goldstein, A. L., and Battistio, J. R. (1970). Isogenic lymphocyte interaction: recognition of self antigens by cells of the neonatal thymus. *Proc. Natl. Acad. Sci. USA*, **67**, 613.

85 von Boehmer, H., Shortman, K., and Adams, P. (1972). Nature of the stimulating cell in the syngeneic and the allogeneic mixed lymphocyte reaction in mice. *J. Exp. Med.*, **136**, 1648.

86 Kyewski, B. A., Shirrmacher, V., and Allison, J. P. (1989). Antibodies against the T cell receptor/CD3 complex interfere with distinct intra-thymic cell–cell interactions *in vivo*: correlation with arrest of cell differentiation. *Eur. J. Immunol.*, **19**, 857.

87 Hirsch, R., Gress, R. E., Pluznik, D. H., Eckhaus, M., and Bluestone, J. A. (1989). Effects of *in vivo* administration of anti-CD3 monoclonal antibody on T cell function in mice. II. *In vivo* activation of T cells. *J. Immunol.*, **142**, 737.

88 Reuff-Juy, D., Liberman, I., Drapier, A.-M., Guillon, J.-C., Leclerc, C., and Cazenave, P.-A. (1991). Cellular basis of the resistance of newborn mice to the pathogenic effects of anti-CD3 treatment. *Int. Immunol.*, **3**, 683.

89 Ceredig, R. and Waltzinger, C. (1990). Neonatal mouse $CD4^+$ mature thymocytes show responsiveness to interleukin 2 and interleukin 7: growth *in vitro* of negatively selected Vβ6- and Vβ-11 expressing CD4+ cells from (C57BL/6 × DBA/2)F1 mice. *Int. Immunol.*, **2**, 869.

90 Riggs, J. E., Hobbs, M. V., and Mosier, D. E. (1992). $CD4^+$ $CD8^-$ Thymocytes from neonatal mice induce IgM production in SCID mice. *J. Immunol.*, **148**, 1389.

91 Adkins, B. and Hamilton, K. (1992). Freshly isolated, murine neonatal T cells produce IL-4 in response to anti-CD3 stimulation. *J. Immunol.*, **149**, 3448.

92 Janeway, C. A. Jr., Jones, B., and Hayday, A. (1988). Specificity and function of T cells bearing γδ receptors. *Immunol. Today*, **9**, 73.

93 Briles, D. E., Forman, C., Hudak, S., and Claflin, J. L. (1982). Anti-phosphorylcholine antibodies of the T15 idiotype are optimally protective against *Streptococcus pneumonia*. *J. Exp. Med.*, **156**, 1177.

94 Feeney, A. J. (1991). Predominance of the prototypic T15 anti-phosphorylcholine junctional sequence in neonatal pre-B cells. *J. Immunol.*, **147**, 4343.

6 The human T cell receptor repertoire

PAUL A. H. MOSS AND JOHN I. BELL

1 Introduction

A central tenet of the clonal selection theory is that T cells express a great diversity of antigen receptors—a broad T cell receptor repertoire. Ever since the genetic organization of the T cell receptor was characterized and found to be analogous to immunoglobulin genes there has been considerable interest in the relative expression of T cell receptor gene segments (1). Although antigenic specificity is determined by the total contribution of both TCR α and β (or γδ) V, D, and J segments as well as N region addition, many groups have focused on overall expression of the individual gene segments. This has led to the concept of the 'V region' and 'J region' repertoire, the former becoming almost synonymous with general TCR repertoire. Nevertheless, although the V region repertoire is only one measure of diversity within T cell populations, the existence of superantigens which stimulate T cells in a Vβ-specific manner validates this type of analysis.

In recent years there has been extensive investigation into human TCR repertoire both in healthy individuals and in disease. Much of the work has been stimulated by the finding of dramatic differences in relative TCR Vβ expression between inbred mouse strains (2). As it was quickly noticed that MHC class II expression influenced the TCR repertoire (3) it was felt that genetically determined variation in human TCR repertoire may provide the explanation for the association of HLA alleles with human disease. Endogenous superantigens (SAgs) have not yet been demonstrated in the human genome although a comprehensive analysis of human repertoire has been hampered by the existence of few monoclonal antibodies to human Vα and Vβ regions. As techniques for studying human TCR expression become increasingly sophisticated there is evidence for heterogeneity within repertoire. How far these differences reflect the effect of superantigens, natural tolerance, or the influence of disease is not yet known. It is not at all clear how far studies of murine repertoire can be extended to models of human repertoire but preliminary analyses of human TCR expression are very intriguing and provide new insights into human T cell immunology.

2 Potential influences on TCR repertoire

2.1 Germline polymorphism

It is becoming clear that many individual TCR gene segments are polymorphic and consequently numerous 'TCR haplotypes' exist (4). The vast majority of this polymorphism is due to allelic variation within V region gene segments (5–10). Such variation is usually the result of a single nucleotide change in the V segment and its significance is unclear. It could represent neutral 'genetic drift' within V regions although heterozygote frequencies within the population suggest a selective advantage for diversity. Deletions of coding Vβ regions are also common (11) although there appear to be few major genetic deletions within the TCR α locus (12).

2.2 Thymic selection

Thymocytes are subject to both positive and negative selection by peptide–MHC complexes within the thymus. If certain V regions are better able to interact with a given MHC allele than others, then different HLA haplotypes might lead to distortions of V segment expression on peripheral T cells. Indeed, HLA identical siblings tend to have a more similar TCR V region repertoire than mismatched siblings or unrelated individuals (13, 14).

2.3 Endogenous superantigens

Mammary tumour viruses that have integrated into the murine genome can cause deletions of a major proportion of the developing thymic T cell population (3, 15, 16) by acting as endogenous superantigens (SAgs). There is, as yet, no clear evidence for similar agents in man despite the fact that a significant proportion of the human genome may be derived from integrated retroviruses. Developing T cells in the rat do not seem subject to deletion by endogenous SAgs (17) and it is possible that the phenomenon may be restricted to mice. Nevertheless, there are suggestions of inherited deletions of Vβ2 (18) and recent reports have shown that more than 20% of adults have low levels of Vβ3 expression on both $CD4^+$ and $CD8^+$ cells (19).

2.4 Natural immunity

It is possible that infection or disease could expand certain T cell populations sufficiently to distort T cell repertoire. Retroviral infections can certainly stimulate extremely high levels of peripheral blood cytotoxic T cells. Environmental superantigens are much more likely to lead to obvious distortions of repertoire and expansions of Vβ2 have already been documented after toxic shock syndrome (20).

2.5 Oligoclonal T cell populations

It is becoming clear that certain populations of $CD8^+$ and $CD4^-CD8^-$ T cells are oligoclonal in peripheral blood (see below). When such expansions are large they will produce distortions in the overall TCR repertoire.

3 The analysis of T cell receptor repertoire

Methods used to analyse TCR repertoire essentially fall into two groups, use of monoclonal antibodies to Vα or Vβ segments or molecular techniques which attempt to quantitate levels of TCR RNA. Neither technique is perfect and a combination of techniques is most satisfactory. Many monoclonal antibodies have now been produced which recognize human Vβ segments and together they cover up to 50% of the expressed Vβ repertoire. There has been less success in the generation of Vα-specific reagents and only three, Vα2.3, Vα12, and Vα23 are so far characterized. These antibodies have been widely applied and have the advantage of speed, ease of use, reproducibility, and unlimited source. However, antibody analysis does not give sequence information and currently can not cover a sizeable portion of V region repertoire.

Molecular techniques can be broadly divided into RNase protection or PCR-based techniques. The former has the advantage of not needing nucleic acid amplification, and thus possible bias, prior to quantitation (21). PCR-based techniques include the use of V region family-specific oligonucleotide primers for multiple parallel amplifications (22, 23), anchored PCR (24, 25), and inverse PCR (26). V region family-specific PCR is the most rapid method but the exponential nature of PCR reaction allows for substantial variation in efficiency to occur depending on the sets of primers used and may account for some of the discrepant results. The latter two techniques have the advantage of quantitation but are slower and more technically demanding to perform (27). PCR-based analysis has been less reproducible from centre to centre than use of antibodies. It is possible that cellular TCR mRNA levels may be distorted by the state of activation of T cell subsets. In addition, some groups expand T cells *in vitro* or activate whole populations with anti-CD3 antibodies to ensure that sufficient RNA is available for detection. This may introduce an additional bias. Nevertheless, when allied with sequencing it gives data on the whole TCR transcript and will have an increasing role to play as the importance of oligoclonal T cell expansions becomes more apparent.

4 T cell receptor repertoire in peripheral blood

4.1 TCR α V region expression

Characterization of TCR Vα usage has been hampered by a shortage of monoclonal antibodies to Vα segments. Until recently, only antibodies to

Vα2.3 and Vα12.1 were available although more are now being reported. There is a similar shortage of antibodies to murine Vα segments and it appears that antibodies to Vα segments are simply more difficult to produce than antibodies to Vβ. The reason for this is unclear but it may be that the Vα segment is less exposed than Vβ in the complete CD3 complex. Expression of Vα2.3 and Vα12.1 in normal individuals shows little variation although the antibodies have been of value in showing increased expression in some cases of sarcoidosis and rheumatoid arthritis (28) respectively.

Anchored PCR has been used to document normal repertoire and reveals significant variation in the relative expression of individual Vα segments (Figure 6.1) (14). Vα2, Vα8, Vα11, Vα14, and Vα23 all contributed over 6% of total repertoire. An analysis of Vα usage in aberrant out-of-frame transcripts showed a similar bias in Vα expression and so it is likely that chromosomal location has a major influence on Vα selection in recombination events. There appears to be little variation in expression between different subjects although identical twins were more similar to each other than outbred individuals showing that there is a genetic influence on Vα expression. No superantigens with specificity for Vα segments have been described although Vα expression does affect the ability of superantigens to activate T cells which share a common Vβ (29). There are now several examples of allelic variation within Vα segments which serve to increase the expressed repertoire (7, 10). Analysis of Vα repertoire will be complicated by the realization that a single T cell may express two different TCR α chains (30, 31).

4.1.1 J region expression

Jα is the most numerous J gene segment found in TCR or immunoglobulin genes and there are approximately 50 murine and up to 61 human segments. The Jα region has now been sequenced in both man and mouse (32) and

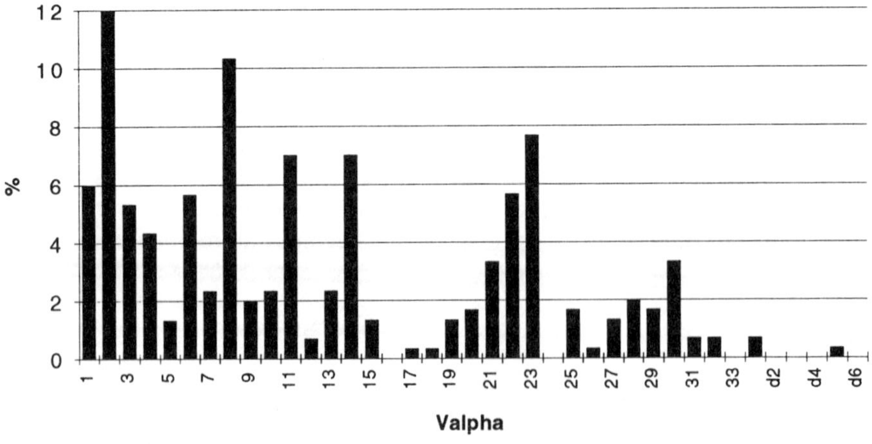

Fig. 6.1 Expression of Vα segments in human peripheral blood (based on reference 14).

reveals that although the general features of the J segment are preserved there has been considerable evolutionary divergence (33). As the Jα region of the TCR transcript is implicated in peptide recognition (34) it is perhaps not surprising that so much genetic variation has occurred in this region. Our own analysis of the relative expression of Jα segments has shown significant differences between individual segments (14). Murine expression of Jα segments appears to be biased during development such that early in ontogeny Jα segments close to the Vα complex are rearranged prior to those more distal to the Vα region. Jα expression in the adult human does not show evidence of a similar phenomenon to any significant extent (Figure 6.2) although this may be apparent from analysis of individual Vα segments (35).

The TCR α gene does not show allelic exclusion at the molecular level and it has been noticed that the two chromosomes within a single cell tend to rearrange to Jα segments that are close to each other in chromosomal location. Although unexplained, this may reflect secondary rearrangements on the TCR α locus which may proceed until the cell is selected by engagement of the TCR. If secondary rearrangements occur sequentially and in parallel on each chromosome then they will stop in similar regions of the Jα complex.

4.2 TCR β V region expression

Of all gene segments within the TCR and immunoglobulin loci, none has attracted so much interest with respect to relative expression as the Vβ segments. The ability of superantigens to activate T cells by binding to Vβ framework residues rather than peptide binding regions and the resultant deletions in murine Vβ repertoire due to the expression of Mtv loci was the spur to the great interest which has developed in TCR repertoire. The influence of Mtv SAgs on murine TCR Vβ expression has been comprehensively reviewed. It remains unclear why superantigens have evolved the ability to activate large numbers of T cells although endogenous retroviruses seem to be dependent on T cell stimulation for infection and dissemination.

Most of the exogenous SAgs are human pathogens and appear to be better at stimulating human rather than murine T cells. There has been considerable interest in finding human correlates of the Mtv loci and although so far there is no direct evidence for human endogenous SAgs (21) there are reports of inherited deletions of both Vβ2 and Vβ3. A bimodal distribution of $V\beta2^+$ T cells in the $CD4^+$ subset of peripheral blood has been reported (18). The range of Vβ2 expression was between 1% and 13% and remained stable over time. Two examples of low level expression of Vβ2 were seen in parents and sibling but this phenotype was not correlated with inheritance of HLA haplotype. Donahue *et al.* have recently reported individuals with low Vβ3 expression on both $CD4^+$ and $CD8^+$ T cells in peripheral blood (19). This bimodal expression may result from two different genetic events. Some families show genetic linkage of low Vβ3 phenotype to the TCR β locus and may exhibit polymorphism in efficiency of TCR recombination (36). In other cases the

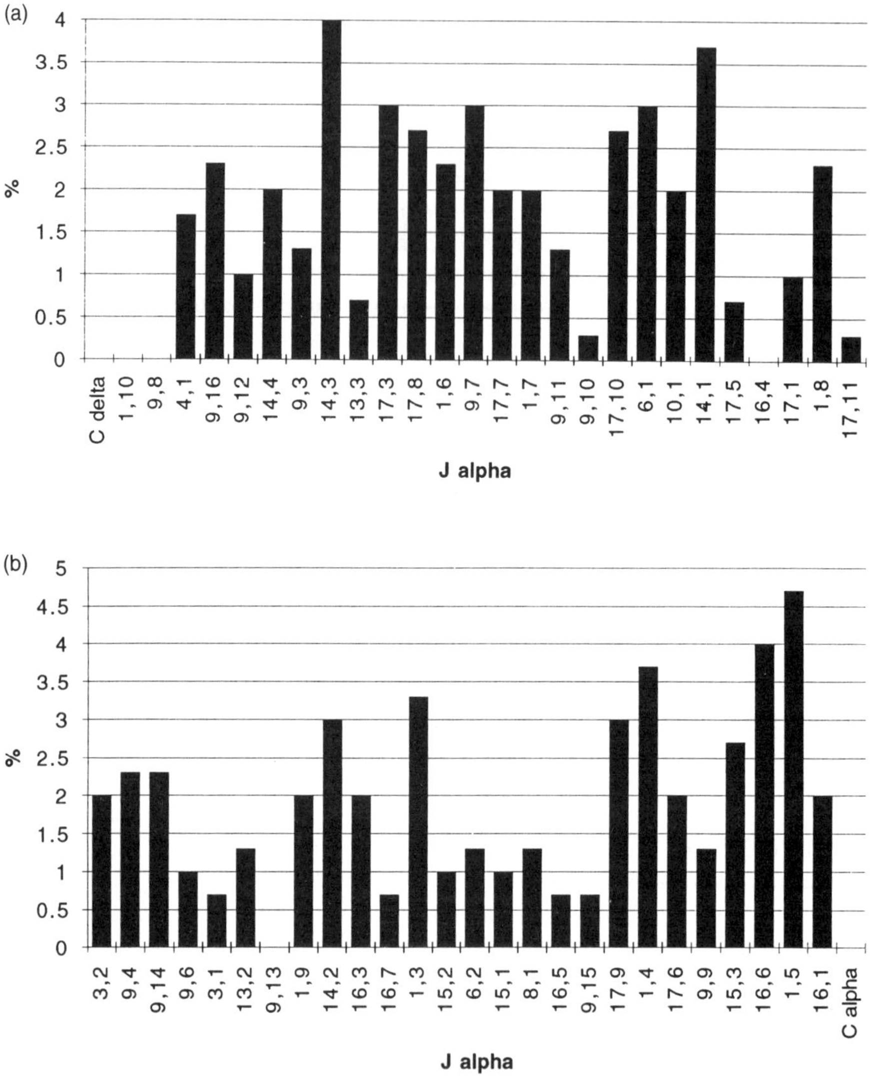

Fig. 6.2 Expression of Jα segments in human peripheral bood (based on reference 14).

effect is not linked to TCR genes and is inherited as a recessive trait in a manner reminiscent of murine endogenous SAgs. Vandekerckhove *et al.* used implantation of fetal liver and thymus into a *scid*–hu mouse to show that polymorphic determinants in the thymus could influence Vβ expression on single-positive cells but no clear Vβ deletions were seen. Distortions in thymic Vβ expression have been seen that were not carried over into peripheral blood (37). A similar pattern had been seen using an RNase protection assay for TCR Vβ RNA in human thymuses (21).

There are now many monoclonal antibodies to human Vβ segments and together they cover up to 50% of the expressed Vβ repertoire. Early reports failed to detect significant variation in expression between individuals although there were differences in the relative expression of different Vβ segments and a genetic influence on Vβ expression seems clear. Expression of a determinant recognized by an antibody to Vβ6.7 fell into three groups and it was realized that the antibody recognized an allelic variant of Vβ6.7, named Vβ6.7a, so that antibody staining reflected homozygous or heterozygous inheritance (6).

Molecular analysis of human Vβ expression confirmed the variation between levels of individual Vβ segments and showed that the Vβ6 family was particularly well represented (38, 39) (Figure 6.3). However, the Vβ6 family contains several members and so this effect could simply reflect the cumulative expression of all loci. The expression profiles determined by both anchor PCR and Vβ family-specific PCR have been reassuringly similar.

Minor differences in Vβ expression are much more easily found using monoclonal antibodies. Akolkar *et al.* used a panel of Vβ-specific antibodies to confirm differences in the human TCR Vβ repertoire and showed that such variation appeared to be related to inheritance of HLA type (40). Despite this, no one has yet shown a simple correlation between inheritance of a particular HLA allele and clear reduction or increase in expression of a particular Vβ allele. Ethnic origin may contribute to variation in TCR β expression (41). A consensus of Vβ expression in peripheral blood derived by anchored PCR is shown in Figure 6.3.

4.2.1 *Jβ expression*

There are 13 functional human Jβ segments grouped as six located 5′ to the Cβ1 segment and seven upstream of Cβ2. There are clear inequalities in the

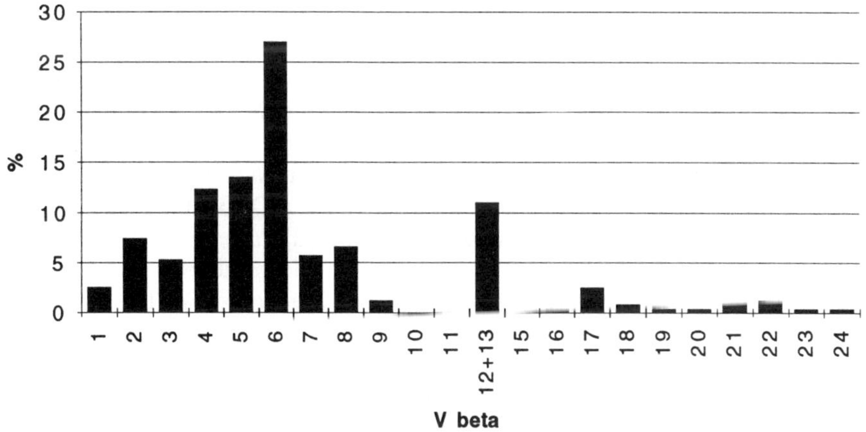

Fig. 6.3 Vβ expression on the surface of human peripheral blood T cells (based on reference 38).

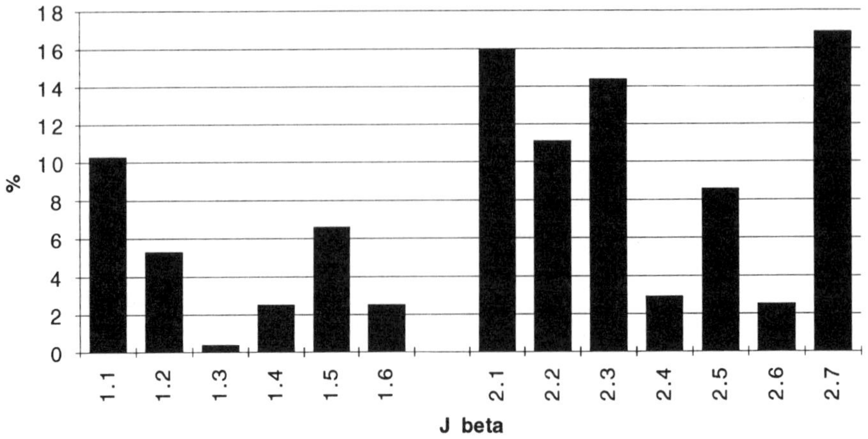

Fig. 6.4 Jβ expression on the surface of human peripheral blood T cells (based on reference 38).

expression of the individual gene segments although this appears to bear little relation to functional selection (Figure 6.4) (38, 37). Similar variation is seen in murine expression (42).

The Cβ2 segment is used approximately twice as commonly as Cβ1. This may reflect a higher probability of recombination to Jβ2 segments rather than Jβ1 although one can not rule out the chance of a secondary recombination to Jβ2 following an out-of-frame join to a Jβ1 segment. Out-of-frame (D)–J–Cβ transcripts also show a bias towards expression of Cβ2 suggesting that the inequality is merely the result of a stochastic process.

4.3 γδ T cells

Although it is now over seven years since the identification of the γδ T cell (43, 44), we are still unclear as to the nature of the antigenic recognition (45–47) or immunological role of this cell. Although the number of gene segments within the TCR γ and δ genes is less than for TCR α or β, the ability of TCR δ transcripts to utilize two in-frame D rearrangements means that there is the potential for enormous diversity within CDR3. It is by no means clear to what extent γδ T cells undergo thymic education and evidence for extrathymic selection has been presented (48). Nevertheless, there is evidence for both positive (49) and negative (50) selection of γδ T cells in the thymus in a manner analogous to αβ cells.

The V region expression of murine γδ T cells reveals a unique and intriguing distribution. Murine γδ T cells appear to undergo an ordered developmental pattern of γδ TCR rearrangement which occurs in association with the localization of γδ T cells with specific TCR rearrangements to different organs. Thus, γδ T cells in the skin of mice predominantly express

$V\gamma3$ gene segments whilst those in gut lymphoid tissue tend to express $V\gamma7$ segments.

Although a similar pattern of developmentally regulated gene expression has not been demonstrated in humans, there is non-random expression of $\gamma\delta$ gene segments in human thymus, blood, and fetal liver. Human fetal thymus has a predominance of $V\delta2$ gene segments whereas after birth the pattern changes to one of predominant $V\delta1$ expression. Although the $V\delta1$ segment remains predominant in thymus and is also found in human large intestine, $V\delta2$ becomes the most commonly expressed germline segment in peripheral blood at the age of six months. Indeed, $\gamma\delta$ T cell expression in peripheral blood is particularly interesting. Based on TCR expression, $\gamma\delta$ cells can be divided into two categories. The larger population, around 70% of cells in most adults, co-expresses $V\gamma9$ and $V\delta2$ gene segments whereas the smaller population tends to express $V\delta1$ paired to a $V\gamma$ segment other than $V\gamma9$. Junctional rearrangements of $V\delta1^+$ cells reveal that there is much more junctional diversity in infants in comparison to adults. This implies that peripheral selection may operate to determine the adult $V\delta1^+$ repertoire (51, 52). $V\delta1$ sequences also predominate in oral epithelium (53).

The fetal liver contains many $\alpha\beta$ and $\gamma\delta$ T cells and the presence of unusual $CD4^+CD8^-$ $\gamma\delta$ cells suggests that fetal liver may be a site of T cell development. The predominant $V\delta$ rearrangement is $V\delta2$, similar to fetal thymus, whereas $V\gamma$ rearrangements are dissimilar (54).

The studies on TCR repertoire described above were performed on whole, unsorted populations of peripheral blood T cells. As so much of the preliminary work has failed to reveal significant variation between individuals research has been extended into the comparison of TCR gene segment expression on well-defined subsets of T cells. This work has already revealed distortions in TCR expression between subsets which may contribute to understanding cellular function.

5 T cell receptor repertoire in T cell subsets

5.1 $CD4^+$ and $CD8^+$ T cells

Separation of peripheral blood T cells into single-positive $CD4^+$ and $CD8^+$ subsets was an obvious place to start the dissection of TCR repertoire. The expression of several $V\beta$- and $V\alpha$-specific monoclonal antibodies was found to vary between $CD4^+$ and $CD8^+$ T cells and a similar finding is obtained from RNase protection assays (21, 55). For most of the MAbs studied the difference was relatively minor, although a twofold difference was seen in several cases. It is likely that such a finding reflects differences in the relative affinity of different framework regions for class I or class II alleles. Positive selection on a class I or class II allele is known to determine a $CD8^+$ or $CD4^+$ T cell respectively. As, for instance, the $V\alpha12.1^+$ gene segment is found predominantly on $CD8^+$ T cells (56) it is likely that $V\alpha12.1^+$ T cells have a higher intrinsic affinity for class I molecules compared to class II. An alternative explanation would be that the affinity for class II molecules is so high that all Vα12.1$^+$ T cells have been deleted during thymic development.

If V region segments do indeed vary in their affinity for HLA alleles one may expect to see predominance of particular V segments in alloreactive T cells which are specific for one MHC allele in the context of many different peptides. Some preference for the use of Vβ4 in the allorecognition of HLA-B27 has been documented (57).

As the TCR repertoire of more and more individuals is studied by V region-specific MAbs we are beginning to see intriguing differences between the TCR expression on CD4$^+$ and CD8$^+$ T cells. The TCR repertoire of CD4$^+$ cells varies little from one person to the next with fairly predictable staining by each MAb. In contrast, staining of CD8$^+$ T cells by each antibody is more variable with a wider distribution of staining seen with each antibody. In several individuals studied there is high percentage staining of CD8$^+$ T cells by one particular MAb and this increase appears to be stable for at least two years. The significance of this expanded population is unclear at present but it may not be genetically determined as there appears to be discordance of TCR β expansion on CD8$^+$ cells from identical twins (58, 59). Preliminary work suggests that the expansion is largely oligoclonal and implies that the CD8$^+$ T cell repertoire may be much more limited than that found on CD4$^+$ cells.

5.2 TCR repertoire in CD3$^+$ large granular lymphocytes

T cells with the membrane phenotype CD8$^+$CD57$^+$ often have large granular lymphocyte morphology and are permanently increased in individuals who have been infected with cytomegalovirus. Their role is unknown although such cells appear to be able to suppress antibody production and the generation of CTL *in vitro*. Recent work has shown that CD8$^+$CD57$^+$ T cells are oligoclonal and that as much as 85% of the total population may be made up from two individual clones (E. Wang *et al.* submitted) (60). As the whole population can comprise 22% of the total CD8$^+$ count, single clones may make up as much as 11% of all CD8$^+$ T cells in peripheral blood. Expansions of large granular lymphocytes (LGLs) are seen in several diseases and it may be that the distortions in TCR V region repertoire which are being reported in such cases reflect their contribution to the whole CD8$^+$ repertoire (61). A notable example is rheumatoid arthritis. CD8$^+$CD57$^+$ expansions have been demonstrated in both the peripheral blood and joints of patients, and clonal expansions of CD8$^+$ T cells bearing Vα12.1 have been described (28).

It is tempting to speculate that the distortions that are being seen in the TCR repertoire of CD8$^+$ cells from normal donors are simply due to expansions of CD57$^+$ bearing cells. However, there is evidence that not all cells within the expansion are CD57$^+$ (P. Moss, unpublished observations) and it may be that CD57$^+$ is simply an activation marker which is present on a subset of the oligoclonal cells.

$CD8^+CD57^+$ T cells represent an increasing proportion of peripheral blood T cells with age although it is not clear whether or not this represents the increasing cumulative seroconversion rate for CMV. Although there is little evidence for changes in Vβ repertoire in ageing mice (62) it has been shown that populations of oligoclonal $CD8^+$ T cells exist in normal elderly humans. Such cells are $CD38^-$ and it has been suggested that these clonal expansions are of no immunological importance but merely represent the T cell equivalent of the benign paraproteinaemia which is so common in elderly people. It remains possible, however, that these oligoclonal populations play a role in immunoregulation. It is noteworthy that LGLs are expanded in conditions associated with B cell dysfunction such as paraproteinaemias, myeloma (63), and rheumatoid arthritis and it is not yet clear whether this expansion represents a T cell response to the B cells or is itself of primary importance. The membrane phenotype of oligoclonal T cells is variable but is usually $CD8^{+high}CD16^{+/-}CD56^{+/-}$ $CD57^{+/-}$. Similar markers are also found on natural killer (NK) cells and an equivalent population of T cells with NK cell markers have been demonstrated in both mice and rats (64–65, 66). In the former case they appear important in preventing the development of B cell autoimmunity. This population may well turn out to have an important role (67), possibly including tumour-directed cytotoxicity (68, 69), and one might speculate that specificity of antigen recognition is similar to that of NK cells but mediated through the TCR. Such cells may also undergo non-classical pathways of development (70). Such speculation awaits testing in cellular and molecular analyses.

5.3 $CD4^-CD8^-$ double-negative cells

The $CD4^-CD8^-$ double-negative (DN) T cell population can contribute up to 7% of all PBLs in healthy donors and up to 50% in certain diseases (71–73). Two recent studies have shown that the population is both oligoclonal and that a single conserved TCR α transcript can be found in virtually all donors studied (74, 75). The origin of DN cells is unclear but murine studies have suggested that they may not be derived from the thymus. Certainly the TCR repertoire of the murine population does not show evidence of either negative or positive selection. Some cells seem to have suppressive properties (76).

The two groups analysed the TCR usage of DN clones derived from healthy donors and went on to look at uncloned DN populations. Analysis of TCR α transcripts showed that a single, invariant Vα24–JαQ TCR α transcript was found in virtually all donors and was preferentially utilized in the DN population. Vα7.2 transcripts were also commonly isolated. The Vβ repertoire showed preferential usage of Vβ2, 8, 11, and 13 and gel electrophoresis revealed that the Vβ transcripts were relatively oligoclonal, rather than reflecting a general stimulation of all cells bearing such framework residues that would be expected if such a bias reflected stimulation by superantigens (75).

By combining limiting dilution analysis with N region oligotyping it was shown that individual clones could survive *in vivo* for four years. One of the clones proliferated in response to *E. coli* suggesting that the extraordinary expansion and survival of these cells may reflect chronic stimulation from gastrointestinal bacteria.

6 T cell receptor repertoire in the gastrointestinal tract and skin

The gastrointestinal tract is a major site of lymphoid development (77) and it appears that the TCR expression of gut-associated T cells may be very different from that found in peripheral blood (78–80). There is considerable evidence for the suggestion that murine T cells may develop in the gut in the absence of thymic maturation and such cells are likely to avoid the influence of thymic selection pressure on repertoire development. Nevertheless, gut-associated T cells appear to undergo some form of selection although details of this are currently unclear. There are suggestions of disordered TCR γδ expression and increased numbers of both αβ and γδ cells in patients with coeliac disease (81–83).

Little work has so far been directed towards the fascinating topic of TCR expression in the T cells found in skin both in health (84, 85) and in the many skin diseases which appear to be mediated by T cells (86, 87). This topic is likely to prove of great interest in the future.

7 The influence of superantigens on T cell repertoire

As noted above, human T cells are susceptible to stimulation from a wide variety of superantigens (22, 88–91) and it might be expected that such agents could distort the overall profile of Vβ expression (92). Although the classical view is that SAgs activate T cells by binding to Vβ framework regions alone it is now clear that other elements of the TCR such as TCR Vα or Jα, can influence the activation of individual T cell clones by SAgs (29, 93).

In vitro analysis revealed that human T cells responded to SAg activation in a Vβ-specific manner similar to that found for murine cells. Following *in vitro* documentation of which Vβ segments were susceptible to stimulation by SAgs such as bacterial exotoxins (89) it was not long before such expansions were detected *in vivo*. Choi *et al.* showed that T cells bearing Vβ2 were expanded following toxic shock syndrome and that the resultant repertoire distortion was present for several months (94).

Several examples of TCR Vβ distortions found in human disease are suggestive of SAg activation. The expansion of $V\beta14^+$ synovial T cell seen in some cases of rheumatoid arthritis may result from SAg activation (95). The TCR distortions seen in the vasculitic condition acute Kawasaki disease may also be secondary to effects of a SAg (96). Polyclonal increases in Vβ2 and Vβ8 bearing T cells have been shown using both PCR and antibody techniques and seem to be marked in the $CD4^+$ population.

8 T cell receptor repertoire in human disease

Few areas in immunology research have shown such consistent problems of irreproducibility as the analysis of T cell receptor repertoire in human disease. To date, no single disease has been analysed in which two studies have observed the same distortions of repertoire. The explanation for these inconsistencies is not obvious, however it may be accounted for by technical difficulties in measuring TCR V regions, choice of patient populations, or the stage of the disease studied.

It would be surprising if substantial homogeneity of TCR V region usage existed in either autoimmune disease or infectious disease. Conservation of V regions is seen in responses to some defined antigens (25), but in these well-controlled circumstances this conservation is far from complete. It is only in exceptional circumstances that a single V region predominates and this is more often true for $CD8^+$ compared to $CD4^+$ clones. During the development of an autoimmune disorder, a range of target antigens are likely to be recognized, as suggested by the host of β cell antigens against which antibodies develop in type I diabetes. Other disorders such as multiple sclerosis (MS) and rheumatoid arthritis (RA) are likely to be accompanied by T cell responses to diverse sets of antigens such as responses to Ig (rheumatoid factor) or heat shock proteins in RA, or responses to myelin basic protein or other myelin proteins in MS. In addition to the development of a range of target antigens during the course of the disease, recruitment of T cells unrelated to disease-associated target antigens are likely to further confuse the interpretation of repertoire. If a restricted response is to be seen, it is likely only to be present early in disease. In extrinsic allergic encephalomyelitis (EAE), where a single T cell clone can initiate disease, a large range of TCRs are seen in the CNS, even at very early stages in the disease, reflecting recruitment of a set of T cells independent of the pathogenic clone.

One argument for the presence of restricted V region representation in human disease is that superantigens may distort the repertoire, increasing the likelihood of self-reactive T cells. One would predict, however, that such superantigen effect, regardless of their theoretical appeal, would create similar distortions in patients studied by different laboratories. This has not proved to be the case. In only one case, Kawasaki's disease, has a putative superantigen been identified and its effects correlated with V region distortions. No similar evidence has been forthcoming in other infections or autoimmune disorders. The techniques used to detect V region distortion are a possible explanation for the inconsistencies between laboratories in the studies. V region usage may also depend on MHC restriction, and in many cases failure to match patients for MHC type limits the interpretation of the data. In some cases twins have been used to overcome these limitations but the variation of TCR usage seen in monozygotic twins makes these analyses difficult.

Four diseases which reflect the confusion pervading the field at present include multiple sclerosis, rheumatoid arthritis, autoimmune thyroid disease, and HIV. In all these diseases TCR V region usage has been implicated in pathogenesis but results vary between laboratories.

8.1 Multiple sclerosis

In multiple sclerosis (MS), interest was provoked by the observation that in EAE a single T cell clone could mediate disease and that different T cell clones showed similar TCR sequences. Interest was strengthened by reports of a genomic linkage between the TCR β locus and development of MS (12). Antibodies to conserved Vβ sequences could alter the course of the disease. Initial data using PCR of MS plaques obtained from post-mortem material suggested that typically only two to four Vα transcripts were observed of which Vα12.1 predominated (97). The lack of information on HLA restriction or quality of RNA from MS brains limited the utility of this study. Attempts to repeat this have proved difficult and attention has drifted to Vβ sequences where Vβ5.2 transcripts, again from MS brains, have shown conserved CDR3 motifs (98). Studies of twins, concordant and discordant for MS, have suggested that discordant twins select different V regions after stimulation with myelin basic protein (MBP) on tetanus toxoid (99). Conservation of V region usage has been clearly identified in MBP-specific clones from both normal individuals and MS patients.

Studies on γδ expression have failed to reveal significant sequence conservation in disease (100) although oligoclonal expansions in early disease have been reported (101).

8.2 Rheumatoid arthritis

In rheumatoid arthritis antibody studies and PCR analysis have suggested a variety of different TCR V region distortions in disease. Genomic linkage is also reported (102, 103). In various studies many different Vα and Vβ segments are either expanded or depleted (95, 104–111). Although there is no agreement on V segment usage, oligoclonal T cells do seem to be obtainable from synovial fluid (112–114) or even blood (28). Such populations have not been seen in γδ subsets (115). Studies on monozygotic twins discordant or concordant for disease showed that rheumatoid arthritis had no clear effect on Vβ expression in peripheral blood (116). These discrepancies in data most probably result from technical problems as well as variable patient selection. Reports of non-random TCR usage by T cells obtained from patients with autoimmune thyroid disease have been similarly difficult to reproduce (117–119).

8.3 HIV infection

Infectious diseases have also not avoided the confusion that prevails in autoimmune conditions. Data suggesting V region distortions in $CD4^+$ populations in individuals with low CD4 counts secondary to HIV infection supported the idea that the virus produced a superantigen. This data was not reproduced in a second study where two other Vβ regions were distorted (Vβ3 and Vβ20) (120) or in a second study of $CD4^+$ TCR repertoire (121). Although there is debate as to the presence of repertoire distortions characterized by monoclonal antibodies rather than PCR (122–124), studies of SIV in rhesus macaques also fail to demonstrate distortions. Twin studies have suggested distortions of repertoire in CD4 cells of the infected twins (Vβ13, Vβ12, and Vβ21). Increased numbers of γδ T cells and expression of the Vδ2 gene segment has been shown in bronchoalveolar lavage cells from an infected donor (125–127) and this correlated with the number of $CD8^+$ T cells in the lavage. Increased numbers of γδ cells are also seen in lavage fluid from patients with pneumocystis pneumonia (127).

Potentially valuable studies on TCR repertoire in other diseases await confirmation (128–133).

The data in all these diseases reflect the difficulty in obtaining consistent results between laboratories. Methodological problems must remain the most likely explanation for these data and no firm conclusions can be drawn until these issues can be resolved. An additional concern is that it is now clear that substantial $CD8^+$ distortions exist in a large number of normal individuals. The presence of oligoclonal T cells can not therefore be assumed to be relevant to disease pathogenesis without comparison to suitably age, sex, and MHC matched controls.

9 Summary

There has been great interest in the study of human TCR receptor expression over the last few years and although few clear messages are yet to emerge it is likely that the field will become increasingly important to our understanding of T cell biology. Much work has gone into standardizing techniques for analysis and this is leading to more powerful and reproducible analyses which combine monoclonal antibodies and molecular techniques. Preliminary observations are pointing towards the possible existence of endogenous superantigens in humans and, if present, these agents may contribute to the genetic susceptibility to disease. Oligoclonal T cells of, as yet, unknown function are common and may have an important role in immunoregulation. Finally, reports of repertoire distortions in human disease are becoming increasingly common and it should not be long before this knowledge can be applied towards potential therapeutic use of immunotherapy.

References

1 Davis, M. M. and Bjorkman, P. J. (1988). T cell antigen receptor genes and T cell recognition. *Nature*, **334**, 395–402.
2 Pullen, A. M., *et al.* (1990). Surprisingly uneven distribution of the T cell receptor V beta repertoire in wild mice. *J. Exp. Med.*, **171**, 49–62.
3 Kappler, J. W., *et al.* (1987). A T cell receptor V beta segment that imparts reactivity to a class II major histocompatibility complex product. *Cell*, **49**, 263–71.
4 Concannon, P., Gatti, R. A., and Hood, L. E. (1987). Human T cell receptor V beta gene polymorphism. *J. Exp. Med.*, **165**, 1130–40.
5 Robinson, M. A. (1989). Allelic sequence variations in the hypervariable region of a T cell receptor beta chain: Correlation with restriction fragment length polymorphism in human families and populations. *Proc. Natl. Acad. Sci. USA*, **86**, 9422–6.
6 Li, Y. X., *et al.* (1990). Allelic variations in the human T cell receptor V beta 6.7 gene products [published erratum appears in *J. Exp. Med.* (1990) **171**, 973]. *J. Exp. Med.*, **171**, 221–30.
7 Cornelis, F., *et al.* (1993). Systematic study of human αβ T cell receptor V segments shows allelic variations resulting in a large number of distinct T cell receptor haplotypes. *Eur. J. Immunol.*, **23**, 1277–83.
8 Reyburn, H., *et al.* (1993). Allelic polymorphism of human T-cell receptor V alpha gene segments. *Immunogenetics*, **38**, 287–91.
9 Gomolka, M., *et al.* (1993). Novel members and germline polymorphisms in the human T-cell receptor Vβ6 family. *Immunogenetics*, **37**, 257–65.
10 Wright, J. A., Hood, L., and Concannon, P. (1991). Human T-cell receptor Vα gene polymorphism. *Hum. Immunol.*, **32**, 277–83.
11 Seboun, E., *et al.* (1989). Insertion/deletion-related polymorphisms in the human T cell receptor beta gene complex. *J. Exp. Med.*, **170**, 1263–70.
12 Barker, P. E., *et al.* (1985). Regional location of T cell receptor gene Ti alpha on human chromosome 14. *J. Exp. Med.*, **162**, 387–92.
13 Gulwani-Akolkar, B., *et al.* (1991). T cell receptor V-segment frequencies in peripheral blood T cells correlate with human leukocyte antigen type. *J. Exp. Med.*, **174**, 1139–46.
14 Moss, P. A. H., *et al.* (1993). Characterization of the human T cell receptor-α chain repertoire and demonstration of a genetic influence on Vα usage. *Eur. J. Immunol.*, **23**, 1153–9.
15 Kappler, J. W., Roehm, N., and Marrack, P. (1987). T cell tolerance by clonal elimination in the thymus. *Cell*, **49**, 273–80.
16 Bill, J., Appel, V. B., and Palmer, E. (1988). An analysis of T-cell receptor variable region gene expression in major histocompatibility complex disparate mice. *Proc. Natl. Acad. Sci. USA*, **85**, 9184–8.
17 Smith, L. R., *et al.* (1992). Vβ repertoire in rats and implications for endogenous superantigens. *Eur. J. Immunol.*, **22**, 641–5.
18 Clarke, G. R., *et al.* (1994). Bimodal distribution of Vβ2CD4+ T cells in human peripheral blood. *Eur. J. Immunol.*, **24**, 837–42.
19 Donahue, J. P., *et al.* (1994). Genetic analysis of low Vβ3 expression in humans. *J. Exp. Med.*, **179**, 1701–6.
20 Choi, Y., *et al.* (1990). Selective expansion of T cells expressing Vβ2 in toxic shock syndrome. *J. Exp. Med.*, **172**, 981–4.

21 Baccala, R., *et al.* (1991). Genomically imposed and somatically modified human thymocyte V beta gene repertoires. *Proc. Natl. Acad. Sci. USA*, **88**, 2908–12.
22 Choi, Y., *et al.* (1989). Interaction of *Staphyloccus aureus* toxin 'superantigens' with human T cells. *Proc. Natl. Acad. Sci. USA*, **86**, 8941–5.
23 Segurado, O. G. and Schendel, D. J. (1993). Identification of predominant T-cell receptor rearrangements by temperature-gradient gel electrophoresis and automated DNA sequencing. *Electrophoresis*, **14**, 747–52.
24 Loh, E. Y., *et al.* (1989). Polymerase chain reaction with single-sided specificity: analysis of T cell receptor delta chain. *Science*, **243**, 217–20.
25 Moss, P. A. H., *et al.* (1991). Extensive conservation of α and β chains of the human T-cell antigen receptor recognizing HLA-A2 and influenza A matrix peptide. *Proc. Natl. Acad. Sci. USA*, **88**, 8987–90.
26 Uematsu, Y. (1991). A novel and rapid cloning method for the T cell receptor variable region sequences. *Immunogenetics*, **34**, 174.
27 Ferradini, L., *et al.* (1993). The use of anchored polymerase chain reaction for the study of large numbers of human T-cell receptor transcripts. *Mol. Immunol.*, **30**, 1143–50.
28 Brenner, M. K. (1992). Peripheral expansions of Vα12 bearing T cells in rheumatoid arthritis. In *Symposium on T cell repertoire*. Arad, Israel.
29 Waanders, G. A., Lussow, A. R., and MacDonald, H. R. (1993). Skewed T cell receptor V alpha repertoire among superantigen reactive murine T cells. *Int. Immunol.*, **5**, 55–61.
30 Padovan, E., *et al.* (1993). Expression of two T cell receptor alpha chains: dual receptor T cells. *Science*, **262**, 422–4.
31 Heath, W. R. and Miller, J. F. (1993). Expression of two alpha chains on the surface of T cells in T cell receptor transgenic mice. *J. Exp. Med.*, **178**, 1807–11.
32 Wilson, R. K., *et al.* (1992). Nucleotide sequence analysis of 95kb near the 3′ end of the murine T-cell receptor α/δ chain locus: strategy and methodology. *Genomics*, **13**, 1198–208.
33 Bougueleret, L. and Claverie, J. M. (1987). Variability analysis of the human and mouse T-cell receptor beta chains. *Immunogenetics*, **26**, 304–8.
34 Jorgensen, J. L., *et al.* (1992). Mapping T-cell receptor-peptide contacts by variant peptide immunization of single-chain transgenics. *Nature*, **355**, 224–30.
35 Dave, V. P., *et al.* (1993). Restricted usage of T-cell receptor V alpha sequence and variable-joining pairs after normal T-cell development and bone marrow transplantation. *Hum. Immunol.*, **37**, 178–84.
36 Posnett, D. N., *et al.* (1994). Level of human TCRBV3S1 Vβ3 expression correlates with allelic polymorphism in the spacer region of the recombination signal sequence. *J. Exp. Med.*, **179**, 1707–11.
37 Jores, R. and Meo, T. (1993). Few V gene segments dominate the T cell receptor beta-chain repertoire of the human thymus. *J. Immunol.*, **151**, 6110–22.
38 Rosenberg, W. R., Moss, P. A. H., and Bell, J. I. (1991). Variation in human T cell receptor V beta and J beta repertoire: analysis using anchor PCR. *Eur. J. Immunol.*, **22**, 541–9.
39 Loveridge, J. A., *et al.* (1991). The genetic contribution to human T cell receptor repertoire. *Immunology*, **74**, 246–50.
40 Akolkar, P. N., *et al.* (1993). Influence of HLA genes on T cell receptor V segment frequencies and expression levels in peripheral blood lymphocytes. *J. Immunol.*, **150**, 2761–73.
41 Geursen, A., *et al.* (1993). Population study of T cell receptor V beta gene usage in

peripheral blood lymphocytes: differences in ethnic groups. *Clin. Exp. Immunol.*, **94**, 201–7.

42 Candeias, S., *et al.* (1991). The Vβ17 T cell repertoire: skewed Jβ usage after thymic selection; dissimilar CDR3s in CD4+ versus CD8+ cells. *J. Exp. Med.*, **174**, 989–1000.

43 Brenner, M. B., *et al.* (1986). Identification of a putative second T-cell receptor. *Nature*, **322**, 145–9.

44 Groh, V., *et al.* (1989). Human lymphocytes bearing T cell receptor gamma/delta are phenotypically diverse and evenly distributed throughout the lymphoid system. *J. Exp. Med.*, **169**, 1277–94.

45 Bluestone, J. A., *et al.* (1988). Structure and specificity of T cell receptor gamma/delta on major histocompatibility complex antigen-specific CD3+, CD4−, CD8− T lymphocytes. *J. Exp. Med.*, **168**, 1899–916.

46 Ciccone, E., *et al.* (1988). Antigen recognition by human T cell receptor gamma-positive lymphocytes. Specific lysis of allogeneic cells after activation in mixed lymphocyte culture. *J. Exp. Med.*, **167**, 1517–22.

47 Fisch, P., *et al.* (1990). Recognition by human V gamma 9/V delta 2 T cells of a GroEL homolog on Daudi Burkitt's lymphoma cells. *Science*, **250**, 1269–73.

48 Parker, C. M., *et al.* (1990). Evidence for extrathymic changes in the T cell receptor gamma/delta repertoire. *J. Exp. Med.*, **171**, 1597–612.

49 Wells, F. B., *et al.* (1993). Phenotypic and functional analysis of positive selection in the γδ T cell lineage. *J. Exp. Med.*, **177**, 1061–70.

50 Dent, A. L., *et al.* (1990). Self-reactive γδ T cells are eliminated in the thymus. *Nature*, **343**, 714–19.

51 Beldjord, K., *et al.* (1993). Peripheral selection of V1+ cells with restricted T cell receptor δ gene junctional repertoire in the peripheral blood of healthy donors. *J. Exp. Med.*, **178**, 121–7.

52 Augustin, A., Kubo, R. T., and Sim, G. K. (1989). Resident pulmonary lymphocytes expressing the gamma/delta T-cell receptor. *Nature*, **340**, 239–41.

53 Pepin, L. F., *et al.* (1993). Preferential V delta 1 expression among TcR gamma/delta-bearing T cells in human oral epithelium. *Scand. J. Immunol.*, **37**, 289–94.

54 Wucherpfennig, K. W., *et al.* (1993). Human fetal liver γδ T cells predominantly use unusual rearrangements of the T cell receptor δ and γ loci expressed on both CD4+CD8− and CD4−CD8 δγ T cells. *J. Exp. Med.*, **177**, 425–32.

55 Grunewald, J., Janson, C. H., and Wigzell, H. (1991). Biased expression of individual T cell receptor V gene segments in CD4+ and CD8+ human peripheral blood T lymphocytes. *Eur. J. Immunol.*, **21**, 819–22.

56 DerSimonian, H., Band, H., and Brenner, M. B. (1991). Increased frequency of T cell receptor Vα12.1 expression on CD8+ T cells: Evidence that Vα participates in shaping the peripheral T cell repertoire. *J. Exp. Med.*, **174**, 639–48.

57 Bragado, R., *et al.* (1990). T cell receptor V beta gene usage in a human alloreactive response. Shared structural features among HLA-B27-specific T cell clones. *J. Exp. Med.*, **171**, 1189–204.

58 Davey, M. P., Meyer, M. M., and Bakke, A. C. (1994). T cell receptor V beta gene expression in monozygotic twins. Discordance in CD8 subset and in disease states. *J. Immunol.*, **152**, 315–21.

59 Hawes, G. E., Struyk, L. and van der Elsen, P. J. (1993). Differential usage of T cell receptor V gene segments in CD4+ and CD8+ subsets of T lymphocytes in monozygotic twins. *J. Immunol.*, **150**, 2033–45.

60 Hingorani, R., *et al.* (1993). Clonal predominance of T cell receptors within the CD8+ CD45RO+ subset in normal human subjects. *J. Immunol.*, **151**, 5762–9.

61 Gorochov, G., *et al.* (1994). Oligoclonal expansion of CD8+ CD57+ T cells with restricted T-cell receptor beta chain variability after bone marrow transplantation. *Blood*, **83**, 587–95.
62 Gonzalez-Quintial, R. and Theofilopoulos A. N. (1992). Vβ gene repertoires in aging mice. *J. Immunol.*, **149**, 230–6.
63 Janson, C. H., *et al.* (1991). Predominant T cell receptor V gene usage in patients with abnormal clones of B cells. *Blood*, **77**, 1776–80.
64 Sarzotti, M., *et al.* (1991). Cloning of murine splenic T lymphocytes and natural killer (NK) cells on filter paper discs: detection of a novel NK/T phenotype. *Eur. J. Immunol.*, **21**, 635–41.
65 Arase, H., *et al.* (1993). Lymphokine-activated killer cell activity of CD4−CD8− TCR alpha beta+ thymocytes. *J. Immunol.*, **151**, 546–55.
66 Brissette, S. C., *et al.* (1994). Characterization and function of the NKR− P1dim/T cell receptor-alpha beta+ subset of rat T cells. *J. Immunol.*, **152**, 388–96.
67 Azuma, M., Phillips, J. H., and Lanier, L. L. (1993). CD28− T lymphocytes. Antigenic and functional properties. *J. Immunol.*, **150**, 1147–59.
68 Ballas, Z. K. and Rasmussen, W. (1993). Lymphokine-activated killer cells. VII. IL-4 induces an NK1.1+CD8 alpha+beta− TCR-alpha beta B220+ lymphokine-activated killer subset. *J. Immunol.*, **150**, 17–30.
69 Halapi, E., *et al.* (1993). Restricted T cell receptor V-beta and J-beta usage in T cells from interleukin-2-cultured lymphocytes of ovarian and renal carcinomas. *Cancer Immunol. Immunother.*, **36**, 191–7.
70 Apasov, S. and Sitkovsky, M. (1993). Highly lytic CD8+, alpha beta T-cell receptor cytotoxic T cells with major histocompatibility complex (MHC) class I antigen-directed cytotoxicity in beta 2-microglobulin, MHC class I-deficient mice. *Proc. Natl. Acad. Sci. USA*, **90**, 2837–41.
71 Londei, M., *et al.* (1989). Definition of a population of CD4−CD8− T cells that express the αβ T cell receptor and respond to interleukins 2, 3 and 4. *Proc. Natl. Acad. Sci. USA*, **86**, 8502.
72 Murison, J. G., *et al.* (1993). Definition of unique traits of human CD4−CD8− alpha beta T cells. *Clin. Exp. Immunol.*, **93**, 464–70.
73 Mieno, M., *et al.* (1991). CD4−CD8− T cell receptor alpha beta T cells: generation of an *in vitro* major histocompatibility complex class I specific cytotoxic T lymphocyte response and allogeneic tumor rejection. *J. Exp. Med.*, **174**, 193–201.
74 Brooks, E. G., *et al.* (1993). Human T-cell receptor (TCR) alpha/beta+ CD4− CD8− T cells express oligoclonal TCRs, share junctional motifs across TCR V beta-gene families, and phenotypically resemble memory T cells. *Proc. Natl. Acad. Sci. USA*, **90**, 11787–91.
75 Porcelli, S., *et al.* (1993). Analysis of T cell antigen receptor (TCR) expression by human peripheral blood CD4−CD8− α/β T cells demonstrates preferential use of several Vβ genes and an invariant TCRα chain. *J. Exp. Med.*, **178**, 1–16.
76 Schmidt, W. I., *et al.* (1993). Homogeneous antigen receptor beta-chain genes in cloned CD4− CD8− alpha beta T suppressor cells. *J. Immunol.*, **151**, 5348–53.
77 Fujihashi, K., *et al.* (1993). Function of alpha beta TCR+ intestinal intraepithelial lymphocytes: Th1- and Th2-type cytokine production by CD4+CD8− and CD4+CD8+ T cells for helper activity. *Int. Immunol.*, **5**, 1473–81.
78 VanKerckhove, C., *et al.* (1992). Oligoclonality of human intestinal intraepithelial T cells. *J. Exp. Med.*, **175**, 57–63.
79 Goodman, T. and Lefrancois, L. (1988). Expression of the gamma-delta T-cell receptor on intestinal CD8+ intraepithelial lymphocytes. *Nature*, **333**, 855–8.
80 Goodman, T. and Lefrancois, L. (1989). Intraepithelial lymphocytes. Anatomical

site, not T cell receptor form, dictates phenotype and function. *J. Exp. Med.*, **170**, 1569–81.

81 Halstensen, T. S. and Brandtzaeg, P. (1993). Activated T lymphocytes in the celiac lesion: non-proliferative activation (CD25) of CD4+ alpha/beta cells in the lamina propria but proliferation (Ki-67) of alpha/beta and gamma/delta cells in the epithelium. *Eur. J. Immunol.*, **23**, 505–10.

82 Kutlu, T., *et al.* (1993). Numbers of T cell receptor (TCR) alpha beta+ but not of TcR gamma delta+ intraepithelial lymphocytes correlate with the grade of villous atrophy in coeliac patients on a long term normal diet. *Gut*, **34**, 208–14.

83 Sturgess, R., *et al.* (1993). Gamma/delta T-cell receptor expression in the jejunal epithelium of patients with dermatitis herpetiformis and coeliac disease. *Clin. Exp. Dermatol.*, **18**, 318–21.

84 Foster, C. A., *et al.* (1990). Human epidermal T cells predominantly belong to the lineage expressing alpha/beta T cell receptor. *J. Exp. Med.*, **171**, 997–1013.

85 Dunn, D. A., *et al.* (1993). T-cell receptor V beta expression in normal human skin. *Proc. Natl. Acad. Sci. USA*, **90**, 1267–71.

86 Fujita, M., *et al.* (1993). Gamma delta T-cell receptor-positive cells in human skin. I. Incidence and V-region gene expression in granulomatous skin lesions. *J. Am. Acad. Dermatol.*, **28**, 46–50.

87 Lewis, H. M., *et al.* (1993). Restricted T-cell receptor V beta gene usage in the skin of patients with guttate and chronic plaque psoriasis. *Br. J. Dermatol.*, **129**, 514–20.

88 Fleischer, B., Schrezenmeier, H., and Conradt, P. (1989). T lymphocyte activation by staphylococcal enterotoxins: role of class II molecules and T cell surface structures. *Cell. Immunol.*, **120**, 92–101.

89 Abe, J., *et al.* (1991). Selective stimulation of human T cells with streptococcal erythrogenic toxins A and B. *J. Immunol.*, **146**, 3747–50.

90 Bowness, P., *et al.* (1992). Clostridium perfringens enterotoxin is a superantigen reactive with human T cell receptor Vβ6.9 and Vβ22. *J. Exp. Med.*, **176**, 893.

91 Tomai, M. A., *et al.* (1991). T cell receptor V gene usage by human T cells stimulated with the superantigen streptococcal M protein. *J. Exp. Med.*, **174**, 285–8.

92 Kotzin, B. L., *et al.* (1993). Superantigens and their potential role in human disease. *Adv. Immunol.*, **54**, 99–166.

93 Vacchio, M. S., *et al.* (1992). Influence of T cell receptor Vα expression on Mls^a superantigen-specific T cell responses. *J. Exp. Med.*, **175**, 1405–8.

94 Choi, Y., *et al.* (1990). Selective expansion of T cells expressing V beta 2 in toxic shock syndrome. *J. Exp. Med.*, **172**, 981–4.

95 Paliard, X., *et al.* (1991). Evidence for the effects of a superantigen in rheumatoid arthritis. *Science*, **253**, 325–9.

96 Abe, J., *et al.* (1993). Characterization of T cell repertoire changes in acute Kawasaki disease. *J. Exp. Med.*, **177**, 791–6.

97 Oksenberg, J. R., *et al.* (1990). Limited heterogeneity of rearranged T-cell receptor V alpha transcripts in brains of multiple sclerosis patients. *Nature*, **345**, 344–6.

98 Oksenberg, J. R., *et al.* (1993). Selection for T-cell receptor V beta-D beta-J beta gene rearrangements with specificity for a myelin basic protein peptide in brain lesions of multiple sclerosis. *Nature*, **362**, 68–70.

99 Utz, U., *et al.* (1993). Skewed T-cell receptor repertoire in genetically identical twins correlates with multiple sclerosis [see comments]. *Nature*, **364**, 243–7.

100 Hvas, J., *et al.* (1993). Gamma delta T cell receptor repertoire in brain lesions of patients with multiple sclerosis. *J. Neuroimmunol.*, **46**, 225–34.
101 Shimonkevitz, R., *et al.* (1993). Clonal expansions of activated gamma/delta T cells in recent-onset multiple sclerosis. *Proc. Natl. Acad. Sci. USA*, **90**, 923–7.
102 de, V. N., *et al.* (1993). A T cell receptor beta chain variable region polymorphism associated with radiographic progression in rheumatoid arthritis. *Ann. Rheum. Dis.*, **52**, 327–31.
103 Ploski, R., Hansen, T., and Forre, O. (1993). Lack of association with T-cell receptor TCRBV6S1*2 allele in HLA-DQA1*0101-positive Norwegian juvenile chronic arthritis patients. *Immunogenetics*, **38**, 444–5.
104 Grom, A. A., *et al.* (1993). Dominant T-cell-receptor beta chain variable region V beta 14+ clones in juvenile rheumatoid arthritis. *Proc. Natl. Acad. Sci. USA*, **90**, 11104–8.
105 Pluschke, G., *et al.* (1991). Biased T cell receptor Vα region repertoire in the synovial fluid of rheumatoid arthritis patients. *Eur. J. Immunol.*, **21**, 2749.
106 Pluschke, G., *et al.* (1993). Analysis of T cell receptor V beta regions expressed by rheumatoid synovial T lymphocytes. *Immunobiology*, **188**, 330–9.
107 Uematsu, Y., *et al.* (1991). The T-cell receptor repertoire in the synovial fluid of a patient with rheumatoid arthritis is polyclonal. *Proc. Natl. Acad. Sci. USA*, **88**, 8534.
108 Broker, B. M., *et al.* (1993). Biased T cell receptor V gene usage in rheumatoid arthritis. Oligoclonal expansion of T cells expressing V alpha 2 genes in synovial fluid but not in peripheral blood. *Arthritis Rheum.* (1993). **36**, 1234–43.
109 Jenkins, R. N., *et al.* (1993). T cell receptor V beta gene bias in rheumatoid arthritis. *J. Clin. Invest.*, **92**, 2688–701.
110 Sottini, A., *et al.* (1991). Restricted expression of T cell receptor Vβ but not Vα gene in rheumatoid arthritis patients. *Eur. J. Immunol.*, **21**, 461.
111 Williams, W. V., *et al.* (1993). Conserved motifs in rheumatoid arthritis synovial tissue T-cell receptor beta chains. *DNA Cell. Biol.*, **12**, 425–34.
112 Stamenkovic, I., *et al.* (1988). Clonal dominance among T-lymphocyte infiltrates in arthritis. *Proc. Natl. Acad. Sci. USA*, **85**, 1179.
113 Cantagrel, A., *et al.* (1993). Clonality of T lymphocytes expanded with IL-2 from rheumatoid arthritis peripheral blood, synovial fluid and synovial membrane. *Clin. Exp. Immunol.*, **91**, 83–9.
114 De, K. F., *et al.* (1993). T cell receptor V beta usage in rheumatoid nodules: marked oligoclonality among IL-2 expanded lymphocytes. *Clin. Immunol. Immunopathol.*, **68**, 29–34.
115 Kohsaka, H., *et al.* (1993). Divergent T cell receptor gamma repertoires in rheumatoid arthritis monozygotic twins. *Arthritis Rheum.*, **36**, 213–21.
116 Kohsaka, H., *et al.* (1993). The expressed T cell receptor V gene repertoire of rheumatoid arthritis monozygotic twins: rapid analysis by anchored polymerase chain reaction and enzyme-linked immunosorbent assay. *Eur. J. Immunol.*, **23**, 1895–901.
117 Iwatani, Y., *et al.* (1993). Intrathyroidal lymphocyte subsets, including unusual CD4+ CD8+ cells and CD3loTCR alpha beta lo/−CD4−CD8− cells, in autoimmune thyroid disease. *Clin. Exp. Immunol.*, **93**, 430–6.
118 McIntosh, R. S., *et al.* (1993). The gamma delta T cell repertoire in Graves' disease and multinodular goitre. *Clin. Exp. Immunol.*, **94**, 473–7.
119 McIntosh, R. S., *et al.* (1993). No restriction of intrathyroidal T cell receptor V alpha families in the thyroid of Graves' disease. *Clin. Exp. Immunol.*, **91**, 147–52.

120 Hodara, V. L., *et al.* (1993). HIV infection leads to differential expression of T-cell receptor V beta genes in CD4+ and CD8+ T cells. *AIDS*, **7**, 633–8.
121 Posnett, D. N., *et al.* (1993). T-cell antigen receptor V beta subsets are not preferentially deleted in AIDS. *AIDS*, **7**, 625–31.
122 Bansal, A. S., *et al.* (1993). HIV induces deletion of T cell receptor variable gene product-specific T cells. *Clin. Exp. Immunol.*, **94**, 17–20.
123 Boyer, V., *et al.* (1993). T cell receptor V beta repertoire in HIV-infection individuals: lack of evidence for selective V beta deletion. *Clin. Exp. Immunol.*, **92**, 437–41.
124 Soudeyns, H., *et al.* (1993). The T cell receptor V beta repertoire in HIV-1 infection and disease. *Semin. Immunol.*, **5**, 175–85.
125 Agostini, C., *et al.* (1994). Gamma delta T cell receptor subsets in the lung of patients with HIV-1 infection. *Cell. Immunol.*, **153**, 194–205.
126 Hermier, F., *et al.* (1993). Decreased blood TcR gamma delta+ lymphocytes in AIDS and p24-antigenemic HIV-1-infected patients. *Clin. Immunol. Immunopathol.*, **69**, 248–50.
127 Kagi, M. K., *et al.* (1993). High proportion of gamma-delta T cell receptor positive T cells in bronchoalveolar lavage and peripheral blood of HIV-infected patients with *Pneumocystis carinii* pneumonias. *Respiration*, **60**, 170–7.
128 De, L. G., *et al.* (1993). T cell receptor heterogeneity in gamma delta T cell clones from intestinal biopsies of patients with celiac disease. *Eur. J. Immunol.*, **23**, 499–504.
129 Diu, A., *et al.* (1993). Limited T-cell receptor diversity in liver-infiltrating lymphocytes from patients with primary biliary cirrhosis. *J. Autoimmun.*, **6**, 611–19.
130 Ferradini, L., *et al.* (1993). Analysis of T cell receptor variability in tumor-infiltrating lymphocytes from a human regressive melanoma. Evidence for *in situ* T cell clonal expansion. *J. Clin. Invest.*, **91**, 1183–90.
131 Forman, J. D., *et al.* (1993). T cell receptor variable beta-gene expression in the normal lung and in active pulmonary sarcoidosis. *Chest*, **103**, 78.
132 Mantegazza, R., *et al.* (1993). Analysis of T cell receptor repertoire of muscle-infiltrating T lymphocytes in polymyositis. Restricted V alpha/beta rearrangements may indicate antigen-driven selection. *J. Clin. Invest.*, **91**, 2880–6.
133 Kay, R. A., *et al.* (1991). An abnormal T cell repertoire in hypergammaglobulinaemic primary Sjo rgren's syndrome. *Clin. Exp. Immunol.*, **85**, 262–4.

7 The role of peptides in positive and negative thymocyte selection

ERIC SEBZDA AND PAMELA S. OHASHI

1 Introduction

A central component of the adaptive immune system is the ability of T cells to respond to foreign antigen while remaining tolerant to self-molecules. T lymphocytes use T cell receptors (TCRs) to recognize peptide antigen presented in the groove of self-major histocompatibility complex (MHC) molecules. An important aspect of this interaction is the TCR specificity—the TCR must be able to bind to the antigen–MHC complex with a relatively high affinity to be effective, but at the same time this TCR can not be allowed to respond to self-peptide–MHC. T cells that react against self-molecules are removed during thymocyte development in a process referred to as negative selection. Additionally, phenomena such as MHC restriction has led to the finding that a second selection process—positive selection—occurs intrathymically. Positive selection skews the TCR response towards antigen recognition in the context of self-MHC. Together, these two forms of selection produce a T cell repertoire capable of recognizing foreign antigen in the context of self-MHC while functionally eliminating the majority of self-reactive T cells.

T cell tolerance is induced through deletion or inactivation of thymocytes that respond against self-determinants (Ramsdell and Fowlkes 1990; Schwartz 1990; Fowlkes and Ramsdell 1993). Clonal anergy functionally inactivates thymocytes whereas clonal deletion physically eliminates immature self-reactive lymphocytes. Support for the concept of clonal deletion came from the finding that mice expressing the MHC class II molecule, H-2E, lacked T cells utilizing the variable element, Vβ17 (Kappler *et al.* 1987*a,b*). In contrast, $V\beta17^+$ T cells were present in H-2E negative mice. Introduction of this MHC molecule into mice normally expressing $V\beta17^+$ T cells resulted in the elimination of these specific lymphocytes. The advent of TCR transgenic mice further demonstrated the process of clonal deletion. TCR transgenic T cells specific for a male antigen were present in female mice but absent in male siblings (von Boehmer 1990). Clonal deletion of TCR transgenic thymocytes in males occurred at an early $CD4^+CD8^+$ double-positive stage of development, prior to high surface expression of the T cell receptor. Other models that examined clonal deletion of superantigen-reactive T cells demonstrated that clonal deletion also occurred at a more mature $CD4^+$ or $CD8^+$ single-positive stage of development (Kappler *et al.* 1988; MacDonald *et al.* 1988; Pullen *et al.* 1988; Abe *et al.* 1988; White *et al.* 1989; Pircher *et al.* 1989; Ohashi *et al.* 1990). Together, these experiments suggest that negative selection is a continuous process that can occur at any thymocyte stage following adequate surface expression of the TCR.

Like negative selection, the process of positive selection was more clearly examined following the generation of TCR transgenic mice (von Boehmer 1994). Using these mice in positively selecting and non-selecting MHC backgrounds, results showed that positive selection required the direct interaction of T cell receptors with particular self-MHC molecules (Teh *et al.* 1988; Kisielow *et al.* 1988; Sha *et al.* 1988; Scott *et al.* 1989; Kaye *et al.* 1989; Berg *et al.* 1989). Positive selection of the thymocytes occurred at the double-positive stage, rescuing these lymphocytes from programmed cell death while they were still functionally and phenotypically immature. This process also terminated RAG-1 and -2 (recombinase-activating genes) expression, thus preventing further rearrangement of the TCR (Turka *et al.* 1991; Brandle *et al.* 1992; Borgulya *et al.* 1992). In this manner, T cells are produced that efficiently recognize foreign antigen presented by self-MHC molecules.

The TCR is directly involved in both forms of selection. How the same TCR can induce either positive or negative selection introduces a conceptual paradox. Several models have tried to address this problem. Most of these hypotheses have incorporated aspects from two opposing models—an altered ligand model and an affinity/avidity model.

2 Altered ligand model

Reports that thymic epithelium induces positive selection while bone marrow-derived cells stimulate negative selection inspired the altered ligand model (Zinkernagel *et al.* 1978; Fink and Bevan 1978; Lo and Sprent 1986; Marrack and Kappler 1987; Marrack *et al.* 1988) (Figure 7.1). This hypothesis suggests that these two populations have a different array of self-peptides bound to their MHC molecules. Deletion of potentially autoreactive lymphocytes was predicted to occur on bone marrow-derived cells expressing a given set of peptide–MHC molecules. This spectrum of peptides may overlap with proteins found in the periphery. In contrast, peptides found on positively selecting epithelial cells are confined to the thymus. As long as mature T lymphocytes do not return to the thymic cortex, there is no risk of self-reactive responses. By assuming that positive and negative selection is induced by mutually exclusive cell populations at defined T cell developmental stages, the altered ligand model explains how the same TCR can be involved in both types of selection.

Several pieces of data argue against an altered peptide model. For example this model suggests that positive selection occurs before negative selection. However, experiments using TCR transgenic mice have shown that thymocytes can be clonally deleted at an early double-positive stage, implying that

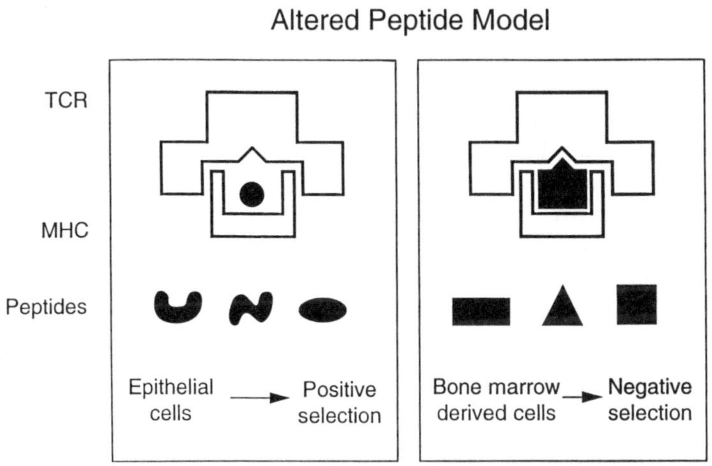

Fig. 7.2 Affinity/avidity model of thymocyte development. Positive and negative selection are governed by avidity thresholds. Thymocytes that can not interact with self-peptide–MHC due to low affinity TCRs are subject to programmed cell death. Thymocytes expressing intermediate affinity receptors for self-peptide–MHC surpass a positive selection threshold and are given a signal to continue T cell maturation. High affinity TCRs are removed when the avidity level for self-peptide–MHC surpasses the negative selection threshold.

volves amino acid substitutions and alterations in the selecting MHC molecule. If the affinity/avidity model is correct, then one would expect minor alterations in TCR–MHC affinity to affect the outcome of thymocyte selection. Again using the 2C TCR transgenic mouse, experiments have demonstrated that changes in the restriction element, H-2K^b, can affect positive and negative selection (Sha *et al.* 1990). More specifically, mutations in MHC residues that directly contact the TCR decreased the amount of positive selection normally observed, possibly by lowering the TCR–MHC affinity. Other H-2K^b mutations induced negative selection. This may be the result of high affinity TCR–MHC interactions. Further experiments using a defined transgenic TCR model also demonstrated that minor alterations in the peptide binding groove led to more efficient positive selection when compared to the natural positively selecting element (Ohashi *et al.* 1993). These data suggest that there is a range of affinities/avidities that could lead to a positive signal for thymocyte development.

Additional experiments have shown that selection can be modified depending upon the MHC concentration. TCR transgenic mice specific for pigeon or moth cytochrome *c* peptide bound to the MHC class II molecule, H-2E^k, were bred into heterozygous MHC backgrounds (Berg *et al.* 1990). Mice heterozygous for the α chain of the H-2E molecule had a reduced ability to positively select transgenic thymocytes, suggesting a direct correlation

A basic postulate of the altered ligand hypothesis is that cortical epithelial cells are the only cells capable of mediating positive selection. However, recent evidence shows that this is not the case. Reconstitution experiments with fetal liver cells suggests that haematopoietic cells can influence positive selection (Bix and Raulet 1992). Likewise, transfected fibroblasts were capable of mediating positive selection (Palowski *et al.* 1993; Hugo *et al.* 1993). Thus, positive selection is not limited to peptide–MHC expressed by thymic epithelial cells.

In summary, although the altered peptide model explains the basic paradox of a single TCR inducing positive or negative selection, very little evidence supports this hypothesis. Indeed, there is a growing body of data refuting basic principles inherent in this model.

3 Affinity/avidity model

The affinity/avidity model argues that positive selection is the result of low–intermediate avidity binding of thymocytes to peptide–MHC whereas high avidity interactions elicit negative selection (Sprent *et al.* 1988). In this sense, there is no distinction between interactions among cortical thymic epithelium and bone marrow-derived cells. During thymocyte development, a small portion of thymocytes expressing a low density of TCRs are positively selected when they bind to self-peptide–MHC on cortical epithelium. These thymocytes have an intermediate avidity for MHC. High avidity thymocytes are deleted at this stage. Thymocytes that are unable to interact with self-peptide–MHC are not selected, and unless further TCR α rearrangement occurs, the cells undergo programmed cell death (Figure 7.2).

TCR density increases as thymocytes mature. In addition, bone marrow-derived cells found primarily in the medulla (e.g. macrophages and dendritic cells), have higher constitutive expression of MHC molecules than thymic epithelium. Therefore, thymocytes that have undergone positive selection due to an intermediate affinity for MHC may still be negatively selected at a later stage due to a high avidity for self-peptide–MHC. Ultimately, T cells with intermediate avidity for self-peptide–MHC are allowed to exit the thymus. T cells with high avidity for self are tolerized before they enter the periphery.

A variety of experiments support the affinity/avidity hypothesis. For example, variations in co-receptor levels can alter the outcome of thymocyte selection (Lee *et al.* 1992; Robey *et al.* 1992). Transgenic thymocytes expressing the 2C TCR were positively selected in H-2^b mice. However, if a CD8 transgene was co-expressed in these mice, the increased surface expression of CD8 resulted in the deletion of thymocytes expressing the 2C transgenic receptor. It is likely that the CD8 transgene increased thymocyte avidity to the point where the negative selection threshold was surpassed.

Another set of experiments which favours the affinity/avidity model in-

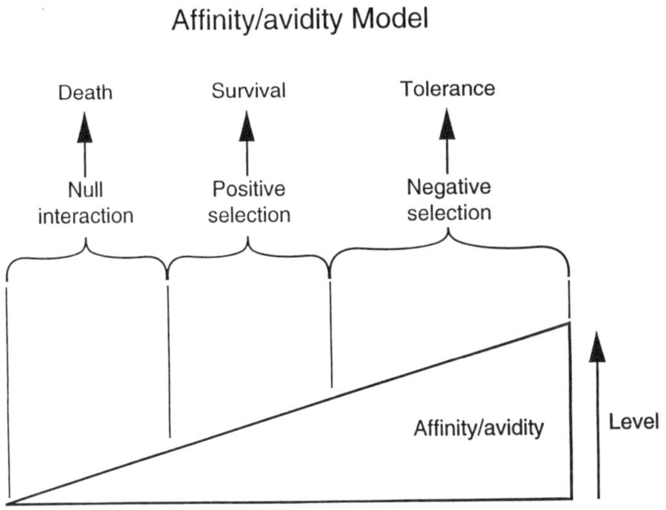

Fig. 7.2 Affinity/avidity model of thymocyte development. Positive and negative selection are governed by avidity thresholds. Thymocytes that can not interact with self-peptide–MHC due to low affinity TCRs are subject to programmed cell death. Thymocytes expressing intermediate affinity receptors for self-peptide–MHC surpass a positive selection threshold and are given a signal to continue T cell maturation. High affinity TCRs are removed when the avidity level for self-peptide–MHC surpasses the negative selection threshold.

volves amino acid substitutions and alterations in the selecting MHC molecule. If the affinity/avidity model is correct, then one would expect minor alterations in TCR–MHC affinity to affect the outcome of thymocyte selection. Again using the 2C TCR transgenic mouse, experiments have demonstrated that changes in the restriction element, H-2K^b, can affect positive and negative selection (Sha *et al.* 1990). More specifically, mutations in MHC residues that directly contact the TCR decreased the amount of positive selection normally observed, possibly by lowering the TCR–MHC affinity. Other H-2K^b mutations induced negative selection. This may be the result of high affinity TCR–MHC interactions. Further experiments using a defined transgenic TCR model also demonstrated that minor alterations in the peptide binding groove led to more efficient positive selection when compared to the natural positively selecting element (Ohashi *et al.* 1993). These data suggest that there is a range of affinities/avidities that could lead to a positive signal for thymocyte development.

Additional experiments have shown that selection can be modified depending upon the MHC concentration. TCR transgenic mice specific for pigeon or moth cytochrome *c* peptide bound to the MHC class II molecule, H-2E^k, were bred into heterozygous MHC backgrounds (Berg *et al.* 1990). Mice heterozygous for the α chain of the H-2E molecule had a reduced ability to positively select transgenic thymocytes, suggesting a direct correlation between H-2E^{k} surface density and positive selection. This experiment showed that positive selection is limited by the avidity of thymocytes for MHC molecules expressed in the thymus. Auphan *et al.* (1992) provided additional evidence demonstrating a direct relationship between negative selection and MHC class I surface density. Together, these experiments show that variations in MHC expression, either qualitative or quantitative, have a direct impact on both positive and negative thymocyte selection.

A potential problem with the affinity/avidity model is the fact that immature thymocytes undergoing selection have only about one-tenth the number of TCR complexes on their cell surfaces as mature T cells. Thymocytes that avoid negative selection in the thymus due to a relatively low affinity interaction with peptide–MHC may potentially have an increased avidity for this peptide in the periphery. However, studies have shown that receptor interactions which were sufficient for clonal deletion in the thymus were unable to induce effector T cell function in the periphery (Pircher *et al.* 1991). This experiment demonstrated that mature T cell activation requires a higher avidity interaction than that required for negative selection. Therefore, the affinity/avidity hypothesis remains a credible model for thymocyte selection.

Evidence in favour of the affinity/avidity model comes from data that suggest that negative selection may occur before positive selection, and that thymic epithelial cells may also induce negative selection. Alterations in co-receptors or antigen–MHC presentation which affect positive and negative selection also support the affinity avidity model of thymocyte selection. Recent experiments using novel gene targeted mice have examined the contribution of peptides to the affinity/avidity between the TCR and MHC during T cell development.

4 Peptides are involved in thymocyte selection

Evidence that peptides are involved in positive selection was obtained from studies using H-2K^{b} class I MHC mutants as antigen. T helper (Th) cells specific for foreign K^{bm6} + self-MHC class II were needed to generate a H-2K^{bm6} allospecific cytolytic T lymphocyte response. Singer *et al.* (1986) demonstrated that Th function was present only in mice expressing both the MHC class II restricting element and the H-2K^{bm6} molecule. This suggested that positive selection of specific Th cells required the presence of a given self-peptide in the context of self-MHC. Other experiments using defined TCR transgenic models combined with MHC mutants containing alterations in the floor of the peptide binding groove also suggested that specific peptides were important in mediating positive selection (Sha *et al.* 1990; Jacobs *et al.* 1990; Ohashi *et al.* 1993).

Further support for peptide involvement in thymocyte selection came from studies in which (H-2^b × H-$2K^b$-mutant)F1 bone marrow cells were injected into lethally irradiated H-2^b or H-$2K^b$-mutant mice (Nikolic Zugic and Bevan 1990). The study focused on the cytotoxic T cell response to the ovalbumin peptide (aa 253–276) since this response is restricted to H-$2K^b$. Their results demonstrated a direct correlation between H-$2K^b$-mutants that could positively select ovalbumin-specific T cells and H-$2K^b$-mutants that could respond to ovalbumin. The data from this study suggested that the self-peptides involved in positive selection were similar or related to the ovalbumin peptide antigen.

Unfortunately, these results do not directly examine the influence of defined self-peptides in positive selection. Experiments have not addressed whether peptide–TCR contact is essential for positive selection. It is possible that peptides merely have an inhibitory effect and prevent thymocyte selection by disturbing necessary TCR–MHC interactions.

Two experimental approaches using gene targeted mice demonstrated the importance of peptides in positive selection. Both groups exploited inherent properties of MHC class I antigen presentation. An MHC class I molecule is a trimeric complex of peptide, heavy chain, and β2-microglobulin (β2M). In the absence of β2M, the MHC class I heavy chain is unstable and only present in an altered conformation on the cell surface (Ljunggren *et al.* 1990; Vitiello *et al.* 1990; Ortiz-Navarrete and Hämmerling 1991). Using β2M deficient mice ($\beta2M^{-/-}$), Hogquist *et al.* (1993) tested the ability of peptides to induce positive selection in fetal thymic organ cultures (Figure 7.3). Previous work had shown that unstable heavy chains were rescued with the addition of exogenous β2M. In addition, if an MHC-binding peptide was supplemented with the β2M, stable MHC class I molecules presenting this peptide were maintained on the cell surface. Ashton-Rickardt *et al.* (1993) used a TAP1 deficient mouse strain to address this same question. The TAP (transporter associated with antigen processing) gene products are involved in peptide transport from the cytoplasm to the endoplasmic reticulum. Mice lacking this gene can not load endogenous peptide into MHC class I molecules. However, class I surface expression can be rescued with the addition of exogenous peptide. Thus both groups could directly test the role of peptides in positive selection. Hogquist *et al.* were able to generate $CD8^+$, TCR^{hi} thymocytes *in vitro*. The population size was dependent upon the complexity of the peptide preparation used. An acid extract of splenocytes or a bacterially synthesized mixture of random peptides was much more effective at stimulating positive selection than a combination of two synthetic peptides. Similar results were obtained using the TAP1 deficient mice. In both studies, peptides that stabilized MHC class I surface expression did not necessarily induce positive selection. Instead, the level of positive selection correlated with the increased peptide complexity suggesting that particular TCR–MHC interactions must occur. Therefore, self-peptides have a specific and essential role in thymocyte selection.

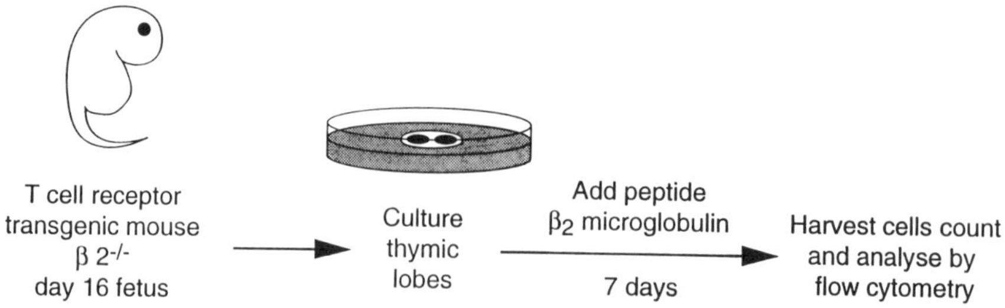

Fig. 7.3 An outline of a fetal thymic organ culture. Thymic lobes are removed from a non-transgenic or TCR transgenic, β2m deficient fetus at day 16 of gestation. The lobes are placed on a filter which floats on media supplemented with β2m and specific peptides. After seven to ten days in culture, during which time the fetal lobes are exposed to media and a humidified environment, the lobes are harvested, counted, and analysed by flow cytometry.

5 Positive and negative thymocyte selection may be induced by defined peptides

Further studies have examined whether a particular peptide could induce the selection of thymocytes expressing a specific T cell receptor. TCR transgenic T cells specific for ovalbumin (aa 257–264) in the context of the MHC class I molecule, H-2K^b, were bred into a H-2^b β2M$^{-/-}$ background (Hogquist *et al.* 1994; Jameson *et al.* 1994). In a positively selecting H-2^b background, thymocytes differentiated towards the CD8$^+$ lineage and expressed high levels of the transgenic TCR. However, in a β2M$^{-/-}$ background which lacked surface expression of normal MHC class I molecules, efficient maturation of the TCR transgenic thymocytes did not occur. A peptide variant which stabilized the MHC class I molecule without interfering with TCR–MHC interactions did not induce positive or negative selection when cultured with exogenous β2M, proving that thymocyte selection requires TCR interaction with specific self-peptides. When high concentrations (2×10^{-5} M) of wild-type peptide were cultured with TCR transgenic β2M$^{-/-}$ fetal lobes in the presence of exogenous β2M, negative selection occurred at the double-positive thymocyte stage. Further analysis was done examining the effects of peptide agonists and antagonists on selection. Agonists are variants of the wild-type peptide antigen that are capable of stimulating a given T cell response. Antagonist peptides contain subtle amino acid variations from the wild-type peptide and functionally inhibit responses to agonists in a specific manner. If similar concentrations of antagonists were substituted for wild-type peptide, positive selection was induced. Peptides that have partial agonist/antagonist characteristics were capable of mediating both forms of selection. Positively selected T cells proliferated in response to wild-type peptide, demonstrating that these lymphocytes were functionally active. Since the positively selecting peptides had antagonist features, their data suggested that positively selecting ligands engage TCRs without leading to a typical mature T cell activation signal.

A similar experiment involved the introduction of transgenic TCRs specific for lymphocytic choriomeningitis virus (LCMV) into a H-2^b $\beta2M^{-/-}$ background (Sebzda *et al.* 1994). These transgenic T cells recognize LCMV glycoprotein 1 (aa 33–41) in the context of the MHC class I molecule, H-$2D^b$. Like the previous mouse model, thymocytes expressed high levels of the transgenic receptor and skewed to the $CD8^+$ lineage in the presence of the restricting element, H-$2D^b$. Lack of β2M caused unstable surface expression of MHC class I molecules resulting in an absence of these lymphocytes. Addition of an irrelevant control peptide that efficiently bound to H-$2D^b$ did not lead to positive selection of the transgenic thymocytes (Sebzda *et al.* unpublished). Culturing TCR $\beta2M^{-/-}$ fetal thymic lobes with low concentrations (10^{-12} M) of the agonist peptide in the presence of exogenous β2M generated a population of $CD8^+$ TCR^{hi} thymocytes. When higher concentrations (10^{-6} M) of p33 were incubated along with β2M, a reduced number of $CD8^+$ thymocytes were observed. In addition, there was a dramatic decrease in $CD4^+CD8^+$ double-positive thymocytes expressing the transgenic TCR. Both of these results are consistent with the notion of negative selection. TCR $\beta2M^{+/-}$ thymic lobes positively selected thymocytes expressing the transgenic TCR. If 10^{-6} M p33 was added to these thymic lobes, there was an absence of TCR transgenic $CD4^+CD8^+$ double-positive and $CD8^+$ thymocytes, demonstrating that high concentrations of this peptide induced negative selection. Together, these results showed that the same peptide can induce both positive and negative selection. Similar results, albeit at different peptide concentrations, were found using LCMV-specific TCR transgenic fetal lobes that were TAP1 deficient (Ashton-Rickardt *et al.* 1994). Other studies have demonstrated that peptide antagonists either prevented the differentiation or induced negative selection of specific T cells (Spain *et al.* 1994; Page *et al.* 1994). These findings show that there is no correlation between antagonist potency and the degree of positive or negative selection. Instead, positive and negative selection may be the respective result of surpassing a low and a high avidity threshold.

The results of these models validate the affinity/avidity hypothesis of thymocyte selection. TCRs that cannot interact with self-peptide–MHC are not selected and die from programmed cell death. Thymocytes with intermediate avidity for self are positively selected. This scenario can be mimicked *in vitro* using high or low concentrations of variant or wild-type peptide respectively (Figure 7.4). A higher avidity interaction, mediated by relatively large amounts of a high affinity peptide, leads to negative selection. Several factors including the inherent TCR affinity for peptide–MHC and the number of TCR–peptide–MHC interactions modify the overall thymocyte avidity for self-peptide–MHC. Additional surface molecules, such as CD4 or CD8 co-receptors, will also alter the selection process.

The next step is to determine how thymocytes respond intracellularly to

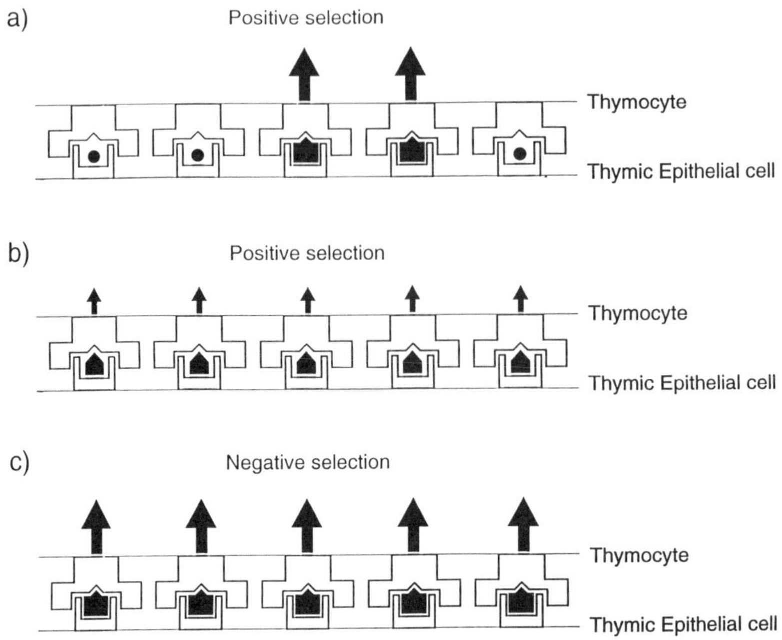

Fig. 7.4 Thymocyte selection is modified by different affinity/avidity interactions. (**a**) Positive selection can be induced with an intermediate avidity TCR–peptide–MHC interaction. Thymocytes that express high affinity TCRs are positively selected *in vitro* if a small number of these receptors interact with peptide–self-MHC. A low number of high affinity TCR interactions surpasses an intermediate avidity threshold, generating a positively selecting signal. (**b**) Alternatively, thymocytes may be positively selected if a large number of low affinity TCRs interact with peptide–MHC. The accumulative effect of many low affinity interactions exceeds the intermediate avidity threshold, and thus provides the necessary positive signal for thymocyte maturation. (**c**) Negative selection can be induced with a high avidity TCR–peptide–MHC interaction. If a substantial number of high affinity TCRs interact with peptide–MHC, negative selection occurs *in vitro*. The negative selection threshold is surpassed and a signal is provided to remove these potentially autoreactive thymocytes.

quantitative differences in TCR–MHC interactions. It is possible that different levels of TCR–antigen–MHC interaction induce the same type of intracellular signal, and that quantitative differences in this signal determine whether positive or negative selection occurs. For instance, a high avidity interaction may lead to extensive TCR–antigen–MHC cross-linking, that could concentrate several protein kinases which have been implicated in intracellular TCR signalling. In contrast, a lower avidity interaction would not activate as many kinases or continue activation for as long a duration of time. Alternatively, different avidity levels may produce qualitative signalling differences. In either case, intermediate avidity interactions would induce weak or incomplete/novel signals which would lead to positive selection.

Higher avidity interactions would result in stronger, extensive intracellular signals which would induce negative selection.

Another question that has been left unanswered involves signalling differences between thymocytes and mature T cells. Experiments have shown that T cell activation requires a higher avidity interaction than negative selection (Pircher *et al.* 1991; Vasquez *et al.* 1992). Differences in intracellular signalling may or may not reflect different pathways. It is possible that the same intracellular machinery that responds to quantitative differences in TCR–MHC interactions in thymocytes is present and functioning in mature T cells. Alternatively, thymocytes may utilize intracellular signalling pathways that vary from mature T lymphocytes. Further experiments are required to resolve this problem.

6 Conclusions

A variety of experiments have demonstrated that peptides are directly recognized by TCRs during thymocyte selection. This selection process appears to be regulated in accordance to an affinity/avidity model. Thymocytes that are unable to surpass a positive selection threshold undergo programmed cell death. In contrast, thymocytes that surpass a negative selection threshold are tolerized. Only thymocytes expressing TCRs with an avidity level between these two extremes are allowed to mature and function outside the thymus. Peptides contribute to the avidity between the thymocyte and stromal cell populations. Future models of thymocyte selection should take into account the relationship between a relatively limited peptide spectrum and a diverse TCR repertoire.

References

Abe, R., Vacchio, M. S., Fox, B., and Hodes, R. J. (1988). Preferential expression of the T-cell receptor V beta 3 gene by Mlsc reactive T cells. *Nature*, **335**, 827–30.

Ashton-Rickardt, P. G., Van Kaer, L., Schumacher, T. N. M., Ploegh, H. L., and Tonegawa, S. (1993). Peptide contributes to the specificity of positive selection of CD8+ T cells in the thymus. *Cell*, **73**, 1041–9.

Ashton-Rickardt, P. G., Bandeira, A., Delaney, J. R., Van Kaer, L., Pircher, H. P., Zinkernagel, R. M., *et al.* (1994). Evidence for a differential avidity model of T cell selection in the thymus. *Cell*, **76**, 651–63.

Auphan, N., Schonrich, G., Malissen, M., Barad, M., Hammerling, G., Malissen, B., *et al.* (1992). Influence of antigen density on degree of clonal deletion in T cell receptor transgenic mice. *Int. Immunol.*, **4**, 541–7.

Berg, L. J., Pullen, A. M., Fazekas de, S. G. B., Mathis, D., Benoist, C., and Davis, M. M. (1989). Antigen/MHC-specific T cells are preferentially exported from the thymus in the presence of their MHC ligand. *Cell*, **58**, 1035–46.

Berg, L. J., Frank, G. D., and Davis, M. M. (1990). The effects of MHC gene dosage and allelic variation on T cell receptor selection. *Cell*, **60**, 1043–53.

Bix, M. and Raulet, D. (1992). Inefficient positive selection of T cells directed by haematopoietic cells. *Nature*, **359**, 330–3.

Bonomo, A. and Matzinger, P. (1993). Thymus epithelium induces tissue-specific tolerance. *J. Exp. Med.*, **177**, 1153–64.

Borgulya, P., Kishi, H., Muller, U., Kirberg, J., and von Boehmer, H. (1991). Development of the CD4 and CD8 lineage of T cells: instruction versus selection. *EMBO J.*, **10**, 913–18.

Borgulya, P., Kishi, H., Uematsu, Y., and von Boehmer, H. (1992). Exclusion and inclusion of α and β T cell receptor alleles. *Cell*, **69**, 529–37.

Brandle, D., Muller, C., Rulicke, T., Hengartner, H., and Pircher, H. (1992). Engagement of the T-cell receptor during positive selection in the thymus down-regulates RAG-1 expression. *Proc. Natl. Acad. Sci. USA*, **89**, 9529–33.

Fink, P. J. and Bevan, M. J. (1978). H-2 antigens of the thymus determine lymphocyte specificity. *J. Exp. Med.*, **148**, 766–75.

Fowlkes, B. J. and Ramsdell, F. (1993). T-cell tolerance. *Curr. Opin. Immunol.*, **5**, 873–9.

Guidos, C. J., Danska, J. S., Fathman, C. G., and Weissman, I. L. (1990). T cell receptor-mediated negative selection of autoreactive T lymphocyte precursors occurs after commitment to the CD4 or CD8 lineages. *J. Exp. Med.*, **172**, 835–45.

Hogquist, K. A., Gavin, M. A., and Bevan, M. J. (1993). Positive selection of CD8+ T cells induced by major histocompatibility complex binding peptides in fetal thymic organ culture. *J. Exp. Med.*, **177**, 1469–73.

Hogquist, K. A., Jameson, S. C., Heath, W. R., Howard, J. L., Bevan, M. J., and Carbone, F. R. (1994). T cell receptor antagonist peptides induce positive selection. *Cell*, **76**, 17–27.

Hugo, P., Kappler, J. W., McCormack, J., and Marrack, P. (1993). Fibroblasts can mediate thymocyte positive selection *in vivo*. *Proc. Natl. Acad. Sci. USA*, **70**, 10335–9.

Hugo, P., Kappler, J. W., Godfrey, D. I., and Marrack, P. C. (1994). Thymic epithelial cell lines that mediate positive selection can also induce thymocyte clonal deletion. *J. Immunol.*, **152**, 1022–34.

Jacobs, H., von Boehmer, H., Melief, C. J. M., and Berns, A. (1990). Mutations in the major histocompatibility complex class I antigen presenting groove affect both negative and positive selection of T cells. *Eur. J. Immunol.*, **20**, 2333–7.

Jameson, S. C., Hogquist, K. A., and Bevan, M. J. (1994). Specificity and flexibility in thymic selection. *Nature*, **369**, 750–2.

Kappler, J. W., Roehm, N., and Marrack, P. (1987*a*). T cell tolerance by clonal elimination in the thymus. *Cell*, **49**, 273–80.

Kappler, J. W., Wade, T., White, J., Kushnir, E., Blackman, M., Bill, J., *et al.* (1987*b*). A T cell receptor Vβ segment that imparts reactivity to a class II major histocompatibility complex product. *Cell*, **49**, 263–71.

Kappler, J. W., Staerz, U. D., White, J., and Marrack, P. (1988). Self tolerance eliminates T cells specific for Mls-modified products of the major histocompatibility complex. *Nature*, **332**, 35–40.

Kaye, J., Hsu, M. L., Sauron, M. E., Jameson, S. C., Gascoigne, N. R., and Hedrick, S. M. (1989). Selective development of CD4+ T cells in transgenic mice expressing a class II MHC-restricted antigen receptor. *Nature*, **341**, 746–9.

Kisielow, P., Teh, H. S., Blüthmann, H., and von Boehmer, H. (1988). Positive selection of antigen-specific T cells in thymus by restricting MHC molecules. *Nature*, **335**, 730–3.

Lee, N. A., Loh, D. Y., and Lacy, E. (1992). CD8 surface levels alter the fate of α β T cell receptor-expressing thymocytes in transgenic mice. *J. Exp. Med.*, **175**, 1013–25.

Ljunggren, H.-G., Stam, N. J., Oehlen, C., Neefjes, J. J., Höglund, P., Heemels, M., *et al.* (1990). Empty MHC class I molecules come out in the cold. *Nature*, **346**, 476–80.

Lo, D. and Sprent, J. (1986). Identity of cells that imprint H-2-restricted T-cell specificity in the thymus. *Nature*, **319**, 672–5.

MacDonald, H. R., Schneider. R., Lees, R. K., Howe, R. C., Acha-Orbea, H., Festenstein, H., *et al.* (1988). T cell receptor Vβ use predicts reactivity and tolerance to Mls^{a}-encoded antigens. *Nature*, **332**, 40–5.

Marrack, P. and Kappler, J. (1987). The T cell receptor. *Science*, **238**, 1073–9.

Marrack, P., Lo, D., Brinster, R., Palmiter, R., Burkly, L., Flavell, R. H., *et al.* (1988). The effect of thymus environment on T cell development and tolerance. *Cell*, **53**, 627–34.

Marrack, P., Ignatowicz, L., Kappler, J. W., Boymel, J., and Freed, J. H. (1993). Comparison of peptides bound to spleen and thymus class II. *J. Exp. Med.*, **178**, 2173–83.

Nikolic Zugic, J. and Bevan, M. J. (1990). Role of self-peptides in positively selecting the T-cell repertoire. *Nature*, **344**, 65–7.

Ohashi, P. S., Pircher, H., Bürki, K., Zinkernagel, R. M., and Hengartner, H. (1990). Distinct sequence of negative or positive selection implied by thymocyte T cell receptor densities. *Nature*, **346**, 861–3.

Ohashi, P. S., Zinkernagel, R. M., Leuscher, I., Hengartner, H., and Pircher, H. (1993). Enhanced positive selection of a transgenic TCR by a restriction element that does not permit negative selection. *Int. Immunol.*, **5**, 131–8.

Ortiz-Navarrete, V. and Hämmerling, G. J. (1991). Surface appearance and instability of empty H-2 class I molecules under physiological conditions. *Proc. Natl. Acad. Sci. USA*, **88**, 3594–7.

Page, D. M., Alexander, J., Snoke, K., Appella, E., Sette, A., Hedrick, S. M., *et al.* (1994). Negative selection of CD4^{+}CD8^{+} thymocytes by T-cell receptor peptide antagonists. *Proc. Natl. Acad. Sci. USA*, **91**, 4057–61.

Palowski, T., Elliot, J. D., Loh, D. Y., and Staerz, U. D. (1993). Positive selection of T lymphoyctes on fibroblasts. *Nature*, **364**, 642–5.

Pircher, H., Bürki, K., Lang, R., Hengartner, H., and Zinkernagel, R. (1989). Tolerance induction in double specific T-cell receptor transgenic mice varies with antigen. *Nature*, **342**, 559–61.

Pircher, H., Hoffmann Rohrer, U., Moskophidis, D., Zinkernagel, R. M., and Hengartner, H. (1991). Lower receptor avidity required for thymic clonal deletion than for effector T cell function. *Nature*, **351**, 482–5.

Pullen, A. M., Marrack, P., and Kappler, J. W. (1988). The T cell repertoire is heavily influenced by tolerance to polymorphic self-antigens. *Nature*, **335**, 796–801.

Ramsdell, F. and Fowlkes, B. J. (1990). Clonal deletion versus clonal anergy: the role of the thymus in inducing self tolerance. *Science*, **248**, 1342–8.

Robey, E. A., Ramsdell, F., and Kioussis, D. (1992). The level of CD8 expression can determine the outcome of thymic selection. *Cell*, **69**, 1089–96.

Salaun, J., Bandeira, A., Khazaal, I., Calman, F., Coltey, M., Coutinho, A., *et al.* (1990). Thymic epithelum tolerizes for histocompatibility antigens. *Science*, **247**, 1471–4.

Schwartz, R. H. (1990). A cell culture model for T lymphocyte clonal anergy. *Science*, **248**, 1349–56.

Scott, B., Blüthmann, H., Teh, H.-S., and von Boehmer, H. (1989). The generation of mature T cells requires interaction of the αβ T cell receptor with major histocompatibility antigens. *Nature*, **338**, 591–3.

Sebzda, E., Wallace, V. A., Mayer, J., Yeung, R. S. M., Mak, T. W., and Ohashi, P. S. (1994). Positive and negative thymocyte selection induced by different concentrations of a single peptide. *Science*, **263**, 1615–18.

Sha, W. C., Nelson, C. A., Newberry, R. D., Kranz, D. M., Russell, J. H., and Loh, D. Y. (1988). Positive and negative selection of an antigen receptor on T cells in transgenic mice. *Nature*, **336**, 73–6.

Sha, W. C., Nelson, C. A., Newberry, R. D., Pullen, J. K., Pease, L. R., Russell, J. H., *et al.* (1990). Positive selection of transgenic receptor-bearing thymocytes by Kb antigen is altered by Kb mutations that involve peptide binding. *Proc. Natl. Acad. Sci. USA*, **87**, 6186–90.

Shortman, K., Vremec, D., and Egerton, M. (1991). The kinetics of T cell antigen receptor expression by subgroups of CD4+8+ thymocytes: Delineation of CD4+8+3^{2+} thymocytes as post-selection intermediates leading to mature T cells. *J. Exp. Med.*, **173**, 323–32.

Singer, A., Mizuochi, T., Munitz, T. I., and Gress, R. E. (1986). Role of self antigens in the selection of the developing T cell repertoire. In *Progress in immunology VI* (ed. B. Cinader and R. G. Miller), pp. 60–6. Academic Press Inc., Orlando, Florida.

Spain, L. M. and Berg, L. J. (1992). Developmental regulation of thymocyte susceptibility to deletion by 'self'-peptide. *J. Exp. Med.*, **176**, 213–23.

Spain, L. M., Jorgensen, J. L., Davis, M. M., and Berg, L. J. (1994). A peptide antigen antagonist prevents the differentiation of T cell receptor transgenic thymocytes. *J. Immunol.*, **152**, 1709–17.

Speiser, D. E., Pircher, H., Ohashi, P. S., Kyburz, D., Hengartner, H., and Zinkernagel, R. M. (1992). Clonal deletion induced by either thymic epithelium or lymphohemopoietic cells at different stages of class I restricted T cell ontogeny. *J. Exp. Med.*, **175**, 1277–83.

Sprent, J., Lo, D., Gao, K. E., and Ron, Y. (1988). T cell selection in the thymus. *Immunol. Rev.*, **101**, 173–90.

Teh, H. S., Kisielow, P., Scott, B., Kishi, H., Uematsu, Y., Blüthmann, H., *et al.* (1988). Thymic major histocompatibility complex antigens and the alpha β T cell receptor determine the CD4/CD8 phenotype of T cells. *Nature*, **335**, 229–33.

Teh, H. S., Kishi, H., Scott, B., Borgulya, P., von Boehmer, H., and Kisielow, P. (1990). Early deletion and late positive selection of T cells expressing a male-specific receptor in T-cell receptor transgenic mice. *Dev. Immunol.*, **1**, 1–10.

Turka, L. A., Schatz, D. G., Oettinger, M. A., Chun, J. J. M., Gorka, C., Lee, K., *et al.* (1991). Thymocyte expression of RAG-1 and RAG-2: termination by T cell receptor cross-linking. *Science*, **253**, 778–81.

Vasquez, N., Kaye, J., and Hedrick, S. M. (1992). *In vivo* and *in vitro* clonal deletion of double-positive thymocytes. *J. Exp. Med.*, **175**, 1307–16.

Vitiello, A., Potter, T. A., and Sherman, L. A. (1990). The role of β2-microglobulin in peptide binding by class I molecules. *Science*, **250**, 1423–6.

von Boehmer, H. (1990). Developmental biology of T cells in T cell-receptor transgenic mice. *Annu. Rev. Immunol.*, **8**, 531–56.

von Boehmer, H. (1994). Positive selection of lymphocytes. *Cell*, **76**, 219–28.

Vukmanovic, S., Jameson, S. C., and Bevan, M. J. (1994). A thymic epithelial cell line induces both positive and negative selection in the thymus. *Int. Immunol.*, **6**, 239–46.

Webb, S. R. and Sprent, J. (1990). Tolerogenicity of thymic epithelium. *Eur. J. Immunol.*, **20**, 2525–8.

White, J., Herman, A., Pullen, A. M., Kubo, R., Kappler, J. W., and Marrack, P. (1989). The Vβ-specific superantigen staphylococcal enterotoxin B: stimulation of mature T cells and clonal deletion in neonatal mice. *Cell*, **56**, 27–35.

Zinkernagel, R. M., Callahan, G. N., Althage, A., Cooper, S., Klein, P. A., and Klein, J. (1978). On the thymus in the differentiation of 'H-2 self-recognition' by T cells: Evidence for dual recognition? *J. Exp. Med.*, **147**, 882–96.

Part II

T cell function

8 T cell activation

DOREEN A. CANTRELL, MANOLO IZQUIERDO PASTOR, KARIN REIF, AND MELISSA WOODROW

1 Introduction

The phrase 'T cell activation' usually refers to the immune activation of mature T lymphocytes in peripheral blood, lymphatics, or tissue. During immune activation T cells undergo a sequence of genetic and phenotypic changes which results ultimately in T cell clonal expansion and the induction of T cell effector function (1). The physiological stimulus that activates T cells is foreign antigen in association with major histocompatibility molecules, presented to the T cell by another cell. Multiple receptors on the surface of the T cell mediate the interaction between the T cell and the antigen-presenting cell. The antigen specificity of the response is dictated by the T cell antigen receptor (TCR) (see Figure 8.1 for structure) (2). However, other receptors play a crucial accessory role and generate amplifying signals that contribute to T cell activation. These include adhesion molecules such as CD2 and LFA1 and their counter ligands LFA3 and ICAM-1, the CD4 and CD8 molecules that contribute to MHC recognition, and molecules such as CD5, CD28, CD44 (3–8). The list of molecules with the potential to participate in the cell activation process is thus long and the relative contributions of the different molecules is not always clear. However, it is undoubtedly relevant that the counter ligands for the T cell accessory molecules may have a restricted tissue distribution that would influence their participation in an immune response. For example, CD5 and CD28 can synergize with the TCR to initiate T cell activation and their respective ligands CD72 and B7/BB1 are found on B lymphocytes. Thus an accessory role for CD5 and CD28 might be expected to occur during co-operative interactions between T cells and B cells.

The T cell activation process can be categorized simplistically as the modulation of cell surface phenotype and the regulation of the secretion of critical cytokines (9). The changes that occur on the T cell surface during an immune response are multiple but it is particularly well characterized that activated T cells synthesize and express new receptor molecules of which the best studied examples are CD69, transferrin receptors, and receptors for cytokines such as the T cell growth factor interleukin-2 (IL-2R). The lymphokines produced by activated T cells include a number of important haematopoetic growth factors including IL-3, GM-CSF, IL-4, and IL-2. T cell growth and clonal expansion is a tightly regulated process requiring the growth factor IL-2, and hence the extent of the T cell proliferative response is determined

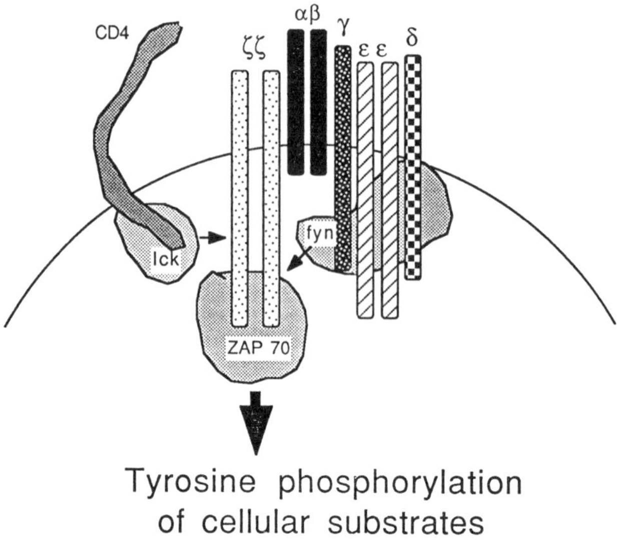

Fig. 8.1 T cell antigen receptor.

by the concentration of IL-2 available to the cell and the level of IL-2R expression on the cell (10, 11). Accordingly, a primary focus in T cell activation studies is the receptors and intracellular signalling mechanisms that control expression of the IL-2 gene and genes encoding the IL-2R. In particular, major breakthroughs in understanding signal transduction in T cells have been achieved because of co-ordinated efforts to link up immediate biochemical events triggered at the cell membrane with the transcription factors that control IL-2 gene expression (1, 9). IL-2 is encoded by a single gene whose transcriptional activity is rapidly increased in activated T cells (12, 13). Several studies have located a 267 bp transcriptional enhancer 5′ of the transcriptional initiation site of the IL-2 gene. Multiple ubiquitous transcription factors such as AP-1, NFkB, Oct-1 interact with this enhancer (14). There are also binding sites for a unique lymphocyte-specific transcriptional complex termed NFAT (nuclear factor of activated T cells) which appears to determine the T cell specificity and inducibility of IL-2 gene expression (15). The study of the signal transduction mechanisms that couple the TCR or accessory molecules to the transcription factors that interact with the IL-2 gene enhancer has become a paradigm of T cell activation.

2 TCR signal transduction

A summary of TCR signal transduction is depicted in Figure 8.2. The immediate consequence of triggering of the TCR is the activation of a protein-tyrosine kinase (PTK) signalling cascade (2). The functioning of this pathway is obligatory for T cell activation as judged by the inhibitory effects of tyrosine kinase inhibitors on TCR induction of IL-2 (16). Three tyrosine kinases have been implicated in T cell activation: $p59^{fyn}$ and ZAP-70 which associate with the cytoplasmic domain of the zeta chain of the TCR (Figure 8.1), and $p56^{lck}$ which because of its physical association with the CD4 and CD8 molecules would complex with the TCR during interaction between the TCR and antigen–MHC (Figure 8.1) (16). Genetic strategies that include the expression of mutated constitutively active or dominant negative kinases by transgenic or transfection protocols have established a vital role for p56 and $p59^{fyn}$ in T cell signalling (17, 18). Recently it has been described that mutations in the ZAP70 gene are associated with defects in TCR signal transduction and the development of severe combined immunodeficiency disease, indicating that ZAP70 also has important functions in T cell activation (19). A discussion of the mechanisms that allow the TCR to regulate the activity of cytosolic PTKs is beyond the scope of the present review. However, the initial step in the activation of *src* family kinases such as $p56^{lck}$ and $p59^{fyn}$ is dephosphorylation of negative regulatory sites of tyrosine phosphorylation (20). One important phosphatase in T cells is CD45, a transmembrane glycoprotein that carries a tyrosine phosphatase within its cytoplasmic domain. The expression of CD45 is essential for TCR stimulation of PTKs and hence T cell activation which suggests some functional coupling between the TCR and CD45 (21).

One substrate for TCR regulated PTK's is a phosphatidylinositol (PtdIns)-specific phospholipase C, PLCγ1 (16). The catalytic activity of PLCγ1 is increased by its tyrosine phosphorylation and thus a PTK coupling mechanism allows the TCR to induce PtdIns metabolism with the subsequent generation of inositol polyphosphates which regulate increases in intracellular calcium and diacylglycerol which stimulates protein kinase C (PKC) (22). A second PTK controlled signalling pathway to originate from the TCR involves the guanine nucleotide binding proteins $p21^{ras}$ (23, 24). The *ras* proteins bind GTP and catalyse its hydrolysis to GDP and a consensus is that the GTP bound form of the protein is the biologically active form (25, 26). In T cells stimulated via the TCR the *ras* proteins rapidly accumulate in the active GTP bound state. Activation of PKC with phorbol esters or diacylglycerols also induces the accumulation of *ras*–GTP complexes. However, PKC does not couple the TCR to $p21^{ras}$ instead, a PKC-independent pathway involving PTK's plays a critical role in TCR regulation of *ras* (27). The realization that and PKC did not couple the TCR to $p21^{ras}$ was significant because it indicated that at least two signalling pathways could originate from the TCR (Figure 8.2). The details of the PTK link between the TCR and $p21^{ras}$ are not completely resolved. The level of active $p21^{ras}$ GTP complexes is determined by a balance of the rate of hydrolysis of bound GTP and the rate of exchange of bound GDP for cytosolic GTP. The GTPase activity of $p21^{ras}$ is controlled by GTPase activating proteins of which two mammalian proteins are known: p120-GAP and neurofibromin (25). Proteins that regulate guanine nucleotide

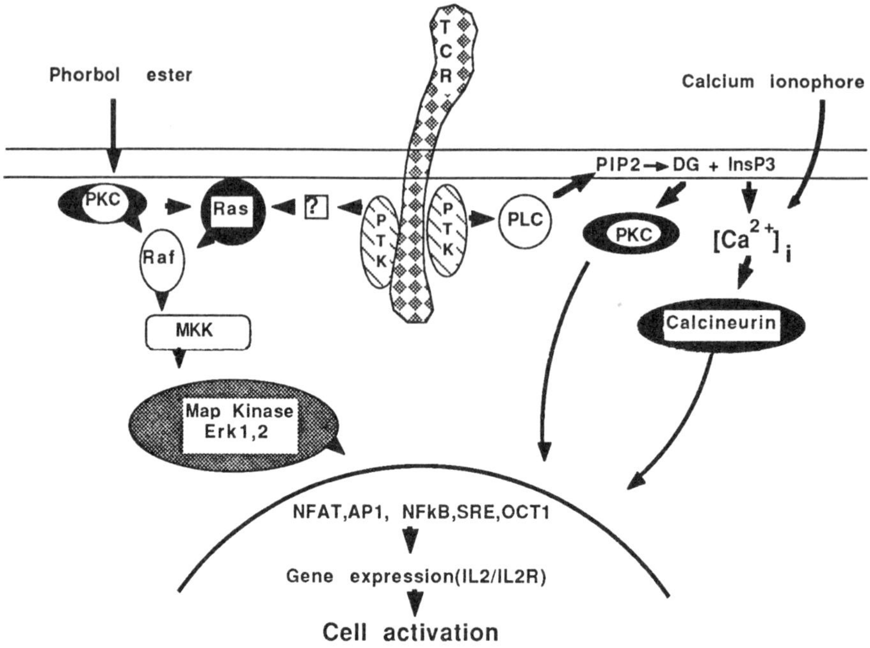

Fig. 8.2 TCR signal transduction.

exchange on *ras* have also been characterized and in mammalian cells they include the homologues of the yeast CDC25 protein and the Drosophila 'Son of sevenless' gene product Sos (28). The alternative mechanisms that could couple the TCR to *ras* are TCR stimulation of $p21^{ras}$ guanine nucleotide exchange or TCR inhibition of *ras*-GAP proteins. To explore the TCR/$p21^{ras}$ coupling pathway, extensive experiments that examined the kinetics of guanine nucleotide metabolism by $p21^{ras}$ were carried out. These studies indicated that in T cells, guanine nucleotide exchange on to $p21^{ras}$ is rapid and unchanged by cell activation. Instead the rate of GTP hydrolysis by *ras* decreases upon TCR triggering allowing *ras* to accumulate in the GTP bound state (29). One interpretation of these data is that TCR coupling to $p21^{ras}$ is mediated by *ras*-GAP proteins and not by guanine nucleotide exchange proteins. This hypothesis is supported by observations that TCR triggering inhibits the activity of *ras*-GAP proteins (29). It is not yet known which GAP protein, i.e. p120-GAP or neurofibromin, is the target for TCR regulation. Similarly, the mechanism that allows the TCR to inhibit *ras*-GAP proteins is unclear. The apparent uncoupling of *ras* guanine nucleotide exchange from receptor activation is a fundamental difference between T cells and other cell types such as fibroblasts. In fibroblasts, growth factors such as the epidermal growth factor receptor (EGFR) regulate $p21^{ras}$ by stimulating *ras* guanine nucleotide exchange (30). The *ras* exchange protein in fibroblasts is the mammalian homologue of the Drosophila Sos gene. Sos is recruited to the activated EGFR cytoplasmic domain and hence to the cell membrane by the adapter protein Grb-2/Sem5 (31a). Studies of Grb2 in T cells (reviewed in ref. 31b) have identified two proteins that when tyrosine phosphorylated can potentially bind to the Grb2 SH2 domain. These are Shc, of which two isoforms of 46 and 52 kDa are expressed in T cells and a membrane located tyrosine phosphoprotein of 36 kDa. In TCR activated cells both Shc and p36 are tyrosine phosphorylated and can bind Grb2 SH2 domains *in vitro*. However, the major tyrosine phosphoprotein found associated with Grb2 SH2 domains in vivo in TCR activated cells is a 36 kDa molecule that is proposed to be an adapter that links the TCR activated PTKs to Grb2/Sos. p36 is only tyrosine phosphorylated in response in TCR triggering and a simplistic analysis would suggest that it is probably a substrate for the kinase ZAP70 which is activated by the TCR. Recently it was observed that p36 also exists in a complex with phospholipase C in TCR-activated cells suggesting that p36 may be an adapter that couples the TCR to both the p21ras and calcium/PKC signalling pathways that originate from the TCR.

The role of Grb2 and Sos in TCR regulation of p21ras is based on correlative data and remains to be proven by genetic studies. Moreover, two molecules in addition to Sos are found complexed to Grb2 SH3 domains in T cells (31b). These two proteins, of 75 and 116 kDa respectively, are constitutively associated with the SH3 domains of Grb2 analogous to the Grb2/Sos association. p75 and p116 are substrates for TCR activated PTKs but nothing else is known regarding the function of these molecules. Nevertheless, the existence of different Grb-2 complexes in T cells suggests that Grb2 may be involved in coupling the TCR to multiple effector pathways.

3 The role of Ca^{2+}, PKC, and $p21^{ras}$ in T cell activation

It has been known for several years that calcium and PKC are important intracellular signals in T cell activation because pharmacological agents that elevate intracellular calcium (calcium ionophores) or stimulate PKC (phorbol esters) can mimic the TCR and induce T cell activation as judged by IL-2 gene expression (32). More recently, it was recognized that $p21^{ras}$ also has a crucial function in T cells. This conclusion was derived from transient transfection protocols that examined the consequences of expressing mutated constitutively active or dominant inhibitory *ras* mutants on T cell activation. The *ras* oncogene p21-V-Ha *ras* is mutated at codon 12 (Ser→Val) and 59 (Ala→Thr). These mutations render the *ras* protein insensitive to negative regulation by *ras*-GAP such that it accumulates in cells in an 'active' GTP bound state. A second *ras* mutant, N17 *ras* is able to block cellular *ras* function by competing for *ras* guanine nucleotide exchange proteins thereby preventing endogenous *ras* activation by stopping exchange of GTP on to cellular *ras* (33). p21-v-Ha *ras* when expressed in Jurkat cells can synergize with a calcium signal to activate the IL-2 gene (34). Contrastingly, expression

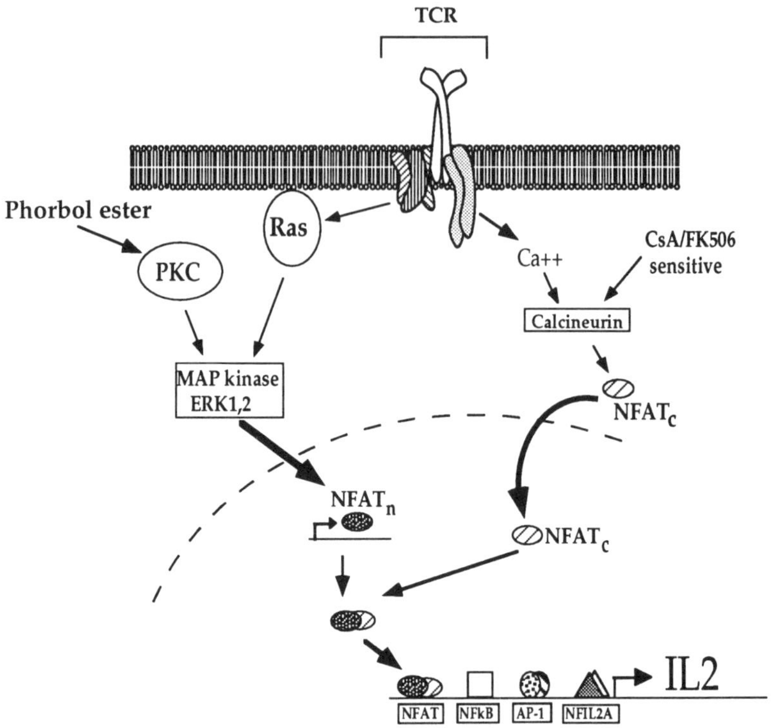

Fig. 8.3 Regulation of NFAT (nuclear factor of activated T cells).

of N17 *ras* inhibits TCR induction of the IL-2 gene which indicates that *ras* function is essential for TCR signal transduction (34).

Multiple transcriptional factors regulate the IL-2 gene but in part *ras* regulation of the IL-2 gene is mediated by *ras* control of the transcriptional factor NFAT (35). NFAT has long been the focus of signal transduction studies in T cells because the immunosuppressive drugs cyclosporin A and FK506, which are widely used therapeutically in transplantation procedures, appear to regulate T cell activation via regulation of NFAT function (14). IL-2 gene expression can not be induced by induction of a single signalling pathway but requires the action of several pathways that integrate at the level of the transcriptional factors that bind to the IL-2 enhancer. Studies of NFAT regulation have revealed how such integration of signals occurs. Hence current knowledge indicates that NFAT is a heterodimer comprising an AP-1 component that is found only in the nucleus of activated T cells and a constitutively expressed cytoplasmic component—a rel family protein, that translocates to the nucleus during T cell activation (Figure 8.3) (36–38). Part of the calcium signalling requirement for NFAT and hence IL-2 expression is that changes in intracellular calcium regulates translocation of the NFAT cytoplasmic component to the nucleus (39). In contrast, PKC and $p21^{ras}$ which are alternative synergistic partners with calcium for NFAT induction are potent, alternative, regulators of kinase cascades that activate AP-1 protein expression and function (40, 41).

For many years it was assumed that the regulatory effects of calcium on T cell activation were mediated by calcium/calmodulin-dependent kinases. However, efforts to understand the mechanism of action of the drugs cyclosporin A and FK506 on NFAT have revealed a crucial role for the calcium/calmodulin-dependent phosphatase, calcineurin in T cells (42–44). FK506 and CyA when bound to their respective immunophilin form a complex with calcineurin thereby sequestering the enzyme and preventing calcineurin function. Calcineurin is composed of two subunits: a catalytic A subunit and a regulatory β subunit. The A subunit has a calcium and calmodulin binding site and when these are deleted the truncated form of the enzyme is constitutively active. The expression by transfection of the truncated 'activated' calcineurin generates a powerful signal that substitutes for calcium and synergizes with either $p21^{ras}$ or PKC signals to activate NFAT and the IL-2 gene (42, 43) (unpublished data). The substrates for calcineurin in T cells are not yet known but are of obvious significance to extending our understanding of T cell signal transduction pathways.

4 The role of $p21^{ras}$ in T cell activation

In fibroblasts and the pheocytochromocytoma PC12 cell line, the transmission of signals from $p21^{ras}$ to the nucleus is proposed to involve the regulation of the activity of MAP kinases or ERKs (mitogen activated protein kinases or extracellular signal regulated kinases) (45, 46). T lymphocytes express at least two MAP kinases ERK1 and ERK2 that are stimulated in response to TCR triggering (47). Two intracellular pathways for ERK2 regulation co-exist in T cells: one mediated by *ras* and the other by PKC. The TCR stimulates both $p21^{ras}$ and PKC but data obtained from transfection studies show that expression of the inhibitory *ras* mutant, N17 *ras* which prevents endogenous *ras* activation suppresses TCR induction of ERK2 (unpublished data). Accordingly it appears that $p21^{ras}$ and not PKC couples the TCR to the regulation of MAP kinases. MAP kinases translocate to the nucleus when activated and their known substrates include transcriptional factors such as ELK-1, c-jun, and c-myc (48–50). Thus one mechanism whereby $p21^{ras}$ could couple the TCR to the nucleus is via stimulation of the ERK kinase cascade.

The activity of the MAP kinases requires their simultaneous phosphorylation on both tyrosine and threonine (51). The phosphorylation and activation of MAP kinase is induced by an activator kinase, MAP kinase kinase (MAPKK). The link between $p21^{ras}$ and the MAPKK is provided by at least one other kinase (MAPKKK). Two candidates for the MAPKKK have been described, the proto-oncogene Raf-1 and a kinase termed MEK-1 (52, 53). Raf-1 can be regulated by at least two pathways: one mediated by $p21^{ras}$ and one controlled by PKC. Raf-1 is apparently activated in T cells by both TCR and

PKC triggering and may provide the link between *ras* or PKC and ERK2 (54). Accordingly, the positioning of *ras* in the mechanisms that couple the TCR and ERK2 provides a framework for positioning other potentially important signalling molecules in the TCR signal transduction pathways.

5 Signal transduction by accessory molecules in T cell activation

It is well established from numerous experimental systems that T cell activation does not occur in response to TCR triggering alone but requires co-operative interactions between the TCR and accessory signalling molecules such as CD2, CD4, CD8, CD5, CD28. In many instances these accessory receptors are in a physical complex with the TCR (e.g. CD2, CD4, CD5) and are therefore able to amplify TCR induced signals by regulating signalling pathways common to the TCR. One accessory molecule that is not in this category is CD28. The cell biology of CD28 indicates that CD28 must regulate at least one signalling pathway that is unique. Hence CD28 can regulate IL-2 and other lymphokine gene expression by both transcriptional and non-transcriptional mechanisms (8). The pathway whereby CD28 induces gene transcription involves PTKs but the pattern of phosphotyrosyl proteins induced by CD28 differs from that induced by the TCR suggesting that CD28 is coupled to different PTKs or tyrosine phosphatases than the TCR. However, the two most striking features of CD28 signalling are firstly that CD28 induces a unique signalling response that stabilizes the mRNA for a variety of lymphokines thereby enhancing their production. Such stabilization of lymphokine mRNA is never observed in response to TCR ligation. Secondly, CD28 induction of T cell activation is not susceptible to inhibition by the immunosuppressive drugs cyclosporin A and FK506 which are potent inhibitors of the TCR activation response. One well established TCR signalling pathway is the pathway linking the TCR to the metabolism of inositol phospholipids. The role of CD28 in regulating inositol lipid metabolism is somewhat controversial. Certain CD28 antibodies stimulate PLC and thus induce inositol lipid hydrolysis. However, when CD28 is triggered by interaction its physiological ligand B7, no activation of PLC occurs. Rather CD28 regulates the activity of a PtdIns-3-kinase thereby regulating cellular levels of D-3 phosphoinositides (55) (S. Ward, personal communication). PtdIns-3-kinase has been implicated as an important growth regulatory molecules for a number of years by virtue of its association with activated or transforming PTKs (56–59). The function of the D-3 phosphoinositides is not known although a recent study revealed the ability of PI (3, 4, 5) P_3 to activate a novel PKC, PKCζ (60). The ability of CD28 to regulate PtdIns-3-kinase probably does not explain the unique functional properties of this receptor since PtdIns-3-kinase stimulation is a common response to triggering of a variety of receptors including the TCR (61). However, PtdIns-3-kinase seems to be an important signal transduction molecule in fibroblasts. Moreover, a recent study has established that in yeast, a target for the immunosuppressive drug rapamycin is a protein with significant homology to the mammalian PtdIns-3-kinase (62). These insights suggest that further investigation into the regulation and function of the PtdIns-3-kinase could be worthwhile because it could clearly be an important amplifying signal for T cell activation.

6 Concluding remarks

So far, T cell activation studies have focused on dissecting the signalling pathways that couple the TCR to the nucleus and the *de novo* regulation of gene expression. Much less is known about the signalling pathways that regulate the changes in T cell surface phenotype that occur during T cell activation, one characteristic response to T cell activation is a change in the affinity of various adhesion molecules for their respective ligand. This response is exemplified by the increase in CD2/LFA-3 or CFA1/ICAM-1 avidity that occurs on activated T cells. Little is known about the intracellular mechanisms that control these changes in cell adhesion properties. The accepted idea is that the TCR can regulate cytoskeleton morphology thereby changing the conformation of different adhesion molecules. However, the signal transduction mechanisms that link the TCR to the cytoskeleton are unknown. It is likely that PTKs are crucial for this aspect of TCR signalling and it is noteworthy that recent studies have identified the proto-oncogene Vav as a substrate for TCR activated PTKs (63, 64). The function of Vav is unknown but it has a region of homology with the dbl oncogene which is a putative guanine nucleotide exchange protein for the rho/rac GTP binding protein (65). In fibroblasts rho and rac are involved in the regulation of membrane morphology and hence control changes in cell shape (66, 67). It will be of interest to determine whether signalling pathways involving GTP binding proteins such as rho/rac are coupled to the TCR via PTKs and whether such molecules play a pivotal role in TCR regulation of cytoskeletal function.

References

1 Crabtree, G. R. (1989). Contingent genetic regulatory events in T lymphocyte activation. *Science,* **243**, 355–61.

2 Weiss, A. (1993). T cell antigen receptor signal transduction: a tail of tails and cytoplasmic protein-tyrosine kinases. *Cell*, **73**, 209–12.

3 Bierer, B. E., Sleckman, B., Ratnofsky, S. E., and Burakoff, S. J. (1989). The role of the CD2, CD4 and CD8 molecules in T cell activation. *Annu. Rev. Immunol.*, **7**, 579–99.

4 Springer, T. (1990). Adhesion receptors of the immune system. *Nature*, **346**, 425–34.

5 Osman, N., Ley, S. C., and Crumpton, M. J. (1992). Evidence for an association between the T cell receptor/CD3 complex and the CD5 antigen in human T lymphocytes. *Eur. J. Immunol.*, **22**, 2995–3000.

6 Davies, A. A., Ley, S. C., and Crumpton, M. J. (1992). CD5 is phosphorylated on tyrosine after stimulation of the T-cell antigen receptor complex. *Proc. Natl. Acad. Sci. USA*, **89**, 6368–72.

7 June, C. H., Ledbetter, J. A., Linsley, P. S., and Thompson, C. B. (1990). Role of the CD28 receptor in T-cell activation. *Immunol. Today*, **11**, 211–16.

8 Schwartz, R. H. (1992). Costimulation of T lymphocytes: the role of CD28,CTLA-4 and B7BB1 in Interleukin2 production and Immunotherapy. *Cell*, **71**, 1065–8.

9 Ullman, K. S., Northrop, J. P., Verweij, C. L., and Crabtree, G. R. (1990). Transmission of signals from the T lymphocyte antigen receptor to the genes responsible for cell proliferation and immune function: the missing link. *Annu. Rev. Immunol.*, **8**, 421–52.

10 Smith, K. A. (1992). Interleukin-2. *Curr. Opin. Immunol.*, **4**, 271–6.

11 Taniguchi, T. and Minami, Y. (1993). The IL-2/IL-2 receptor system: a current overview. *Cell*, **73**, 5–8.

12 Durand, D. B., Bush, M. R., Morgan, J. G., Weiss, A., and Crabtree, G. R. (1987). A 275 base pair fragment at the 5′ end of the interleukin 2 gene enhances expression from a heterologous promoter in response to signals from the T cell antigen receptor. *J. Exp. Med.*, **165**, 395–407.

13 Durand, D. B., Shaw, J.-P., Bush, M. R., Replogle, R. E., Belagaje, R., and Crabtree, G. R. (1988). Characterization of antigen receptor response elements within the interleukin-2 enhancer. *Mol. Cell. Biol.*, **8**, 1715–24.

14 Emmel, E. A., Verweij, C. L., Durand, D. B., Higgins, K. M., Lacy, E., and Crabtree, G. R. (1989). Cyclosporin A specifically inhibits function of nuclear proteins involved in T cell activation. *Science*, **246**, 1617–20.

15 Verweij, C. L., Guidos, C., and Crabtree, G. R. (1990). Cell type specificity and activation requirements for NFAT-1 (nuclear factor of activated T-cells) transcriptional activity determined by a new method using transgenic mice to assay transcriptional activity of an activated nuclear factor. *J. Biol. Chem.*, **265**, 15788–95.

16 Klausner, R. D. and Samelson, L. E. (1991). T cell antigen receptor activation pathways: the tyrosine kinase connection. *Cell*, **64**, 875–8.

17 Appleby, M. W., Gross, J. A., Cooke, M. P., Levin, S. D., Qian, X., and Perlmutter, R. M. (1992). Defective T cell receptor signaling in mice lacking the thymic isoform of p59fyn. *Cell*, **70**, 751–63.

18 Cooke, M. P., Abraham, K. M., Forbusch, K. A., and Perlmutter, R. M. (1991). Regulation of T cell receptor signalling by a src family protein tyrosine kinase (p59 fyn). *Cell*, **65**, 281–91.

19 Chan, A. C., Kadlecek, T. A., Elder, M. E., Filipovich, A. H., Kuo, W. L., Iwashima, M., Parslow, T. G., and Weiss, A. (1994). ZAP70 deficiency in an autosomal recessive form of severe combined immunodeficiency. *Science*, **264**, 1599–601.

20 Veillette, A. and Davidson, D. (1992). Src-related protein tyrosine kinases and T-cell receptor signalling. *Trends Genet.*, **8**, 61–6.

21 Koretzky, G. A., Picus, J., Schultz, T., and Weiss, A. (1991). Tyrosine phosphatase CD45 is required for T-cell antigen receptor and CD2-mediated activation of a protein tyrosine kinase and interleukin 2 production. *Proc. Natl. Acad. Sci. USA*, **88**, 2037–41.

22 Weiss, A. and Imboden, J. B. (1987). Cell surface molecules and early events involved in human T lymphocyte activation. *Adv. Immunol.*, **41**, 1–38.

23 Downward, J., Graves, J. D., Warne, P. H., Rayter, S., and Cantrell, D. A. (1990). Stimulation of $p21^{ras}$ upon T-cell activation. *Nature*, **346**, 719–23.

24 Izquierdo, M., Downward, J., Graves, J. D., and Cantrell, D. A. (1992). Role of Protein Kinase C in T-cell antigen receptor regulation of p21ras: evidence that two p21ras regulatory pathways coexist in T cells. *Mol. Cell. Biol.*, **12**, 3305–12.

25 Downward, J. (1992). Regulation of $p21^{ras}$ by GAPs and guanine nucleotide exchange proteins in normal and oncogenic cells. *Curr. Opin. Genet. Dev.*, **2**, 13–18.

26 Satoh, T., Nafakuku, M., and Kaziro, Y. (1992). Function of Ras as a molecular switch in signal transduction. *J. Biol. Chem.*, **267**, 2414–52.

27 Izquierdo, M., Downward, J., Leonard, W. J., Otani, H. and Cantrell, D. A. (1992). IL-2 activation of p21ras in murine myeloid cells transfected with human IL-2 receptor beta chain. *Eur. J. Immunol.*, **22**, 817–21.

28 Downward, J. (1992). Exchange rate mechanisms. *Nature*, **358**, 282–3.

29 Graves, J. D., Downward, J., Rayter, S., Warne, P., Tutt, A. L., Glennie, M., *et al.* (1991). CD2 antigen mediated activation of the guanine nucleotide binding proteins p21ras in human T lymphocytes. *J. Immunol.*, **146**, 3709–12.

30 Buday, L. and Downward, J. (1993). Epidermal growth factor regulates the exchange rate of guanine nucleotides on p21ras in fibroblasts. *Mol. Cell. Biol.*, **13**, 1903–10.

31a Buday, L. and Downward, J. (1993). Epidermal growth factor regulates p21ras through the formation of a complex of receptor, Grb2 adapter protein and Sos nucleotide exchange factor. *Cell*, **73**, 611–20.

31b Izquierdo Pastor, M., Reif, K., and Cantrell, D. A. (1995). The regulation and function of p21ras during T cell activation and growth. *Immunol. Today*, **16**, 159–164.

32 Berry, N. and Nishizuka, Y. (1989). Protein kinase C and T cell activation. *Eur. J. Biochem.*, **89**, 205–14.

33 Medema, R. H., Vries-Smits, A. M. M., van der Zon, G. C. M., Maassen, J. A., and Bos, J. L. (1993). Ras activation by insulin and epidermal growth factor through enhanced exchange of guanine nucleotides on p21ras. *Mol. Cell. Biol.*, **13**, 155–62.

34 Rayter, S., Woodrow, M., Lucas, S. C., Cantrell, D., and Downward, J. (1992). p21ras mediates control of IL2 gene promoter function in T cell activation. *EMBO J.*, **11**, 4549–56.

35 Woodrow, M., Rayter, S., Downward, J., and Cantrell, D. A. (1993). p21ras function is important for T cell antigen receptor and protein kinase C regulation of nuclear factor of activated cells. *J. Immunol.*, **150**, 1–9.

36 Northrop, J. P., Ullman, K. S., and Crabtree, G. R. (1993). Characterisation of the nuclear and cytoplasmic components of the lymphoid-specific nuclear factor of activated T cells (NFAT) complex. *J. Biol. Chem.*, **268**, 22917–25.

37 Boise, L. H., Petryniak, B., Mao, X., June, C. H., Wang, C., Lindsten, C. B. *et al.* (1993). The NFAT-1 DNA binding complex in activated T cells contains Fra-1 and Jun-B. *Mol. Cell. Biol.*, **113**, 1911–20.

38 Rao, A. (1994). NF-ATp: a transcription factor required for the co-ordinate induction of several cytokine genes. *Immunol. Today*, **15**, 274–81.

39 Flanagan, W. M., Corthésy, B., Bram, R. J., and Crabtree, G. R. (1991). Nuclear association of a T-cell transcription factor blocked by FK-506 and cyclosporin A. *Nature*, **352**, 803–7.

40 Smeal, T., Binetruy, B., Mercola, D. A., Birrer, M., and Karin, M. (1991). Oncogenic and transcriptional cooperation with Ha-Ras requires phosphorylation of c-Jun on serines 63 and 73. *Nature*, **354**, 494–6.

41 Hunter, T. and Karin, M. (1992). The regulation of transcription by phosphorylation. *Cell*, **70**, 375–87.
42 Clipstone, N. A. and Crabtree, G. R. (1992). Identification of calcineurin as a key signalling enzyme in T-lymphocyte activation. *Nature*, **357**, 695–7.
43 O'Keefe, S. J., Tamura, J., Kincaid, R. L., Tocci, M. J. and O'Neill, E. A. (1992). FK-506 and Cs-A-sensitive activation of the interleukin-2 promoter by calcineurin. *Nature*, **357**, 692–4.
44 Schreiber, S. L. and Crabtree, G. R. (1992). The mechanism of action of cyclosporin A and FK506. *Immunol. Today*, **13**, 136–42.
45 Leevers, S. J. and Marshall, C. J. (1992). Activation of extracellular signal-regulated kinase, ERK2, by p21ras oncoproteins. *EMBO J.*, **11**, 569–74.
46 Leevers, S. J. and Marshall, C. J. (1992). MAP kinase regulation—the oncogene connection. *Trends Cell. Biol.*, **2**, 283–6.
47 Whitehurst, C. E., Boulton, T. G., Cobb, M. H., and Geppert, T. G. (1992). Extracellular signal-regulated kinases in T cells. Anti-CD3 and 4beta-Phorbol 12-Myristate 13-Acetate-induced phosphorylation and activation. *J. Immunol.*, **148**, 3230–7.
48 Pulverer, B. J., Kyriakis, J. M., Avruch, J., Nikolakaki, E., and Woodgett, J. R. (1991). Phosphorylation of c-jun mediated by MAP kinases. *Nature*, **353**, 670.
49 Seth, A., Gonzalez, F. A., Gupta, S., Radenm, D. L., and Davis, R. J. (1992). Signal transduction within the nucleus by mitogen-activated protein kinase. *J. Biol. Chem.*, **34**, 24796–804.
50 Marais, R., Wynne, J., and Treisman, R. (1993). The SRF accessory protein ELK-1 contains a growth factor-regulated transcriptional activation domain. *Cell*, **73**, 381–93.
51 Nishida, E. and Gotoh, Y. (1993). The MAP kinase cascade is essential for diverse signal transduction pathways. *Trends Biol. Sci.*, **18**, 128–31.
52 Howe, L. R., Leevers, S. J., Gomez, N., Nakielney, S., Cohen, P., and Marshall, C. J. (1992). Activation of the MAP kinase pathway by the protein kinase raf. *Cell*, **71**, 335–42.
53 Lange-Carter, C. A., Pleiman, C. M., Gardner, A. M., Blumer, K. J. and Johnson, G. L. (1993). A divergence in the MAP Kinase regulatory network defined by MEK Kinase and Raf. *Science*, **260**, 315–19.
54 Siegel, J. N., Klausner, R. D., Rapp, U. R., and Samelson, L. E. (1990). T cell antigen receptor engagement stimulates c-raf phosphorylation and induces c-raf kinase activity via a protein kinase C-dependent pathway. *J. Biol. Chem.*, **265**, 18472–80.
55 Stephens, L. R., Hughes, K. T., and Irvine, R. F. (1991). Pathway of phosphatidylinositol (3,4,5)-triphosphate synthesis in activated neutrophils. *Nature*, **351**, 33–9.
56 Otsu, M., Hiles, I., Gout, I., Fry, M. J., Ruiz-Larrea, F., Panayotou, G. *et al.* (1991). Characterization of two 85 kD proteins that associate with receptor tyrosine kinases, middle-T/pp60c-src complexes, and PI3-kinase. *Cell*, **65**, 91–104.
57 Escobedo, J. A., Navankasattusas, S., Kavanaugh, W. M., Milfay, D., Fried, V. A., and Williams, L. T. (1991). cDNA cloning of a novel 85 kD protein that has SH2 domains and regulates binding of PI 3-kinase to the PDGF beta-receptor. *Cell*, **65**, 75–82.
58 Escobedo, J. A., Kaplan, D. R., Kavanaugh, W., Turck, C. W., and Williams, L. R. (1991). A phosphatidylinositol 3-kinase binds to platelet derived growth factor receptor sequence containing phosphotyrosine. *Mol. Cell. Biol.*, **11**, 1125–32.
59 Cantley, L. C., Auger, K. R., Carpenter, C., Duckworth, B., Graziani, A., Kapeller, R., *et al.* (1991). Oncogenes and signal transduction. *Cell*, **64**, 281–302.

60 Nakanishi, H., Brewer, K. A., and Exton, J. H. (1993). Activation of the zeta isozyme of Protein kinase C by phosphatidylinositol 3,4,5-triphosphate. *J. Biol. Chem.*, **268**, 13–16.
61 Ward, S. G., Ley, S. C., Macphee, C., and Cantrell, D. A. (1992). Regulation of D-3 phosphoinositides during T cell activation via the T cell antigen receptor/CD3 complex and CD2 antigens. *Eur. J. Immunol.*, **22**, 45–9.
62 Kunz, J., Henriquez, R., Schneider, U., Deuter-Reinhard, M., Movva, N., and Hall, M. N. (1993). Target of rapamycin in yeast, TOR2, is an essential phosphatidylinositol kinase homologue required for G1 progression. *Cell*, **73**, 585–96.
63 Margolis, B., Hu, P., Katzav, S., Li, W., Oliver, J. M., Ullrich, A., *et al.* (1992). Tyrosine phosphorylation of vav proto-oncogene product containing SH2 domain and transcription factor motifs. *Nature*, **356**, 71–4.
64 Bustelo, X. R., Ledbetter, J. A., and Barbacid, M. (1992). Product of the vav proto-oncogene defines a new class of tyrosine protein kinase substrates. *Nature*, **356**, 68–74.
65 Adams, J. M., Houston, H., Allen, J., Lints, T. and Harvey, R. (1992). The hematopoietically expressed vav proto-oncogene shares homology with the dbl GDP-GTP exchange factor, the bcr gene and a yeast gene (CDC24) involved in cytoskeletal organization. *Oncogene*, **7**, 611–18.
66 Ridley, A. J. and Hall, A. (1992). The small GTP binding protein rho regulates the assembly of focal adhesions an actin stress fibers in response to growth factors. *Cell*, **70**, 389–99.
67 Ridley, A. J., Paterson, H. F., Johnston, C. L., Diekman, D., and Hall, A. (1992). The small GTP binding protein rac regulates growth factor induced membrane ruffling. *Cell*, **70**, 401–10.

9 *Src*-related kinases and their receptors in T cell activation

JANICE C. TELFER, OTTMAR JANSSEN, K. V. S. PRASAD, MONIKA RAAB, ANTONIO DA SILVA, AND CHRISTOPHER E. RUDD

1 Introduction

T cell activation is mediated by ligation of the TCRζ–CD3 complex in conjunction with the CD4 and CD8 co-receptors. Co-ligation of CD4 or CD8 with the TCRζ–CD3 complex dramatically potentiates activation (Rudd *et al.* 1989). The TCRζ–CD3 complex binds to antigen in the form of peptide bound to the polymorphic cleft of major histocompatibility (MHC) class I or II proteins, as expressed by antigen-presentation cells (Townsend and Bodmer 1989). CD4 and CD8 binds to invariant regions of the major histocompatibility complex (MHC) antigens class II and I, respectively (Doyle and Strominger 1987; Norment *et al.* 1988). The TCRζ–CD3 complex and CD4/CD8 antigens presumably bind to the same MHC molecules during antigen recognition. The accessory function played by CD4 and CD8 enhances proliferation signals in cases of low affinity binding to antigen, or low antigen abundance. The binding of CD4 and CD8 to thymic MHC class II and I during T cell development is critical for the establishment of the functional dichotomy of the T cell population into $CD4^+CD8^-$ helper T cells recognizing antigen in the context of MHC class II and $CD8^+CD4^-$ cytotoxic T cells recognizing antigen in the context of MHC class I.

Structurally, CD4 (a single chain at 52 kDa) and CD8 (a 34 kDa α chain and a 32 kDa β chain) are members of the immunoglobulin gene superfamily. The two N-terminal domains of CD4 have been determined by X-ray crystallography (Ryu *et al.* 1990; Wang *et al.* 1990). CD4 possesses a lateral protrusion from the N-terminal immunoglobulin domain, derived from an extra β-pleated sheet structure. Overall, CD4 appears to form an extended rod-like structure. Binding to MHC class II involves an extended region on the face of the outer first two domains of CD4 (Clayton *et al.* 1989; Lamarre *et al.* 1989). By contrast, CD8 has a single N-terminal immunoglobulin-like domain (Leahy *et al.* 1992). CD8 is expressed either as a disulfide bonded αα homodimer or a disulfide bonded αβ heterodimer. CD8 binds to the α3 membrane proximal domain of MHC class I proteins (Salter *et al.* 1989; Sanders *et al.* 1991).

T cell activation is accompanied by a number of early events that include Ca^{2+} influx, inositol lipid turnover, and the activation of protein kinase C. Tyrosine phosphorylation of intracellular substrates is amongst the earliest of signals preceding phosphotidylinositol hydrolysis by phospholipase C and calcium flux (Klausner and Samelson 1991). Pharmacological inhibition of T cell tyrosine kinase activity will also block activation of phospholipase C via stimulation of the T cell receptor suggesting that tyrosine kinase activity is necessary for the generation of the second messengers IP_3 and DAG (June *et al.* 1990; Mustelin *et al.* 1990). A major unsolved problem has been to uncover the mechanism by which T cell tyrosine kinases regulate T cell activation. T cells express three $p60^{src}$-related protein tyrosine kinases: $p56^{lck}$, $p59^{fyn(T)}$, and $p60^{yes}$. *Src*-related kinases are entirely intracellular, being linked to the inner face of the plasma membrane by a myristic moiety. The role of T cell *src*-related kinases in signal transduction first became apparent with the finding that $p56^{lck}$ interacts non-covalently with the T cell receptors CD4 and CD8, endowing these receptors with the enzymatic capacity of the transmembrane tyrosine kinase receptors (Rudd *et al.* 1988; Veillette *et al.* 1988; Barber *et al.* 1989; Rudd 1991). Subsequently, the TCRζ–CD3 complex has been found to associate with $p59^{fyn(T)}$ (Samelson *et al.* 1990, 1992). The interleukin-2 receptor (IL-2R) (Hatakeyama *et al.* 1991) and CD2 (Bell *et al.* 1992) have been found to associate with $p56^{lck}/p59^{fyn(T)}$. An interaction between the TCR ζ chain and the tyrosine kinase ZAP-70 has also been described (Chan *et al.* 1992). As will be outlined, each of these receptor-associated kinases has the capacity to phosphorylate and associate with additional signalling enzymes, forming a signalling complex with the capacity to generate a complex cascade of signals.

2 The structure of *src*-related protein-tyrosine kinases

The *src* family of kinases is comprised of a group of eight tyrosine kinases with a characteristic structure. These include the prototype kinase $pp60^{src}$, as well as $p56^{lck}$, $p59^{fyn}$, $p53/56^{lyn}$, $p55^{blk}$, $p60^{yes}$. Each of the *src* kinases possesses a highly conserved protein-tyrosine kinase domain [*src* homology 1 domain (SH1)] of about 300 amino acids. This protein-tyrosine kinase domain is a feature shared by a related family of receptor-tyrosine kinases such as the platelet-derived growth factor (PDGF-R), the fibroblast growth factor receptor (FGF-R), the colony stimulating factor receptor (CSF-1-R), the insulin receptor, and others. Except for the shared kinase domain, the structure of *src*-related kinases and receptor-tyrosine kinases diverge substantially. Features of the *src* family that are not found in receptor-tyrosine kinase receptors include an N-terminal sequence for the attachment of myristic acid and receptor binding, two globular *src* homology 2 and 3 domains (SH2 and SH3), and a carboxyl regulatory region. Figure 9.1 illustrates the basic organization of the *src*-related kinases ($p56^{lck}$, $p59^{fyn}$) that have been reported to associate with surface receptors.

The N-terminus is linked to a myristic group, a modification that is required for membrane attachment and localization (Cross *et al.* 1985; Kamps *et al.*

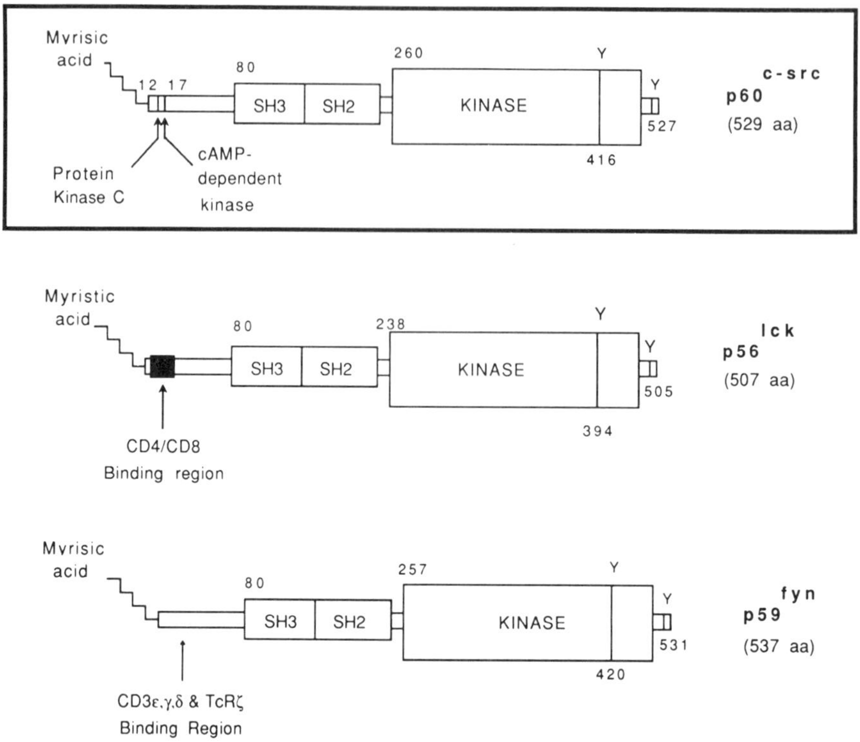

Fig. 9.1 The structure of *src*-related protein-tyrosine kinases that associate with surface receptors. Structure of $p60^{src}$ and other members of the family of non-receptor kinases ($p56^{lck}$, $p59^{fyn(T)\ and\ (B)}$). $p60^{src}$ serves as the prototype for other *src*-related kinases which bind to surface receptors. Each kinase is modified by the addition of a myristic acid moiety to the NH_2-terminus allowing an interaction with the lipid bilayer of the plasma membrane. The NH_2 region (70 to 80 residues) contains sequences largely unique to each of the *src*-related kinases. Sites for serine phosphorylation include a protein kinase C site (Ser-12) and cAMP-dependent kinase site (Ser-17) in $p60^{src}$. Phosphorylation sites of $p56^{lck}$, Ser-42 and Ser-59 may act as substrates for protein kinase C and MAP-2 kinase, respectively. The site of interaction of $p56^{lck}$ with CD4 and CD8 has been mapped within residues 10 to 30 of the NH_2-terminal domain. Likewise, $p59^{fyn(T)}$ binding to CD3 γ,δ,ε and TCR ζ is mediated by the first ten residues. The NH_2-terminal domain is followed by SH3 and SH2 domains which share moderate homology between all *src*-related kinases. Similar domains have been identified in a variety of non-catalytic intracellular molecules. SH3 may bind to proline-rich motifs, while the SH2 domains bind to phosphotyrosine residues located next to hydrophobic residues. Next is the kinase domain (SH1) which is highly conserved and possesses an autophosphorylation (Tyr-416 for $pp60^{src}$; Tyr-394 for $p56^{lck}$). The C-terminal portion of each kinase possesses a negative regulatory site (Tyr-527 for $p60^{src}$; Tyr-505 for $p56^{lck}$; Tyr-531 for $p59^{fyn}$).

1985). This is followed by a small N-terminal region of approximately 70–80 amino acids which varies between $p56^{lck}$, $pp60^{src}$ and other *src* family members. Depending on the kinase, the N-terminal region includes sites that associate with the surface receptors CD4 and CD8 (Shaw *et al.* 1989, 1990; Turner *et al.* 1990), sites of serine phosphorylation by protein kinase C (Ser-12 for $pp60^{src}$), and cAMP-dependent kinases (Ser-17 for $pp60^{src}$) (Gould *et al.* 1985; Gentry *et al.* 1986; Patschinsky *et al.* 1986; Hirota *et al.* 1988). Two serine phosphorylation sites within $p56^{lck}$ have been mapped to positions 42 and 59 by site-directed mutagenesis (Winkler *et al.* 1993). Phosphorylation of peptides corresponding to these sites suggests that Ser-42 and Ser-59 act as substrates for protein kinase C and MAP-2 kinase, respectively (Winkler *et al.* 1993). In the case of $p56^{lck}$, the N-terminal region also possesses cysteine residues at positions 3 and 5 that influence the localization of the kinase to the plasma membrane (Turner *et al.* 1990).

The conserved SH2 and SH3 domains lie downstream of the amino-terminus (Figure 9.1). Both domains are found in other proteins, either in non-enzymatic molecules ('adaptors'), or in enzymes (Pawson and Gish 1992). SH2 domains of about 100 amino acids are found in many intracellular polypeptides including phospholipase Cγ (PLCγ) (Stahl *et al.* 1988; Suh *et al.* 1988), *ras*GAP (GTPase activating protein) (McCormick 1989), the regulatory subunit (p85) of phosphatidylinositol-3-kinase (PI-3-kinase) (Escobedo *et al.* 1991*a*; Otsu *et al.* 1991; Skolnik *et al.* 1991), and others (Pawson and Gish 1992). SH2 domains in adaptors appear designed to link tyrosine phosphorylated signalling proteins and receptors which lack SH2 domains. The SH2 domain binds to phosphotyrosine and adjacent hydrophobic sequences in the cytoplasmic tails of growth factor receptors (i.e. PDGF-R, CSF-1-R) and in $pp60^{src}$ (Kaplan *et al.* 1987, 1990; Margolis *et al.* 1989; Molloy *et al.* 1989). Nuclear magnetic resonance (NMR) analysis of *abl*-SH2 and crystallographic analysis of $pp60^{src}$ SH2 show a spherical domain containing a hydrophobic pocket with residues capable of binding to one peptide at a time (Overduin *et al.* 1992; Waksman *et al.* 1992, 1993; Eck *et al.* 1993). The overall structure is formed by two antiparallel sheets surrounded by two alpha helices. The phosphotyrosine extends into the pocket interacting with a conserved Arg residue (*src* Arg-175) which forms an ion pair with the phosphate. Other residues stabilize the interaction by hydrogen bonding and amino–aromatic group interactions.

Cantley and colleagues have defined the optimal binding sequences for different SH2 domains, illustrating the importance of adjacent residues in defining the specificity of binding (Songyang *et al.* 1993). SH2 domains of *src*-related kinases bind optimally to the pTyr–Glu–Glu–Ile (pTyr + 1–3 = YEEI), with diminished affinities of binding to substituted residues at pY + 1 = D, T, Q; pY + 2 = N, Y, D, Q; pY + 3 = M, L, V. Each of the *src*-related kinases' (*src, fyn, lck*) SH2 domains differ slightly in their affinities for individual residues. The dissociation constant of binding of the *lck*-SH2 domain to a phosphopeptide containing the YEEI motif (EPQpYEEIPIYL) is in the 1 nM range (Payne et al. 1993). Crystal structure of this SH2–peptide high affinity complex reveals a second conserved pocket within the SH2 domain for the pY + 3 residue toward the C-terminus (Eck *et al.* 1993; Waksman *et al.* 1993). Other SH2 domains of phosphatidylinositol-3-kinase (PI-3-kinase) (YDHP) and PLCγ (N-terminal = YLEL, C-terminal = YVIP) exhibit different specificities for phosphotyrosine-containing sequences. Consistent with this, phosphotyrosine sites within the PDGF-R such as Tyr-751 bind to PI-3-kinase, but poorly if at all to *ras*GAP or PLCγ (Coughlin *et al.* 1989; Kazlauskas *et al.* 1990; Morrison *et al.* 1990; Escobedo *et al.* 1991*b*). In the case of *src* kinases, deletion or mutations within this region of $pp60^{src}$ alters its association with intracellular proteins (Reynolds *et al.* 1989; Wang and Parson 1989; Wendler and Boschelli 1989). The SH2 domain (but not the SH3 domain) is required for oncogenic transformation by constitutively active $p56^{lck}$ (F-505) (Veillette *et al.* 1992).

The SH3 motif is a small domain of approximately 50 amino acids found in various proteins including PLCγ (Stahl *et al.* 1988; Suh *et al.* 1988), $p47^{gag\text{-}crk}$ (Mayer *et al.* 1988), α-spectrin (Sahr *et al.* 1990), myosin-IB, fodrin, and a yeast actin binding protein (Rodaway *et al.* 1989; Drubin *et al.* 1990). SH3 domains seem to function as independent units, found in various locations within polypeptides, and often in tandem with SH2 domains. The crystal structure of this domain shows a compact β-barrel of five antiparallel β-strands that form a hydrophobic pocket (Musacchio *et al.* 1992; Yu *et al.* 1992). Cicchetti and colleagues have identified an *abl*-SH3 binding protein 3BP-1 with a proline-rich hydrophobic binding motif (PPPLPPV) (Cicchetti *et al.* 1992). A consensus sequence XPXXPPPXXP is found in other SH3 binding proteins such as the murine *limb deformity* (ld) proteins, formins, and a subtype of the muscarinic acetylcholine receptor (Ren *et al.* 1993). SH3 interactions have been implicated in the regulation of a number of events including *src* kinase binding to PI-3-kinase (Prasad *et al.* 1993*a*), and the regulation of *ras* guanine nucleotide exchange (Olivier *et al.* 1993; Simon *et al.* 1993). Prasad and co-workers have shown binding of the fyn SH3 domain to the p85 subunit of PI-3-kinase (Prasad *et al.* 1993*a*). The SH2 domain of the *drk* protein in Drosophila binds the tyrosine kinase sevenless, while its SH3 domain can bind to the proline-rich C-terminus of Son of sevenless (Sos), a ras guanine nucleotide-releasing protein (GNRP) (Simon *et al.* 1991, 1993; Olivier *et al.* 1993). This interaction provides a means to couple protein-tyrosine kinases to $p21^{ras}$ activation.

More C-terminal than the SH2 and SH3 domains is the highly conserved kinase domain of about 300 amino acids [also called the *src* homology domain 1 (SH1)] that exhibits approximately 80% homology between c-*src*, c-*yes*, c-*fgr*, *fyn*, *lyn*, *lck*, *hck*, and *tkl*. This region has been defined by homologies with other kinases, mutational effects (Bryant and Parsons 1984; Snyder *et al.* 1985; Kamps and Sefton 1986), and by the fact that proteolytic fragments of this region possess kinase activity (Brugge and Darrow 1984). The region has a glycine-rich ATP binding site with a key lysine residue that interacts with

ATP (*src* K-295; *lck* K-273) (Kamps *et al.* 1984). Included in the kinase domain is the autophosphorylation site, the primary site of *in vitro* phosphorylation (Smart *et al.* 1981; Casnellie *et al.* 1982; Patschinsky *et al.* 1982). Although wild-type kinase is not readily phosphorylated at this site *in vivo*, constitutively active transforming versions of the kinase (i.e. $pp60^{v\text{-}src}$) are phosphorylated at this site (Smart *et al.* 1981; Patschinsky *et al.* 1982).

A conserved tyrosine residue in a consensus motif that is involved in the regulation of kinase activity is at the C-terminus of *src* kinases (Figure 9.1). Tyr-527 of $pp60^{c\text{-}src}$ and Tyr-505 of $p56^{lck}$ serve as the main regulator of tyrosine phosphorylation *in vivo* (Cooper *et al.* 1986). Transforming $pp60^{v\text{-}src}$ lacks this terminal tyrosine, and mutation of this tyrosine induces constitutive kinase activity (Cartwright *et al.* 1987; Kmiecik and Shalloway 1987; Piwnica-Worms *et al.* 1987; Reynolds *et al.* 1987). Phosphorylation of this residue inhibits the catalytic activity, while dephosphorylation stimulates kinase activity. The interaction of middle T antigen of polyoma virus with the *src* kinases $pp60^{c\text{-}src}$, $p59^{fyn}$, and $pp60^{c\text{-}yes}$ activates the kinase by preventing phosphorylation at this site (Cartwright *et al.* 1986; Cheng *et al.* 1989). Phosphorylation of this kinase regulatory site is mediated by the cytosolic tyrosine kinase $p50^{csk}$ (**c**-terminal **s**rc **k**inase) (Partanen *et al.* 1990; Nada *et al.* 1991), while dephosphorylation may be mediated by protein-tyrosine phosphatases (PTPases) such as CD45 (Mustelin *et al.* 1989; Ostergaard *et al.* 1989; Ostergaard and Trowbridge 1990).

3 $p56^{lck}$ binding to CD4 and CD8

Src kinases lack transmembrane and extracellular regions, making them unable to interact directly with the extracellular environment. Their role in signal transduction and growth control became apparent with the finding that $p56^{lck}$ could interact with the cellular receptors, CD4 and CD8 (Rudd *et al.* 1988; Veillette *et al.* 1988; Barber *et al.* 1989; Rudd 1991). The interaction of $p56^{lck}$ with CD4 and CD8 was shown by co-precipitation and *in vitro* kinase labelling. Between 30% to 90% of $p56^{lck}$ is stably associated with CD4 or CD8, depending on the T cell line examined. CD4 and CD8 cytoplasmic regions share a region of limited similarity encoded by the CD4 sequence KKT**CQCP**HRFQKT and the CD8 sequence RRV**CKCP**RPVVKS designating a 13 residue motif (K/RK/RX**CXCP**XXXXKT/S) (Figure 9.2 I). This sequence is located midway in the cytoplasmic domain of CD4 (residues 419 to 431) and proximal to the lipid bilayer in the CD8 α chain (residues 190 to 203). Within this sequence is a conserved **CXCP** motif as well as five basic amino acids that alternate with the hydrophobic sequences on either side of a possible alpha helix (Figure 9.2 II). Although the exact molecular basis of the interaction is unknown, mutation of the cysteine residues disrupts $p56^{lck}$ binding (Shaw *et al.* 1989, 1990; Turner *et al.* 1990). The cysteine requirement may differ slightly for CD4 and CD8. Mutation of either cysteine (C) elimin-

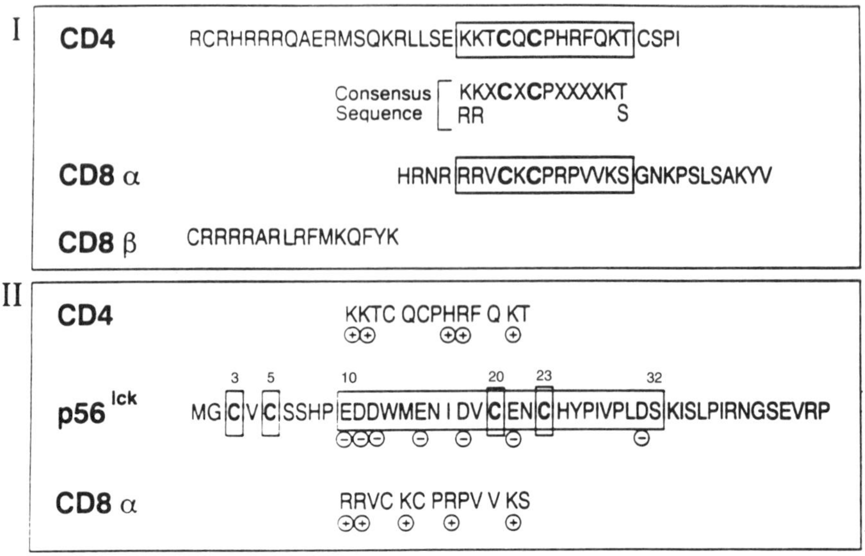

Fig. 9.2 Comparison of CD4 and CD8 cytoplasmic domains (p56^{lck} binding motif) and the N-terminal region of p56^{lck}. (I) CD4 and CD8 α cytoplasmic regions share a sequence of limited homology which includes **CXCP** binding motif for p56^{lck}. A larger consensus region is designated by the CD4 residues (KT**C**Q**CP**HRFQKT) and the CD8 α sequences (RRV**C**K**CP**RPVVKS), providing the consensus sequence K/RK/RX**C**X**CP**XXXXKT/S. The CD8 β chain, which lacks this sequence, does not bind to p56^{lck}. (II) CD4 and CD8 cysteine residues within the binding site are flanked by positively charged amino acids. In contrast, the p56^{lck} cysteines are flanked by negatively charged amino acids, suggesting an ionic interaction between CD4, CD8, and p56^{lck}.

ated complex formation of p56^{lck} with CD4. CD8 α chain may require the mutation of both cysteines in order to disrupt the complex (Turner *et al.* 1990). The cysteine residues do not undergo covalent bonding (Rudd *et al.* 1988).

Other amino acids within the cytoplasmic sequences may directly or indirectly stabilize the interaction. This can be inferred from the observation that the deletion of C-terminal residues following the **CXCP** resulted in diminished levels of precipitable p56^{lck} (Turner *et al.* 1990). In addition, while the CD8 α V**CKCP**R motif fused to the VSV G viral protein associated with p56^{lck}, the analogous sequence from CD4 was unable to support complex formation (Shaw *et al.* 1990). Other T cell antigens possessing the **CXCP** sequence include the 4-1BB antigen (Kwon *et al.* 1987), and NK1.1 antigen (Giorda *et al.* 1990). In the case of NK1.1 the motif is oriented in reverse (Giorda *et al.* 1990).

Residues 10 to 32 in p56^{lck} bind to the cytoplasmic sequences of CD4 and CD8 (Shaw *et al.* 1989; Turner *et al.* 1990) (Figure 9.2 II). This region of p56^{lck} differs from other *src* family members. Within this sequence, mutation of cysteines at positions 20 and 23 abrogates the binding to CD4 and CD8 (Turner *et al.* 1990). Whether these cysteines mediate direct interactions, or play an indirect role by affecting the conformation of the kinase has not been established. Crystallographic analysis should resolve this issue. A direct interaction involving cysteine residues in both the receptor and kinase argue for the possible role of metal ions in the interaction. CXXC sequences are found in several proteins which undergo tetrahedral formation with metals. These include the metallothionin protein which binds Zn^{2+} (Furey *et al.* 1986), and the Tat protein of the human immunodeficiency virus which binds Cd^{2+} and Zn^{2+} (Frankel *et al.* 1988). This region is also rich in negatively charged amino acids which could potentially form ionic bonds with alternating basic residues within the CD4 and CD8 binding sequence (Figure 9.2 II). Ionic interactions do not appear to be required for complex formation, but may play a role in regulating kinase activity.

4 The TCRζ–CD3–$p59^{fyn}$ association

Besides $p56^{lck}$, T cells express the *src*-related protein-tyrosine kinases, $p59^{fyn(T)}$ and $p60^{yes}$. $p59^{fyn}$ is expressed in most cell types; however, as a result of mutually exclusive splicing, $p59^{fyn}$ exists as two isoforms (Cooke and Perlmutter 1989). The more prevalent form, $p59^{fyn(B)}$, is highly expressed in brain and other cell types, while $p59^{fyn(T)}$ is expressed in T cells. The splicing occurs within exon 7 in a region at the beginning of the kinase domain. Samelson and co-workers first showed that $p59^{fyn(T)}$ can co-precipitate with the TCRζ–CD3 complex (Samelson *et al.* 1990, 1992). The TCRζ–CD3 complex itself is composed of the TCR αβ chains, the CD3 γ, δ, and ε chains, and the ζ/η/Fcγ chains (Baniyash *et al.* 1988). Associated $p59^{fyn}$ has been identified in only a limited number of detergents (such as digitonin or Brij 96) that appear to maintain the integrity of weak protein–protein interactions. Chemical cross-linking and co-capping studies have verified the existence of the complex (Gassmann *et al.* 1992; Sarosi *et al.* 1992). Some 20% of *fyn* can be detected in association with the receptor, depending on the conditions of isolation. A surprisingly small 2–4% of total TCRζ–CD3 is *fyn*-associated. In a few cell lines, an additional form of *fyn* at 72 kDa also associates with the complex and differs from $p59^{fyn(T)}$ by the presence of additional sites of serine/threonine phosphorylation (da Silva and Rudd 1993).

As with $p56^{lck}$ binding to CD4, the N-terminal region of $p59^{fyn}$ mediates binding to the receptor (Gauen *et al.* 1992). The first ten amino acids of *fyn* are sufficient to mediate binding to the TCR ζ chain, a kinase region which is normally associated with myristoylation and membrane localization. The molecular mechanism regulating the association remains unclear. This region is unrelated to other receptor-associated *src*-related kinases such as $pp60^{src}$, $p56^{lck}$, $p56^{lyn}$, or $p55^{blk}$. Consistent with this, neither $p56^{lck}$ nor $pp60^{src}$ bound

Human CD3γ	P	N	D	Q	L	Y	Q	P	L	K	D	R	E	D	D	Q	*	Y	S	H	L	Q	G	N
Human CD3δ	R	N	D	Q	V	Y	Q	P	L	R	D	R	D	D	A	Q	*	Y	S	H	L	G	G	N
Human CD3ε	V	P	N	P	D	Y	E	P	I	R	K	G	Q	R	D	L	*	Y	S	G	L	*	*	*
Human ζ chain (1)	G	Q	N	Q	L	Y	N	E	L	N	L	G	R	R	E	E	*	Y	D	V	L	D	K	R
Human ζ chain (2)	P	Q	E	G	L	Y	N	E	L	Q	K	D	K	M	A	E	A	Y	S	E	I	G	M	K
Human ζ chain (3)	G	H	D	G	L	Y	Q	G	L	S	T	A	T	K	D	T	*	Y	D	A	L	H	M	Q
Mouse MB1	E	D	E	N	L	Y	E	G	L	N	L	D	D	C	S	M	*	Y	E	D	I	S	R	C
Mouse B29	E	E	D	H	T	Y	E	G	L	N	I	D	Q	T	A	T	*	Y	E	D	I	V	T	L
Rat Fcε γ chain	P	D	D	R	L	Y	E	E	L	H	V	Y	S	P	I	*	*	Y	S	A	L	E	D	T
Rat Fcε β chain	K	S	D	A	V	Y	T	G	L	N	T	R	N	Q	E	T	*	Y	E	T	L	K	H	E
Human CD5	H	V	D	N	E	Y	S	Q	P	P	R	N	S	R	L	S	A	Y	P	A	L	E	G	V
Lck Tyr 394	I	E	D	N	E	Y																		
Fyn Tyr 420	I	E	D	N	E	Y																		

Fig. 9.3 Conserved Y-X_{11}-Y-X-X-L motifs within the cytoplasmic domains of the CD3 γ,δ,ε, TCR ζ, MB1, B29, and CD5 antigens. The CD3 γ,δ,ε, TCR ζ, MB1, B29, and CD5 antigens share the Y-X_{11}-Y-X-X-L activation motif. Each of these antigens are phosphorylated on tyrosine residues as a result of receptor ligation. The Y-X-X-L sequence is of particular importance for CD3 ε function. In the case of CD5, the Y-X-X-L submotif has been replaced by the Y-X-X-P submotif. CD5 possesses a set of residues next to the first Y residue (DNEY) which is homologous to the autophosphorylation site within $p56^{lck}$ and $p59^{fyn}$.

ζ or CD3 subunits, although the transfer of the *fyn* N-terminal motif to *src* conferred binding to TCR ζ (Gauen *et al.* 1992). The binding sequence possesses two Cys residues, the proximal Cys being conserved in $p56^{lck}$ and $p56^{lyn}$, but not $pp60^{src}$ or $p55^{blk}$. In the case of $p56^{lck}$, the first Cys has been implicated in membrane localization (Turner *et al.* 1990). $p59^{fyn(T)}$ associates with the various CD3 γ, δ, and ε chains as well as the phosphorylated TCR ζ (Gauen *et al.* 1992). As first pointed out by Reth (1989), each of these chains share a common motif defined by the sequence Y-X-X-L/I-X_{7-8}-Y-X-X-L (191) (Figure 9.3). It is unclear as to whether the binding of $p59^{fyn}$ to these various chains requires phosphorylation of the tyrosine residues. Although historically, tyrosine phosphorylation of the CD3 γ, δ, and ε chains has been difficult to detect, a low level of tyrosine phosphorylation has recently been reported (Qian *et al.* 1993).

Although the stoichiometry of the *fyn*(T)–TCRζ–CD3 interaction appears quite low, antibody-induced TCRζ/CD3 stimulation of $p59^{fyn(T)}$ activity can be detected (da Silva *et al.* 1992; Tsygankov *et al.* 1992). By contrast, $p56^{lck}$ was not consistently activated, although this may vary between cell lines (Veillette *et al.* 1989; da Silva *et al.* 1992; Tsygankov *et al.* 1992). $p59^{fyn(T)}$ stimulation was accompanied by a dramatic increase in the phosphorylation of $p59^{fyn(T)}$ associated p120/130 molecule (da Silva *et al.* 1992). Transfection of $p59^{fyn(T)}$ in fibroblasts has indicated that p120/130 interacts directly with $p59^{fyn}$, and therefore associates indirectly with the TCRζ–CD3 complex. The *fyn-* SH2 domain binds to the p120/130 polypeptides. Although the identity of p120/130 is unknown, it appears to associate with $p59^{fyn}$ and not $p56^{lck}$ (da Silva *et al.*, submitted).

5 Role of CD4/CD8–p56^{lck} and TCRζ–CD3–p59$^{fyn(T)}$ in T cell activation

The biological functions of p56^{lck} and p59^{fyn} are linked to the functions of their receptors. As previously outlined, CD4, CD8, and TCRζ–CD3 regulate T cell growth by the recognition of antigen in the form of peptide presented by major histocompatibility (MHC) class I or II antigens (Townsend and Bodmer 1989; Unanue and Cerottini 1989). A functional linkage between the TCRζ–CD3 complex and p59$^{fyn(T)}$ has been demonstrated most aptly in transgenic mice. Over-expression of p59^{fyn} is correlated with enhanced Ca^{2+} influx and proliferation (Cooke *et al.* 1991). By contrast, expression of the dominant negative form of kinase inactive p59^{fyn} inhibited these events. Similarly, mice generated by homologous recombination that lack either p59$^{fyn(T)}$ (Appleby *et al.* 1992), or p59$^{fyn(T+B)}$ (Stein *et al.* 1992) are partially defective in TCRζ–CD3-mediated signalling. The signalling defect was most pronounced in thymocytes, although peripheral T cells also showed a decrease in Ca^{2+} mobilization and proliferation ($>$ 50%) to anti-CD3 plus phorbal ester. p59$^{fyn(T)}$ may therefore be most important in signalling during development in a subpopulation of thymocytes. In mature splenic T cells, the response to alloantigen was either only partially affected, or unaffected by the loss of p59^{fyn} (Appleby *et al.* 1992; Stein *et al.* 1992). In terms of proliferation *per se*, p59^{fyn} appears to be partially required in T cell signalling. Redundancy may exist in the ability of p56^{lck} to compensate for p59^{fyn}. However, individual T cells may differ in the dependency on distinct *src*-related kinases and the nature of effector functions linked to individual kinases may be found to differ. A differential role for the p59$^{fyn(T)}$ isoform has been suggested by the use of constitutively active forms of the kinase in the transfection of T cell clones. The transforming form of p59$^{fyn(T)}$, but not p59$^{fyn(B)}$, has been reported to potentiate IL-2 production (Davidson *et al.* 1992).

Constitutively activated forms of p56^{lck} (F-505) can also augment TCR-induced tyrosine phosphorylation and IL-2 production, even in the absence of CD4 and CD8 (Abraham *et al.* 1991). In one instance, the SH2 domain, but not the SH3 domain was essential for TCR-induced tyrosine phosphorylation; however, both were essential for TCR-induced lymphokine production (Caron *et al.* 1992). In another study, T cells transfected with the constitutively activated form of *lck* (F-505) produce significantly higher amounts of IL-2 in the absence of antigen stimulation. In this case, the *lck* SH2 domain, but not the *lck* SH3 domain, appeared to be necessary for increased lymphokine production (Luo and Sefton 1992).

An obvious role for the CD4–p56^{lck} and CD8–p56^{lck} complexes during antigen presentation would be to introduce p56^{lck} to the TCRζ–CD3 complex. Within a multimolecular complex, TCR ζ, CD5, p59^{fyn}, and PLCγ could serve as potential substrates of the kinase domain (Samelson *et al.* 1986, 1992; Burgess *et al.* 1992; Davies *et al.* 1992). Co-receptor function of CD4 and

CD8 requires associated p56^{lck} (Sleckman *et al.* 1988; Zamoyska *et al.* 1989; LeTourneur *et al.* 1990; Chalupny *et al.* 1991; Glaichenhaus *et al.* 1991). Forms of CD4 that lack the cytoplasmic tail, or that lack cysteines within the *lck* binding motif are profoundly defective in costimulatory function. In certain cases, the requirement for p56^{lck} can be overcome by increasing the concentration of antigen, or increasing the density of CD4 or CD8 on the cell surface (Sleckman *et al.* 1988; Zamoyska *et al.* 1989). An enhanced response to low antigen concentration is of importance given the fact that physiological concentrations of antigen are likely to be quite limited (Harding and Unanue 1990). Using a CD4-dependent T cell hybridoma, Glaichenhaus and co-workers demonstrated that the CD4–p56^{lck} interaction potentiated responses to antigen by 50- to 100-fold (Glaichenhaus *et al.* 1991). A smaller contribution was made by CD4–MHC adhesion, independent of p56^{lck}. p56^{lck} is therefore necessary to signalling in antigen stimulated T cells which require CD4/CD8 co-receptor function.

Peripheral T cells from *lck* negative transgenic mice are also profoundly defective in T cell signalling by antigen. Although these mice exhibit a block in thymic differentiation, there exists a small population of T cells in the periphery (Molina *et al.* 1992). Given the caveat that aberrant differentiation may alter signalling, p56^{lck} appears necessary for signalling by antigen. Interestingly, the same cells could be stimulated by anti-CD3 ligation suggesting that antibody may be capable of stimulating cells via an alternative pathway, perhaps employing p59$^{fyn(T)}$. Further support for an essential role of p56^{lck} in the activation of certain T cells has come from the study of a mutant of the Jurkat T cell line. Straus and Weiss (1992) have analysed a mutant T cell tumour (JCaM1) which lacks p56^{lck} expression and is defective in response to TCRζ–CD3 ligation. Both the induction of tyrosine phosphorylation and Ca^{2+} release was dramatically impaired (Straus and Weiss 1992). Expression of p56^{lck} restored the defect. The signalling defect did not involve, in an obvious way, TCRζ–CD3 ligation in the context of the CD4–p56^{lck} complex. The cells did not express CD8 and expressed very low levels of CD4. Although low levels of CD4–p56^{lck} might constitutively associate with TCRζ–CD3, as found in other cells (Burgess *et al.* 1991), this study introduced the possibility that p56^{lck} may function independently of CD4.

p56^{lck} and p59$^{fyn(T)}$ could generate an intracellular signalling cascade via protein-tyrosine kinase activity, by SH3 binding to PI-3-kinase, or by SH2 binding to phosphotyrosine labelled substrates (Figure 9.4). In a scenario involving the tyrosine kinase domain, TCR ζ phosphorylation may act as the first step in a multistep process that recruits downstream molecules. One candidate is the SH2 carrying, non-myristoylated tyrosine kinase ZAP-70 which is related to the *syk* family of tyrosine kinases (Chan *et al.* 1992). Unlike p56^{lck} and p59$^{fyn(T)}$, ZAP-70 associates with the TCR ζ chain only after receptor cross-linking. Further, the association in Cos cells requires the co-expression of either p59^{fyn} or p56^{lck}. TCR ζ possesses three repeats of a

Y-X-X-L-X_{7-8}-Y-X-X-L/I motif found in a variety of receptor phosphoproteins (Figure 9.3). p56^{lck}/p59^{fyn} are likely to directly phosphorylate the Y-X-X-L-X_{7-8}-Y-X-XL/I motif, which in turn recruits SH2 carrying proteins such as ZAP-70. TCR ζ dimerization alone is capable of generating intracellular signals (Irving and Weiss 1991; Romeo and Seed 1991). ZAP-70 binding to the Y-X-X-L-X_{7-8}-Y-X-X-L/I motif in ζ is required for ζ transduced Ca^{2+} mobilization and tyrosine phosphorylation (Irving *et al.* 1993). The C-terminal motif is essential. Expression of multiple repeats of the motif in chimeric constructs results in enhanced signalling suggesting that the redundancy of the motifs with the TCR ζ chain may act to amplify the signal. Both SH2 domains within ZAP-70 are required suggesting tandem binding to the two tyrosine residues within the Y-X-X-L-X_{7-8}-Y-X-XL/I motif. In this sense, TCRζ–ZAP-70 binding is somewhat analogous to recruitment of multiple SH2 binding proteins by the EGF-R and PDGF-R. The major difference is that the initial phosphorylation event is mediated by a separate kinase, rather than by autophosphorylation. It is likely that the CD3 γ,δ,ε chains utilize a similar mechanism to recruit downstream targets. These chains appear to be capable of generating distinct kinase-mediated signals (LeTourneur and Klausner 1992; Wegener *et al.* 1992), which may lead to the phosphorylation of a distinct, but overlapping spectrum of proteins (LeTourneur and Klausner 1992).

Situations of abundant ligand, or high affinity between antigen and the TCR αβ chains, may suffice to directly activate p59^{fyn} activity and thereby phosphorylate TCR ζ. Under conditions where the amount of ligand is limiting, or where the binding affinity of the TCRζ–CD3 complex for ligand is low, engagement may be insufficient to allow stimulation of p59$^{fyn(T)}$, thereby necessitating the co-ligation of CD4–p56^{lck} or CD8–p56^{lck}. Alternatively, since only a small per cent of TCRζ–CD3 complexes bind to p59$^{fyn(T)}$, CD4–p56^{lck} may be required to phosphorylate TCR ζ in p59$^{fyn(T)}$ negative complexes. On a basic level, p56^{lck} could be conceived as a back-up system for p59^{fyn}. However, the system will almost certainly involve subtle differences in the kinase activities of p56^{lck} and p59$^{fyn(T)}$ towards specific tyrosine residues within the TCR ζ and CD3 chains. Different affinities of *lck* SH2 and *fyn* SH2 for binding the various phosphotyrosines should allow for the selection of different substrates. In the case of the EGF-R and FGF-R, intracellular substrates such as PLCγ and PI-3-kinase have varying affinities for separate Tyr-P sites on the receptor kinases (Eck *et al.* 1993; Payne *et al.* 1993; Songyang *et al.* 1993; Waksman *et al.* 1993).

Other T cell antigens and receptors may also impinge on this process. The stimulatory properties of CD2, Thy-1, and Ly-6 in T cells are known to depend on TCR ζ expression (Breitmeyer *et al.* 1987; Frank *et al.* 1990; Moingeon *et al.* 1992). These antigens can induce TCR ζ phosphorylation and may associate with *src*-related kinases (Frank *et al.* 1990; Stefanova *et al.* 1991; Bell *et al.* 1992; Moingeon *et al.* 1992). CD5, a member of the macrophage scavenger receptor family (Krieger 1992), shares a number of characteristics with the TCR ζ chain. Both structures can be co-purified with the

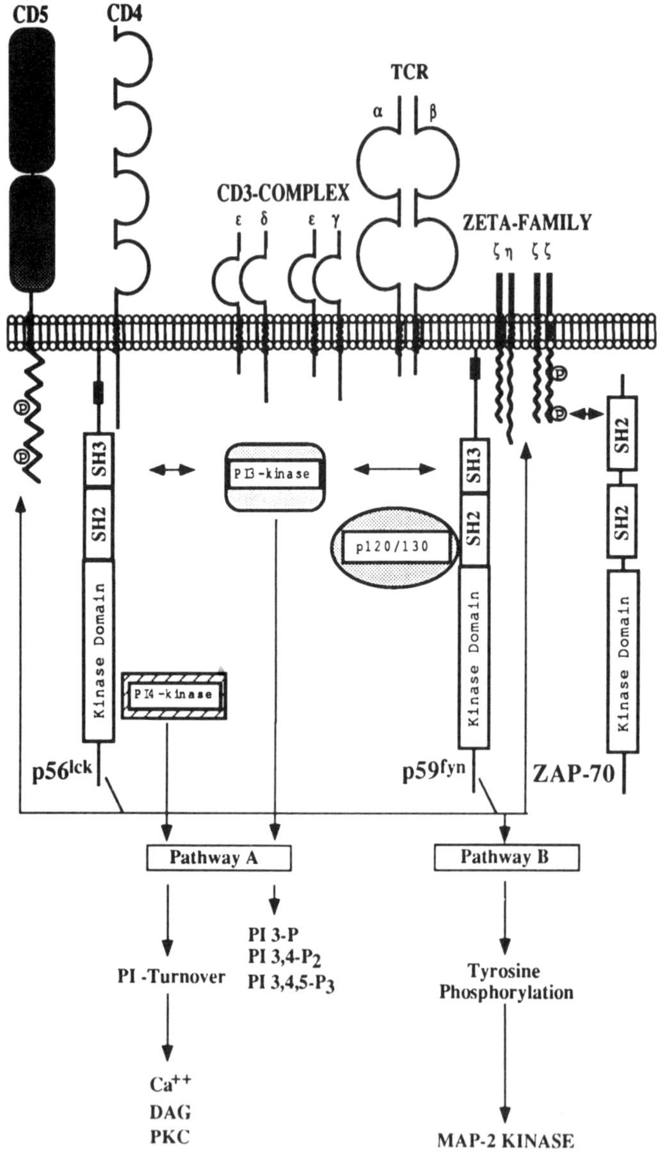

TCR–CD3 complex in mild detergents and act as substrates for a protein-tyrosine kinase (Beyers *et al.* 1992; Burgess *et al.* 1992; Davies *et al.* 1992). As seen in Figure 9.3, CD5 possesses a version of the Y-X_{11}-Y-X-X-L motif. The main difference is that a proline is substituted for the leucine in the Y-X-X-L submotif found in the TCR ζ, CD3, MB1, B29, and Fc ε chains. Mutational analysis has shown the Y-X-X-L sequence to be of particular functional importance in CD3 ε (LeTourneur and Klausner 1992). This sug-

Fig. 9.4 Signalling cascade mediated by CD4–p56^{lck} and TCRζ/CD3–p59$^{fyn(T)}$; protein-tyrosine and lipid kinase-mediated pathways. The CD4, CD5, and TCRζ–CD3 structures, their associated kinases (p56^{lck}, p59$^{fyn(T)}$, ZAP-70) and SH2/SH3 binding proteins (p120/130 and PI-3-kinase) are depicted. One signalling pathway is mediated by the protein-tyrosine kinase domain of p56^{lck} and p59$^{fyn(T)}$ and the phosphorylation of TCR ζ, CD5, and possibly the CD3 subunits. Co-localization and ligation of CD4–p56^{lck}, CD5, TCRζ–CD3–p59$^{fyn(T)}$, and CD45 (not shown) stimulates the kinase activity leading to the tyrosine phosphorylation of substrates. TCR ζ phosphorylation at the activation motif Y-X-X-L-X_{7-8}-Y-X-X-L/I leads to the recruitment of SH2 carrying proteins such as ZAP-70. Similarly, CD5 possesses a version of the Y-X_{11}-Y-X-X-L activation motif with residues next to the first Y residue (DNEY) that are homologous to the autophosphorylation site in p56^{lck} and p59^{fyn}. CD4–p56^{lck}-mediated tyrosine kinase activity appears correlated to the activation of MAP-2 kinase (ERK) (pathway B). An alternate pathway is one mediated by PI-3- and PI-4-kinases, leading to PI turnover (Ca^{2+} release, DAG, and PKC activation) as well as the generation of PI-3-P, PI-3,4-P_2, PI-3,4,5-P_3 (pathway A). Pathways A and B may/could function in conjunction with each other. Alternatively, the demonstration that *lck* and *fyn* SH3 domains mediate binding to PI-3-kinase introduces the possibility that PI-3-kinase binding and function may operate independently of *lck* and *fyn* tyrosine kinase activity.

---

gests the possibility that CD5 may recruit an intracellular mediator distinct from the TCRζ–CD3 chains. CD5 also possesses a set of residues next to the first Y residue (DNEY) which is homologous to the autophosphorylation site of p56^{lck} and p59^{fyn} (see Figure 9.3). CD5 is amongst the most rapidly phosphorylated substrates within the T cell ($T_{1/2}$ = 20 sec) (Burgess *et al.* 1992) (Figure 9.4). This signalling event may be tied to its ability to bind to the B cell antigen CD72 (Van de Velde *et al.* 1991).

CD45 negative mutants have also provided indirect evidence implicating protein-tyrosine kinase activity and phosphorylation in the regulation of T cell activation. These cells are defective in T cell signalling and in anti-CD4 and CD3 induction of substrate phosphorylation (Pingel and Thomas 1989; Koretzky *et al.* 1990, 1991). The defect may be linked to the depressed activity of p56^{lck} and p59^{fyn} in these cells (Mustelin and Altman 1990; Volarevic *et al.* 1990; Weiss *et al.* 1991; Mustelin *et al.* 1992; Shiroo *et al.* 1992). Tyrosine kinase inhibitors also inhibit T cell function (June *et al.* 1990; Mustelin *et al.* 1990). The requirement for tyrosine kinase and phosphatase activity appears restricted to early events of activation. Transfection with the muscarinic receptor, a G protein-dependent receptor, can bypass the defect and induce activation-related events (Koretzky *et al.* 1990). Another way of bypassing the absence of CD45 is to co-ligate CD4–p56^{lck} and the TCRζ–CD3 complex, which causes the phosphorylation of PLCγ (Deans *et al.* 1992). This suggests the possibility that *trans* phosphorylation between *fyn* and *lck* may suffice to activate one or both kinases, and to overcome the low basal level of activity imposed by the absence of the phosphatase.

Aside from the tyrosine kinase domain, $p56^{lck}$ and $p59^{fyn}$ possess SH2 and SH3 domains which could transfer signals by recruitment of downstream molecules. Importantly, the use of *lck* and *fyn* SH2, SH3, and SH2–SH3 fusion proteins have shown that CD4–p56lck and TCRζ–CD3–$p59^{fyn(T)}$ utilize a unique mechanism to recruit PI-3-kinase (Prasad *et al.* 1993*a*,*b*,*c*). Unlike receptor-tyrosine kinases which recruit PI-3-kinase by the binding of the SH2 domain of p85 to autophosphorylation sites within the cytoplasmic domain of the kinase, $p56^{lck}$ and $p59^{fyn(T)}$ bind to PI-3-kinase via their SH3 domains. Binding to PI-3-kinase is further modified by the presence of an adjacent SH2 region suggesting an ability of the SH2 domain to influence SH3 recognition of its substrate. Limited SH2 recognition of PI-3-kinase was also noted, but some 50-fold less than precipitated by the SH3 domain. The *fyn* and *lck* SH3 interaction with PI-3-kinase represents a distinct mechanism by which *src*-related kinases bind molecules and regulate activity.

Identification of the SH3 domain PI-3-kinase interaction underscores a major difference between conventional receptor-tyrosine kinases and the CD4–$p56^{lck}$ and CD8–$p56^{lck}$ system. In the case of the PDGF-R, PI-3-kinase binding is entirely dependent on growth factor binding to the receptor. The ensuing autophosphorylation then creates binding sites for the SH2 domains of various downstream targets such as PLCγ and PI-3-kinase. By contrast, SH3 binding does not obviously involve phosphotyrosine residues. Instead, SH3 domains bind to proline-rich consensus motifs within the 3BP-1, 3BP-2, formin, Sos, and the rat m4 muscarinic acetylcholine receptor (Cicchetti *et al.* 1992; Olivier *et al.* 1993; Ren *et al.* 1993; Simon *et al.* 1993). The *lck* SH3 interaction with PI-3-kinase may provide a tyrosine phosphorylation-independent signalling mechanism in co-receptor function. In this sense, it is intriguing that the co-receptor function of CD4–$p56^{lck}$ can be partially retained with the combined disruption of the kinase and SH2 domains (Xu and Littman 1993). PI-3-kinase has been found crucial for mitogenesis initiated by platelet-derived growth factor (PDGF) (Fantl *et al.* 1992; Valius and Kazlauskas 1993). Two potential proline-rich binding sites exist within the p85 subunit, but not within the p110 subunit. PI-3-kinase, like the *drk–Sos* interaction (Olivier *et al.* 1993; Simon *et al.* 1993), represents a case of a SH3 interaction with a key signalling protein, underlining the importance of the SH3 domain in CD4–$p56^{lck}$ signalling. It remains to be determined whether other mechanisms exist to recruit PI-3-kinase. Controversy exists as to whether tyrosine phosphorylation of PI-3-kinase occurs under physiological conditions (Hu *et al.* 1992; Kavanaugh *et al.* 1992; McGlade *et al.* 1992).

As seen in Figure 9.4, receptor-associated $p59^{fyn(T)}$ is organized in discrete functional cassettes that regulate both binding to the receptor complex and to the downstream targets. While the N-terminal region of *fyn* binds to the CD3 γ, δ, ε chains and the TCR ζ chain, *fyn* SH2 can bind p120/130, and *fyn* SH3 binds PI-3-kinase. Both PI-3-kinase and p120/130 co-purify with receptor-free *fyn* and the TCRζ–CD3 complex. $p59^{fyn(T)}$ may therefore introduce PI-3-kinase and p120/130 into the TCRζ–CD3 prior to receptor ligation. As in the case of CD4–$p56^{lck}$, the fact that *fyn* SH3 mediates binding to PI-3-kinase within TCRζ–CD3–$p59^{fyn(T)}$ suggest that PI-3-kinase is recruited independently of *fyn* tyrosine kinase activity and phosphorylation. Signals derived from the SH3 PI-3-kinase interaction may therefore be distinct from signals generated by the protein-tyrosine kinase domain. Consistent with this, anti-CD3 induced stimulation of PI-3-kinase has been reported to occur independently of tyrosine phosphorylation (Ward *et al.* 1992). By contrast, PLCγ may associate with the TCRζ–CD3 complex and as such be regulated by tyrosine phosphorylation (Dasgupta *et al.* 1992).

Besides PI-3-kinase, CD4–$p56^{lck}$ is associated with PI-4-kinase activity, at levels greater than PI-3-kinase (Prasad *et al.* 1993*b,c*). PI-4-kinase is part of the classical PI pathway, responsible for the generation of PI-4-P and PI-4,5-P_2 for PI turnover. PI-4-P and PI-4,5-P_2 can be acted upon by PI-3-kinase to generate PI-3,4-P_2 and PI-3,4,5-P_3. To date, the EGF-R is the only other receptor to have been reported to associate with PI-4-kinase (Cochet *et al.* 1991). As in the case of PI-3-kinase, antibody-induced CD4 cross-linking caused an increase in CD4 precipitable PI-4-kinase activity (Prasad *et al.* 1993*b, c*). Further work is required to determine whether the enzyme associates with the kinase, or with the CD4 cytoplasmic domain.

A further connection between *src* kinases and inositol turnover is evident with the finding of phospholipase Cγ (PLCγ) associated with $p56^{lck}$. Using TrpE–PLCγ1 fusion proteins, Weber *et al.* (1992) provide evidence that the N-terminal SH2 domain of PLCγ binds to $p56^{lck}$ (Weber *et al.* 1992). PLCγ catalyses the hydrolysis of PI-4,5-P_2 to generate inositol-1,4,5-triphosphate (IP_3) and diacylglycerol (DAG), thus mobilizing intracellular Ca^{2+} and stimulating protein kinase C (Rhee *et al.* 1989). The SH2 binding region of PLCγ binds to phosphotyrosine residues within autophosphorylated EGF-R and PDGF-R (Anderson *et al.* 1990; Margolis *et al.* 1990). T cell activation leads to the tyrosine phosphorylation and activation of PLCγ (Mustelin *et al.* 1990; Weiss *et al.* 1991). TCRζ–CD3 engagement promotes the association of PLCγ with CD4–$p56^{lck}$, providing an example of the modulation of the CD4–$p56^{lck}$ complex via ligation of TCRζ–CD3.

CD4–$p56^{lck}$ and TCRζ–CD3–$p59^{fyn}$ complexes have the potential to induce numerous downstream events. Candidates for mediators include a p32 GTP binding protein, $p72^{raf}$, and a p110 Raf-1 related polypeptide, which associate with the CD4–$p56^{lck}$ complex (Telfer and Rudd 1991; Prasad and Rudd 1992; Thompson *et al.* 1992). The p32 protein was recognized by an antisera to the consensus GTP binding region (Gly–X_4–Gly–Lys) of the heterotrimeric G proteins. GTP binding and hydrolytic activity are associated with the CD4 complex and [α-^{32}P]GTP covalently labels p32 in UV-mediated cross-linking experiments (Telfer and Rudd 1991). Further studies will be required to determine the structure of p32 and whether p32 can hydrolyse GTP. The serine/threonine kinase $p72^{raf}$ was first identified as the transforming gene of the murine sarcoma virus 3611 (Jansen *et al.* 1984; Li *et al.* 1991). $p72^{raf}$ can be tyrosine phosphorylated and may associate in low amounts with the PDGF-R and other receptors (Morrison *et al.* 1989; Li *et al.* 1991). $p21^{ras}$ and $pp60^{v\text{-}src}$ appear to operate synergistically in the activation of downstream Raf-1 (Williams *et al.* 1992).

Growth factor activation of receptor-tyrosine kinases induces increased GTP occupancy on $p21^{ras}$, the activation of Raf-1, MAP kinase kinase, MAP-2 kinase (ERKs), S6 kinase, and others (Cheng *et al.* 1989; Downward *et al.* 1990; Hunter 1991; Li *et al.* 1991). Anti-TCRζ–CD3 cross-linking and constitutively active $pp60^{v\text{-}src}$ will activate $p21^{ras}$ (Smith *et al.* 1986; Downward 1990; Downward *et al.* 1990). Receptor-associated *lck* and *fyn* could theoretically regulate $p21^{ras}$ by a mechanism similar to the receptor kinase-mediated *drk–Sos* connection (Olivier *et al.* 1993; Simon *et al.* 1993). Further downstream, anti-CD3 activates the serine kinase MAP-2 kinase, an event that is augmented by CD4 co-ligation, and inhibited by anti-CD4 pre-ligation (Nel *et al.* 1990). MAP-2 kinase activity is regulated by the combined effects of serine and tyrosine phosphorylation. In T cell clones requiring CD4 expression for a functional response, the loss of CD4 is correlated with the loss of the phosphorylation and activation of MAP-2 kinase (Ettehadieh *et al.* 1992).

6 Other kinase–receptor interactions

Since the identification of the CD4/CD8–$p56^{lck}$ and TCRζ–CD3–$p59^{fyn}$ complexes, a variety of other receptor interactions with *src* family members have been described (Burkhardt *et al.* 1991; Hatakeyama *et al.* 1991; Rudd 1991; Yamanashi *et al.* 1991; Bell *et al.* 1992). *Src*-related kinases may therefore play a variety of functional roles in receptor signalling systems. In the case of T cells, reported interactions include associations between the 4-1BB antigen and $p56^{lck}$ (Kim *et al.* 1993), the interleukin-2 growth factor receptor (IL-2R), and $p56^{lck}/p59^{fyn(T)}$ (Hatakeyama *et al.* 1991), CD2 and $p56^{lck}/p59^{fyn(T)}$ (Bell *et al.* 1992), and phosphatidylinositol (GPI) anchored proteins and $p56^{lck}$/other kinases (Stefanova *et al.* 1991). The major difference between these interactions and the CD4–*lck*, CD8-*lck*, and TCRζ–CD3-*fyn* associations is their relative degree of promiscuity in binding multiple *src* family members. Interactions such as CD2–$p56^{lck}$ and IL–2R–$p56^{lck}$ have been reported to withstand extraction in harsher detergents, similar to CD4–$p56^{lck}$. The biological role of these other receptor–kinase interactions in intracellular signalling remains to be clarified.

6.1 $p59^{lck}$, $p59^{fyn}$, and the IL-2 receptor

Although the initial binding of antigen to the TCRζ–CD3 receptor results in T cell proliferation, a second receptor system, the interleukin-2 receptor

(IL-2R) serves as an obligatory second signal allowing progression to DNA synthesis (Smith 1988). The lymphokine interleukin-2 (IL-2) binds its receptor in an autocrine loop. The IL-2R is composed of two subunits, the IL-2R α chain (p55) and a β chain (p70–75) that bind IL-2 with low and intermediate affinities (k_D 10 nM and 1 nM). Formation of a IL-2R αβ heterodimer creates a high affinity version of the receptor (k_D 10 pM). Signalling occurs by the β chain containing intermediate and high affinity receptors, but not the low affinity homodimer (Hatakeyama *et al.* 1989). A recently cloned third chain γ (p64) associates with the β chain and participates in the formation of high and intermediate affinity receptors (Takeshita *et al.* 1992). Both α and β chains are integral type 1 proteins with cytoplasmic tails which possess no homology with kinases (Leonard *et al.* 1984). The length of the cytoplasmic tail of the IL-2 β is 286 residues, while that of the IL-2 α chain is only 13 amino acid residues. The β chain cytoplasmic region can be divided into a 'serine-rich' region and an 'acidic region'. Deletion analysis has shown the 'serine-rich' region (but not the 'acidic region') to be key to IL-2-mediated growth control (Hatakeyama *et al.* 1990). IL-2 binding induces tyrosine phosphorylation, thereby suggesting linkage to an intracellular kinase (Horak *et al.* 1991).

Hatakeyama *et al.* have described an interaction between $p56^{lck}$, $p59^{fyn}$, and the IL-2 β chain (Hatakeyama *et al.* 1991). Only 0.5–1% of intracellular *lck* associates with the receptor which (assuming 1:1 stoichiometry) could occupy as many as 10–30% of receptors. Mapping studies reveal that the presence of the 'acidic region' is crucial to complex formation (Hatakeyama *et al.* 1991). Within this region reside two Tyr residues (Tyr-355 and Tyr-358) that may serve as substrates for the kinase. Conversely, the N-terminal half of the kinase domain of $p56^{lck}$ is required for association with IL-2R β. Although $p56^{lck}$ and $p59^{fyn}$ may not be required for signalling by the IL-2R (Mills *et al.* 1992), the mystery over the assignment of kinase binding to a seemingly non-essential region of IL-2 β may be resolved with the demonstration that there exist two pathways, one leading to the induction of c-*jun* and c-*fos*, and another involving the induction of c-*myc*. Abrogation of $p56^{lck}$ binding interferes with c-*jun* and c-*fos*, but not c-*myc* induction (Shibuya *et al.* 1992).

6.2 $p56^{lck}$, $p59^{fyn}$, and the CD2 antigen

CD2 is a pan T cell antigen which binds to a structurally-related antigen, LFA-3, and plays a key role in a variety of T cell functions including activation-related lymphokine release and cytotoxicity (Meuer *et al.* 1984; Springer *et al.* 1987). Antibodies to distinct epitopes on human CD2 can activate T cells, an event dependent on the expression of the TCR ζ chain (Meuer *et al.* 1984; Moingeon *et al.* 1992). The binding of the natural ligand for CD2 fails to directly stimulate, instead acting as a costimulatory signal (Hunig *et al.* 1987). A proline/histidine-rich region within the cytoplasmic tail is required for signalling via this receptor (Chang *et al.* 1990; Beyers *et al.* 1991). Bell and co-workers (1992) have reported that both $p56^{lck}$ and $p59^{fyn}$ can be co-purified with rat CD2 in a variety of detergents including NP-40 (Bell *et al.* 1992). Specificity was shown in that the association could only be detected in $CD53^+$ subset of thymocytes, despite the fact that CD2 and $p56^{lck}$ are expressed at equivalent levels in both the $CD53^+$ and $CD53^-$ subsets. The interaction therefore appears to be developmentally regulated, by unknown mechanisms. Furthermore, the association is resistant to alkylating agents, arguing that the molecular mechanism of association differs from the CD4-$p56^{lck}$ interaction. CD2 ligation has also been reported to stimulate $p56^{lck}$ activity (Danielian *et al.* 1991). Whether $p56^{lck}$ can account for the stimulatory function of CD2 remains unclear.

7 Summary

Like the structurally similar transmembrane receptor kinases such as PDGF-R and EGF-R, the CD4/CD8–$p56^{lck}$ and TCRζ–CD3–$p59^{fyn(T)}$ complexes have the capacity to recruit downstream signalling molecules, and as such, have the potential to regulate divergent branches of a signalling cascade initiated by engagement of the CD4/CD8 and TCRζ–CD3 complexes. One signalling pathway may involve the kinase domain and its ability to phosphorylate targets such as CD5, TCR ζ, and PLCγ. Tyrosine phosphorylation can act directly to regulate the activity of the enzymes such as PLCγ. Alternatively, the phosphorylation of specific sites within the TCR ζ chain can recruit targets through phosphotyrosine–SH2 domain binding interactions. Another pathway involves the ability of *src* kinase SH3 domains to bind to PI-3-kinase. Other potential downstream targets include PI-4-kinase, GTP binding proteins, Raf-1 serine/threonine kinase, and MAP-2 kinase. The mechanism of binding and the means of regulation of these proteins by CD4/CD8–$p56^{lck}$ and the TCRζ/CD3–$p59^{fyn}$ complexes remains to be determined. Nevertheless, these enzymes and pathways represent many of the known cellular signal transducers, suggesting that the CD4/CD8–$p56^{lck}$ and TCRζ–CD3–$p59^{fyn}$ complexes are powerful switch points at the initiation of activation signals in T cells.

References

Abraham, N., Miceli, M. C., Parnes, J. R., and Veillette, A. (1991). Enhancement of T-cell responsiveness by the lymphocyte-specific tyrosine protein kinase p56lck. *Nature*, **350**, 62–6.

Anderson, D., Koch, C. A., Grey, L., Ellis, C., Moran, M. F., and Pawson, T. (1990). Binding of SH2 domains of phospholipase C gamma 1, GAP, and Src to activated growth factor receptors. *Science*, **250**, 979–82.

Appleby, M., Gross, J., Cooke, M., Levin, S., Qian, X., and Perlmutter, R. (1992). Defective T cell receptor signaling in mice lacking the thymic isoform of $p59^{fyn}$. *Cell*, **70**, 751–63.

Baniyash, M., Hsu, V. W., Seldin, M. F., and Klausner, R. D. (1988). The isolation and characterization of the murine T-cell antigen receptor zeta chain gene. *J. Biol. Chem.*, **263**, 9874–8.

Barber, E. K., Dasgupta, J. D., Schlossman, S. F., Trevillyan, J. M., and Rudd, C. E. (1989). The CD4 and CD8 antigens are coupled to a protein-tyrosine kinase ($p56^{lck}$) that phosphorylates the CD3 complex. *Proc. Natl. Acad. Sci. USA*, **86**, 3277–81.

Bell, G. M., Bolen, J. B., and Imboden, J. B. (1992). Association of src-like protein tyrosine kinases with the CD2 cell surface molecule in rat T lymphocytes and natural killer cells. *Mol. Cell. Biol.*, **12**, 5548–54.

Beyers, A. D., Davis, S. J., Cantrell, D. A., Izquierdo, M., and Williams, A. F. (1991). Autonomous roles for the cytoplasmic domains of the CD2 and CD4 T-cell surface antigens. *EMBO J.*, **10**, 377–85.

Beyers, A. D., Spruyt, L. L., and Williams, A. F. (1992). Molecular associations between the T-lymphocyte antigen receptor complex and the surface antigens CD2, CD4, or CD8, and CD5. *Proc. Natl. Acad. Sci. USA*, **89**, 2945–9.

Breitmeyer, J. B., Daley, J. F., Levine, H. B., and Schlossman, S. F. (1987). The T11 (CD2) molecule is functionally linked to the T3/Ti T cell receptor in the majority of T cells. *J. Immunol.*, **139**, 2899–905.

Brugge, J. S. and Darrow, D. (1984). Analysis of the catalytic domain of phosphotransferase activity of two avian sarcoma virus-transforming proteins. *J. Biol. Chem.*, **259**, 4550–7.

Bryant, D. L. and Parsons, J. T. (1984). Amino acid alterations within a highly conserved region of the Rous sarcoma virus src gene product pp60src inactivate tyrosine protein kinase activity. *Mol. Cell. Biol.*, **4**, 862–6.

Burgess, K. E., Odysseos, A. D., Zalvan, C., Druker, B. J., Anderson, P., Schlossman, S. F., *et al.* (1991). Biochemical identification of a direct physical interaction between CD4:$p56^{lck}$ and Ti(TcR)/CD3 complex. *Eur. J. Immunol.*, **21**, 1663–8.

Burgess, K. E., Yamamoto, M., Prasad, K. V. S., and Rudd, C. E. (1992). CD5 serves as a tyrosine substrate in a receptor complex including TcRζ, $p56^{lck}$ and $p59^{fyn}$. *Proc. Natl. Acad. Sci. USA*, **89**, 9311–15.

Burkhardt, A. L., Brunswick, M., Bolen, J. B., and Mond, J. J. (1991). Anti-immunoglobulin stimulation of B lymphocytes activates src-related protein-tyrosine kinases. *Proc. Natl. Acad. Sci. USA*, **88**, 7410–14.

Caron, L., Abraham, N., Pawson, T., and Veillette, A. (1992). Structural requirements for enhancement of T-cell responsiveness by the lymphocyte-specific tyrosine protein kinase p56lck. *Mol. Cell. Biol.*, **12**, 2720–9.

Cartwright, C. A., Kaplan, P. L., Cooper, J. A., Hunter, T., and Eckhart, W. (1986). Altered sites of tyrosine phosphorylation in pp60c-src associated with polyoma virus middle tumor antigen. *Mol. Cell. Biol.*, **6**, 1562–70.

Cartwright, C. A., Eckhart, W., Simon, S., and Kaplan, P. L. (1987). Cell transformation by pp60c-src mutated in the carboxy-terminal regulatory domain. *Cell*, **49**, 83–91.

Casnellie, J. E., Harrison, M. L., Hellstrom, K. E., and Krebs, E. G. (1982). A lymphoma protein with an *in vitro* site of tyrosine phophorylation homologous to that in pp60src. *J. Biol. Chem.*, **257**, 13877–9.

Chalupny, N. J., Ledbetter, J. A., and Kavathas, P. (1991). Association of CD8 with p56lck is required for early T cell signalling events. *EMBO J.*, **10**, 1201–7.

Chan, A. C., Iwashima, M., Turck, C. W., and Weiss, A. (1992). ZAP-70: A 70 Kd protein-tyrosine kinase that associates with the TCRζ chain. *Cell*, **71**, 649–62.
Chang, H.-C., Moingeon, P., Pedersen, R., Lucich, J., Stebbins, C., and Reinherz, E. L. (1990). Involvement of the PPPGHR motif in T-cell activation via CD2. *J. Exp. Med.*, **172**, 351–5.
Cheng, S. H., Harvey, R. W., Piwnica-Worms, H., Espino, P. C., Roberts, T. M., and Smith, A. E. (1989). Mechanism of activation of complexed pp60-src by the middle T antigen of polyomavirus. *Curr. Top. Microbiol. Immunobiol.*, **144**, 109–20.
Cicchetti, P., Mayer, B. J., Thiel, G., and Baltimore, D. (1992). Identification of a protein that binds to the SH3 region of Abl and is similar to Bcr and GAP-rho. *Science*, **257**, 803–6.
Clayton, L. K., Sieh, M., Pious, D. A., and Reinherz, E. L. (1989). Identification of human CD4 residues affecting class II MHC virus HIV-1 gp120 binding. *Nature*, **339**, 548–51.
Cochet, C., Fihol, O., Payrastre, B., Hunter, T., and Gill, G. N. (1991). Interaction between the epidermal growth factor receptor and phosphoinositide kinases. *J. Biol. Chem.*, **266**, 637–44.
Cooke, M. P. and Perlmutter, R. M. (1989). Expression of a novel form of the fyn proto-oncogene in hematopoietic cells. *New Biol.*, **1**, 66–74.
Cooke, M. P., Abraham, K. M., Forbush, K. A., and Perlmutter, R. M. (1991). Regulation of T cell receptor signaling by a *src* family protein-tyrosine kinase ($p59^{fyn}$). *Cell*, **65**, 281–91.
Cooper, J. A., Gould, K. L., Cartwright, C. A., and Hunter, T. (1986). Tyr527 is phosphorylated in pp60src: implications for regulation. *Science*, **231**, 1431–4.
Coughlin, S. R., Escobedo, J. A., and Williams, L. T. (1989). Role of phosphatidylinasitol kinase in PDGF receptor signal transduction. *Science*, **243**, 1191–4.
Cross, F. R., Garber, E. A., Pellman, D., and Hanafusa, H. (1985). A short sequence in the p60src N terminus is required for p60src myristoylation and membrane association and for cell transformation. *Mol. Cell. Biol.*, **4**, 1834–42.
da Silva, A. J. and Rudd, C. E. (1993). A 72Kd fyn-related polypeptide ($p72^{fyn\text{-}R}$) binds to the antigen-receptor/CD3 (TcR/CD3) complex. *J. Biol. Chem.*, **268**, 16537–43.
da Silva, A. J., Yamamoto, M., Zalvan, C. H., and Rudd, C. E. (1992). Engagement of the TcR/CD3 complex stimulates $p59^{fyn(T)}$ activity: detection of associated proteins at 72 and 120/130KD. *Mol. Immunol.*, **29**, 1417–25.
Danielian, S., Fagard, R., Alcover, A., Acuto, O., and Fischer, S. (1991). The tyrosine kinase activity of p56lck is increased in human T-cells activated via CD2. *Eur. J. Immunol.*, **21**, 1967–70.
Dasgupta, J. D., Granja, C., Druker, B., Lin, L. L., Yunis, E. J., and Relias, V. (1992). Phospholipase C-gamma 1 association with CD3 structure in T-cells. *J. Exp. Med.*, **175**, 282–8.
Davidson, D., Chow, L. M., Fournel, M., and Veillette, A. (1992). Differential regulation of T-cell antigen responsiveness by isoforms of the src-related tyrosine protein kinase p59fyn. *J. Exp. Med.*, **175**, 1483–92.
Davies, A. A., Ley, S. C., and Crumpton, M. J. (1992). CD5 is phosphorylated on tyrosine after stimulation of the T-cell antigen receptor complex. *Proc. Natl. Acad. Sci. USA*, **89**, 6369–72.
Deans, J. P., Kanner, S. B., Torres, R. M., and Ledbetter, J. A. (1992). Interaction of CD4: Ick with the T cell receptor/CD3 complex induces early signaling events in the absence of CD45 tyrosine phosphatase. *Eur. J. Immunol.*, **22**, 661–8.
Downward, J. (1990). The ras superfamily of small GTP-binding proteins. *Trends Biochem. Sci.*, **15**, 469–72.

Downward, J., Graves, J. D., Warne, P. H., Rayter, S., and Cantrell, D. A. (1990). Stimulation of p21 ras upon T-cell activation. *Nature*, **346**, 719–23.

Doyle, C. and Strominger, J. L. (1987). Interaction between CD4 and class II MHC molecules mediates cell adhesion. *Nature*, **330**, 256–9.

Drubin, D. G., Mulholland, J., Zhu, Z. M., and Botstein, D. (1990). Homology of a yeast actin-binding protein to signal transduction proteins and myosin-I. *Nature*, **343**, 288–90.

Eck, M. J., Shoelson, S. E. and Harrison, S. C. (1993). Recognition of a high affinity phosphotyrosil peptide by the Src homology-2 domain of $p56^{lck}$. *Nature*, **362**, 87–91.

Escobedo, J. A., Navankasattusas, S., Kavanaugh, W. M., Milfay, D., Fried, V. A., and Williams, L. T. (1991*a*). cDNA cloning of a novel 85 kd protein that has SH2 domains and regulates binding of PI3-kinase to the PDGF beta-receptor. *Cell*, **65**, 75–82.

Escobedo, J. A., Kaplan, D. R., Kavanaugh, W. M., Turck, C. W., and Williams, L. T. (1991*b*). A phosphatidylinositol-3 kinase binds to platelet-derived growth factor receptors through a specific receptor sequence containing phosphotyrosine. *Mol. Cell. Biol.*, **11**, 1125–32.

Ettehadieh, E., Sanghera, J. S., Peclech, S. L., Hess-Bienz, D., Watts, J., Shastri, N. *et al.* (1992). Tyrosyl phosphorylation and activation of MAP kinases by p56lck. *Science*, **255**, 853–5.

Fantl, W. J., Escobedo, J. A., Martin, G. A., Turck, C. W., del, R. M., McCormick, F., *et al.* (1992). Distinct phosphotyrosines on a growth factor receptor bind to specific molecules that mediate different signaling pathways. *Cell*, **69**, 413–23.

Frank, S. J., Niklinska, B. B., Orloff, D. G., Mercep, M., Ashwell, J. D., and Klausner, R. D. (1990). Structural mutations of the T cell receptor ζ chain and its role in T cell activation. *Science*, **249**, 174–7.

Frankel, A. D., Bredt, D. S., and Pabo, C. O. (1988). Tat protein from human immunodeficiency virus forms a matal-linked dimer. *Science*, **240**, 70–3.

Furey, W. F., Robbins, A. H., Clancy, L. L., Winge, D. R., Wang, B. C., and Stout, C. D. (1986). Crystal structure of Cd, Zn metallothionein. *Science*, **231**, 704–10.

Gassmann, M., Guttinger, M., Amrien, K. E., and Burn, P. (1992). Protein tyrosine kinase p59fyn is associated with the T-cell receptor CD3 complex in functional human lymphocytes. *Eur. J. Immunol.*, **22**, 283–6.

Gauen, L. K. T., Kong, A.-N. T., Samelson, L. E., and Shaw, A. S. (1992). $p59^{fyn}$ tyrosine kinase associates with multiple T-cell receptor subunits through its unique amino-terminal domain. *Mol. Cell. Biol.*, **12**, 5438–46.

Gentry, L. E., Chaffin, K. E., Shoyab, M., and Purchio, A. F. (1986). Novel serine phosphorylation of pp60c-src in intact cells after tumor promoter treatment. *Mol. Cell. Biol.*, **6**, 735–8.

Giorda, R., Rudert, W. A., Vavassori, C., Chambers, W. H., Hiserodt, J. C., and Trucco, M. (1990). NKR-PI, a signal transduction molecule on natural killer cells. *Science*, **249**, 1298–300.

Glaichenhaus, N., Shastri, N., Littman, D. R., and Turner, J. M. (1991). Requirement for association of $p56^{lck}$ with CD4 in antigen-specific signal transduction in T cells. *Cell*, **64**, 511–20.

Gould, K. L., Woodgett, J. R., Cooper, J. A., Buss, J. E., Shalloway, D., and Hunter, T. (1985). Protein kinase C phosphorylates pp60src at a novel site. *Cell*, **42**, 849–57.

Harding, C. V. and Unanue, E. R. (1990). Quantitation of antigen presenting cell MHC class II/peptide complexes necessary for T-cell stimulation. *Nature*, **346**, 574–6.

Hatakeyama, M., Tsudo, M., Minamoto, S., Kono, T., Doi, T., Miyata, T., *et al.* (1989). Interleukin-2 receptor beta chain gene: generation of three receptor forms by cloned human alpha and beta chain cDNA's. *Science*, **244**, 551–6.

Hatakeyama, M., Mori, H. M., Doi, T., and Taniguchi, T. (1990). A restricted cytoplasmic region of IL-2 receptor beta chain is essential for growth signal transduction but not ligand binding and internalization. *Cell*, **59**, 837–45.

Hatakeyama, M., Kono, T., Kobayashi, N., Kawahara, A., Levin, S. D., Perlmutter, R. M., *et al.* (1991). Interaction of the IL-2 receptor with the src-family kinase p56lck: identification of novel intermolecular association. *Science*, **252**, 1523–8.

Hirota, Y., Kato, J., and Takeyak, T. (1988). Substitution of Ser-17 of pp60c-src: biological and biochemical characterization in chicken embryo fibroblasts. *Mol. Cell. Biol.*, **8**, 1826–30.

Horak, I. D., Gress, R. E., Lucas, P. J., Horak, E. M., Waldmann, T. A., and Bolen, J. B. (1991). T-lymphocyte interleukin 2-dependent tyrosine protein kinase signal transduction involves the activation of p56lck. *Proc. Natl. Acad. Sci. USA*, **88**, 1996–2000.

Hu, P., Margolis, B., Skolnik, E. Y., Lammers, R., Ullrich, A., and Schlessinger, J. (1992). Interaction of phosphatidylinositol 3-kinase-associated p85 with epidermal growth factor and platelet-derived growth factor receptors. *Mol. Cell. Biol.*, **12**, 981–90.

Hunig, T., Tiefenthaler, G., Meyer zum Buschenfeld, K. M., and Meuer, S. C. (1987). Alternative pathway activation of T-cells by binding of CD2 to its cell-surface ligand. *Nature*, **326**, 298–301.

Hunter, T. (1991). Cooperation between oncogenes. *Cell*, **64**, 249–53.

Irving, B. A. and Weiss, A. (1991). The cytoplasmic domain of the T cell receptor ζ chain is sufficient to couple to receptor-associated signal transduction pathways. *Cell*, **64**, 891–901.

Irving, B. A., Chan, A. C., and Weiss, A. (1993). Functional characterization of a signal transducing motif present in the T cell antigen receptor ζ chain. *J. Exp. Med.*, **177**, 1093–103.

Jansen, H. W., Lurz, R., Blister, K., Bonner, T. I., Mark, G. E., and Rapp, U. R. (1984). Homologous cell-derived oncogenes in avian carcinoma virus MH2 and murine sarcoma virus 3611. *Nature*, **307**, 281–4.

June, C. H., Fletcher, M. C., Ledbetter, J. A., Schieven, G. L., Siegel, J. N., Phillips, A. F., *et al.* (1990). Inhibition of tyrosine phosphorylation prevents T-cell receptor-mediated signal transduction. *Proc. Natl. Acad. Sci. USA*, **87**, 7722–6.

Kamps, M. P. and Sefton, B. M. (1986). Neither arginine nor histidine can carry out the function of lysine-295 in the ATP-binding site of pp60src. *Mol. Cell. Biol.*, **6**, 751–7.

Kamps, M. P., Taylor, S. S., and Sefton, B. M. (1984). Direct evidence that oncogenic tyrosine kinases and cyclic AMP-dependent protein kinase have homologous ATP-binding sites. *Nature*, **310**, 589–92.

Kamps, M. P., Buss, J. E., and Sefton, B. M. (1985). Mutation of NH2-terminal glycine of p60src prevents both myristolation, and morphological transformation. *Proc. Natl. Acad. Sci. USA*, **82**, 4625–8.

Kaplan, D. R., Whitman, M., Schaffhausen, B., Pallas, D. C., White, M., Cantley, L., *et al.* (1987). Common elements in growth factor stimulation and oncogenic transformation: 85 kD phosphoprotein and phosphatidylinositol kinase activity. *Cell*, **50**, 1021–9.

Kaplan, D. R., Morrison, D. K., Wong, G., McCormick, F., and Williams, L.T. (1990). PDGF beta-receptor stimulates tyrosine phosphorylation of GAP and association of GAP with a signalling complex. *Cell*, **61**, 125–33.

Kavanaugh, W. M., Klippel, A., Escobedo, J. A., and Williams, L. T. (1992). Modification of the 85-kilodalton subunit of phosphatidylinositol-3 kinase in platelet-derived growth factor-stimulated cells. *Mol. Cell. Biol.*, **12**, 3415–24.

Kazlauskas, A., Ellis, C., Pawson, T., and Cooper, J. A. (1990). Binding of GAP to activated PGDF receptors. *Science*, **247**, 1578–81.

Kim, Y.-J., Zhou, Z., Pollok, K. E., Shaw, A., Bolen, J. B., Fraser, M., *et al.* (1993). Novel T cell antigen 4-1BB associates with the protein tyrosine kinase $p56^{lck}$. *J. Immunol.*, **151**, 1255–62.

Klausner, R. D. and Samelson, L. E. (1991). T cell antigen receptor activation pathways: the tyrosine kinase connection. *Cell*, **64**, 875–8.

Kmiecik, T. E. and Shalloway, D. (1987). Activation and suppression of pp60c-src transforming ability by mutation of its primary sites of tyrosine phosphorylation. *Cell*, **49**, 65–73.

Koretzky, G. A., Picus, J., Thomas, M. L., and Weiss, A. (1990). Tyrosine phosphatase CD45 is essential for coupling T-cell antigen receptor to the phosphatidyl inositol pathway. *Nature*, **346**, 66–8.

Koretzky, G. A., Picus, J., Schultz, T., and Weiss, A. (1991). Tyrosine phosphatase CD45 is required for T-cell antigen receptor and CD2-mediated activation of a protein tyrosine kinase and interleukin 2 production. *Proc. Natl. Acad. Sci. USA*, **88**, 2037–41.

Krieger, M. (1992). Molecular flypaper and atherosclerosis: structure of the macrophage scavenger receptor. *Trends Biochem. Sci.*, **17**, 141–6.

Kwon, B. S., Kim, G., Prystowsky, D., Lancki, D., Sabath, D., Pan, J., *et al.* (1987). Isolation and characterization of multiple species of T lymphocytes subset cDNA clones. *Proc. Natl. Acad. Sci. USA*, **84**, 2896–900.

Lamarre, D., Ashkenazi, A., Fleury, S., Smith, D. H., Sekaly, R. P. and Capron, D. J. (1989). The MHC-binding and gp120-binding functions of CD4 are separable. *Science*, **245**, 743–6.

Leahy, D. J., Axel, R., and Hendrickson, W. A. (1992). Crystal structure of a soluble form of the human T cell coreceptor CD8 at 26 Å resolution. *Cell*, **68**, 1145–62.

Leonard, W. J., Depper, J. M., Crabtree, G. R., Rudikoff, S., Pumphrey, J., Robb, R. J., *et al.* (1984). Molecular cloning and expression of cDNAs for the human interleukin-2 receptor. *Nature*, **311**, 626–31.

LeTourneur, F. and Klausner, R. D. (1992). Activation of T cells by a tyrosine kinase activation domain in the cytoplasmic tail of CD3. *Science*, **255**, 79–82.

LeTourneur, F., Gabert, J., Cosson, P., Blanc, D., Davoust, J., and Malissen, B. (1990). A signaling role for the cytoplasmic segment of the CD8α chain detected under limiting stimulatory conditions. *Proc. Natl. Acad. Sci. USA*, **87**, 2339–43.

Li, P., Wood, K., Mamon, H., Haser, W., and Roberts, T. (1991). Raf-1; a kinase currently without a cause but not lacking in effects. *Cell*, **64**, 479–84.

Luo, K. and Sefton, B. (1992). Activated lck tyrosine protein kinase stimulates antigen-independent interleukin-2 production in T-cells. *Mol. Cell. Biol.*, **12**, 4724–32.

Margolis, B., Rhee, S. G., Felder, S., Mervic, M., Lyall, R., Levitzki, A., *et al.* (1989). EGF induces tyrosine phosphorylation of phospholipase C-II: a potential mechanism for EGF receptor signalling. *Cell*, **57**, 1101–7.

Margolis, B., Li, N., Koch, A., Mohammadi, M., Hurwitz, D. R., Zilberstein, A., *et al.* (1990). The tyrosine phosphorylated carboxyterminus of the EGF receptor is a binding site for GAP and PLC-gamma. *EMBO J.*, **9**, 4375–80.

Mayer, B. J., Hamaguchi, M., and Hanafusa, H. (1988). A novel viral oncogene with structural similarity to phospholipase C. *Nature*, **332**, 272–5.

McCormick, F. (1989). Ras GTPase activating protein: signal transmitter and signal terminator. *Cell*, **56**, 5–8.

McGlade, C. J., Ellis, C., Reedijk, M., Anderson, D., Mbamalu, G., Reith, A. D., *et al.* (1992). SH2 domains of the p85 alpha subunit of phosphatidylinositol 3-kinase regulate binding to growth factor receptors. *Mol. Cell. Biol.*, **12**, 991–7.

Meuer, S. C., Hussey, R. E., Fabbi, M., Fox, D., Acuto, O., Fitzgerald, A., *et al.* (1984). An alternative pathway of T-cell activation: a functional role for the 50 kd TII sheep erythrocyte receptor protein. *Cell*, **36**, 897–906.

Mills, G. B., Arima, N., May, C., Hill, M., Schmandt, R., Li, J., *et al.* (1992). Neither the LCK nor the FYN kinases are obligatory for IL-2-mediated signal transduction in HTLV-I-infected human T cells. *Int. Immunol.*, **4**, 1233–43.

Moingeon, P., Lucich, J. L., McConkey, D. L., Letourneur, F., Malissen, B., Kochan, J., *et al.* (1992). CD3 zeta dependence of the CD2 pathway of activation in T lymphocytes and natural killer cells. *Proc. Natl. Acad. Sci. USA*, **89**, 1492–6.

Molina, T. J., Kishihara, K., Siderovski, D. P., van, E. W., Narendran, A., Timms, E., *et al.* (1992). Profound block in thymocyte development in mice lacking p56lck. *Nature*, **357**, 161–4.

Molloy, C. J., Bottaro, D. P., Fleming, T. P., Marshall, M. S., Gibbs, J. B., and Aaronson, S. A. (1989). PDGF induction of tyrosine phosphorylation of GTPase activating protein. *Nature*, **342**, 711–14.

Morrison, D. K., Kaplan, D. R., Escobedo, J. A., Rapp, U. R., Roberts, T. M., and Williams, L. T. (1989). Direct activation of the serine/threonine kinase activity of Raf-1 through tyrosine phosphorylation by the PDGF beta-receptor. *Cell*, **58**, 649–57.

Morrison, D. K., Kaplan, D. R., Rhee, S. G., and Williams, L. T. (1990). Platelet-derived growth factor (PDGF)-dependent association of phospholipase C-gamma with the PDGF receptor signaling complex. *Mol. Cell. Biol.*, **10**, 2359–66.

Musacchio, A., Noble, M., Pauptit, R., Wierenga, R., and Saraste, M. (1992). Crystal structure of a src-homology 3 (SH3) domain. *Nature*, **359**, 851–5.

Mustelin, T. and Altman, A. (1990). Dephosphorylation and activation of the T cell tyrosine kinase pp56lck by the leukocyte common antigen (CD45). *Oncogene*, **5**, 809–13.

Mustelin, T., Coggeshall, K. M., and Altman, A. (1989). Rapid activation of the T-cell tyrosine protein kinase pp56lck by the CD45 phosphotyrosine phosphatase. *Proc. Natl. Acad. Sci. USA*, **86**, 6302–6.

Mustelin, T., Coggeshall, K. M., Isakov, N., and Altman, A. (1990). T cell antigen receptor-mediated activation of phospholipase C requires tyrosine phosphorylation. *Science*, **247**, 1584–7.

Mustelin, T., Pessa-Morikawa, T., Autero, M., Gassman, M., Anderson, L. C., Gahmberg, P. G., *et al.* (1992). Regulation of the p59fyn protein tyrosine kinase by the CD45 phosphotyrosine phosphatase. *Eur. J. Immunol.*, **22**, 1173–8.

Nada, S., Okada, M., MacAuley, A., Cooper, J. A., and Nakagawa, H. (1991). Cloning of a complementary DNA for a protein-tyrosine kinase that specifically phosphorylates a negative regulatory site of $p60^{c\text{-}src}$. *Nature*, **351**, 69–72.

Nel, A. E., Pollack, S., Landreth, G., Ledbetter, J. A., Hultin, L., Williams, K., *et al.* (1990). CD3-mediated activation of MAP-2 kinase can be modified by ligation of the CD4 receptor. Evidence for tyrosine phosphorylation during activation of this receptor. *J. Immunol.*, **145**, 971–9.

Norment, A. M., Salter, R. D., Parham, P., Engelhard, V. H., and Littman, D. R. (1988). Cell-cell adhesion mediated by CD8 and MHC class I molecules. *Nature*, **336**, 79–81.

Olivier, J. P., Raabe, T., Henkemeyer, M., Dickson, B., Mbamalu, G., Margolis,

B., *et al.* (1993). A Drosophila SH2-SH3 adaptor protein implicated in coupling the Sevenless tyrosine kinase to an activator of ras guanine nucleotide exchange, Sos. *Cell*, **73**, 179–91.

Ostergaard, H. L., Shackelford, D. A., Hurley, T. R., Johnson, P., Hyman, R., Sefton, B. M., *et al.* (1989). Expression of CD45 alters phosphorylation of the lck-encoded tyrosine protein kinase in murine lymphoma T-cell lines. *Proc. Natl. Acad. Sci. USA*, **86**, 8959–63.

Ostergaard, H. L. and Trowbridge, I. S. (1990). Coclustering CD45 with CD4 or CD8 alters the phosphorylation and kinase activity of p56lck. *J. Exp. Med.*, **172**, 347–50.

Otsu, M., Hiles, I., Gout, I., Fry, M. J., Ruiz, L. F., Panayotou, G., *et al.* (1991). Characterization of two 85 kd proteins that associate with receptor tyrosine kinases, middle-T/pp60c-src complexes, and PI3-kinase. *Cell*, **65**, 91–104.

Overduin, M., Rios, C. B., Mayer, B. J., Baltimore, D., and Cowburn, D. (1992). Three-dimensional solution structure of the src homology 2 domain of c-abl. *Cell*, **70**, 697–704.

Partanen, J., Makela, T. P., Alitalo, R., Lehvaslaiho, H., and Alitalo, K. (1990). Putative tyrosine kinases expressed in K-562 human leukemia cells. *Proc. Natl. Acad. Sci. USA*, **87**, 8913–17.

Patschinsky, T., Hunter, T., Esch, F. S., Cooper, J. A., and Sefton, B. M. (1982). Analysis of the sequence of amino acids surrounding sites of tyrosine phosphorylation. *Proc. Natl. Acad. Sci. USA*, **79**, 973–7.

Patschinsky, T., Hunter, T., and Sefton, B. M. (1986). Phosphorylation of the transforming protein of Rous sarcoma virus: direct demonstration of phosphorylation of serine 17 and identification of an additional site of tyrosine phosphorylation in pp60-src of Prague Rous sarcoma virus. *J. Virol.*, **59**, 73–81.

Pawson, T. and Gish, G. (1992). SH2 and SH3 domains: from structure to function. *Cell*, **71**, 359–62.

Payne, G., Sholeson, S. E., Gish, G., Pawson, T., and Walsh, C. T. (1993). Kinetics of $p56^{lck}$ and $p60^{src}$ Src homology 2 domain binding to tyrosine-phosphorylated peptides determined by a competition assay or surface plasmon resonance. *Proc. Natl. Acad. Sci. USA*, **90**, 4902–6.

Pingel, J. T. and Thomas, M. L. (1989). Evidence that the leukocyte-common antigen is required for antigen-induced T lymphocyte proliferation. *Cell*, **58**, 1055–65.

Piwnica-Worms, H., Saunders, K. B., Roberts, T. M., Smith, A. E. and Cheng, S. H. (1987). Tyrosine phosphorylation regulates the biochemical and biological properties of pp60src. *Cell*, **49**, 75–82.

Prasad, K. and Rudd, C. (1992). A raf-1-related p110 polypeptide associates with the CD4-pp56lck complex in T cells. *Mol. Cell. Biol.*, **12**, 5260–7.

Prasad, K. V. S., Janssen, O., Kapeller, R., Raab, M., Cantley, L. C., and Rudd, C. E. (1993*a*). Src-homology 3 domain of protein kinase p59fyn mediates binding to phosphatidylinositol 3-kinase in T cells. *Proc. Natl. Acad. Sci. USA*, **90**, 7366–70.

Prasad, K. V. S., Kapeller, R., Janssen, O., Duke-Cohan, J. S., Repke, H., Cantley, L. C., *et al.* (1993*b*). Phosphatidylinositol 3-kinase and phosphatidylinositol 4-kinase binding to the CD4-p56lck complex: p56lck SH3 domain mediates binding to PI-3 kinase. *Mol. Cell. Biol.*, **13**, 7708–17.

Prasad, K. V. S., Kapeller, R., Janssen, O., Repke, H., Cantley, L. C., and Rudd, C. E. (1993*c*). Regulation of CD4-p56lck associated phosphatidylinositol 3-kinase (Pl 3-kinase) and phosphatidylinositol 4-kinase (PI 4-kinase). *Phil. Trans. R. Soc. Biol.*, **342**, 35–42.

Qian, D., Griswold-Prenner, I., Rosner, M. R., and Fitch, F. W. (1993). Multiple components of the T cell antigen receptor complex become tyrosine-phosphorylated upon activation. *J Biol. Chem.*, **6**, 4488–93.

Ren, R., Mayer, B. J., Cicchetti, P., and Baltimore, D. (1993). Identification of a ten-amino acid proline-rich SH3 binding site. *Science*, **259**, 1157–61.

Reth, M. (1989). Antigen receptor tail clue. *Nature*, **338**, 383–4.

Reynolds, A. B., Vila, J., Lansing, T. J., Potts, W. M., Weber, M. J., and Parsons, J. T. (1987). Activation of the oncogenic potential of the avian cellular src protein by specific structural alteration of the carboxy terminus. *EMBO J.*, **6**, 2359–64.

Reynolds, A. B., Kanner, S. B., Wang, H. C. R., and Parsons, J. T. (1989). Stable association of activated pp60src with two tyrosine-phosphorylated cellular proteins. *Mol. Cell. Biol.*, **9**, 3951–8.

Rhee, S. G., Suh, P. G., Ryu, S.-H., and Lee, S. Y. (1989). Studies of inositol phospholipid-specific phospholipase C. *Science*, **244**, 545–50.

Rodaway, A. R. F., Sternberg, M. J. E., and Bentley, D. L. (1989). Similarity in membrane proteins [letter]. *Nature*, **342**, 624.

Romeo, C. and Seed, B. (1991). Cellular immunity to HIV activated by CD4 fused to T-cell or Fc receptor polypeptides. *Cell*, **64**, 1037–46.

Rudd, C. E. (1991). CD4,CD8 and the TcR/CD3 complex: a novel class of protein-tyrosine kinase receptor. *Immunol. Today*, **11**, 400–5.

Rudd, C. E., Trevillyan, J. M., Dasgupta, J. D., Wong, L. L., and Schlossman, S. F. (1988). The CD4 receptor is complexed in detergent lysates to a protein-tyrosine kinase (pp58) from human T lymphocytes. *Proc. Natl. Acad. Sci. USA*, **85**, 5190–4.

Rudd, C. E., Anderson, P., Morimoto, C., Streuli, M., and Schlossman, S. F. (1989). Molecular interactions, T-cell subsets and a role of the CD4/CD8:p56^{lck} complex in human T-cell lymphocytes. *Immunol. Rev.*, **111**, 225–66.

Ryu, S. E., Kwong, P. D., Trunch, A., Porter, T. G., Artos, J., Rosenberg, M., *et al.* (1990). Crystal structure of an HIV-binding recombinant fragment of human CD4. *Nature*, **348**, 419–26.

Sahr, K. E., Laurila, P., Kotula, L., Scarpa, A. L., Coupal, E., Leto, T. L., *et al.* (1990). The complete cDNA and polypeptide sequences of human erythroid alpha-spectrum. *J. Biol. Chem.*, **265**, 4434–43.

Salter, R. D., Norment, A. M., Chen, B. P., Clayberger, C., Krensky, A. M., Littman, D. R., *et al.* (1989). Polymorphism in the alpha 3 domain of HLA-A molecules affects binding to CD8. *Nature*, **338**, 345–7.

Samelson, L. E., Patel, M. D., Weissman, A. M., Harford, J. B., and Klausner, R. (1986). Antigen activation of murine T cells induces tyrosine phosphorylation of a polypeptide associated with the T cell antigen receptor. *Cell*, **46**, 1083–90.

Samelson, L. E., Phillips, A. F., Luong, E. T., and Klausner, R. D. (1990). Association of the fyn protein-tyrosine kinase with the T-cell antigen receptor. *Proc. Natl. Acad. Sci. USA*, **87**, 4358–62.

Samelson, L. E., Egerton, M., Thomas, P. M., and Wange, R. L. (1992). The T-cell antigen receptor tyrosine kinase pathway. *Adv. Exp. Med. Biol.*, **323**, 9–16.

Sanders, S. K., Fox, R. O., and Kavathas, P. (1991). Cell-cell adhesion mediated by CD8 and human histocompatibility leukocyte antigen G, a nonclassical major histocompatibility complex class I molecule on cytotrophoblasts. *J. Exp. Med.*, **174**, 371–9.

Sarosi, G. A., Thomas, P. M., Egerton, M., Phillips, A. F., Kim, K. W., Bonvini, E., *et al.* (1992). Characterization of the T-cell antigen receptor-p60fyn protein tyrosine kinase association by chemical cross-linking. *Int. Immunol.*, **4**, 1211–17.

Shaw, A. S., Amrein, K. E., Hammond, C., Stern, D. F., Sefton, B. M., and Rose, J. K. (1989). The lck tyrosine protein kinase interacts with the cytoplasmic tail of the CD4 glycoprotein through its unique amino-terminal domain. *Cell*, **59**, 627–36.

Shaw, A. S., Chalupny, J., Whitney, A., Hammond, C., Amrein, K. E., Kavathas, P., *et al.* (1990). Short related sequences in the cytoplasmic domains of CD4 and CD8 mediate binding to the amino-terminal domain of p56lck tyrosine protein kinase. *Mol. Cell. Biol.*, **10**, 1853–62.

Shibuya, H., Yoneyama, M., Ninomiya-Tsuji, J., Matsumoto, K., and Taniguchi, T. (1992). IL-2 and EGF receptors stimulate the hematopoietic cell cycle via different signalling pathways: demonstration of a novel role for c-myc. *Cell*, **70**, 57–67.

Shiroo, M., Goff, L., Shivan, E., and Alexander, D. (1992). CD45 tyrosine phosphatase activated p59fyn couples the T cell antigen receptor to pathways of deacylglycerol production, protein kinase C activation and calcium flux. *EMBO J.*, **11**, 4887–97.

Simon, M. A., Bowtell, D. D. L., Dodson, G. S., Laverty, T. R., and Rubin, G. M. (1991). Ras1 and a putative guanine nucleotide exchange factor perform crucial steps in signaling by the sevenless protein tyrosine kinase. *Cell*, **67**, 701–16.

Simon, M. A., Dodson, G. S., and Rubin, G. M. (1993). An SH3-SH2-SH3 protein is required for p21Ras1 activation and binds Sevenless and Sos proteins *in vitro*. *Cell*, **73**, 169–77.

Skolnik, E. Y., Margolis, B., Mohammadi, M., Lowenstein, E., Fischer, R., Drepps, A., *et al.* (1991). Cloning of PI3 kinase-associated p85 utilizing a novel method for expression/cloning of target proteins for receptor tyrosine kinases. *Cell*, **65**, 83–90.

Sleckman, B. P., Peterson, A., Foran, J. A., Gorga, J. C., Kara, C. J., Strominger, J. L., *et al.*, (1988). Functional analysis of a cytoplasmic domain-deleted mutant of the CD4 molecule. *J. Immunol.*, **141**, 49–54.

Smart, J. E., Opperman, H., Czernilofsky, A. P., Purchio, A. F., Erickson, R. L., and Bishop, J. M. (1981). Characterization of sites for tyrosine phosphorylation in the transforming protein of Rous sarcoma virus (pp60v-src) and its normal cellular homologue (pp60c-src). *Proc. Natl. Acad. Sci. USA*, **78**, 6013–17.

Smith, K. A. (1988). Interleukin-2: inception, impact, and implications. *Science*, **240**, 1169–76.

Smith, M. R., DeGudicibus, S. J., and Stacey, W. (1986). Requirement for c-ras proteins during viral oncogene transformation. *Nature*, **320**, 540–3.

Snyder, M. A., Bishop, J. M., McGrath, J. P., and Levinson, A. D. (1985). A mutation at the ATP-binding site of pp60v-src abolishes kinase activity, transformatiom, and tumorigenicity. *Mol. Cell. Biol.*, **5**, 1772–9.

Songyang, Z., Shoelson, S. E., Chaudhuri, M., Gish, G., Pawson, T., Haser, W. G., *et al.* (1993). SH2 domains recognize specific phosphopeptide sequences. *Cell*, **72**, 767–78.

Springer, T. A., Dustin, M. L., Kishimoto, T. K., and Marlin, S. D. (1987). The lymphocyte function-associated LFA-1, CD2, and LFA-3 molecules: cell adhesion receptors of the immune system. *Annu. Rev. Immunol.*, **5**, 223–52.

Stahl, M. L., Ferenz, C. R., Kelleher, K. L., Kriz, R. W., and Knopf, J. L. (1988). Sequence similarity of phospholipase C with the non-catalytic region of src. *Nature*, **332**, 269–72.

Stefanova, I., Horejsi, V., Ansotegui, I. J., Knapp, W., and Stockinger, H. (1991). GPI-anchored cell-surface molecules complexed to protein tyrosine kinases. *Science*, **254**, 1016–19.

Stein, P. L., Lee, H. M., Rich, S., and Soriano, P. (1992). pp59fyn mutant mice display differential signaling in thymocytes and peripheral T cells. *Cell*, **70**, 741–50.

Straus, D. and Weiss, A. (1992). Genetic evidence for the involvement of the lck tyrosine kinase in signal transduction through the T cell antigen receptor. *Cell*, **70**, 585–93.

Suh, P.-G., Ryu, S. H., Moon, K. H., Suh, H. W., and Rhee, S. G. (1988). Inositol phospholipid-specific phospholipase C: complete cDNA and protein sequences. *Proc. Natl. Acad. Sci. USA*, **85**, 5419–23.

Takeshita, T., Asao, H., Ohtani, K., Ishii, N., Kumaki, S., Tanaka, N., *et al.* (1992). Cloning of the gamma chain of the IL-2 receptor. *Science*, **257**, 379–82.

Telfer, J. C. and Rudd, C. E. (1991). A 32Kd GTP-binding protein associated with CD4:p56^{lck} and CD8:p56^{lck} T cell receptor complexes. *Science*, **254**, 439–41.

Thompson, P. A., Ledbetter, J., Rapp, U. R., and Bolen, J. B. (1992). The Raf-1 serine-threonine kinase is a substrate for the p56lck protein tyrosine kinase in T cells. *Cell Growth Different.*, **2**, 609–17.

Townsend, A. and Bodmer, H. (1989). Antigen recognition by class I restricted T lymphocytes. *Annu. Rev. Immunol.*, **7**, 601–24.

Tsygankov, A., Broker, B., Fargnoli, J., Ledbetter, J., and Bolen, J. (1992). Activation of tyrosine kinase p60^{fyn} following T cell antigen receptor cross-linking. *J. Biol. Chem.*, **267**, 18259–62.

Turner, J. M., Brodsky, M. H., Irving, B. A., Levin, S. D., Perlmutter, R. M., and Littman, D. R. (1990). Interaction of the unique N-terminal region of tyrosine kinase p56^{lck} with cytoplasmic domains of CD4 and CD8 is mediated by cystein motifs. *Cell*, **60**, 755–65.

Unanue, E. R. and Cerottini, J. C. (1989). Antigen Presentation. *FASEB J.*, **3**, 2496–502.

Valius, M. and Kazlauskas, A. (1993). Phospholipase γ 1 and phosphatidylinositol 3 kinase are the downstream mediators of the PDGF receptor's mitogenic signal. *Cell*, **73**, 321–34.

Van de Velde, H., von Hoegen, I., Luo, W., Parnes, J. R., and Thielemans, K. (1991). The T/B cell antigen, CD5, and the B-cell surface protein, CD72, form a pair of interacting receptors. *Nature*, **351**, 662–5.

Veillette, A., Bookman, M. A., Horak, E. M., and Bolen, J. B. (1988). The CD4 and CD8 T cell surface antigens are associated with the internal membrane tyrosine-protein kinase p56^{lck}. *Cell*, **55**, 301–8.

Veillette, A., Bolen, J. B., and Bookman, M. A. (1989). Alterations in tyrosine protein phosphorylation induced by antibody-mediated cross-linking of the CD4 receptor of T lymphocytes. *Mol. Cell. Biol.*, **9**, 4441–6.

Veillette, A., Caron, L., Fournel, M., and Pawson, T. (1992). Regulation of the enzymatic function of the lymphocyte-specific tyrosine protein kinase p56lck by the non-catalytic SH2 and SH3 domains. *Oncogene*, **7**, 971–80.

Volarevic, S., Burns, C. M., Sussman, J. J., and Ashwell, J. D. (1990). Intimate association of Thy-1 and the T cell antigen receptor with the CD45 tyrosine phosphatase. *Proc. Natl. Acad. Sci.*, **87**, 7085–9.

Waksman, G., Kominos, D., Robertson, S. C., Pant, N., Baltimore, D., Birge, R. B., *et al.* (1992). Crystal structure of the phosphotyrosine recognition domain SH2 of v-src complexed with tyrosine-phosphorylated peptides [see comments]. *Nature*, **358**, 646–53.

Waksman, G., Shoelson, S. E., Pant, N., Cowburn, D., and Kuriyan, J. (1993). Binding of a high affinity phosphotyrosil peptide to the src SH2 domain: crystal structures of the complexes and peptide-free forms. *Cell*, **72**, 779–90.

Wang, H. C. and Parson, J. T. (1989). Deletions and insertions within an amino-terminal domain of pp60v-src inactivate transformation and modulate membrane stability. *J. Virol.*, **63**, 291–302.

Wang, J., Yan, Y., Garrett, T. P. J., Tarr, G. E., Husain, Y., Reinherz, E., *et al.*

(1990). Atomic structure of a fragment of human CD4 containing two immunoglobulin-like domains. *Nature*, **348**, 411–18.

Ward, S. G., Reif, K., Ley, S., Fry, M. J., Waterfield, M. D., and Cantrell, D. A. (1992). Regulation of phosphoinositide kinases in T cells. Evidence that phosphatidylinositol 3-kinase is not a substrate for T cell antigen receptor-regulated tyrosine kinases. *J. Biol. Chem.*, **267**, 23862–9.

Weber, J. R., Bell, G. M., Han, M. Y., Pawson, T., and Imboden, J. B. (1992). Association of the tyrosine kinase LCK with phospholipase C-gamma 1 after stimulation of the T cell antigen receptor. *J. Exp. Med.*, **176**, 373–9.

Wegener, A.-M. K., Letourneur, F., Hoeveler, A., Brocker, T., Luton, F., and Malissen, B. (1992). The T cell receptor/CD3 complex is composed of at least two autonomous transduction modules. *Cell*, **68**, 83–95.

Weiss, A., Koretzky, G., Schatzman, R. C., and Kadlecek, T. (1991). Functional activation of the T-cell antigen receptor induces tyrosine phosphorylation of phospholipase C-γ1. *Proc. Natl. Acad. Sci. USA*, **88**, 5484–8.

Wendler, P. A. and Boschelli, F. (1989), Src homology 2 domain deletion mutants of p60v-src do not phosphorylate cellular proteins of 120–150 kDa. *Oncogene*, **4**, 231–6.

Williams, N. G., Roberts, T. M., and Li, P. (1992). Both p21ras and pp60v-src are required, but neither alone is sufficient to activate the Raf-1 kinase. *Proc. Natl. Acad. Sci. USA*, **89**, 2922–6.

Winkler, D. G., Park, I., Kim, T., Payne, N. S., Walsh, C. T., Strominger, J. L. *et al.* (1993). Phosphorylation of Ser-42 and Ser-59 in the N-terminal region of the tyrosine kinase $p56^{lck}$ *Proc. Natl. Acad. Sci. USA*, **90**, 5176–80.

Winkler, D. G., Park, I., Payne, N. S., Walsh, C., and Strominger, J. L. (1993). *J. Cell. Biochem.*, **17A**, 260.

Xu, H. and Littman, D. R. (1993). A kinase-independent function of LCK in potentiating antigen-specific T cell activation. *Cell*, **74**, 633–43.

Yamanashi, Y., Kakiuchi, T., Mizuguchi, J., Yamamoto, T., and Toyoshima, K. (1991). Association of B cell antigen receptor with protein tyrosine kinase Lyn. *Science*, **251**, 192–4.

Yu, H., Rosen, M. K., Shin, T. B., Seidel-Dugan, C., Brugge, J. S., and Schreiber, S. L. (1992). Solution structure of the SH3 domain of src and identification of its ligand-binding site. *Science*, **258**, 1665–8.

Zamoyska, R., Derham, P., Gorman, S. D., von Hoegen, P., Bolen, J. B., Veillette, A., *et al.* (1989). Inability of CD8α' polypeptides to associate with $p56^{lck}$ correlates with impaired function *in vitro* and lack of expression *in vivo*. *Nature*, **342**, 278–81.

10 Transgenesis and the T cell receptor

ANDREW L. MELLOR

1 Introduction

In 1988 two groups reported that they had created transgenic mice using productively rearranged TCR genes derived from murine $CD8^+$ T cell clones (Kisielow *et al.* 1988*a*; Sha *et al.* 1988*a*). This achievement represented a significant watershed in immunological research since the technical barrier posed by the enormous structural heterogeneity of the T cell repertoire in normal mice had been circumvented. The advent of TCR transgenic (TCR–Tg) mice allowed the fate of thymocytes and T cells expressing particular TCR molecules to be followed with relative ease using TCR-specific monoclonal antibodies. Thus, entirely new opportunities for studying thymocyte development and selection as well as the role of T cells in a wide range of immunological phenomena were opened up. The potential utility of TCR–Tg mice became widely appreciated in the wake of the initial reports of their creation. Since then, TCR–Tg mice have been generated in many laboratories and have been used to study many aspects of T cell biology. It is my intention in this chapter to describe how TCR–Tg mice are made and how they can be used in research to illustrate their pivotal importance in contemporary research into T cell development and function. In doing so it is not my intention to give details of results and interpretations of experiments involving the use of TCR–Tg mice since many recent and comprehensive reviews of this field are available—see chapters in *Immunological Reviews*, volumes 122 (1991); 133 (1993); 135 (1993). Rather, it is my intention to provide a practical review of advantages and disadvantages of using TCR–Tg mice as well as a brief overview of the practical applications which are of most interest to immunologists.

2 Background

Transgenesis is a powerful method for studying gene expression *in vivo* and for studying phenotypic effects of transgene expression in mice. Detailed reviews of how to make, identify, and analyse transgene expression in Tg mice are available (Hogan *et al.* 1986; Grosveld and Kollias 1992; Murphy and Carter 1993). Methods for creating Tg mice were developed in the early 1980s. Injection of purified DNA into fertilized mouse oocytes is the most frequently used method for introducing new genetic material into the germline of mice although other methods, involving transfection of totipotent embryonic stem (ES) cells, have also been used successfully. This ES cell route is obligatory for creating 'knock-out' Tg mice which have lost gene function as a result of homologous recombination into a particular gene. Control elements necessary for regulating gene transcription and expression can be identified in Tg mice provided that large enough segments of cloned DNA are available to allow 'properly regulated' (during development and tissue-specific) expression of the transgene in mice. Large constructs are then subjected to progressive deletion of DNA until proper regulation of transgene expression no longer occurs. The study of gene expression in thymocytes and T cells is, in itself, a major activity in the field of T cell biology which has benefited from the application of transgenesis techniques. However, the main reason why Tg mice are now so frequently used in immunological research is the potential for creating mice which express a single novel protein of immunological interest on a particular genetic background from inbred laboratory strains of mice. By comparing biological phenotypes of Tg and matched non-Tg mice, newly acquired functions can be rigorously assessed and unequivocally attributed as consequences of transgene expression. TCR–Tg mice have already been used to unravel some of the mysteries surrounding the development and function of thymocytes and T cells. Initially, TCR–Tg mice were made with very specific goals in mind; namely to help resolve the protracted controversies surrounding the cellular basis for immunological tolerance and repertoire selection.

TCR–Tg mice are usually made by microinjecting fertilized mouse oocytes with DNA containing TCR genes and transferring the injected oocytes back to foster mothers. However, there are two special considerations in the case of TCR genes. First, two genes, TCR α and TCR β genes are required to create unique TCRs. This issue is resolved by co-injecting DNA constructs containing TCR α and TCR β chain genes. In general, DNA fragments injected into oocytes are ligated to each other before they become stably integrated into mouse chromosomal DNA as long concatamers. This ensures that TCR α and TCR β genes are closely linked and are inherited as a single genetic locus in which TCR α and TCR β genes are co-expressed. Secondly, TCR genes exist in germline and productively rearranged configurations. Unless one is interested in studying the processes of TCR gene rearrangement (Chapter 15) or allelic exclusion (see Chapter 16) there is little point in using TCR genes in the germline configuration as transgenes since they will be subject to rearrangement and allelic exclusion as are endogenous TCR genes during thymocyte development. Consequently, transgenic TCRs will be expressed on a low percentage of T cells. However, use of productively rearranged TCR transgenes has a profound effect on the TCR repertoire since TCR transgenes prevent rearrangement of endogenous TCR genes due to allelic exclusion. If allelic exclusion is efficient, nearly all thymocytes and T cells will express a single TCR clonotype and the resultant T cell repertoire will be, essentially, monospecific. The frequency of T cells expressing a particular TCR clonotype in normal mice has been estimated to be in the range of 10^{-4} to 10^{-6}. Thus, the advantage of using TCR–Tg mice is enormous since it becomes possible to follow the fate of thymocytes expressing particular TCR clonotypes under different circumstances.

3 Generation of TCR–Tg mice

In this section I will discuss practical factors which should be considered when designing a strategy for making TCR–Tg mice (Figure 10.1).

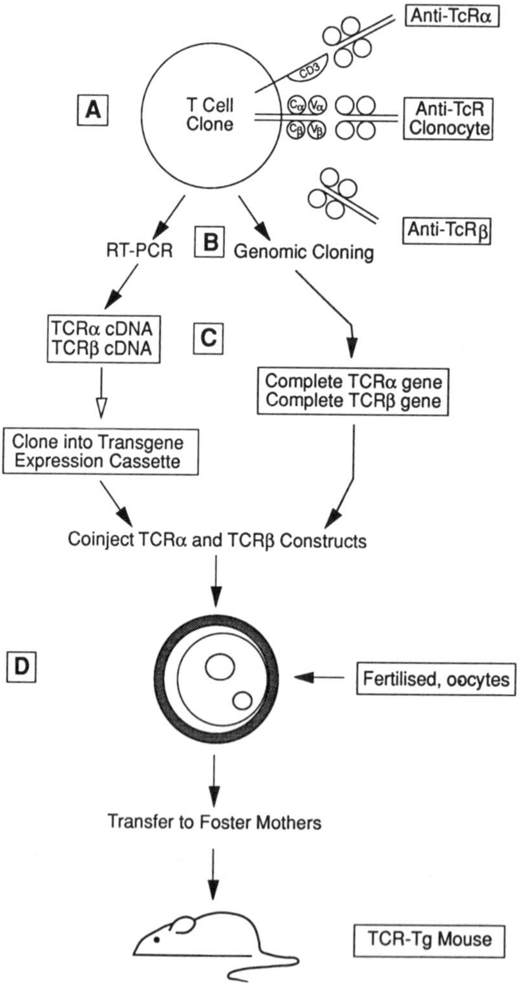

Fig. 10.1 From T cell clone to TCR transgenic mice. See text for discussion.

3.1 Source of DNA constructs for transgenesis

DNA containing TCR genes in productively rearranged configurations can be obtained from only one source, cloned T cells (Figure 10.1A). The choice of T cell clone is dictated by research goals; often the MHC and/or antigen specificity of interest is the key factor influencing choice of T cell clone. Usually, this will also dictate whether the T cell clone is a $CD4^+$ helper-type or a $CD8^+$ cytotoxic T cell clone. It is possible to use T cell clones from other species (e.g. humans, Viney *et al.* 1992; Rothe *et al.* 1993), but other factors must then be considered since mouse T cells expressing xeno-TCRs may not develop and/or function properly due to suboptimal interaction with:

(1) murine MHC molecules;

(2) other cell surface structures required for optimal surface expression of the TCR–CD3 complex;

(3) intracellular molecules involved in signal transduction.

Several other points should be considered before isolating TCR genes from T cell clones. First, it may be useful to know which TCR V region gene segments are present in the TCR expressed by the T cell clone since thymocytes expressing certain thymocyte subsets which express particular TCR Vβ gene segments are subject to superantigen-mediated elimination irrespective of the nature of the Vα segment used. Many TCR Vβ chains and a few TCR Vα chains can be discriminated by staining T cell clones with monoclonal antibodies which recognize specific TCR V region epitopes. Secondly, the availability of anti-TCR clonotypic monoclonal antibodies is very useful for studies on TCR–Tg mice. Although antibodies specific for TCR Vβ or TCR Vα chains can also be used to detect thymocytes and T cells expressing TCR transgenes this is not ideal since transgenic TCR Vβ chains may associate with TCR Vα chains expressed as a result of rearrangements of endogenous TCR Vα genes. Thus, it is highly desirable to have anti-TCR clonotypic reagents available. Consequently, it is worth trying to generate such reagents when TCR–Tg mice are to be used for long-term studies.

Once an appropriate T cell clone is available one of two methods are used to isolate DNA containing productively rearranged TCR α and TCR β genes from chromosomal DNA of the T cell clone (Figure 10.1B). Entire TCR genes can be cloned directly by genomic cloning procedures. TCR genes, particularly TCR α genes in mice, can be spread over as much as 30–40 kb of DNA. Consequently, a technique which allows large pieces of DNA to be cloned (e.g. cosmid cloning) should be used. The alternative, and now more frequently used procedure, involves cDNA synthesis and PCR amplification (RT–PCR) of TCR sequences using RNA isolated from T cell clones. After cDNA synthesis a panel of PCR primers which define unique sequences at the 5′ end of all known TCR Vα and TCR Vβ regions is used to identify the TCR V regions present in the rearranged genes (Casanova *et al.* 1991). PCR can then be used to sequence and to clone TCR cDNA into appropriate vectors or DNA cassettes for transgene expression. The advantage of this strategy is that it is relatively fast and does not involve construction of a cloned DNA library. However, transcriptional control elements and other regulatory elements essential for correct gene expression must be provided from another source since cDNA amplified by PCR from T cell RNA will not contain these elements.

3.2 Expression of TCR transgenes

It is important that transgenes are expressed in appropriate cell types in transgenic mice. Ideally, the best way to achieve this is to incorporate transcriptional control elements from the actual gene of interest into the DNA for injection (Figure 10.1C). However, CD3 expression in the same cells is obligatory for surface expression of TCR αβ heterodimers (see Chapter 16). In effect, this limits cell surface expression of TCR chains to cells expressing CD3 (i.e. thymocytes and T cells), even if the TCR transgenes are expressed in other cells. This fact has been exploited by investigators who have used promoters derived from human CD2 (Mamalaki *et al.* 1992) or murine MHC class I genes (Pircher *et al.* 1989*a*,*b*) to drive expression of TCR transgenes. In practice, patterns of TCR transgene expression in thymocytes and T cells which are identical to the normal pattern of TCR regulation are difficult to achieve since there are nearly always subtle variations in timing and/or levels of expression of TCR molecules in TCR–Tg mice even when genomic constructs from murine TCR genes are used. In addition, TCR expression in non-Tg mice is normally regulated by rearrangement of the TCR α chain genes (see Chapters 2 and 5). Consequently, surface expression in TCR–Tg mice occurs earlier since the rate-limiting step is expression of CD3 polypeptides.

Genomic DNA obtained directly from chromosomal DNA of a T cell clone should include control elements in the cloned DNA flanking the coding regions of each TCR gene if the cloned DNA segment is large enough. Enhancer and other locus control elements which are required for tissue-specific, position-independent, copy number-dependent transgene expression may be some distance from structural genes and may not be included in cloned DNA. Thus, enhancer elements may have to be inserted into TCR constructs to ensure that correct regulation of transcriptional activity occurs in mice. For TCR–Tg mice, enhancers need only be inserted into one gene construct since co-injected DNA fragments integrate into mouse chromosomal DNA as a single ligated unit allowing enhancer elements to operate on both transgenes. Whether one has included all relevant control elements in the transgene constructs can only be assessed empirically by injecting the DNA fragment into oocytes and determining the pattern of transgene transcription and expression. Initially, TCR–Tg mice were made using this approach (Kisielow *et al.* 1988*a*; Sha *et al.* 1988*a*). When these experiments were successful, it was possible to use the same TCR gene constructs as cassettes into which new TCR structural genes could be placed by recombination and replacement of original TCR coding regions. In this way control elements required for correct transgene expression could be reused in new TCR gene constructs. Several expression cassettes based on cloned TCR genes are now available. In principle, new TCR transgenes can be generated quickly and routinely via the RT–PCR route and inserted into TCR expression cassettes. However, this procedure can be complicated for several reasons:

(1) if either TCR V region cannot be identified using the two sets of PCR primers;

(2) more than one rearranged Vα or Vβ region is detected by PCR, necessitating identification of which V region is required for correct specificity;

(3) recombination of DNA from the cassette and new TCR cDNA may not be straightforward and depends on the availability of restriction enzyme sites.

These problems can be resolved by:

(1) making a total cDNA library or using PCR protocols which require knowledge of DNA sequence from only one end of DNA segments to be amplified (anchored PCR);

(2) sequencing of all PCR products obtained and, if necessary, use of all cloned TCR sequences for transfection or transgenesis to establish which combination gives the required specificity;

(3) multiple cloning steps or engineering of restriction enzyme sites to facilitate cloning.

In general, the following points should be considered carefully. TCR cDNA obtained by RT–PCR should be:

(1) full-length or, at least, contain the entire V–(D)–J regions;

(2) preferably contain restriction enzyme sites for rapid cloning at each end;

(3) include signals necessary for polyadenylation at the 5′ end of RNA transcripts if not provided in the expression cassette (and vice versa if they are provided).

Expression cassettes should:

(1) have restriction enzyme cloning sites for cDNAs adjacent to the transcriptional promoter;

(2) include all control elements for correct transcription and RNA processing (promoter, enhancers, tissue-specific regulatory elements, at least one intron);

(3) have unique restriction enzyme sites so that vector (plasmid) DNA can be separated from cloned eukaryotic DNA prior to oocyte injection.

3.3 Mouse production

Frequently, genetic background is of critical importance in immunological research because of the effect that polymorphic gene products have on shaping the repertoire of lymphocyte receptor specificity. Hence the genetic background should be carefully considered when making TCR–Tg mice (Figure 10.1D) for use in long-term projects. In the past, the majority of Tg mice have been made using second generation (F2) oocytes from two laboratory inbred mouse strains with different genetic backgrounds (Hogan *et al.* 1986; Grosveld and Kollias 1992; Murphy and Carter 1993). This means that each transgenic founder mouse possesses a unique set of background genes inherited as segregating sets of genes derived from both parental strains and can be homozygous, hemizygous, or null for alleles from each parental strain. This may not present a great problem since MHC haplotype and, in some cases superantigen genes, exert the most prominent influence on T cell repertoire selection (see below); heterogeneity at these loci can often be resolved in a single generation by judicious choice of mating partners. Thus, in the example outlined in Figure 10.2B, a founder mouse hemizygous for the transgene (+/−) and MHC haplotype (k/b) can be used to generate transgenic offspring homozygous for either MHC haplotype (k/k or b/b) by mating to non-Tg (−/−) partners of either MHC haplotype (k/k or b/b). On average, a quarter of the mice born of such matings will be transgenic on a homozygous MHC haplotype. However, TCR–Tg lineages derived from a single founder mouse and maintained by repeated backcrossing to partners from a particular inbred strain will not acquire 'inbred' genetic backgrounds until the 12th and subsequent backcross generation. This can take up to two years to achieve and, where influences from background genes must be eliminated completely as in studies on tissue transplantation, this can be a major problem. One solution is to generate Tg mice using 'inbred' oocytes where both parents are from one inbred strain; pure breeding inbred Tg lines can then be generated in only two generations (Figure 10.2A). This is feasible for several inbred strains including CBA/Ca (Sponaas *et al.* 1994) and C57BL strains (Mamalaki *et al.* 1992). Other inbred strains can be more difficult to use for transgenesis due to lower yields of fertilized oocytes and/or lower survival rates of oocytes following injection and culture *in vitro*. One indication of the potential of an inbred strain for transgenesis is the ease with which they can be bred as a colony. In general, the ease of using a given strain for transgenesis correlates with the ease of breeding the colony. Some, apparently identical, inbred strains breed better in some Institutions than others so difficulties in using a particular strain in one Institution may be less severe in another.

For TCR–Tg mice, the influence of genetic background can be profound due, in particular, to two sets of genes: MHC genes and endogenous superantigen (SAg) genes (see Acha-Orbea, Chapter 11). Proteins encoded by MHC and SAg genes interact directly with TCR molecules. Although the specificity of TCRs for MHC (+ peptide) ligands depends jointly on the TCR

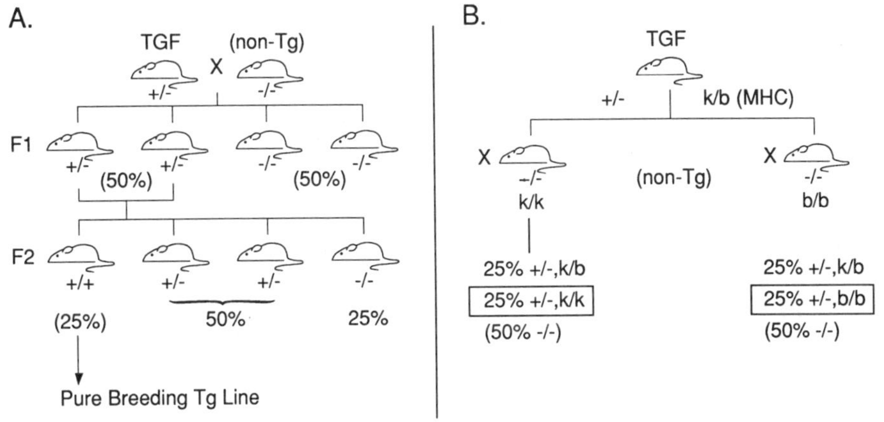

Fig. 10.2 A. Breeding transgenic mice to homozygosity. If founder mice and non-transgenic partners are from the same inbred strain a pure breeding transgenic line can be established after the second generation. B. Resolving hemizygosity at a single background locus (e.g. MHC) by choice of appropriate partners for breeding. See text for details.

α and TCR β chains (Hilyand *et al.*, Chapter 18), interactions between TCRs and SAgs usually involve the TCR β chain only (Acha-Orbea, Chapter 11). Thus, profound alterations to the fate of thymocytes and T cells can occur depending on the MHC and SAg genes inherited and the specificity of the clonotype TCR encoded by the transgenes (see Section 4.1). Often these effects can be predicted in advance since the MHC and antigen specificities as well as the TCR Vβ family of the TCR expressed by a T cell clone is known prior to making TCR–Tg mice. Consequently, undesirable interactions between transgenic TCR molecules and endogenous MHC and SAg gene products can be avoided by shrewd choice of inbred strain for oocyte production. However, technical limitations on the choice of strain for oocyte production may limit options such that unfavourable TCR–ligand interactions can not be avoided. In these cases, TCR–Tg mice must be made on an inappropriate genetic background and bred on to a more appropriate background by repeated backcrosses and selection of those TCR–Tg mice which do not inherit inappropriate MHC or SAg gene alleles. Even if unfavourable TCR interactions with MHC and/or SAgs can be anticipated and avoided, unanticipated background gene effects can influence thymocyte and T cell development or function in TCR–Tg mice. These effects can be circumvented completely only if background genes of mice used for T cell cloning and for oocyte production are matched perfectly and this is rarely possible.

Several types of genetic background in which genes required for TCR gene rearrangement are defective are now available. *Scid* mice have no T or B cells due, it is thought, to a defective gene coding for a DNA ligase required after TCR gene rearrangement has occurred (Chapter 15). However, this is a relatively leaky mutation and better options are now available as *RAG* gene 'knock-out' transgenic mice in which *RAG-1* or *RAG-2* genes have been inactivated via ES cell-mediated gene targeting (Chapter 15). Backcrossing of TCR transgenes on to these genetic backgrounds can be worthwhile since endogenous TCR gene rearrangements can not take place. This is particularly useful when anti-clonotypic reagents are not available.

TCR–Tg mice are immunodeficient since allelic exclusion mediated by the TCR transgenes ensures that the normal heterogeneity of the T cell repertoire is almost completely suppressed. This can lead to health problems which are peculiar to TCR–Tg mice. Thus, facilities for the production and breeding of TCR–Tg mice should be as clean as possible to minimize health risks. This means that barrier facilities should be used where possible and access should be kept to an absolute minimum. In extreme cases, some researchers generate Tg mice under specific pathogen-free (SPF) conditions; in this case injection apparatus must be housed in the SPF facility.

3.4 Genotype analysis and breeding TCR–Tg lines

Transgenic mice are identified by genotype analysis using probes specific for DNA contained in the transgene which is not present in mouse chromosomal DNA. Several genotyping methods are available. Tissue samples for DNA preparation from potential TCR–Tg mice are obtained either by tail biopsy of neonates (as early as seven days after birth) or by bleeding from the tail vein (from about four weeks after birth). Mice are toe or ear tagged at the time of sampling for subsequent identification. Until recently, Tg mice were identified in most laboratories by digesting DNA samples with restriction enzymes and carrying out Southern blot/*in situ* hybridization analysis to identify transgenes. Probes for this procedure were obtained from TCR V or C region sequences (which also hybridize to endogenous mouse TCR genes) or from DNA present in the expression cassette incorporated into the transgene. Now, most transgenes are detected using PCR methods which are relatively quick and simple and do not require the use of radioactive probes. However, PCR methods can not be used to determine whether Tg mice are homozygous or heterozygous whereas this is possible, but not foolproof, using the more laborious Southern blot/hybridization procedure described above.

Tg mice identified immediately after oocyte injection should be used to found TCR–Tg lineages by breeding to non-Tg mice of an appropriate strain; offspring of these matings are then typed and used for further breeding or experiments. It is advisable to select more than one transgenic founder mouse for each DNA construct because transmission of the transgene from founder to offspring is not guaranteed and transgene expression may vary due to differences in copy number and integration site position in individual founder mice. Provided that founder mice are not mosaic with respect to TCR transgenes, half of their offspring should inherit the TCR transgenes (Figure 10.2A). However, partial mosaicism often leads to a reduced frequency of transgene transmission in the first generation; in some cases no offspring inherit the transgene from the founder mouse, presumably because germline cells of the founder mouse do not contain stably integrated transgenes. Occasionally, transgene inheritance does not stabilize for the first or even the second generation; this possibility should be considered before selecting mice of only one lineage for experimental use. Instability of transgene inheritance can occur on relatively rare occasions when founder mice have two, or more, chromosomal sites where transgenes are integrated; segregation at these separate loci can lead to genotypic and, potentially, to phenotypic changes from one generation to the next. Alternatively, illegitimate recombination within the cluster of linked transgenes at a single integration site can generate phenotypic instability from one generation to the next. When TCR–Tg mice are to be used in a long-term project it is advisable to breed them to homozygosity and freeze morulae so that the line can be resurrected at a later date, should this become necessary if existing colonies are lost due to breeding failure or infection.

3.5 Phenotypic analysis of TCR–Tg mice

Analyses of transgene expression can begin when founder mice are old enough for blood samples to be taken without putting the survival of the mouse at risk. However, it is somewhat safer to postpone this until first generation offspring are available if patience allows. Peripheral blood T cells should express TCR molecules encoded by TCR transgenes if the DNA construct has been made properly. However, due to the genetic mosaicism of some founder mice, only a low percentage of T cells in founder mice may express transgenic TCR molecules even when the DNA construct is expressed appropriately. This problem will disappear when Tg mice from subsequent generations are analysed since all T cells should contain TCR transgenes in such mice. The simplest method to detect TCR expression is to use anti-clonotypic (anti-TCR) or anti-TCR V region monoclonal antibodies, when available, to carry out cytofluorimetric analyses of blood lymphocytes. Several analyses can be carried out using lymphocytes extracted from a few hundred microlitres of mouse blood and, as a consequence, allowing TCR expression to be assessed on T cells (e.g. CD3, Thy-1) or T cell subsets (e.g. CD4, CD8). More detailed analyses can be carried out when TCR–Tg mice are available for sacrifice so that thymocytes, as well as peripheral T cells, can be obtained. It is important to establish whether *all* thymocytes and T cells express TCR–Tg molecules since there may be problems with transgene stability or developmental blocks on thymocyte development in some genetic backgrounds if this is not observed; clear interpretation of results obtained from such TCR–Tg mice may not be possible in these circumstances. The effect of age on TCR–Tg expression can also be profound as determined by phenotypic analysis. In our experience, the proportion of TCR–Tg^+ T cells declines with age; this seems to be a general observation based on informal discussions with others. However, this does not lead to detectable reductions in the strength of functional responses with age in our experience.

3.6 Functional analyses of TCR–Tg mice

Assays to determine functional characteristics of T cells and thymocytes from TCR–Tg mice can be carried out as soon as mice are available for experiments *in vivo* or for sacrifice so that *in vitro* assays for T cell function can be set-up. Spleen or lymph node cells are the usual source of peripheral T cells for functional assays *in vitro*, although thymocytes can also be tested because the frequency of mature TCR–Tg$^+$ thymocytes which have acquired effector functions but which have not yet left the thymus is very high in TCR–Tg mice. Because of the increase in frequency of T cells with the antigen specificity conferred by TCR–Tg molecules, T cells from TCR–Tg mice mount vigorous responses to cells expressing appropriate ligand molecules both *in vivo* and *in vitro*. This means that the responses measured are magnified in TCR–Tg relative to non-transgenic control mice. For example, tissue grafts expressing appropriate ligands may be rejected faster and immunity to infection and autoimmunity may be apparent much earlier in TCR–Tg mice. In addition, mixed lymphocyte cultures may produce measurable T cell responses (cytokine release, proliferation, and/or cytolytic activity) in shorter periods and protocols have to be modified to take account of this. We find that T cells from TCR–Tg mice can be treated as if they were T cell clones when used in functional assays *in vitro*. However, the functional status of the peripheral T cell repertoire in TCR–Tg mice is not entirely predictable. For example, the full range of anticipated effector functions may not be manifested in TCR–Tg mice where TCR genes are derived from a cytotoxic (usually CD8$^+$) T cell clone; this may arise due to the absence of non-specific help for T cell responses in TCR–Tg mice which do not develop a full repertoire of T cells (Auphan *et al.* 1992*a*; Gaugler *et al.* 1993). In addition, T cells from some TCR–Tg mice only respond in the presence of IL-2, as if they enter a quiescent state *in vivo*. Lack of functional responses from TCR–Tg mice may come about because of background gene effects which, for example, lead to the induction of a state of anergy in the peripheral T cell pool.

4 TCR–Tg mice in immunological research

In this section I will review the main areas of research which have benefited from the use of TCR–Tg mice to illustrate their utility in the immunology field. Since the number of studies in which TCR–Tg have been used has grown exponentially since 1988 I will only point out specific examples to illustrate major applications.

The main impetus which lead to the generation of the first TCR–Tg mice was the desire to resolve two of the central mysteries in immunology, namely:

1. How self-reactive T lymphocytes are prevented from causing autoimmunity or, putting the problem another way, how self-tolerance is imposed on the T cell repertoire.
2. How the T cell repertoire is selected to have a preference for recognition of foreign antigens in the context of self-MHC molecules.

Before the production of TCR–Tg mice, the major barrier to progress on these issues arose from the sheer complexity of the T cell repertoire in terms of the heterogeneity, in normal individuals, of TCR structures expressed by T cells. This meant that there was no method sensitive enough to detect lymphocytes expressing one particular TCR structure (clonotype). In contrast, TCR–Tg mice express identical TCR structures on nearly all thymocytes and T cells facilitating studies designed to assess the fate and functional status of thymocytes and T cells expressing particular TCR clonotypes. TCR–Tg mice also facilitate studies on thymocyte and T cell differentiation since transgenes are expressed as part of the normal programme of gene expression (Chapter 2), if one assumes that thymocyte and T cell differentiation is not perturbed by expression of TCR transgenes. This is not entirely justified as TCR transgenes do not have to be rearranged during thymocyte differentiation and TCR molecules appear on the cell surface as soon as CD3 is expressed; i.e. they appear at an earlier stage compared to normal mice since surface expression is normally delayed by TCR α gene rearrangement.

Immunologists have also realized that TCR–Tg mice are a very useful source of T cells for structural and functional analyses of immunological phenomena in attempts to define these processes in terms of the cellular and molecular interactions which are involved. In effect T cells from peripheral lymphoid organs of TCR–Tg mice can be used experimentally as surrogates for T cell clones with the same specificity, although care must be exercised initially in determining the functional status of the peripheral T cell pool of such mice.

4.1 Background effects

Since TCR molecules enable thymocytes and T cells to interact with MHC–peptide complexes and superantigens expressed on other cells, the genetic background of TCR–Tg mice is a critical factor determining the fate of thymocytes and T cells. Three basic patterns of thymocyte development can be discerned for each transgenic TCR specificity depending on the nature of the interactions between TCR molecules and self-MHC–peptide complexes and superantigens expressed on cells in the thymus (Figure 10.3). More subtle effects on the pattern of thymocyte development in TCR–Tg mice have also been described, particularly in cases where the pattern of self-ligand expression has been modified (see Section 4.2). In order to interpret the fate of thymocytes in TCR–Tg mice it is necessary to understand some aspects of thymocyte development in normal mice. In non-Tg mice, four predominant thymocyte subsets can be distinguished using monoclonal antibodies against CD4 and CD8 (Chapters 2 and 3). About 5–10% of thymocytes are immature $CD8^-CD4^-$ thymocytes which differentiate into $CD8^+CD4^+$ thymocytes which are by far the most abundant subset (70–80% of thymocytes). After this stage, thymocytes are still regarded as immature but they can, following selection processes involving interaction with MHC class II (Berg *et al.* 1989*a*) or class I molecules (Scott *et al.* 1989), differentiate into either $CD8^-CD4^+$ or $CD8^+CD4^-$ thymocytes, respectively. These two populations of more mature thymocytes constitute about 10% and 5% of thymocytes, respectively; and are precursors of helper or of cytotoxic T cells, respectively, in the periphery. Two processes, both of which involve TCR–MHC interactions, shape the repertoire of thymocytes during this developmental process. First, self-reactive clones which interact 'inappropriately' (e.g. with high affinity) with MHC or superantigens expressed on thymic cells are either eliminated (negative selection) or rendered functionally unresponsive and, secondly, thymocytes which interact 'appropriately' (e.g. lower affinity) with cells expressing self-MHC molecules are actively selected and mature further until they emerge from the thymus as T cells (positive selection). Of course, one of the most intriguing aspects of these processes is why interactions between TCR molecules and self-MHC can lead to such profoundly different outcomes for thymocytes. TCR–Tg mice are in the forefront of research aimed at trying to understand the cellular and molecular basis of these selection processes. If TCR–MHC interactions do not take place thymocytes are thought to die (via programmed cell death) *in situ*, at the $CD8^+CD4^+$ stage. Experiments in which TCR–Tg mice were used played a significant role in demonstrating that these events take place during thymocyte development and selection. Many reviews detailing relevant immunological questions at issue and the role of TCR–Tg mice in helping to resolve them, are available for readers who wish to delve more deeply (see *Immunological Reviews*, volumes 122 and 135).

4.1.1 Selecting backgrounds

If selecting MHC and/or peptide ligands are expressed on thymic cells increased numbers of $CD8^+CD4^+$ thymocytes survive programmed cell death and mature into T cells in TCR–Tg mice compared to control non-Tg mice (Figure 10.3). Examples of what happens during thymocyte development, as defined by cytofluorimetric analyses of CD4 and CD8 expression on thymocyte subsets is shown in Figure 10.4. These data are derived from experiments with BM3.6 TCR–Tg mice in which the TCR–Tg molecules confer specificity for $H\text{-}2K^b$ and are selected on $H\text{-}2K^k$ (Table 10.1, and Sponaas *et al.* 1994). However, the extent of these effects can differ widely for different TCR–Tg mice. In principle, thymocyte development on a selecting background leads to a situation in which relative proportions of the two immature thymocyte

Table 10.1 TCR–Tg mice

TCR–Tg line (promoter)[a]	Specificity			Selection	Reference
	Allo-MHC	MHC–antigen	Additional		
(αβ)	–	$H\text{-}2D^b$/H-Y	–	$H\text{-}2D^b$	Kisielow *et al.* 1988*a,b*
2C	$H\text{-}2L^d$	–	–	$H\text{-}2K^b$	Sha *et al.* 1988*a,b*
KB5C20	$H\text{-}2K^b$	–	–	$H\text{-}2K^k$	Schönrich *et al.* 1991
BM3.6	$H\text{-}2K^b$	–	–	$H\text{-}2K^k$	Sponaas *et al.* 1994
F3	$H\text{-}2K^b$	–	H-2E/MtvSag	?	Morahan *et al.* 1991
P14	–	$H\text{-}2D^b$/LCMV GP	$Mls\text{-}1^a$	$H\text{-}2D^b$	Pircher *et al.* 1989*a,b*
(MHC I)					
K	–	$H\text{-}2K^k$/SV40-L-T	$Mls\text{-}1^a$	$H\text{-}2K^k$	Geiger *et al.* 1992
F5	–	$H\text{-}2D^b$/flu NP	H-2E/MtvSag	$H\text{-}2D^b$	Mamalaki *et al.* 1992
(hCD2)					
DO.10	–	$H\text{-}2A^d$/OVA	–	$H\text{-}2A^d$	Murphy *et al.* 1990
2B4,	–	$H\text{-}2E^k$/cyt c	$Mls\text{-}2^a/3^a$	$H\text{-}2E^k$	Berg *et al.* 1989*a,b*
AND	–	,,		,,	Kaye *et al.* 1989
4B2A1	–	$H\text{-}2E^d$/Igλ	–	$H\text{-}2E^d$	Bogen *et al.* 1992
vir-2-15	–	$H\text{-}2E^d$/flu HA	–	$H\text{-}2E^d$	Swat *et al.* 1992

[a] *Transcriptional promoters used are derived from TCR genes except when indicated.*

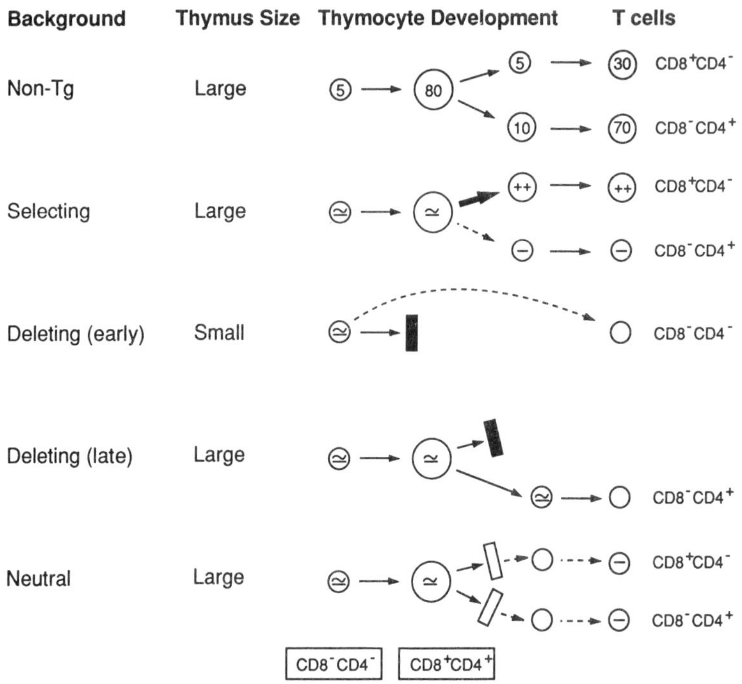

Fig. 10.3 The effect of genetic background on thymocyte development in TCR transgenic mice. Numbers in circles refer to the approximate percentage of each thymocyte or T cell subpopulation as a proportion of total thymocytes or T cells. Actual proportions can vary depending on mouse strain, age, and health status. The meaning of other symbols are described in the text. Bars represent complete (filled bars) or leaky (open bars) blocks on thymocyte development.

subsets are not much changed compared to non-Tg controls (shown by the $\simeq$ symbol in Figure 10.3) whereas selection favours development of either $CD8^-CD4^+$ or $CD8^+CD4^-$ thymocytes since a single TCR molecule usually only interacts with one MHC molecule (class II or class I respectively). In the example outlined in Figure 10.3, the TCR–Tg molecule expressed on all thymocytes confers recognition of a self-MHC class I molecule as an appropriate selecting ligand and, hence, development of $CD8^+CD4^-$ thymocytes and T cells is favoured. Skewing (++) in the relative proportion of $CD8^+$ $CD4^-$ thymocytes and T cells which develop in TCR–Tg mice compared to non-Tg control mice will be observed at the expense of the $CD8^-CD4^+$ subsets which will appear in relatively reduced proportions (−). In practice, skewing into only one of the more mature thymocyte subsets on selecting backgrounds is relative, not absolute, and T cells of both subsets can usually be found in the periphery of TCR–Tg mice even though the normal ratios differ (e.g. Figure 10.4). In addition, the relative proportion of $CD8^+CD4^+$ thymocytes observed can also be much reduced in some TCR–Tg

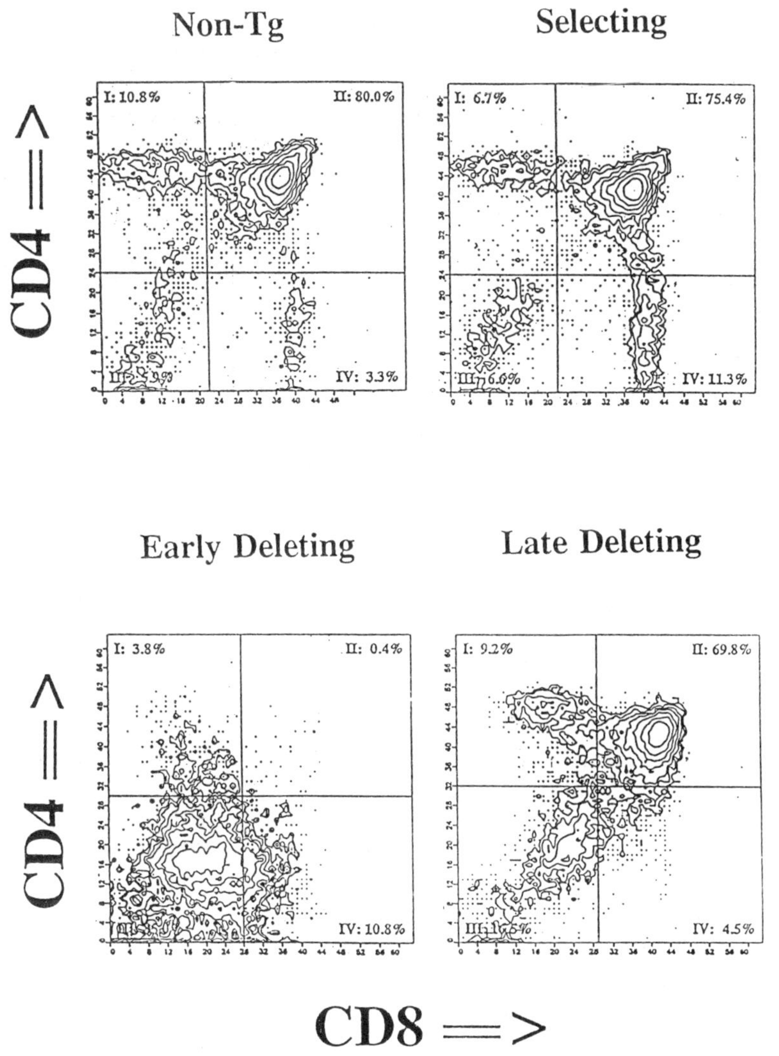

Fig. 10.4 Cytofluorimetric analyses of thymocytes from non-transgenic (CBA) mice and from TCR–Tg mice on various genetic backgrounds (Sponaas *et al.* 1994). Thymocyte subpopulations are identified by staining with monoclonal antibodies directed against murine CD8 and CD4 molecules. The TCR–Tg molecules expressed by the TCR–Tg mice used in these examples confers recognition of MHC class I molecules (H-$2K^b$) and is selected by H-$2K^k$ molecules; note the increase in proportion of $CD8^+CD4^-$ thymocytes (*bottom right* quadrant) on a selecting background, the total absence of $CD8^+CD4^+$ thymocytes (*top right* quadrant) in the early deleting background, and the absence of $CD8^+CD4^-$ only in the late deleting background. See text for more details.

mice due to the efficiency of selection into one of the mature subsets. Skewing in the proportions of $CD8^-CD4^+$ and $CD8^+CD4^-$ thymocytes and T cells which develop is usually observed in TCR–Tg mice, although the extent of this is variable for different lines of TCR–Tg mice. Since development of only a single type of mature thymocyte subset is generally not observed in TCR–Tg mice the question arises as to how thymocytes are selected into the inappropriate subset (i.e. how do thymocytes expressing TCR molecules conferring recognition of MHC class I molecules develop into $CD8^-CD4^+$ thymocytes ?). There are several possible explanations for this phenomenon. Positive selection may be:

(1) relatively independent of co-receptor interactions;

(2) there may be cross-reactivity with other MHC molecules of a different class;

(3) allelic exclusion could be inefficient such that new TCR specificities are created by association of TCR α or TCR β chains encoded by productive rearrangements of endogenous TCR genes and transgenic TCR chains.

The last hypothesis seems likely in some cases because skewing is significantly increased when TCR transgenes are backcrossed on to genetic backgrounds with defective recombination of TCR genes (*scid* and *RAG-1*, *RAG-2*, or TCR gene knock-out mice). In contrast, co-receptor interactions appear to be essential for positive selection (Robey *et al.* 1992; Fung-Leung *et al.* 1993; Schönrich *et al.* 1993).

TCR–Tg mice have also been used in attempts to define the precise sequence of events which takes place during thymocyte differentiation and selection. For example, by devising appropriate cytofluorimetric analyses the level of TCR expression on immature, and both types of mature, thymocyte subsets can be compared. In most TCR–Tg mice the level of TCR is almost always higher on mature, compared to immature, thymocytes. However, it may be wrong to assume that this increase in TCR level comes about as a direct consequence of selection mediated through TCR–MHC interactions. Put simply, the complex sequence of events from the immature ($CD8^+CD4^+$) to the mature selected ($CD8^-CD4^+$ or $CD8^+CD4^-$) stage is now thought to proceed as follows:

(1) decrease in the level of expression of CD8 or CD4;

(2) increase in TCR level;

(3) interaction with selecting MHC ligand but only if the appropriate co-receptor molecule is still expressed at high level;

(4) loss of expression of one co-receptor.

Nevertheless, there are a number of contentious issues which are still debated in the context of thymocyte selection (for example see Matzinger 1993; von Boehmer *et al.* 1993; Hugo *et al.* 1993) not the least of which is whether co-receptor down-regulation is a random or a selective event in itself, which would affect profoundly the sequence of events outlined above. Moreover, the increase in TCR expression is seen, by some, as a prelude to, not part of, the process of positive selection mediated by self-MHC. For example, increased TCR expression may occur as a result of MHC-independent increases in expression of CD3 chains which limit the level of TCR expression in $CD8^+CD4^+$ thymocytes; this could explain why TCR up-regulation is observed in TCR–Tg mice in which non-TCR expression cassettes are used (Mamalaki *et al.* 1992; Pircher *et al.* 1989*a,b*). In addition, there are very important, and still outstanding, questions concerning the role of peptides complexed with selecting MHC molecules in the selection process. What is clear is that TCR–Tg mice have been, and are continuing to be, extremely useful in the search for answers to these issues.

4.1.2 Deleting backgrounds

Tolerogens expressed on cells in the thymus usually mediate thymocyte elimination (deletion) via negative selection (see Kisielow *et al.* 1991; Lo *et al.* 1991). In TCR–Tg mice this can lead to a profound reduction in the absolute number of thymocytes which has an obvious manifestation in reduction of thymus size (Figure 10.3). Tolerogenic ligands can be MHC molecules, endogenous *Mtv* superantigens which associate with self-MHC class II molecules (e.g. *Mls*), or peptides which associate with particular MHC molecules (Table 10.1). These ligands mediate deletion of immature thymocytes following interaction between TCR molecules and their cognate ligands expressed on other thymic cells. However, the timing of deletion can vary so that large numbers of $CD8^+CD4^+$ either do, or do not, develop in the thymus (Figures 10.3 and 10.4). The first experimental demonstrations of thymocyte deletion in TCR–Tg mice involving male-specific antigen, H-Y (Kisielow *et al.* 1988*a*) or $H\text{-}2L^d$ (Sha *et al.* 1988*a*) as deleting ligands were of the early deletion phenotype in which deletion occurs before differentiation to the $CD8^+CD4^+$ stage, effectively blocking their appearance in the thymus. Early deletion is often, although not exclusively, observed when the deleting ligand is either an MHC class I molecule or a peptide which complexes with a self-MHC class I molecule. However, it has been argued that early deletion is artefactual and comes about as a consequence of early expression of transgenic TCR molecules in TCR–Tg mice allowing TCR–MHC interactions to take place at an earlier developmental stage than non-Tg mice (Berg *et al.* 1989*b*). In addition, the extent of deletion depends on the level of expression of the deleting ligand (Auphan *et al.* 1992*b*). In contrast, there are no clear examples, to date, of early deletion occurring when a MHC class II molecule, or a MHC class II-associated peptide, act as the deleting ligand although fewer studies have been carried out with MHC class II ligands (see below). Artefact or not, deletion at an early stage usually prevents subsequent development of both subsets of mature thymocytes and T cell subsets ($CD8^-CD4^+$ and $CD8^+CD4^-$) in TCR–Tg mice, even though the deleting ligand is a MHC class I peptide complex. Presumably, this occurs because both mature subsets pass through the $CD8^+CD4^+$ stage. Late deletion has been observed in several TCR–Tg systems, especially when the expression pattern of the deleting ligand in thymus is modified (Hämmerling *et al.* 1991; Sponaas *et al.* 1994). Late deletion is characterized by a large thymus with relatively normal proportions of immature thymocytes but deletion of only one mature thymocyte subset and absence of the corresponding mature T cell subset. In the examples outlined in Figures 10.3 and 10.4, late deletion mediated by a MHC class I ligand prevents development of $CD8^+CD4^-$ thymocytes but some $CD8^-CD4^+$ thymocytes still develop. In some TCR–Tg mice, where deletion is mediated by a self-MHC class I ligand, this pattern of thymocyte development is observed and substantial pools of $CD8^-CD4^+$ T cells, which also express TCR–Tg molecules, accumulate (e.g. Sponaas *et al.* 1994). As discussed above, selection of these cells may occur as a consequence of rearrangement and expression of endogenous TCR genes.

Most TCR–Tg mice which express TCR molecules conferring recognition of MHC class II molecules involve specific peptides as the deleting ligand (Table 10.1). Thus, deletion only takes place when the peptide is supplied by administration *in vivo*. However, an unexpected cross-reaction with $H\text{-}2A^s$ molecules allowed one group to assess the effect of MHC class II deleting ligands on thymocyte fate (Vasquez *et al.* 1992). They reported that deletion occurred but at a late stage in development since $CD8^+CD4^+$ thymocytes appeared, albeit in reduced numbers, whereas $CD8^-CD4^+$ thymocytes failed to mature. This could represent a real difference between MHC class I and MHC class II-mediated patterns of deletion since MHC class I molecules are expressed by all thymic cells (including thymocytes themselves) whereas MHC class II molecules are expressed by fewer thymic cells, in absolute terms. However, the TCR–$H\text{-}2A^s$ interaction may be relatively weak and it is possible that a stronger interaction could produce earlier deletion. Administration of peptides has also been tried in this and other TCR–Tg systems for MHC class II-mediated deletion (Vasquez *et al.* 1992). In most cases, the deletion stage seems to be late since $CD8^+CD4^+$ are often observed in relatively large numbers (Berg *et al.* 1989*b*; Vasquez *et al.* 1992). TCR molecules which recognize MHC class II ligands are also subject to deletion when endogenous *Mtv* encoded superantigens are expressed by cells in the thymus. In general, these ligands bring about deletion in TCR–Tg mice but at a late stage in thymocyte development as $CD8^+CD4^+$ thymocytes appear in relatively large numbers (Berg *et al.* 1989*b*; Kisielow *et al.* 1991; Lo *et al.* 1991; Vasquez *et al.* 1992).

Often, $CD8^-CD4^-$ T cells expressing TCR–Tg molecules accumulate in relatively large numbers when deletion occurs in the thymus of TCR–Tg mice (Russell *et al.* 1990; von Boehmer *et al.* 1991; Wilson *et al.* 1992; Sponaas *et al.* 1994). T cells of this type could develop via two independent pathways:

(1) directly from immature $CD8^-CD4^-$ thymocytes which do not require positive selection for survival;

(2) from more mature thymocytes (or T cells) expressing only one co-receptor which down-regulate CD8, or CD4, expression as a result of interaction with self-MHC class I, or MHC class II, ligands.

This raises an important technical issue which should be taken into account when interpreting data obtained from experiments involving TCR–Tg mice. Careful analysis of cytofluorimetric data is essential in order to determine whether thymocytes of particular subsets are deleted or survive but express reduced levels of TCR, CD8, or CD4 molecules as a result of interactions with tolerogens (Schönrich *et al.* 1991). Deletion can often affect only a proportion of thymocytes in TCR–Tg mice. For example, thymocytes expressing high levels of TCR molecules can be deleted whereas those expressing lower levels mature as T cells which continue to express reduced levels of TCR molecules (Teh *et al.* 1989). This has been interpreted as evidence that thresholds for tolerance induction exist such that deletion only occurs when high affinity interactions between thymocytes and tolerizing cells take place (Auphan *et al.* 1992*a*). In contrast, lower affinity interactions involving identical TCR molecules and cognate MHC ligands are below the tolerance threshold for deletion and do not contribute to autoimmune responses in the periphery, perhaps because of their inability to be activated *in vivo*.

4.1.3 Neutral backgrounds

In theory, it should be possible to identify genetic backgrounds for TCR–Tg mice in which thymocytes are not subject to negative or positive selection. In such environments thymocytes expressing TCR–Tg molecules encounter no tolerizing or selecting ligands during their passage through the thymus. Consequently, thymocytes develop as far as the immature $CD8^+CD4^+$ stage are not deleted but mature no further due to lack of positive selection (Figure 10.3). Neutral backgrounds are very useful as controls for TCR transgenes expressed on deleting or selecting backgrounds. In practice, TCR transgenes expressed on a variety of different genetic backgrounds usually produce patterns of thymocyte development which suggest that negative or positive selection is taking place to some extent. Negative selection can occur when transgenic TCR molecules cross-react with tolerogens expressed on the new background. These cross-reactivities with MHC, superantigen, or other peptide antigens are unpredictable and each time TCR transgenes are placed on new genetic backgrounds it is important to assess whether selection events are shaping the pattern of thymocyte development or not. Positive selection can occur as a result of rearrangement of endogenous TCR genes which allow some thymocytes to recognize selecting ligands. Evidence in support of this latter explanation has been obtained since mature thymocyte and T cell subsets largely disappear when TCR–Tg mice are backcrossed on to genetic backgrounds where rearrangements of endogenous TCR genes are prevented. However, subtle changes to the structure of selecting ligands or the level or pattern of their expression can have either profound or negligible effects on the pattern of thymocyte development in TCR–Tg mice (Sha *et al.* 1990; Auphan *et al.* 1992*b*; Ohashi *et al.* 1993*a*; Sponaas *et al.* 1994).

4.2 Tolerance

One of the most important contributions to immunological research made possible by resorting to the use of TCR–Tg mice is resolution of the controversy surrounding the mechanism of self-tolerance induction during thymocyte development. Studies involving the use of TCR–Tg mice provided, for the first time, unequivocal evidence that a state of specific immunological unresponsiveness could be induced by clonal deletion of thymocytes (Kisielow *et al.* 1988*a*; Sha *et al.* 1988*a*). Many factors which are important in the negative selection process have been identified by observations on thymocyte development in TCR–Tg mice (see *Immunological Reviews*, Vol. 122, 1991). For example, requirements for co-receptor (Knobloch *et al.* 1992; Schönrich *et al.* 1993; Fung-Leung *et al.* 1993), MHC (Auphan *et al.* 1992*a*,*b*; Sponaas *et al.* 1994), and peptide (Iwabuchi *et al.* 1992; Mamalaki *et al.* 1992) expression during negative selection have been investigated. TCR–Tg mice have facilitated these studies since they provide a means of avoiding problems with the normal complexity of TCR expression in developing thymocytes.

Co-receptor interactions are differentially required during negative selection in TCR–Tg mice (Schönrich *et al.* 1993; Fung-Leung *et al.* 1993). This suggests that the affinity of interaction between TCR and cognate ligand is not always sufficient to mediate negative selection. Complementary studies in which MHC class I ligands are mutated (Ingold *et al.* 1991) or were recombined (Aldrich *et al.* 1991) to eliminate CD8 binding sites also demonstrate that co-receptors are necessary for some, but not all, thymocyte selection events. This illustrates an important point about TCR–Tg mice; results which are obtained in a single system should not be extrapolated to other thymocytes which express TCR molecules conferring the same nominal specificities for deleting or selecting ligands as specific requirements during selection events may vary depending on the affinity of TCR for cognate MHC–peptide ligand in each case.

TCR–Tg mice have been used in other studies on tolerance induction in situations where self-antigen expression is modified. Many investigators have concentrated on finding out how tolerance is induced in situations where potential T cell tolerogens, such as those listed in Table 10.1, are expressed exclusively in the extrathymic environment (see *Immunological Reviews*, Vol. 133). Frequently, this goal has been achieved by expressing genes coding for potential tolerogens in Tg mice using transcriptional promoter elements derived from genes which are not expressed in the thymus. Mating TCR–Tg mice to Tg mice which express the cognate tolerogen (Tol–Tg) recognized by transgenic TCR molecules allows the fate and functional status of thymocytes and T cells in (TCR–Tg × Tol–Tg) double transgenic mice to be readily assessed as a means of determining tolerance status and mode of tolerance induction. In this way, several non-deletional mechanisms by which tolerance is induced and maintained have been described (Hämmerling *et al.* 1991, 1993; Miller and Heath 1993). In one case, peripheral T cells were rendered unresponsive to cognate tolerogen (H-2K^b) expressed in the periphery via mechanisms involving reduction in the level of surface expression of TCR and CD8 co-receptor molecules on T cells which, as a result, become unresponsive to the tolerogen (Schönrich *et al.* 1991, 1992). In another case, T cell tolerance to skin grafts was induced by cognate ligands (H-2K^b) expressed exclusively on pancreatic β-islet cells but thymocytes were not eliminated and T cells expressed normal levels of TCR and CD8 molecules and yet were unresponsive to antigen *in vivo* and *in vitro* (Morahan *et al.* 1991). Perhaps the most extreme case occurred when the cognate ligand (LCMV-GP) was again expressed by pancreatic β-islet cells; in this case T cells developed normally, and mounted a vigorous response after viral infection of the mice which led to extremely rapid onset of diabetes (Ohashi *et al.* 1991). Thus, 'tolerance' is maintained passively presumably because β-islet cells failed to activate resting T cells. In other cases, altering the pattern of expression of a tolerogenic molecule leads to tolerance induction in the thymus with deletion of only a proportion of thymocytes expressing TCR–Tg molecules. However, subtle modification to the phenotype of thymocytes does take place as they develop in the thymus and this can be manifested in modulation of the levels of surface TCR, CD8, or CD4 molecules which correlates with the induction of a state of unresponsiveness to the tolerogen.

That tolerance is induced in the extrathymic compartment can be demonstrated by comparing thymocyte development and function in (TCR–Tg × Tol–Tg) mice and in control TCR–Tg mice. If expression of the tolerogen has no measurable effect on thymocyte development and acquisition of effector functions it is reasonable to assume that tolerance, if it is acquired *in vivo* is not induced as a result of tolerogen expression in the thymus. This has become an important issue since many transgenes which were designed not to express in the thymus do so when expression is assessed carefully (Miller and Heath 1993). T cells from TCR–Tg mice also provide a very sensitive means for assessing immunologically relevant expression of tolerogens which can not be detected by other means. For example, cells and tissues from MHC Tg mice can be used to stimulate responses from responder T cells from TCR–Tg mice; such assays can be very sensitive and, obviously, specific.

In vitro systems entailing the use of TCR–Tg mice have also been developed to carry out more detailed investigations into the cellular mechanisms and the factors involved in negative selection. TCR–Tg mice can be used as a source of well-defined thymocyte subsets based on CD4, CD8, TCR, and CD3 expression and their relative size. CD8$^+$CD4$^+$ thymocytes from TCR–Tg mice are subject to negative selection when incubated with a variety of cells expressing appropriate ligands *in vitro* (Iwabuchi *et al.* 1992; Pircher *et al.* 1992, 1993; Speiser *et al.* 1992). The relative importance of thymocyte stage, nature of the cell presenting tolerogens to thymocytes, and the effect of structural alterations to the tolerogen on efficiency of negative selection can be assessed readily in these *in vitro* systems.

4.3 Thymocyte selection

Thymocytes must undergo positive selection before they can mature into peripheral T cells. Characteristic changes which take place during positive selection have been observed in TCR–Tg mice on selecting backgrounds (see *Immunological Reviews*, Vol. 135, 1993). Thus, thymocytes which undergo selection often express increased levels of TCR and CD8 (e.g. see Figure 10.4), if selected by MHC class I molecules or CD4 if selected by MHC class II co-receptor molecules. In addition, thymocytes stop expression of either CD4 or CD8, respectively. TCR–Tg have been widely used in studies aimed at defining the sequence of steps which constitute positive selection and the role of factors such as peptide, levels and pattern of MHC expression, and the role of co-receptors. In particular, two models were put forward to explain how expression of one of the co-receptors might come about during positive selection. In the instructive model, TCR–MHC interactions dictate whether CD4 or CD8 would be switched off whereas proponents of the stochastic model envisage that the loss of expression of one co-receptor gene is essentially random and TCR-mediated selection takes place at a later stage and is only successful if the appropriate co-receptor molecules are still expressed; this implies that positive selection will be co-receptor dependent (see below). Support for both hypotheses has been obtained with reference to TCR–Tg mice in conjunction with CD4 or CD8 knock-out mice (Robey *et al.* 1992; Davis *et al.* 1993). Most workers in the field now believe that the evidence for the stochastic model is more convincing (see *Immunological Reviews*, Vol. 135, for detailed discussion of this issue). As discussed in Sections 4.1.1 and 4.1.2, the importance of co-receptor interactions during positive and negative selection appears to differ. CD8 or CD4 co-receptor interactions during positive selection are necessary in the majority of cases tested to date whereas at least some thymocytes can be eliminated in the absence of co-receptor. This suggests that more stringent requirements are necessary for interactions which lead to positive rather than negative selection. This is also borne out in experiments to determine whether positive selection can be mediated by different types of cells expressing selecting ligand. Thymocytes themselves, even when they express increased levels of appropriate selecting MHC molecules in Tg mice, are very poor mediators of positive selection *in vitro* despite being effective mediators of negative selection (Schönrich *et al.* 1993). Whereas most cells are able to mediate negative selection, few types of cell can mediate positive selection of immature $CD8^+CD4^+$ thymocytes in TCR–Tg mice *in vivo* or in fetal thymus organ culture. Cells capable of mediating positive selection reside exclusively in the radioresistant compartment of the thymus and are thought to be cells of epithelial origin although mesenchymal cells may also play an indirect role in the selection process (Anderson *et al.* 1993). Indeed, epithelial cells and fibroblasts injected intrathymically into TCR–Tg mice are able to affect positive selection. TCR–Tg mice provide a very versatile and convenient source of thymocytes with well-defined specificities which can be enriched for thymocytes at particular developmental stages. Reconstitution of fetal thymus organ cultures with thymocytes from TCR–Tg mice provides an experimental system with which to study the factors which influence positive selection in cellular and molecular terms.

4.4 Immune responses

As mentioned above, immune responses measured *in vivo* and *in vitro* are enormously amplified in TCR–Tg mice due to the vast increase in the frequency of T cells with a single specificity. This facilitates studies designed to assess the relative contribution of functionally distinct T cell subsets in immunological processes, such as immunity to infectious agents, autoimmunity, tissue graft rejection, and in mouse model systems in which specific tolerance can be induced by manipulation of the mature T cell repertoire. The immunological status of the T cell repertoire has been studied in great detail for some TCR–Tg mice. These studies reveal that T cells in TCR–Tg mice seem to differentiate into functional states *in vivo* which are unique, but stably inherited by all mice of a particular TCR–Tg lineage. For example, co-receptor dependence during effector responses is determined by the TCR clonotype since the same dependence is manifested in the repertoire of TCR–Tg mice (Gaugler *et al.* 1993). T cells from some TCR–Tg mice will readily produce effector responses *in vitro* whereas T cells from other TCR–Tg mice require considerable encouragement, such as exogenous IL-2 as well as a period of stimulation, before they express effector functions in response to cognate ligands (e.g. Mamalaki *et al.* 1992). T cells from such mice are probably in a resting state *in vivo* and may even enter an anergic state which can usually be reversed *in vitro*. However, some TCR–Tg mice have T cell repertoires from which effector responses which might be expected can not be generated. It should be remembered that TCR–Tg mice are almost certainly immunodeficient for all functional T cell subsets and due to the lack of co-operation from other types of T cells not present in TCR–Tg mice. Consequently, it is wise to investigate how best to generate functional responses from TCR–Tg mice so that the functional status of T cells can be assessed.

4.5 Autoimmunity and immunity

Failure to induce tolerance is thought to lead to autoimmunity. However, autoimmune responses are characterized by being weak and developing progressively over a protracted period. Consequently, they are hard to detect, at least in the initial stages prior to the onset of pathological damage to tissues. This is also the case in some mouse models which develop autoimmunity as a consequence of expression of antigens encoded by transgenes. However, progression of autoimmune disease can be accelerated greatly if

TCR–Tg mice are used in such systems. For example, mice which take several weeks to develop autoimmune diabetes due to expression of transgenes in pancreatic β-islet cells can develop autoimmunity in only a few days when crossed on to a TCR–Tg background (Ohashi *et al.* 1991, 1993*b*). However, slow progression to an autoimmune state has also been described for TCR–Tg mice (Geiger *et al.* 1992). Increases in the strength of immune responses *in vivo* to foreign antigens, whether they are viral antigens, allo-MHC, or minor histocompatibility antigens expressed on tissue grafts or immunizing cells, is also observed in TCR–Tg mice.

4.6 γδ TCR mice

I have concentrated on discussing TCR–Tg mice which express TCR molecules of the αβ type. Similar approaches have been used to investigate aspects of γδ T cell biology (see Chapter 4). Several groups have generated γδ TCR–Tg mice and have used them to study the development of γδ thymocytes and their relationship to the separate αβ thymocyte lineage (Hedrick and Dent 1990; Tatsumi *et al.* 1993). The role of this minor subset of T cells in immunological phenomena is still unclear but studies involving the use of TCR–Tg mice have demonstrated that γδ thymocytes develop along pathways which are separate but very similar to the developmental pathway that TCR αβ thymocytes follow. Thus, they are subject to programmed cell death (Dent *et al.* 1993), are selected on non-MHC, self-ligands (Wells *et al.* 1991, 1993), and are tolerized by self-ligands (Bonneville *et al.* 1990).

4.7 TCR gene knock-out mice

Another class of transgenic mice which are proving to be extremely useful in unravelling the complexities of T cell biology are mice with defective TCR genes generated as a result of gene targeting in ES cells. TCR α, TCR β, TCR γ, and TCR δ genes have all been inactivated using appropriate gene targeting vectors to modify the constant regions necessary for TCR gene expression. Thymocytes fail to develop in mice which are homozygous for defective TCR β genes. However, large numbers of $CD8^+CD4^+$ thymocytes develop in mice with defective TCR α genes although medullary regions fail to differentiate in the thymus of such mice (Philpott *et al.* 1992). Further investigations revealed that differentiation of $CD8^-CD4^-$ into $CD8^+CD4^+$ thymocytes is mediated by surface expression of homodimeric TCR β chains on the surface of thymocytes which signal differentiation (Groettrup *et al.* 1992; Groettrup and von Boehmer 1993). Interestingly, thymocytes expressing TCR αβ heterodimers do develop in mice which are homozygous for defective TCR γ or TCR δ genes. This has been interpreted as evidence that thymocytes expressing the two different forms of TCR develop along pathways which are independent in thymus.

5 Summary

In summary, TCR–Tg mice are very convenient and versatile resources for investigating cellular interactions which bring about immunological phenomena involving T cells. They also provide the means to understand the molecular events and factors which influence the efficacy and specificity of these interactions. The near monoclonality of their T cell repertoires permit detailed investigation of thymocyte development and selection in designated backgrounds and also facilitates enrichment of thymocytes and T cells at particular stages of development and differentiation for use in studies on T cell development. T cells from TCR–Tg are useful in functional studies aimed at defining the relative contribution of different T cell subsets during immune responses. The continued growth in the number of different TCR–Tg mice and the growing trend to devise experimental systems involving other types of transgenic mice, such as 'knock-out' mice, will provide a wealth of new information and possibilities for further investigation of immunological phenomena in the future.

References

Aldrich, C. J., Hammer, R. E., Jones-Youngblood, S., Koszinowski, U., Hood, L., Stroynowski, I., *et al.* (1991). Negative and positive selection of antigen-specific cytotoxic T lymphocytes affected by the alpha 3 domain of MHC I molecules. *Nature*, **352**, 718–21.

Anderson, G., Jenkinson, E. J., Moore, N. C., and Owen, J. J. (1993). MHC class II-positive epithelium and mesenchyme cells are both required for T-cell development in the thymus. *Nature*, **362**, 70–3.

Auphan, N., Jezo-Bremond, A., Schonrich, G., Hammerling, G., Arnold, B., Malissen, B., *et al.* (1992*a*). Threshold tolerance in H-2Kb-specific TCR transgenic mice expressing mutant H-2Kb: conversion of helper-independent to helper-dependent CTL. *Int. Immunol.*, **4**, 1419–28.

Auphan, N., Schonrich, G., Malissen, M., Barad, M., Hammerling, G., Arnold, B., *et al.* (1992*b*). Influence of antigen density on degree of clonal deletion in T cell receptor transgenic mice. *Int. Immunol.*, **4**, 541–7.

Berg, L. J., Fazekas de St. Groth, B., Pullen, A. M., and Davis, M. M. (1989*a*). Phenotypic differences between αβ versus β T-cell receptor transgenic mice undergoing negative selection. *Nature*, **340**, 559–62.

Berg, L. J., Pullen, A. M., Fazekas de St. Groth, B., Mathis, D., Benoist, C., and Davis, M. M. (1989*b*). Antigen/MHC-specific T cells are preferentially exported from the thymus in the presence of their MHC ligand. *Cell*, **58**, 1035–46.

Bogen, B., Gleditsch, L., Weiss, S., and Dembic, Z. (1992). Weak positive selection of transgenic T cell receptor-bearing thymocytes: importance of major histocompatibility complex class II, T cell receptor and CD4 surface molecule densities. *Eur. J. Immunol.*, **22**, 703–9.

Bonneville, M., Ishida, I., Itohara, S., Verbeek, S., Berns, A., Kanagawa, O., *et al.* (1990). Self-tolerance to transgenic gamma delta T cells by intrathymic inactivation. *Nature*, **344**, 163–5.

Casanova, J. L., Romero, P., Widmann, C., Kourilsky, P., and Maryanski, J. L. (1991). T cell receptor genes in a series of class I major histocompatibility complex-restricted cytotoxic T lymphocyte clones specific for a *Plasmodium berghei* nonapeptide: implications for T cell allelic exclusion and antigen-specific repertoire. *J. Exp. Med.*, **174**, 1371–83.

Davis, C. B., Killeen, N., Crooks, M. E., Raulet, D., and Littman, D. R. (1993). Evidence for a stochastic mechanism in the differentiation of mature subsets of T lymphocytes. *Cell*, **73**, 237–47.

Dent, A. L., Matis, L. A., Bluestone, J. A., and Hedrick, S. M. (1993). Evidence for programmed cell death of self-reactive gamma delta T cell receptor-positive thymocytes. *Eur. J. Immunol.*, **23**, 2482–7.

Ferrick, D. A., Ohashi, P. S., Wallace, V., Schilham, M., and Mak, T. W. (1989). Thymic ontogeny and selection of alpha beta and gamma delta T cells. *Immunol. Today*, **10**, 403–7.

Fung-Leung W. P., Wallace, V. A., Gray, D., Sha, W. C., Pircher, H., Teh, H. S., *et al.* (1993). CD8 is needed for positive selection but differentially required for negative selection of T cells during thymic ontogeny. *Eur. J. Immunol.*, **23**, 212–16.

Gaugler, B., Schmitt-Verhulst, A. M., and Guimezanes, A. (1993). Evaluation of functional heterogeneity in the CD8 subset with T cells from T cell receptor-transgenic mice. *Eur. J. Immunol.*, **23**, 1851–8.

Geiger, T., Gooding, L. R., and Flavell, R. A. (1992). T-cell responsiveness to an oncogenic peripheral protein and spontaneous autoimmunity in transgenic mice. *Proc. Natl. Acad. Sci. USA*, **89**, 2985–9.

Groettrup, M. and von Boehmer, H. (1993). T cell receptor beta chain dimers on immature thymocytes from normal mice. *Eur. J. Immunol.*, **23**, 1393–6.

Groettrup, M., Baron, A., Griffiths, G., Palacios, R., and von Boehmer, H. (1992). T cell receptor (TCR) beta chain homodimers on the surface of immature but not mature alpha, gamma, delta chain deficient T cell lines. *EMBO J.*, **11**, 2735–45.

Grosveld, F. and Kollias, G. (ed.) (1992). *Transgenic animals*. Academic Press, London.

Hammerling, G. J., Schonrich, G., Momburg, F., Auphan, N., Malissen, M., Malissen, B., *et al.* (1991). Non-deletional mechanisms of peripheral and central tolerance: studies with transgenic mice with tissue-specific expression of a foreign MHC class I antigen. *Immunol. Rev.*, **122**, 47–67.

Hammerling, G. J., Schonrich, G., Ferber, I., and Arnold, B. (1993). Peripheral tolerance as a multi-step mechanism. *Immunol. Rev.*, **133**, 93–104.

Hedrick, S. M. and Dent, A. (1990). A model for gamma delta T-cell development: rearranged gamma- and delta-chain genes incorporated into the germline of mice. *Res. Immunol.*, **141**, 588–92.

Hogan, B., Constantini, F., and Lacy, E. (1986). *Manipulating the mouse embryo.* Cold Spring Harbor, New York.

Hugo, P., Kappler, J. W., and Marrack, P. C. (1993). Positive selection of TCRαβ thymocytes: Is cortical thymic epithelium an obligatory participant in the presentation of MHC protein? *Immunol. Rev.*, **135**, 133–56.

Ingold, A. L., Landel, C., Knall, C., Evans, G. A., and Potter, T. A. (1991). Co-engagement of CD8 with the T cell receptor is required for negative selection. *Nature*, **352**, 721–3.

Iwabuchi, K., Nakayama, K., McCoy, R. L., Wang, F., Nishimura, T., Habu, S., *et al.* (1992). Cellular and peptide requirements for *in vitro* clonal deletion of immature thymocytes. *Proc. Natl. Acad. Sci. USA*, **89**, 9000–4.

Kaye, J., Hsu, M. L., Sauron, M. E., Jameson, S. C., Gascoigne, N. R., and Hedrick, S. M. (1989). Selective development of CD4+ T cells in transgenic mice expressing a class II MHC-restricted antigen receptor. *Nature*, **341**, 746–9.

Kisielow, P., Bluethmann, H., Staerz, U., Steinmetz, M., and von Boehmer, H. (1988*a*). Tolerance in T cell receptor transgenic mice involves deletion of non-mature CD4+CD8+ thymocytes. *Nature*, **333**, 742.

Kisielow, P., Teh, H. S., Bluthmann, H., and von Boehmer, H. (1988*b*). Positive selection of antigen-specific T cells in thymus by restricting MHC molecules. *Nature*, **335**, 730–3.

Kisielow, P., Swat, W., Rocha, B., and von Boehmer, H. (1991). Induction of immunological unresponsiveness *in vivo* and *in vitro* by conventional and super-antigens in developing and mature T cells. *Immunol. Rev.*, **122**, 69–85.

Knobloch, M., Schonrich, G., Schenkel, J., Malissen, M., Malissen, B., Schmitt-Verhulst, A. M., *et al.* (1992). T cell activation and thymic tolerance induction require different adhesion intensities of the CD8 co-receptor. *Int. Immunol.*, **4**, 1169–74.

Lo, D., Freedman, J., Hesse, S., Brinster, R. L., and Sherman, L. (1991). Peripheral tolerance in transgenic mice: tolerance to class II MHC and non-MHC transgene antigens. *Immunol. Rev.*, **122**, 87–102.

Mamalaki, C., Norton, T., Tanaka, Y., Townsend, A. R., Chandler, P., Simpson, E., *et al.* (1992). Thymic depletion and peripheral activation of class I major histocompatibility complex-restricted T cells by soluble peptide in T-cell receptor transgenic mice. *Proc. Natl. Acad. Sci. USA*, **89**, 11342–6.

Matzinger, P. (1993). Why positive selection? *Immunol. Rev.*, **135**, 81–118.

Miller, J. F. and Heath, W. R. (1993). Self-ignorance in the peripheral T cell pool. *Immunol. Rev.*, **133**, 131–50.

Morahan, G., Hoffmann, M. W., and Miller, J. F. A. P. (1991). A nondeletional mechanism of peripheral tolerance in T-cell receptor transgenic mice. *Proc. Natl. Acad. Sci. USA*, **88**, 11421–5.

Murphy, D. and Carter, D. A. (ed.) (1993). *Transgenesis techniques*. Humana Press, Totowa, NJ.

Murphy, K. M., Heimberger, A. B., and Loh, D. Y. (1990). Induction by antigen of intrathymic apoptosis of CD4+CD8+TCRlo thymocytes *in vivo*. *Science*, **250**, 1720–3.

Ohashi, P. S., Oehen, S., Buerki, K., Pircher, H., Ohashi, C. T., Odermatt, B., *et al.* (1991). Ablation of 'tolerance' and induction of diabetes by virus infection in viral antigen transgenic mice. *Cell*, **65**, 305–17.

Ohashi, P. S., Zinkernagel, R. M., Leuscher, I., Hengartner, H., and Pircher, H. (1993*a*). Enhanced positive selection of a transgenic TCR by a restriction element that does not permit negative selection. *Int. Immunol.*, **5**, 131–8.

Ohashi, P. S., Oehen, S., Aichele, P., Pircher, H., Odermatt, B., Herrera, P., *et al.* (1993*b*). Induction of diabetes is influenced by the infectious virus and local expression of MHC class I and tumor necrosis factor-alpha. *J. Immunol.*, **150**, 5185–94.

Philpott, K. L., Viney, J. L., Kay, G., Rastan, S., Gardiner, E. M., Chae, S., *et al.* (1992). Lymphoid development in mice congenitally lacking T cell receptor alpha beta-expressing cells. *Science*, **256**, 1448–52.

Pircher, H., Burki, K., Lang, R., Hengartner, H., and Zinkernagel, R. M. (1989*a*). Tolerance induction in double specific T-cell receptor transgenic mice varies with antigen. *Nature*, **342**, 559–61.

Pircher, H., Mak, T. W., Lang, R., Ballhausen, W., Rüedi, E., Hengartner, H., *et* al. (1989*b*). T cell tolerance to Mls[a] encoded antigens in the T cell receptor Vβ8.1 chain transgenic mice. *EMBO J.*, **8**, 719–27.

Pircher, H., Muller, K. P., Kyewski, B. A., and Hengartner, H. (1992). Thymocytes can tolerize thymocytes by clonal deletion *in vitro*. *Int. Immunol.*, **4**, 1065–9.

Pircher, H., Brduscha, K., Steinhoff, U., Kasai, M., Mizuochi, T., Zinkernagel, R. M., *et al.* (1993). Tolerance induction by clonal deletion of CD4+8+ thymocytes *in vitro* does not require dedicated antigen-presenting cells. *Eur. J. Immunol.*, **23**, 669–74.

Robey, E. A., Fowlkes, B. J., Gordon, J. W., Kioussis, D., von Boehmer, H., Ramsdell, F., *et al.* (1991). Thymic selection in CD8 transgenic mice supports an instructive model for commitment to a CD4 or CD8 lineage. *Cell*, **64**, 99–107.

Robey, E. A., Ramsdell, F., Kioussis, D., Sha, W., Loh, D., Axel, R., *et al.* (1992). The level of CD8 expression can determine the outcome of thymic selection. *Cell*, **69**, 1089–96.

Rothe, J., Ryser, S., Mueller, U., Steinmetz, M., and Bluethmann, H. (1993). Functional expression of a human TCR beta gene in transgenic mice. *Int. Immunol.*, **5**, 11–17.

Russell, J. H., Meleedy-Rey, P., McCulley, D. E., Sha, W. C., Nelson, C. A., and Loh, D. Y. (1990). Evidence for CD8-independent T cell maturation in transgenic mice. *J. Immunol.*, **144**, 3318–25.

Schonrich, G., Kalinke, U., Momburg, F., Malissen, M., Schmitt-Verhulst, A. M., Malissen, B., *et al.* (1991). Down-regulation of T cell receptors on self-reactive T cells as a novel mechanism for extrathymic tolerance induction. *Cell*, **65**, 293–304.

Schonrich, G., Momburg, F., Malissen, M., Schmitt-Verhulst, A. M., Malissen, B., Hammerling, G. J., *et al.* (1992). Distinct mechanisms of extrathymic T cell tolerance due to differential expression of self antigen. *Int. Immunol.*, **4**, 581–90.

Schonrich, G., Strauss, G., Muller, K. P., Dustin, L., Loh, D. Y., Auphan, N., *et al.* (1993). Distinct requirements of positive and negative selection for selecting cell type and CD8 interaction. *J. Immunol.*, **151**, 4098–105.

Scott, B., Bluthman, H., Teh, H. S., and von Boehmer, H. (1989). The generation of mature T cells requires interaction of the alpha beta T-cell receptor with major histocompatibility antigens. *Nature*, **338**, 591–3.

Sha, W. C., Nelson, C. A., Newberry, R. D., Kranz, D. M., Russell, J. H., and Loh, D. Y. (1988*a*). Selective expression of an antigen receptor on CD8-bearing T-lymphocytes in transgenic mice. *Nature*, **335**, 271–4.

Sha, W. C., Nelson, C. A., Newberry, R. D., Kranz, D. M., Russell, J. H., and Loh, D. Y. (1988*b*). Positive and negative selection of an antigen receptor on T cells in transgenic mice. *Nature*, **336**, 73–6.

Sha, W. C., Nelson, C. A., Newberry, R. D., Pullen, J. K., Pease, L. R., Russell, J. H., *et al.* (1990). Positive selection of transgenic receptor-bearing thymocytes by Kb antigen is altered by Kb mutations that involve peptide binding. *Proc. Natl. Acad. Sci. USA*, **87**, 6186–90.

Speiser, D. E., Pircher, H., Ohashi, P. S., Kyburz, D., Hengartner, H., and Zinkernagel, R. M. (1992). Clonal deletion induced by either radioresistant thymic host cells or lymphohemopoietic donor cells at different stages of class I-restricted T cell ontogeny. *J. Exp. Med.*, **175**, 1277–83.

Sponaas, A.-M., Tomlinson, P. D., Antoniou, J., Auphan, N., Langlet, C., Malissen, B., *et al.* (1994). Induction of tolerance to self MHC class I molecules expressed under the control of milk protein or β-globin gene promoters. *Int. Immunol.*, **6**, 277–87.

Swat, W., Dessing, M., Baron, A., Kisielow, P., and von Boehmer, H. (1992).

Phenotypic changes accompanying positive selection of CD4+CD8+ thymocytes. *Eur. J. Immunol.*, **22**, 2367–72.

Tatsumi, Y., Pena, J. C., Matis, L., Deluca, D., and Bluestone, J. A. (1993). Development of T cell receptor-gamma delta cells. Phenotypic and functional correlations of T cell receptor-gamma delta thymocyte maturation. *J. Immunol.*, **151**, 3030–41.

Teh, H. S., Kishi, H., Scott, B., and von Boehmer, H. (1989). Deletion of autospecific T cells in T cell receptor (TCR) transgenic mice spares cells with normal TCR levels and low levels of CD8 molecules. *J. Exp. Med.*, **169**, 795–806.

Vasquez, N. J., Kaye, J., and Hedrick, S. M. (1992). *In vivo* and *in vitro* clonal deletion of double-positive thymocytes. *J. Exp. Med.*, **175**, 1307–16.

Viney, J. L., Prosser, H. M., Hewitt, C. R., Lamb, J. R., and Owens, M. J. (1992). Generation of monoclonal antibodies against a human T cell receptor beta chain expressed in transgenic mice. *Hybridoma*, **11**, 701–13.

von Boehmer, H., Kirberg, J., and Rocha, B. (1991). An unusual lineage of alpha/beta T cells that contains autoreactive cells. *J. Exp. Med.*, **174**, 1001–8.

von Boehmer, H., Swat, W., and Kisielow, P. (1993). Positive selection of immature αβ T cells. *Immunol. Rev.*, **135**, 67–80.

Wells, F. B., Gahm, S. J., Hedrick, S. M., Bluestone, J. A., Dent, A., and Matis, L. A. (1991). Requirement for positive selection of gamma delta receptor-bearing T cells. *Science*, **253**, 903–5.

Wells, F. B., Tatsumi, Y., Bluestone, J. A., Hedrick, S. M., Allison, J. P., and Matis, L. A. (1993). Phenotypic and functional analysis of positive selection in the gamma/delta T cell lineage. *J. Exp. Med.*, **177**, 1061–70.

Wilson, A., Pircher, H., Ohashi, P., and MacDonald, H. R. (1992). Analysis of immature (CD4−CD8−) thymic subsets in T-cell receptor alpha beta transgenic mice. *Dev. Immunol.*, **2**, 85–94.

11 Superantigens and tolerance

HANS ACHA-ORBEA

1 Introduction

The immune system has the difficult task of distinguishing self-antigens which should be spared from immune attack from foreign potentially harmful antigens which should be eliminated before they can become harmful. This distinction results in tolerance to self-antigens. The clonally distributed T cell receptors (TCR) can form billions of different highly specific recognition structures, many of which could react with self-antigens. Most self- and foreign antigens recognized by T cells are processed short peptides presented by self-major histocompatibility complex (MHC) class I and class II molecules.

Tolerance is achieved in part through elimination of autoreactive cells during the maturation and differentiation of the immune system. The thymus plays an important role in establishing tolerance in the T lymphocyte compartment. Elimination of autoreactive cells during thymic maturation is by clonal deletion but, in addition, several mechanisms exist to eliminate or inactivate mature autoreactive T cells after they leave the thymus. Three other basic mechanisms of tolerance have been described:

- peripheral deletion
- induction of unresponsiveness (anergy)
- immune regulation (suppression)

CD4 molecules increase the affinity of interaction between TCR and MHC class II molecules whereas CD8 molecules have a comparable function in their interaction with MHC class I molecules. Only a minority of cells in the body, including classical antigen-presenting cells, express MHC class II molecules. Antigens generated in the absence of MHC class II molecules can not activate $CD4^+$ T cells. In contrast, $CD8^+$ T cells recognize antigens in the context of MHC class I molecules expressed on the surface of the majority of cells in the body (see Chapter 3).

Antigens presented by cells which are not professional antigen-presenting cells generally fail to induce a primary immune response, rather inducing T cell anergy. To induce an immune response, secreted lymphokines and other interactions between receptors and co-receptors expressed on the T cell and APC are crucial in addition to cognate interactions between the TCR and MHC molecule bearing the antigenic epitope.

Recent studies have provided insight into the basic mechanisms of tolerance induction although much has to be learned. Four important approaches have been key to our current understanding of immune tolerance:

1. Bone marrow chimeras: irradiation of mice leads to a rapid loss of bone marrow-derived radiosensitive cells. Amongst those are T and B cells, macrophages, and dendritic cells. By injecting (semi) allogeneic T cell depleted bone marrow into such irradiated mice, T cell maturation in a foreign thymic environment can be studied.
2. Thymus transplants: similar analyses as above.
3. Transgenic mice expressing TCR molecules of known specificity express these receptor structures on the majority of T cells due to allelic exclusion of the endogenous TCR (see Chapter 10). Introducing selecting or non-selecting MHC haplotypes or antigens into such mice or transferring transgenic T cells into a naïve host allows the analysis of changes in a marked T cell compartment.
4. Superantigens interact with a large percentage of T cells in naïve mice. Since superantigens interact with TCR Vβ elements of the TCR, superantigen-reactive cells can be analysed with antibodies specific for TCR Vβ regions.

Each of the experimental systems has its advantages and drawbacks but overall each has led to similar conclusions.

This review will cover superantigens alone, discussing a selection of the most important experiments. Neither tolerance in the B cell compartment nor in the other major T cell subset, TCR $\gamma\delta^+$ T cells, will be covered. After introducing the current knowledge about superantigens, I will show how they have been used as tools to learn more about immune response and tolerance.

2 An introduction to superantigens

2.1 General

Classically, antigen recognition occurs through interaction of small processed peptides bound to self-major histocompatibility complex molecules which are recognized by the highly diverse parts of the TCR. Variable (V), junctional (J), diversity (D), and random deletions and insertions of nucleotides in N regions during rearrangement all contribute to classical peptide–MHC recognition (for review see reference 1). A typical estimate of the frequency of T cells specific for any particular peptide antigen is in the order of one in 10^6. In contrast, superantigens have the capability to stimulate a large percentage of T lymphocytes. The high frequency of superantigen interaction is explicable following the finding that they interact with a T cell receptor (TCR) Vβ segment which represents only a minor part of TCR diversity (see Figure 11.1) (2–10). Other parts of the highly diverse TCR are of less importance for interaction with superantigen in comparison with classical peptide antigen recognition although preferential usage of particular TCR Vα as well as Jβ elements have been detected in superantigen-reactive T cells (11–14) (for a comparison of general features of bacterial, viral, and retroviral superantigens see Table 11.1). There are in the order of 25 Vβ segments in the

Table 11.1 Properties of superantigens

	Bacterial	Retroviral	Viral
Stimulation of superantigen-reactive T cell	+	+/–	+
Clonal deletion	+	+	+
Induction of unresponsiveness	+	+	?
Class II dependence for presentation	+/–[a]	+	+
Binding to Ig	+	?	+
Induction of cytotoxic response	+	–	?
Amino acid homology[b]	Low	Low	Lo
Role in infection	?	+	?

[a] Recently it has been described that some bacterial superantigen can be presented in mice lacking the classical MHC class II molecules I-A and I-E (142).
[b] Homology between the different families of superantigens.

mouse and each is expressed in 1% to 15% of peripheral T cells. Superantigens can interact with one or several TCR Vβ chains following presentation by MHC class II molecules but, in contrast with classical peptide antigens, they can not be presented by MHC class I molecules (for review see references 15–19). Superantigen recognition breaks the rule of MHC restriction observed with classical peptide recognition since most alleles of MHC class II and even xenogeneic class II molecules can present superantigens to a given T cell (20, 21). The frequency of responding cells is about 10% of mature peripheral T cells for a single superantigen (2, 3, 5, 10, 22) which can be distinguished from mitogens, since these stimulate all T cells. In contrast, classical peptide antigens interact with an estimated 10 000-fold lower frequency of T cells. For a comparison between classical peptide antigen and superantigen recognition see Table 11.2. Recently it was shown that some γδ T cells can interact with bacterial superantigens as well (23, 24). Superantigens are presented exclusively by MHC class II molecules to T cells by

Table 11.2 Comparison between classical antigen and superantigen recognition

	Peptide antigen	Bacterial superantigen	Retroviral superantigen
Presenting molecule	Class I or class II	Class II	Class II
Parts of TCR important for recognition	Vα,N, Jα,Vβ,N,Dβ, Jβ	Vβ	Vβ
Frequency of responsive cells	$1/10^4$–$1/10^6$	1/3–1/10	1/3–1/10
Requirement for antigen processing	Yes	No	No (partial?)
MHC restriction	Yes	No	No

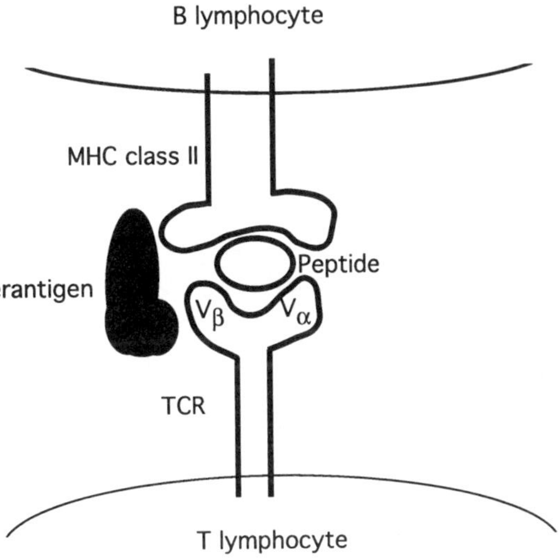

Fig. 11.1 Peptide and superantigen interaction with TCR and MHC molecules.

cross-linking TCR Vβ with MHC class II molecules [see Figure 11.1 (10, 25–27), and Chapter 19].

Superantigens have very strong influences on the immune system and therefore represent key tools for the investigation of immune response and tolerance. After TCR Vβ-specific monoclonal antibodies became available, it became possible to monitor the fate of superantigen-reactive T cells *in vitro* and *in vivo*. Depending on the time in the development of a T cell, the dose and place of encounter with superantigen, the following outcomes can be observed: Encounter with superantigen of superantigen-reactive T cells in the thymus of neonatal mice leads to clonal deletion (2–5, 10), with 50–99% of superantigen-responsive T cells being deleted depending on the type of endogenous or the dose of exogenous superantigens used (4, 5, 10, 28). In addition to neonatal clonal deletion, interaction with superantigens in adult mice can lead to deletion of the superantigen-reactive cells by a combination of peripheral and thymic deletion (29–31). Encounter with superantigens can result in a strong stimulation of the immune response (10, 15, 32–34). Surprisingly this immune response declines shortly after the initial encounter and thereafter the cells become unresponsive (anergic) (29, 35–42). Unresponsiveness has also been observed in the absence of a significant superantigen response following injection of very low doses of exogenous superantigen or superantigen expressing cells (30, 43). A summary of the possible fates of T cells after antigen interaction is shown in Figure 11.2 and the fates of T cells after superantigen encounter in Figure 11.3. Superantigen encounter does not always lead to stimulation, anergy, and deletion. Affinities between the three interacting molecules, dosage, the microenvironment, stage of maturation, and other factors may influence the outcome. Inhibition of proliferation and

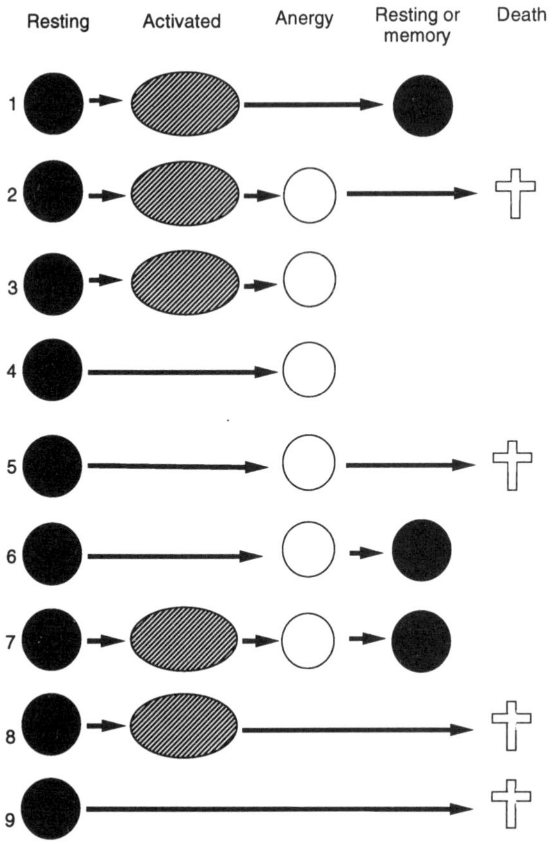

Fig. 11.2 Hypothetical pathways for immune response, anergy, and death of T cells. When T cells become activated by antigen–MHC complexes on antigen-presenting cells several outcomes are possible: (1) T cells become activated blasts and differentiate into memory cells or become small resting T cells again. (2) After activation anergy is induced and the cells die. (3) Same as (2) but the T cells survive as 'anergic' cells. (4) T cells become anergic without activation and survive as 'anergic' cells. (5) Same as (4) but the T cells die. (6) T cells become anergic and revert to small resting T cells. (7) Same as (6) but T cells go through an activation stage. (8) T cells become activated and die. (9) T cells die without activation and anergization.

IL-2 production following *in vitro* stimulation with either superantigen or after cross-linking of TCR Vβ molecules with monoclonal antibodies can be used as a readout of anergy, when cells become incapable of mounting a response on re-encounter with superantigen. The molecular events leading to this inhibition of lymphokine secretion are still not clear (44–46).

Superantigens are the products of various micro-organisms: bacteria, retroviruses, and viruses. The general properties of these superantigens, an overview of the nature of the micro-organisms producing them, and their life cycles in their natural host will be discussed below.

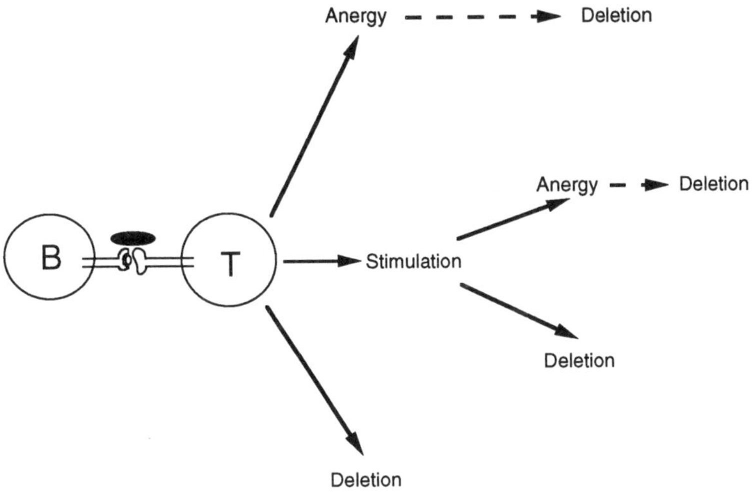

Fig. 11.3 Superantigen-induced T–B interaction. The T cell can directly or indirectly reach all the stages of stimulation, anergy, and deletion. The dashed arrows show possible but not exclusive outcomes.

2.2 Bacterial superantigens

Bacterial superantigens are the causative agents of food poisoning and toxic shock syndrome (47). It has been known for a long time that bacteria can produce 'T cell mitogens' (48, 49). It was recently discovered that these exotoxins are not mitogens stimulating all T cells but superantigens stimulating subpopulations of T cells identified by their TCR Vβ expression (10, 34, 50). They have been isolated from *Staphylococci* (Staphylococcal enterotoxins SEA, B, C1–3, D, E, toxic shock syndrome toxin TSST-1) (10, 18, 34, 50), *Streptococci* (51–55), *Mycoplasma* (56–58), as well as a molecularly undefined product from *Yersinia* (59). Some of the clinical symptoms attributed to interaction with superantigens include toxic shock syndrome, arthritis, and food poisoning. The detailed mechanisms leading to food poisoning ingestion are not known but it seems that symptoms in mice after injection are dependent on the presence of superantigen-reactive T cells (60). Superantigens have been shown to cause tissue damage in human fetal small intestinal epithelium due to T cell activation (61).

Humans are much more sensitive to bacterial superantigens than rodents but T cells with homologous TCR Vβ sequences react in both species (see Table 11.3). Lethal toxic shock in mice can be induced, however, in D-galactosamine sensitized mice (62). T or monocyte stimulation with superantigens leads to production of large amounts of lymphokines and to release of leukotrienes from mast cells (62–71). Superantigens are effective at very low concentrations, they can induce a strong response in the nanomolar range which makes them the most potent 'mitogens' (49, 72). This potency of super-

Table 11.3 TCR specificities of bacterial superantigens

Bacterial toxin	Mouse Vβ	Human Vβ
SEA	1,3,10,11,12,17	1.1,5.3,6.3,6.4,6.9,7.3,7.4,9.1
SEB	7,8.1,8.2,8.3	3,12,14,15,17,20
SEC1	7,8.2,8.3,11	12
SEC2	8.2,10	12,13.2,14,15,17,20
SEC3	7,8.2	5,12,13.2
SED	3,7,8.3,11,17	5,12
SEE	11,15,17	5.1,6.1–6.4
ExFT	10,11,15	2
TSST-1	15,16	2
SPE-A	8.2	2,12,14,15
SPE-B	ND	8
SPE-C	ND	1,2,5.1,10
pepM5	ND	2,4,8
YPS	3,6,7,8.1,9,11	ND
MAM	1,3,6,5.1,8.1–3,16	3,5.1,7.1, 8.2,10,12.2,16,17.1,20

(The results summarized from references 18, 128, 222, 226)

antigens may cause problems in their identification since minute contaminations of preparations can lead to wrong conclusions about specificity (55, 73).

The bacterial superantigens characterized to date have molecular weights of 22–66 kDa (18, 71). Not all have been molecularly characterized yet but the comparison between the different staphylococcal superantigens reveals groups of proteins with sequence homology. SED and SEE are 71% and 92% homologous to SEA, respectively. When this group is compared with SEB and SEC, the homology drops to 30% (18, 71). Homologies to other bacterial superantigens can be even lower. Recent studies indicated that several of the enterotoxin genes in *Staphylococcus* are encoded by phage (74), plasmid, or transposable elements (for review see reference 71). They may have developed by convergent evolution by similar selective pressures.

The bacterial superantigens can bind with high affinity to MHC class II molecules (75–81). Affinities for human class II molecules are in the range of 10^{-5}–10^{-7} M (82). Affinities for mouse MHC class II molecules are lower. Bacterial superantigens do stimulate the normally class II restricted $CD4^+$ lymphocyte subset but more surprisingly are able to stimulate the usually MHC class I restricted $CD8^+$ T cells comparably even though superantigens are presented exclusively by MHC class II molecules (20, 83).

2.3 Retroviral superantigens

When it became clear that the genes encoding all the known endogenous superantigens mapped close to MMTV proviral loci, it did not take long to

Table 11.4 TCR specificities of MMTV superantigens

Vβ	*Mtv* and MMTV	Presenting molecule
2	MMTV(C4), MMTV(), *Mtv*-DDO	I-E
3	-1, -3, -6, -13, -27, -44, -MAI	I-E, I-A
5	-6, -9, -44,	I-E
6	-7, -43, -44, 53, MMTV(SW), MMTV(JYG)	I-E > I-A
7	-7, -43, 53, MMTV(SW), MMTV(JYG)	I-E > I-A
8.1	-7, -43, -44, 53, MMTV(SW), MMTV(JYG)	I-E > I-A
9	-7, -43, -44, 53, MMTV(SW), MMTV(JYG)	I-E > I-A
11	-8, -9, -11	I-E
12	? (possibly -8, -9, -11)	I-E
14	-2, MMTV(C3H), MMTV(GR)	I-E ≫ I-A
15	MMTV(C3H), MMTV(GR)	I-E ≫ I-A
16	?	I-E
17a1	-3, -8, -9	I-E (Mtv-3 with I-A)
17a2	-6, MAI	I-E
17a(cz)	-1, -3, -6, 13, -27, -44	I-E
19a	?	
20	?	

The results summarized from references 101, 180, 225. Infectious viruses are named MMTV and endogenous proviral loci *Mtv*-.
Partial deletion is induced by: *Mtv-1, -8, 13, -27, -44.*

show that these superantigens are actually encoded by MMTV (84–88). The TCR Vβ specificities of the MMTV superantigens so far characterized are summarized in Table 11.4. The products of some of these endogenous proviruses were originally described as minor lymphocyte stimulating (Mls) antigens by Festenstein some 20 years ago (32). He discovered that a large proportion of Mls-responsive cells is stimulated in a non-MHC restricted fashion when T cells from mice lacking a particular Mls antigen were mixed with B cells expressing an Mls antigen. It was clear that the genes encoding these proteins were dominant and, surprisingly, several different Mls loci were described each characterized by a stimulatory Mls allele and a null allele not able to stimulate a T cell response of MHC matched stimulator cells (15). Several such +/− bi-allelic systems were defined in the common laboratory mouse strains and mapped to different chromosomal localizations.

MMTV proviruses arose by integration of infectious MMTV into germ cells and are inherited like normal host genes. In contrast, infectious MMTV is maternally transmitted from mother to offspring via milk during the first two weeks of life before their stomach contents acidify (for review see references 16, 17). Laboratory as well as wild mice have several MMTV provirus integrations in their genome. These loci are called *Mtv-1* to *−53*, with more being defined. In most laboratory mice two to eight endogenous *Mtv* loci are found

(89). In wild mice the number differs depending on where the mice were caught. It ranges from zero to more than ten. Most of the endogenous proviruses have lost the capability of producing infectious virus particles. Most, with the exception of *Mtv-17*, which has a defect in its promoter, produce superantigens (90). These superantigens are encoded in an open reading frame (*orf*) in the long terminal repeat of the virus (84–88). The function of MMTV-*orf* has been a puzzle to retrovirologists. They had difficulties ascribing a function to this potential protein which they could not detect in normal or tumour cells. It was thought that this protein had an effect on gene regulation, with both positive and negative gene regulation being described (91, 92). Since viral genes often have more than one function, it will be interesting to see if functions other than their superantigenic properties can be found for MMTV-*orf*. Very little information about the biochemical nature of the *orf* gene product is available (93, 94). Blocking studies with monoclonal anti-MMTV-*orf* antibodies, however, clearly established that the carboxyl-terminus of pORF is accessible on the cell surface. The different MMTVs are about 90–95% homologous in their amino acid sequence (95). Although the expressed protein of MMTV-*orf* is not yet characterized, based on the cDNA sequences several key features of the potential protein(s) can be postulated (see Figure 11.4):

1. In the carboxyl-terminal 10–30 amino acids a very high level of polymorphism is detected. This polymorphism correlates perfectly with the TCR Vβ specificity (17, 87, 88). So far six groups with different TCR Vβ specificities have been defined (95 and unpublished results). For each group the sequences of one to six different loci or infectious particles have been defined and few differences within groups have been detected in most instances. Comparison of MMTVs which are evolutionary distant can help define the key TCR Vβ interaction residues (97–99). Recently the first *Mtv-* loci with specificity for T cells found in two different groups have been defined and the few observed changes in the carboxyl-terminal MMTV-*orf* sequence will allow pin-pointing key residues for TCR interaction (100, 101).
2. Five in-frame potential start codons (ATG) in the amino-terminal half of all the MMTV-*orf* sequences have been found.
3. 30 amino acids after the first potential start codon, just after the second ATG, a highly hydrophobic sequence is detected which could serve as a transmembrane region or a signal peptide, depending on which ATG is being used. Transfection experiments indicated that using the longest possible MMTV-*orf* DNA led to best superantigen presentation results, DNA starting at the second ATG was weaker and shorter fragments were completely unable to present a superantigen (86). *In vitro* translation experiments in the presence of microsomal membranes indicated the potential of producing a type II transmembrane glycoprotein with the longest mRNA molecule (86, 102, 103). These proteins are inserted into

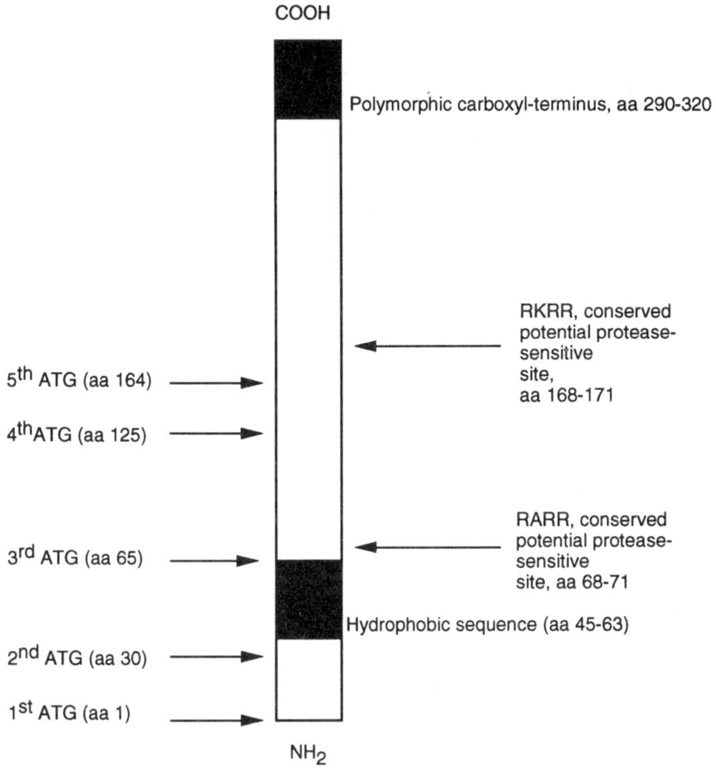

Fig. 11.4 Potential MMTV superantigen protein.

the membrane with opposite orientation to classical proteins which would agree with the finding that the carboxyl-terminal polymorphism correlates with TCR Vβ specificity.

4. Two highly conserved potential protease sensitive sites (RXRR) are found between the transmembrane domain and the middle of the molecule and these might be cleaved during transport or after surface expression. There are indications that MMTV superantigens are processed into shorter fragments (94, 99). But it is clear that MMTV superantigens are not peptides since peptides containing the carboxyl-terminal 40 amino acids were unable to elicit a superantigen response (unpublished results).

Based on older experiments as well as experiments which only became possible after the discovery of superantigens encoded by MMTV, the following scenario for infection can be proposed (Figure 11.5).

The virus is taken up through milk starting at birth and comes into contact with lymphocytes in the Peyer's patches (O. Karapetian, A. N. Shakhov, J. P. Kraehenbuhl, H. Acha-Orbea. Retroviral infection of neonatal Peyer's patch lymphocytes. The mouse mammary tumor virus model. *J. Exp. Med.*

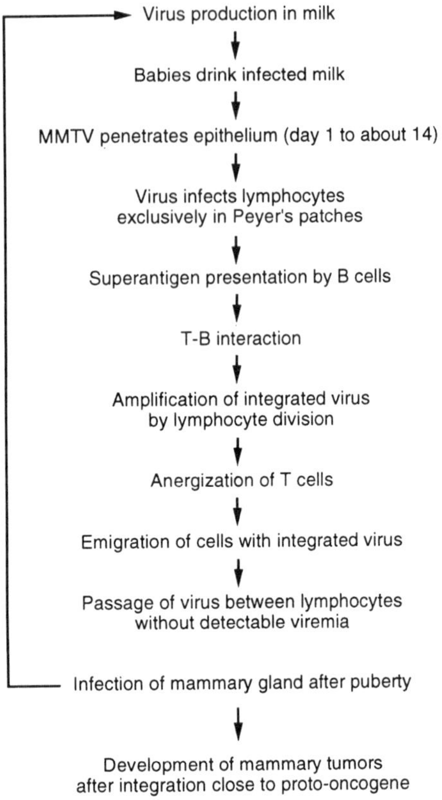

Fig. 11.5 Life cycle of MMTV.

180, ISM 1994). Lymphocytes are infected and as soon as the immune system matures sufficiently, B cells present superantigen to superantigen-reactive T cells (28, 96, 104). From experiments using adult mice it was clear that infection can be detected exclusively in B cells and that a strong augmentation of the superantigen expression by B cells occurs (96, 105). This results in differentiation and an increase in the number of cells with an integrated MMTV provirus. A continuous and vigorous immunologic stimulation (which would be harmful for the host) is prevented by anergization and deletion of the responsive T cells. The expression of superantigen therefore allows much more efficient infection (105). We recently showed that in newborn mice the virus infects lymphocytes exclusively in Peyer's patches, the lymph nodes of the small intestine (O. Karapetian, A. N. Shakhov, J. P. Kraehenbuhl, H. Acha-Orbea, *loc. cit.*). The infection of lymphocytes represents a crucial step in the life cycle of the virus. Mice lacking MMTV superantigen-responsive T cells or mice having an incomplete immune system are not able to propagate the virus efficiently (104–106). Later in life, the exogenous (but not the endogenous) virus is able to be transferred between lymphocyte subsets (30).

Viremia with MMTV has never been observed, the only place where free virus can be detected is in milk. This situation permits a continuous low level of infection of lymphocytes. After puberty and under the influence of hormones the virus can be transferred from lymphocytes to the mammary gland epithelium where virus replication occurs with secretion of infectious virus particles into the milk. This is the only place where free virus particles have been observed. When the mothers transmit the virus to their offspring, the life cycle is completed (for review see (107)). This is schematically shown in Figure 11.5. As the name implies, MMTV viruses are the major causative agent of mammary tumours in mice. Integration close to a mouse proto-oncogene can activate it and induce the first steps in carcinogenesis (108–113).

It has been suggested that murine leukaemia virus expresses a superantigen (114). In these experiments mice lacking H–2E expression (see below) were infected with a mixture of defective and intact MuLV. These mice develop an immunodeficiency syndrome similar to AIDS. It was shown that lymphoma cells from such mice stimulate T cells expressing Vβ5 and 11. There remains a problem in interpreting this study to attribute superantigen production to MuLV since these infected mice also express an endogenous superantigen which has been shown to interact with these TCR populations. In H-2E negative mice, it does not lead to deletion of superantigen-reactive T cells but recently we have shown that H-2E-dependent superantigens can be presented in the context of H-2A albeit in a much less efficient way (16). Therefore, only experiments in mice lacking these endogenous superantigens will provide conclusive evidence about the existence of MuLV encoded superantigens.

During the last year two reports have appeared in the scientific literature raising the possibility that the human immunodeficiency virus (HIV) encodes a superantigen (115, 116). The first analysed the expression of mRNA of the different TCR Vβ chains in the peripheral blood of patients at late stages of AIDS. Several TCR Vβ elements were strongly reduced in late stage patients, after the drop in the $CD4^+$ T cell population. Semi-quantitative polymerase chain reaction was used to quantitate mRNA expression in unseparated blood isolates. Analyses on purified $CD4^+$ T cell populations are required to address this question more directly since it is possible that the TCR Vβ chains which were reduced are underrepresented in the $CD8^+$ T cell population.

In another study it was shown that freshly isolated T cell lines expressing TCR Vβ6.7, 8, 17 or TCR Vβ12 could be infected similarly with HIV but that after addition of antigen-presenting cells the latter had an approximately 100-fold higher HIV titre than the former three suggesting a superantigen effect in HIV amplification. More needs to be done to convincingly show that HIV encodes a superantigen. If it does, it opens new strategies for vaccination and would make the MMTV system a model system of choice for new vaccination studies directed at the superantigen effects. In the case of MMTV superantigen, the answer to the question of why these viruses have adopted a strategy of superantigenic stimulation of T cells seems answered. The superantigen allows a more efficient infection of the host lymphocytes which seems crucial for the survival of the virus.

2.4 Viral superantigens

In another study it has been shown that rabies virus encodes a superantigen in its nuclear protein (117). Recombinant protein showed strong binding to MHC class II molecules and stimulated T cells expressing Vβ12 in human and Vβ6 (the closest murine homologue) in mouse. It will be interesting to determine the effect of this superantigen on the efficiency of infection with rabies virus. Once this virus reaches the nerve cells, it is no longer accessible to the immune system. In addition, it will be interesting to see if and how this superantigen plays a role in the vaccines shown to provide protection from rabies infection. The results of these experiments should also provide insights into the possible role of superantigens in vaccination.

2.5 Roles in infection

The biological role of MMTV superantigens has now become clearer. The superantigen-induced amplification of the infection is required for productive infection. In the absence of superantigen-reactive T cells, the life cycle of MMTV is interrupted (104–106).

The biological role of bacterial superantigens is much less clear. The diversity of proteins with the same function points to a strong selection to produce such superantigens. Several reports indicated that superantigen responses might suppress a specific immune response (37, 38, 118–120). It was recently shown that injection with SEB leads to a suppression of a classical peptide-specific immune response by as much as 80% (60). One can speculate that expression of superantigens helps the bacteria to reduce the immune response of the host and to establish a more productive infection. More experiments are required to confirm this hypothesis.

2.6 Potential roles in autoimmunity

In a normal individual many potentially autoreactive lymphocytes persist. They are kept in check by ill-defined tolerance mechanisms. Since superantigens trigger T cells (and B cells) polyclonally, independent of their original antigen specificity, they might induce a chain reaction leading to autoimmunity. One key question is whether superantigen encounter invariably leads to induction of anergy or if appropriate antigen presentation following superantigen stimulation can rescue cells from anergy and apoptosis. Several recent studies showed that mice which are close to development of autoimmunity can be made sick by treatment with superantigens which stimulate the autoimmune T cells (120*a*,*b*). Both activation *in vivo* and *in vitro* were effective. Several studies addressed the role of superantigens in autoimmunity. In one, mice with a lymphoproliferative syndrome (lpr and gld mice) were investigated.

These mice show a large thymus-dependent amplification of lymphocyte subsets leading to lymphadenopathy characterized by Thy-1^+, Ly-1^+, $CD8^-$, $CD4^-$, $B220^+$, Pgp-1^+ T cells. They express a normal TCR repertoire and therefore undergo thymic deletion of superantigen-reactive T cells (121, 122). A parallel situation may exist in patients with Kawasaki disease who have a large increase of Vβ2 T cells in peripheral blood (123, 124) and in rheumatoid arthritis patients, in whom an increase of T cells expressing Vβ14 was found in joints in contrast to a decrease in the periphery. These last results have to be taken with caution because quantitation has been performed by PCR analysis and the differences were not bigger than twofold (125). Experimentally, unresponsiveness in autoreactive T cells by exogenous superantigen has been induced in autoimmune mice and these mice were protected from autoimmunity (126, 127).

Future experiments will be needed to explore the nature of the association between superantigens and autoimmunity.

2.7 Role of MHC class II in superantigen presentation

Much experimental evidence suggests that superantigens have to bind to MHC class II molecules to activate T cells. Cross-linking of superantigen in the absence of MHC class II molecules does not lead to T cell stimulation. Single point mutations in the α helix of MHC class II molecules clearly indicated that single amino acid substitutions in the peptide binding groove can abolish classical peptide presentation. In contrast, superantigen recognition was not affected by these changes. Changing amino acid residues outside the peptide binding groove affected superantigen binding moderately whereas peptide binding was not affected (79). Different MHC class II alleles or isotypes have different affinities for staphylococcal superantigens. Several recent reviews cover these experiments (82, 128). TSST, for example, binds more weakly to DP molecules and to mouse H-2E molecules and amongst mouse H-2A molecules only weak binding was observed to I-A^b. SEA was shown to bind strongly to practically all HLA-DR molecules except DRB4*0101. In addition SEA requires Zn^{2+} ions for superantigen function which allowed the mapping of the histidine residues important for binding (129) (see also Chapter 19).

MMTV superantigens in general interact better with H-2E than H-2A molecules. A whole group of endogenous and exogenous MMTV superantigens have been defined which do not induce clonal deletion when H-2E molecules are absent. At least one of these shows, however, a very slow deleting capacity with particular H-2A haplotypes which indicates an isotype preference but in no way an isotype specificity (16). Considering the superantigens which induce an immune response in the context of H-2A molecules, a clear hierarchy of presentation is seen:

$$A^k > A^b > A^d > A^s >> A^q.$$

Mice expressing H-2A^q (in the absence of H-2E) show neither clonal deletion nor MMTV superantigen stimulation. Both bacterial and MMTV superantigen function can be efficiently blocked with anti-class II monoclonal antibodies. Continuous exposure to anti-class II H-2E monoclonal antibodies *in vivo* from birth leads to a rescue of superantigen-reactive T cells from deletion (130) and those expressing the superantigen-reactive TCR are not tolerant to the superantigen. Stopping the continuous antibody treatment leads to a rapid elimination of the monoclonal antibody from the system. After about 31 days, complete peripheral deletion of the superantigen-reactive cells is observed.

A very recent experiment addressed the question whether MMTV and bacterial superantigens compete for binding to MHC class II molecules. The conclusion was that some (SEA and SEB) compete with an MMTV-*orf* polypeptide whereas TSST does not (131).

2.8 Superantigen interaction with TCR and MHC

One bacterial superantigen, staphylococcal enterotoxin B (SEB), has recently been crystallized (132). Introduction of mutations in the SEB coding sequence already previously allowed localization of important MHC and the TCR interaction residues (133). Two adjacent amino acids have been found in SEA which are involved in TCR and MHC class II interaction respectively (134). By generating hybrid molecules between SEA and SEE, which are highly homologous but differ in TCR specificity, it was possible to map two adjacent amino acids which are involved in TCR interaction (135). For all the superantigens analysed, the affinity for the TCR seems to be much lower than for MHC molecules and they need to be presented by MHC class II molecules (128, 136).

2.9 TCR and superantigen interaction

The sequence of natural variants as well as mutagenesis of TCR Vβ molecules have allowed identification of few amino acids in the TCR Vβ chain important for MMTV superantigen interaction (137–141). The general conclusion is that the predicted CDR4 loop is interacting with the superantigen, more specifically a hypervariable region between residues 70–74 of the TCR β chain (see also Chapter 19).

The overall conclusions of these structural studies suggest that unprocessed superantigen molecules can interact with the lateral side of the TCR and MHC class II molecules. From the available results, it seems clear that similar parts of the TCR are involved in interaction with different superantigens but that different binding sites may exist on MHC class II molecules

2.10 Stimulation

One of the key features of superantigens is their capacity to induce a strong proliferative response of T cells expressing the superantigen-reactive TCR Vβ chains. For this stimulation to occur they must associate with MHC class II molecules on antigen-presenting cells. However, there are other, so far not clearly defined, presenting molecules different from MHC class II H-2A and H-2E in the mouse and which also can present some of the bacterial superantigens (142). The staphylococcal superantigens have a strong binding affinity to MHC class II molecules. No binding to MHC class I molecules has been detected. The binding affinities to human MHC class II molecules are in the order of 10^{-5} to 10^{-7} M which resembles an affinity constant for an average antibody molecule (75–77, 82). The affinities for mouse MHC class II molecules are one to two orders of magnitude lower. Different superantigens have different preferences for DR, DP, or DQ isotypes in humans. A general feature of the so far characterized bacterial superantigens is reactivity to MHC molecules in different species including mouse, rat, and man. Using this superantigen bridge, for example, mouse T cells can be stimulated with human antigen-presenting (MHC class II expressing) cells. In general, staphylococcal superantigens have a much higher binding affinity to human than to mouse MHC class II molecules.

T cells interacting normally with MHC class II molecules express the CD4 surface molecules and these increase the affinity of contact between effector and stimulator cells. CD8 T cells (cytotoxic cells) usually interact with antigen presented by MHC class I. Bacterial superantigens have a stronger interaction with CD4 T cells but they can induce a very strong CD8 superantigen-mediated cytotoxic response even across species barriers. Both bacterial as well as viral superantigens bind to human and mouse MHC class II molecules and little MHC restriction in this interaction is observed. There is, however, a hierarchy in presenting capability of the different MHC class II isotypes and alleles which goes from strong presentation to lack of presentation. Recently it was shown in transfectants that human MHC class II molecules can present MMTV superantigens to human as well as mouse superantigen-reactive T cells (143, 144). Most of the superantigens so far described stimulate both $CD4^+$ and $CD8^+$ T cells expressing particular TCR Vβ elements (20, 34, 117). Cytotoxicity, which is readily detectable following stimulation with bacterial superantigens, however, has so far not been found with MMTV superantigens (66, 145).

Injection of SEB into mice leads to a strong and transient activation of $CD4^+$ as well as $CD8^+$ T cells. The CD8 cells are highly cytotoxic towards class II expressing target cells in the presence of SEB between two and three days after injection in *ex vivo* (83). Thereafter cytotoxicity drops to background levels. It is not clear why these cells do not cause tissue damage *in vivo*.

Amongst the MMTV superantigens, differences in their stimulation capacity were noted and a whole series of weakly or non-stimulating MMTV superantigens were discovered. These differences in stimulating capacity were found both *in vitro* and *in vivo*. MMTV superantigens induced a much weaker but clearly detectable response in $CD8^+$ cells as compared to $CD4^+$ cells. *In vitro* stimulation with Mls-1^a (*Mtv-7*) expressing cells resulted in clear stimulation of superantigen-reactive $CD8^+$ $V\beta6^+$ T cells (146–150). Measuring the efficiencies of presentation by different MHC class II haplotypes suggested a lower affinity for $CD8^+$ T cells than for $CD4^+$ T cells. Surprisingly, a twofold increase of CD8 T cell stimulation was observed after blocking the CD8 molecules with monoclonal antibodies (150, 151). Similar observations were made with bacterial enterotoxins (20).

Mls-1^a (*Mtv-7*) expression leads to clonal deletion of T cells expressing TCR Vβ 6, 7, 8.1, and 9. Analysis of the responding cells after injection of superantigen expressing B cells revealed a clear hierarchy of the responses in these four T cell populations: Vβ6>Vβ7>Vβ8.1>Vβ9 (152).

There have been contradictory results following *in vivo* injection with MMTV superantigen expressing cell populations. Webb and collaborators showed in neonatal mice that transfer of $CD8^+$ T cells from MMTV superantigen expressing mice are more efficient at induction of clonal deletion than B or $CD4^+$ T cells. Much smaller numbers of CD8 T cells are required to induce deletion. In addition they showed that such $CD8^+$ T cells are capable of inducing a strong stimulation of superantigen-reactive T cells in the Mls-1^a negative host (148, 153). The capacity of inducing deletion by $CD8^+$ T cells was confirmed by Waanders *et al.* (30). In adult mice, in contradiction to these results, no detectable stimulation was observed by using the same mouse strains. The route of injection and the purification of the relevant subsets were, however, different (30, 153).

3 Superantigens in peripheral and thymic tolerance

Experiments with bone marrow chimeras, thymus transplantations, TCR transgenic mice, fetal thymic organ culture, as well as superantigens, have been important for our current understanding of thymic and peripheral tolerance. Each experimental system has its advantages and disadvantages, some of which are listed below.

1. TCR transgenic mice: TCR transgenic mice express the transgenic receptor on the majority of TCR expressing immature and mature T cells. By breeding with MHC congenic mice, MHC haplotypes corresponding to their original restriction element or to other MHC haplotypes can be obtained. This allows one to address questions about positive selection. Introduction of the antigen for which the receptor is specific permits analysis of tolerance induction by clonal deletion or peripheral mechanisms. Double transgenic mice can be created which express a self-antigen exclusively in the periphery (not in the thymus) and tolerance induction to peripheral antigens can be

studied. One of the major disadvantages is the expression of TCR at early differentiation stages of the thymocytes. In addition, due to utilization of different DNA constructs, different transgenes can be expressed at different stages of development and different percentages of transgenic TCR expressing mature lymphocytes are found. The recent finding that a large percentage of peripheral T cells can express two functional TCR α chains complicates the analysis of thymic selection in normal mice.

2. Bone marrow chimeras: Elimination of radiosensitive cells (B, T, macrophages, dendritic cells) by lethal irradiation and injection of semi-allogenic or allogeneic T cell depleted bone marrow allows the analysis of maturation of T cells in a foreign environment. In many instances, lethal irradiation spares some of the 'radiosensitive' cells which can result in partial chimerism.

3. Thymus transplantations: Transplantation of a foreign thymus into normal or nude mice allows complementary analyses as described in bone marrow chimeras. Often bone marrow-derived cells are transferred with the thymus transplants.

4. Fetal thymic organ culture: Thymic differentiation can be easily manipulated by addition of factors or cells to the organ. Disadvantages are that it is not clear whether fetal thymus *in vitro* behaves like the adult thymus *in vivo*.

5. Superantigens: Analysis of thymic maturation in a normal environment without manipulation. This allows stimulation, deletion, and anergy in the thymus and the periphery to be studied for effectively polyclonal T cell responses. Disadvantages: It is still not clear how similar superantigen responses are to classical antigen responses (see below).

3.1 Thymic selection

One of the major sites for tolerance induction in the T cell compartment is the thymus. Here, T cells undergo differentiation and the differentiation steps can be followed by changes in their surface markers. A large amount of literature exists on the events in maturation and they are reviewed in references 154, 155, and covered in Chapter 2 by Crispe and Schatz in this volume. In this review, only the major subsets will be considered. Two of the commonly used markers are CD4 and CD8. CD4 is expressed by the major peripheral mature T cell population which interacts with antigens presented by MHC class II molecules. CD8 is found on the other major peripheral T cell subset which interacts with peptides presented by MHC class I molecules. These molecules have at least two functions: on the one hand they increase the affinity of the interaction with MHC class I and class II molecules, respectively, and on the other hand they are able to transmit signals to the cell. In the thymus all combinations of these two molecules are found. Cells expressing both are called double-positive thymocytes; cells expressing none, double-negative; and cells expressing one of the two, single-positive or

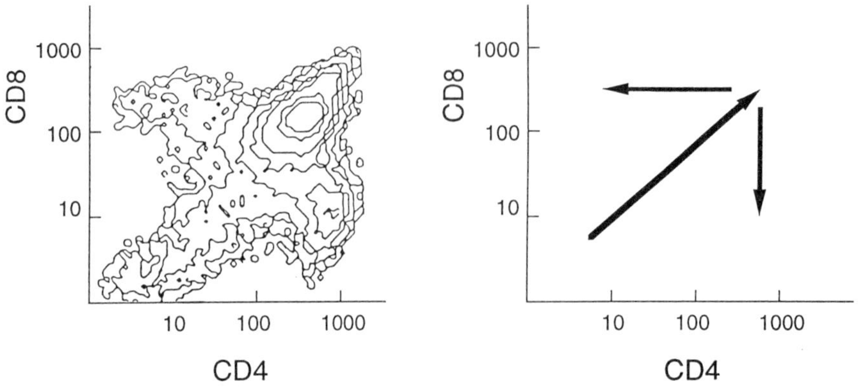

Fig. 11.6 FACS profile of a normal thymus. Thymocytes of six-week-old mice were stained with FITC-labelled anti-CD8 and phycoerythrin-labelled anti-CD4 (*left*). The differentiation pathway of thymocytes from $CD4^-8^-$ to $CD4^+8^+$ and finally to $CD4^+8^-$ or $CD4^-8^+$ populations is shown to the *right*.

mature T cells. Thymocytes differentiate from $CD4^-CD8^-$ precursors to $CD4^+8^+$ double-positive immature thymocytes in the cortex and finally to $CD4^+8^-$ or $CD4^-8^+$ single-positive T cells in the medulla (see Figure 11.6). This is a much simplified view; in reality these subsets are much more complex. For example, the double-negative population contains besides the precursor cells several different more mature populations. Single-positive cells have been selected for export to the periphery where they make up the mature T cell repertoire. Only about 1–2% of the double-positive cells survive this selection process. During this differentiation two major selection events determine the cells which are allowed to mature. These selection events have been called positive selection and clonal deletion. From blocking studies *in vivo* with monoclonal antibodies directed at TCR, MHC class I, MHC class II, CD4, as well as CD8 clearly indicated that the interaction of these molecules is required for positive selection and clonal deletion to occur (156–160). Despite the large numbers of experiments, different possibilities exist still as to how positive and negative selection operate: see Chapters 2 and 7.

These observations leave us with the paradox that maturing T cells can receive signals which make them survive or die. For the time being three major (not exclusive) interpretations have been proposed:

1. The affinity of the TCR determines whether positive selection or clonal deletion occurs. This might happen at any differentiation stage. The affinity of the T cell determines the fate in selection: cells with a high affinity for self-antigens (autoreactive T cells) die during maturation whereas cells able to interact with self-MHC (and peptide?) with weak affinity are positively selected. Therefore, the useful cells (which can interact with self-MHC molecules) would be positively selected and the harmful (autoreactive) would

be eliminated. The cells which undergo neither of the two selection events would die in the thymus (for review see reference 154).

2. The thymocyte changes during maturation from a cell which reacts to TCR interaction by survival to a cell which will die upon TCR interaction. Therefore the maturation stage of the thymocyte determines the outcome of selection.

3. The microenvironment (interleukins, co-receptors) determines whether interaction leads to survival or death. It is also possible that different microenvironments present different peptides to the maturing T cells. Since immature double-positive T cells express about ten times fewer TCR molecules on the cell surface, an interaction at this stage of maturation would be of much lower affinity (for review see references 161–164). This is supported by the fact that thymic cortical epithelial cells are incapable of presenting antigens to T cell clones (165) or by T cell hybridomas reacting with thymic epithelial cells but not with other cells (166). In addition, because thymocytes change during differentiation, interaction may transmit different signals to the cell at different stages of development. One of these changes during differentiation leads to an up-regulation of TCR molecules by about a factor of ten when double-positive thymocytes are compared to single-positives (167). Depending on the maturation stage of thymocytes, interaction with superantigens can lead to either deletion or stimulation. Immature cells are deleted whereas mature cells are stimulated to proliferate (168, 169).

Originally it was thought that different cells are responsible for these selection events. Radioresistant thymic cortical epithelial cells were thought to mediate positive selection whereas radiosensitive bone marrow-derived cells such as macrophages, dendritic cells, and B cells were important for clonal deletion. More recent results indicate, however, that any MHC expressing cell can induce clonal deletion and also that intrathymic injection of many different cell types including fibroblasts can induce positive selection (170). A simplified schema of thymic selection is shown in Figure 11.7. Some of the results obtained are discussed below in more detail.

3.2 Clonal deletion

Due to expression of endogenous MMTV superantigens or after injection of bacterial superantigens, the general observation has been that in normal mice holes in the mature T cell repertoire exist and that such holes are generated during T cell maturation in the thymus (154, 155). Genomic copies of TCR Vβ elements encode the expected frequencies of T cells expressing these receptors in the immature thymocytes in the cortex, but a near complete lack of particular Vβ expressing T cells is found among the mature thymocytes in the medulla and among the mature T cells in the periphery in mice expressing the corresponding MMTV. A sharp reduction of these superantigen-reactive T cells is seen at the cortico-medullary junction in the thymus (171). Selection

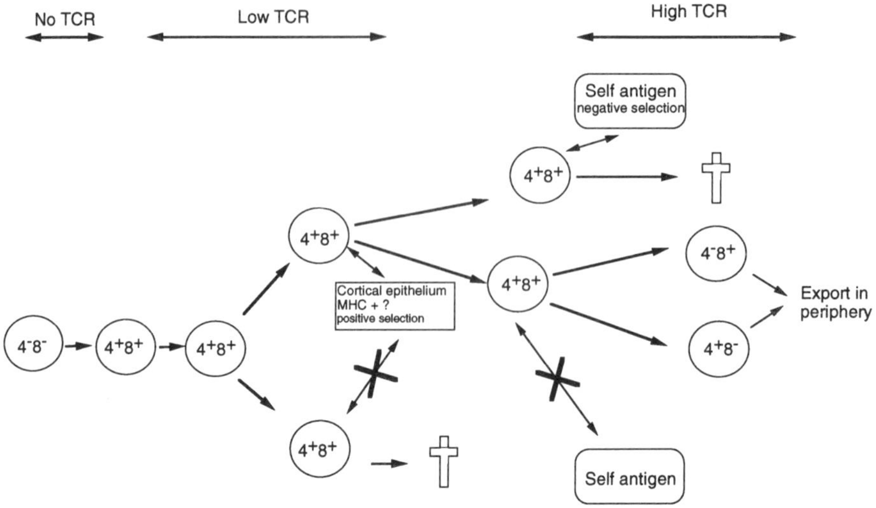

Fig. 11.7 Simplified schema of thymic T cell differentiation. Crossed out arrows indicate lack of interaction.

events in TCR transgenic mice with receptors cross-reactive with a nominal peptide antigen (H-Y or viral peptide) and a superantigen (exogenous SEB or endogenous MMTV) take place at different localizations in the thymus (164, 172). In both cases it was clear that deletion of self-reactive cells directed at peptide antigens or superantigens happened at different stages of thymocyte differentiation. For the two classical peptide antigens analysed, deletion was readily detectable in the cortex amongst the immature double-positive thymocytes whereas superantigen-induced clonal deletion appeared later at the cortico-medullary junction and was only measurable at the single-positive stage. In another TCR transgenic mouse model it has been shown that class II restricted antigen-specific TCR molecules cross-reacting with MMTV superantigens are deleted at the double-positive stage of development in TCR $\alpha\beta$ transgenic mice whereas β only transgenic mice revealed deletion at a later stage of development comparable to the previous two above (173). These differences could be due to:

1. Different affinities of the analysed clonal TCR for classical antigens and for superantigens.
2. Different levels of MHC class II expression (cortical epithelial cells express less MHC class II antigens than medullary epithelial or dendritic cells).
3. Preferential expression of superantigens at the cortico-medullary junction and therefore different cells inside the thymus inducing deletion.

These observations clearly indicate that clonal deletion might occur at different stages of thymocyte maturation.

The TCR forms a complex with a signal transducing CD3 molecule (see Chapter 17). After cross-linking of CD3 with monoclonal antibodies, rapid death of practically all the double-positive thymocytes is observed. Mature T cells are activated to proliferate under these conditions. Interestingly, cross-linking of the TCR with anti-TCR monoclonal antibodies spares about half the thymocytes from death (174). Although different antibodies cannot be compared directly, these results indicate that double-positive thymocytes with different sensitivities to TCR cross-linking exist. It has been suggested that it is these cells which survive a first step of positive selection. The two populations differ in their capability to mobilize intracellular Ca^{2+} after stimulation: only the cells able to mobilize intracellular Ca^{2+} die after anti-TCR antibody interaction. In mature thymocytes, as well as peripheral T cells, anti-TCR, as well as anti-CD3 cross-linking, leads to mobilization of intra-and extracellular Ca^{2+} and to activation (174).

Since B cells are the best MMTV superantigen-presenting cells (175, 176), the effects of B cell depletion after continuous anti-IgM antibody injection from birth has been analyzed on the deletion of the superantigen-reactive T cells (177, 178). The results of these studies show that a drastic reduction of this deletion process can be caused by depletion of the majority of B cells. These experiments have been carried out with the endogenous superantigens *Mtv-7* and with *Mtv-9*. With the strong superantigen Mls-1 (*Mtv-7 orf*) a clear difference in this blocking activity was seen in mice expressing poor superantigen-presenting MHC haplotypes as compared with good presenters. In the poor presenters ($H\text{-}2^b$, E^-) the blocking of clonal deletion was efficient and the cells in the periphery were not tolerant to their own Mls-1 antigens (177). In good presenting MHC haplotypes deletion was not blockable and the superantigen-reactive T cells were anergic (177, 178). For *Mtv-9*, a clear increase of the superantigen-reactive T cells was observed after B cell depletion (178). It seems that the near complete depletion of B cells is enough to partially block the clonal deletion in the case of weak superantigens (or MHC molecules with weak presentation capabilities).

Mice expressing endogenous superantigens clonally delete both CD4 and CD8 T cells (5). Injection of anti-CD4 but not anti-CD8 monoclonal antibodies *in vivo* into neonatal mice leads to a partial rescue of the normally deleted $V\beta6^+$ $CD8^+$ T cells from deletion. This led to the interpretation that negative selection occurs at the $CD4^+8^+$ (double-positive) stage of T cell maturation and that it is the CD4 molecule which provides the affinity required for induction of clonal deletion (159, 160).

The kinetics of superantigen-mediated clonal deletion is strongly influenced by the MHC class II haplotype expressed. In addition, different endogenous superantigens induce different deletion kinetics in the superantigen-reactive T cell populations. For example, $Mls\text{-}1^a$ (*Mtv-7*) leads to a near complete deletion of Vβ6 expressing T cells within ten days after birth. $Mls\text{-}3^a$ (*Mtv-6*) is even faster in the deletion of Vβ3 expressing T cells. On the other hand, *Mtv-9* which is the strongest Vβ11 deleting superantigen known requires over six weeks until it reaches similar levels of deletion (87, 179), and others such as *Mtv-1*, *-8*, *-13*, *-27*, *-44* lead only to partial deletion (for review see reference 180). In addition, their affinities of TCRs for a particular superantigen might influence the kinetics of clonal deletion (152). We have recent evidence that one of the parameters determining the speed of deletion is the site of integration of the provirus. Depending on the site of integration, different amounts of mRNA can be produced which in turn can lead to different levels of superantigen on the cell surface (Shakhov *et al.* submitted).

In the case of injection of cells carrying the endogenous superantigen, or after injection of infectious MMTV, a life-long low level chimerism or infection is established which leads to a life-long presence of the superantigen and deletion of superantigen reactive T cells (28, 30).

Similar experiments have been performed with bacterial superantigens. Repeated injections of newborn mice leads to partial thymic deletion of the superantigen-reactive T cells (10). In adult mice, after a single injection of the superantigen SEB, a strong induction of an immune response is followed by long-lasting partial deletion of the superantigen-reactive T cells (31, 41, 42, 181). Repeated injections of low doses of SEC or MMTV superantigen expressing cells leads to a complete deletion of the superantigen-reactive T cells (43). Even four months after SEB injection the levels of the superantigen-responsive T cells remain partially deleted (182). It is not clear whether this is due to maintenance of SEB at low concentrations in specific localizations (such as follicular dendritic cells which are known to retain intact antigens for many months) or what else might cause this sustained partial (50%) deletion. After two to four months, the cells are no longer unresponsive. Injection of SEB into mice led to the conclusion that very low superantigen doses are required for clonal deletion to occur in the thymus. Doses 100 times lower are required to induce clonal deletion of immature thymocytes than for stimulation of mature T cells (183).

Bone marrow chimera experiments clearly indicate that different cells in the thymus are responsible for negative and positive selection. The hierarchy of MMTV superantigen presentation capabilities of different MHC class II haplotypes correlated with induction of clonal deletion or anergy (184–186). Transplantation of a thymus from an *Mtv-7* (Mls-1^a) donor into nude mice which do not develop a thymus did not lead to deletion of superantigen-reactive cells when donor lymphocytes were eliminated before transplantation. When donor lymphocytes were present, an initial deletion was observed and this faded out later in life, most likely when the donor-derived cells were lost (187). These results clearly indicate that it is the bone marrow-derived cells which are responsible for clonal deletion of MMTV superantigen-reactive T cells.

3.3 Peripheral deletion

Does clonal deletion occur exclusively in the thymus? Although discrepancies exist in the previously published results, the answer to this question is clearly no.

Several observations indicated that, in the absence of a thymus, peripheral deletion does not occur. For example, several groups reported that after thymectomy within the first three days of life T cells specific for an endogenous superantigen can be found amongst peripheral T cells (188, 189). In addition, after intravenous injection of *Mtv* congenic cells, no deletion was observed in normal mice (39).

T cells which have interacted with bacterial or MMTV superantigens are highly sensitive to treatments with either cyclosporin A or cortisone (190, 191). These agents accelerate the deletion of these cells *in vivo*. Although the mechanisms of action of these reagents is not completely understood, both inhibit lymphokine secretion by T cells. It is therefore possible that cells require specific lymphokines or growth factors after interaction with superantigens.

Conflicting results concerning injection of MMTV superantigen expressing lymphocytes into adult mice have been reported (29, 30, 39). In all cases induction of unresponsiveness (see below) was observed but the discrepancy was in the induction of clonal deletion. Most likely the differences were due to the types of cells injected, the route, dose of injection, and to the establishment of chimerism in the mice analysed. Different forms of treatment might preferentially lead to deletion or unresponsiveness. Using superantigen congenic cells usually leads to life-long chimerism whereas injection of cells from strains of mice differing at multiple minor histocompatibility loci leads more often to rejection of the injected cells. To summarize the published results, the following generalizations can be made:

1. Intravenous injection of MMTV superantigen congenic cells can lead to initial expansion and then to induction of unresponsiveness followed in some but not all reports by a deletion of the superantigen-reactive cells. In thymectomized mice deletion can be observed.
2. Injection of similar cell doses subcutaneously leads to a fast and near complete life-long deletion of the superantigen-reactive cells. In residual cells unresponsiveness can be observed (see below). This deletion is clearly not thymus-dependent.

Webb *et al.* made the interesting observation that MMTV superantigen congenic B cells, $CD4^+$ T cells, as well as $CD8^+$ T cells are capable of inducing deletion in either newborn mice or adult, thymectomized mice (29, 148, 153). In the newborn but not in adult mice the $CD8^+$ cells were actually much more efficient at inducing this effect. Waanders *et al.* extended this observation but could not reproduce the finding that $CD4^+$ cells can induce deletion (30). Another difference between these two studies was that Webb *et al.* repeatedly found an induction of a superantigen response after injection of purified $CD8^+$ cells whereas Waanders *et al.* never found such a response. The different purification schemes and routes of injection might explain these discrepancies. Webb *et al.* used a depletion approach for purifying the CD8 cells whereas Waanders *et al.* used a positive selection approach. In addition, the former group injected the cells intravenously whereas the second subcutaneously. Direct comparison of the different purification and injection schemes are required to clarify this point.

3.4 Positive selection

Positive selection is thought to select T cells which have the potential to efficiently interact with peptides bound to self-MHC molecules. Cells which are unable to interact with self-MHC molecules would be useless. During the last years it has become clearly established that this selection event is important in T cell differentiation. Where and when positive selection occurs, however, is a matter of discussion.

Mice expressing H-2E molecules have generally a higher level of expression of T cells expressing TCR Vβ6 in the $CD4^+$ subset than mice lacking H-2E expression (192), and TCR transgenic mice only export sufficient numbers of T cells expressing the transgenic receptor when a positively selecting MHC haplotype is expressed on the thymic epithelium (for review see reference 164). In addition, it has been found that genes outside the MHC are important for the level of positive selection (193): different strains expressing the same MHC molecules and the same genomic TCR Vβ haplotype can have drastic differences in the expression levels of TCR Vβ17a expressing T cells. It will be interesting to identify these genes; they might give clues about the mechanisms. Expression of MHC class II H-2E molecules leads to a faster deletion in $Mls\text{-}1^a$ mice. It is thought that this isotype confers a higher affinity for most of the endogenous superantigens. Recently it has been suggested that MMTV superantigens might be responsible for positive selection of whole Vβ populations (194) but in this area different groups have conflicting results (101, 180).

In bone marrow chimera experiments, the question about which cell in the thymus is responsible for positive and negative selection has been addressed in studies which indicate that different thymic cells might be responsible for these two selection events. For endogenous superantigens, radiosensitive bone marrow-derived cells were found to be responsible (184) probably because B cells are the major presenting (and therefore clonally deleting) cells for endogenous MMTV superantigens. Positive selection can be attributed to radioresistant thymic cells.

Several MHC haplotypes exist which do not express MHC class II H-2E molecules (such as q, s, b) due to a deletion in the I-Eα promoter. Therefore transgenic mice can be used to direct H-2E expression into different cell types in the thymus. Using transgenes, in which specific promoter regions were excised, it was possible to direct expression to the cortical thymic epithelium (with expression in about 1% of medullary dendritic cells) or exclusively the medullary epithelial cells, dendritic cells, and macrophages. It has been observed that in normal $Mls\text{-}1^a$ negative mice H-2E expression leads to a higher level of TCR Vβ6 expression (195–197). The effect on positive selection of T cells by H-2E expression were then analysed. The general conclusion of these studies is that expression of H-2E on the cortical epithelial cells was required for positive selection to occur.

3.5 Where and when do positive selection and clonal deletion occur?

The results in the literature are not conclusive about the order of these selection events: they depend on the type of cell expressing the antigen responsible for clonal deletion as well as the affinity of the TCR to determine the stage of development when deletion occurs. For MMTV superantigens, for example, this would most likely be the B cells. Positive selection does not become apparent until the single-positive stage in normal mice, therefore it is very difficult to see at exactly what stage it occurs. Large numbers of double-positive thymocytes are produced every day independently of any selection events and this makes it impossible to detect the small percentage of positively selected cells in normal mice. In TCR transgenic mice, positive selection becomes apparent at the double-positive stage.

3.6 Neonatal and adult tolerance

A key observation for tolerance development has been the observation that during the neonatal period it is quite easy to induce tolerance to foreign cells as well as to soluble proteins. In general, continuous presence of antigen is required to maintain a long-lasting tolerance. Several observations suggested that newly matured T cells are easy to tolerize. In mice the window for tolerance induction in the neonatal period is between fetal life and one to two days after birth. A second important observation has been that during early life the immune system is not yet capable of mounting an efficient specific immune response. In addition, potentially autoreactive T cells still leave the thymus and can be detected in the periphery.

MMTV superantigens are wonderful tools for dissecting the early immune response in mice. As described above, easily measurable T as well as B cell responses can be induced in adult animals. Since MMTV infects mice during the first two weeks of life and the effect of the superantigen is required for productive infection to occur, we started analysing the neonatal immune response. MMTV is taken up through milk and has to traverse first the intestinal mucosa to come into contact with the host immune system. Therefore the mucosal lymph nodes can be analysed early in life for presence of the superantigen response. Using this approach we were able to show that the mucosal immune system reaches (at least) partial maturity two to three days after birth. Experiments performed in parallel show that there is a long delay in the maturation of the peripheral immune response in which, around the age of weaning, there is a detectable B and T cell response. Using these tools, it should now become possible to dissect further the neonatal incapability of mounting an efficient immune response and to find the reasons for this deficiency.

3.7 Anergy

Anergy has been defined as a state of the T cell in which it is no longer able to respond to a second challenge with antigen or after TCR cross-linking with anti-TCR monoclonal antibodies. It has been shown that many anergic cells subsequently die of programmed cell death (apoptosis) *in vivo*.

After induction of a superantigen-response *in vivo* with bacterial or MMTV superantigens the responsive T cells are rendered unresponsive (anergic). Different states of anergy can be distinguished:

1. For stimulation to occur it is not enough to provide TCR cross-linking. Other signals and factors are required to provide a productive stimulation (198–203), for example, expression of H-2E molecules on β cells of the pancreas can lead to anergization of H-2E restricted T cells in the presence of antigen (204). If other stimuli are not occurring, the cell becomes anergic, losing the capacity to secrete IL-2 and to proliferate after a second stimulation.
2. Another mechanism is clonal exhaustion (205). By over-stimulation the cells are lost.
3. Repeated antigen stimulation can lead to a down-regulation of the TCR. These cells still can produce IL-2 but can not be induced to proliferate (206–209).

The induction of anergy by superantigens is most likely a combination of the first two states above (29, 31, 39, 41, 42). TCR transgenic mice with a TCR expressing an Mls-1^a-reactive Vβ chain express the superantigen-reactive TCR but these cells are rendered anergic (210).

From the published results it seems clear that not just one state of anergy exists (for review see reference 211). Several experiments indicated that anergy can be broken by addition of exogenous IL-2 whereas others found that IL-2 has no effect. It has been suggested that anergic cells represent a differentiation stage of T cells and that our way of looking at immune responses *in vitro* might mistake this differentiation stage for unresponsiveness (212).

3.8 Professional and non-professional antigen-presenting cells

For many classical antigen recognition systems, it has been shown that dendritic cells and macrophages are very capable antigen-presenting cells. On the other hand, the ability of B cells but not T cells to stimulate a response to MMTV superantigen (then still called Mls antigens) was already noted in 1974 (15, 213). B cells were shown by several groups to be poor presenters in primary immune responses (214–216). The older literature is filled with contradictory results about MMTV superantigen-presenting cells. Several reports indicated a role of B cells (175, 176, 213, 217, 218), macrophages (118), as well as dendritic cells (219, 220), but in more recent studies it has been shown that dendritic cells and macrophages are very poor MMTV superantigen-presenting cells. The best presenting cells are clearly B cells. Mazda *et al.* show that B cells and dendritic cells are required for deletion in fetal thymic organ culture. Dendritic cells alone were not effective (219). In another recent study, Inaba *et al.* showed that thymic dendritic cells are able to induce clonal anergy of the superantigen-reactive T cells but not deletion (220). There are still discrepancies between the findings of different groups but the common ground is that B cells are the major MMTV superantigen-presenting cells. For bacterial superantigens the situation *in vitro* is much clearer. Any MHC class II expressing cell is capable of presenting bacterial superantigens to T cells.

For MMTV superantigens, the best presenting cells are non-professional antigen-presenting cells, the B cells. Therefore it seems likely that the combination of the presenting cell for a primary immune response with the very high frequency of responding T cells drives the immune response towards anergy and deletion. Although it is too early to say how good a model for classical immune responses superantigens represent, several recent publications have suggested that in classical antigen recognition similar mechanisms can be observed (205, 221, 222).

For bacterial superantigens the situation *in vivo* is not clear. It will be important to define whether B cells play a predominant role for presentation *in vivo* or if other cells are responsible.

3.9 Suppression

A large amount of literature exists about Mls antigens and suppression of the immune response. Injection of Mls-1^a cells into Mls-1^b mice results in absence of a secondary response. Mls antigens are incapable of inducing graft-versus-host disease (GvH) or splenomegaly. Interestingly, injection of Mls-1^b mice with Mls-1^a splenocytes results in protection from GvH disease after challenge with H-2 compatible or non-compatible cells (38). Even allo-or classical antigen reactivities are decreased after such treatments (for review see reference 15). These phenomena were attributed to suppressor cells. More recently, it has been shown that injection of Mls-1^a cells into Mls-1^b mice results in initial activation of the superantigen-responsive T cells, followed by a loss in their capability to produce IL-2 and to proliferate shortly thereafter. This state of anergy results in disappearance of the superantigen-reactive T cells from the repertoire by thymic as well as peripheral deletion mechanisms. It is likely that putative suppressive effects were due to this rapid induction of long-lasting anergy and deletion.

Why do low doses of superantigen bearing cells lead to deletion of the superantigen-reactive T cells in the absence of a measurable stimulation?

Under these conditions, life-long chimerism is reported (30). High doses of MMTV injections subcutaneously lead to a very strong immune response in the draining lymph nodes but such a response has never been observed at later times in non-draining lymph nodes (28). In the absence of 'suppression', I would have expected to see a visible amplification of the immune response at later time points. Is it possible that the two so apparently different and difficult to explain mechanisms of anergy and suppression are the same? Many experiments have suggested that small numbers of adoptively transferred 'suppressor' cells can influence the majority of cells specific for a specific antigen such that they become 'tolerant' (223).

3.10 Lymphokines

Many of the pathological effects of bacterial superantigens have been ascribed to the over-production of lymphokines. Cross-linking of MHC class II as well as cross-linking of TCR have been implicated in lymphokine production. Shortly after injection of bacterial superantigens IL-1, IL-2, IL-3, IL-4, IL-5, IL-10, TNF, interferon-γ, TGF-β are produced some of which can be detected as soon as two hours after injection (62, 64, 65, 67–70, 224).

3.11 Outlook

Despite the fact that superantigens have taught us a lot about the immune response and tolerance, there is a long way to go to understand the immune system. Some of the key directions for the future are to understand what anergy is and how the different states of anergy and immune response can be manipulated. The results of these studies will have a great impact for tolerization in allergy, transplantation, and autoimmunity, as well as a better understanding of immune deficiencies and ways to induce reversion to normal immune responses (as in the case of AIDS).

References

1 Jorgensen, J. L., Reay, P. A., Ehrlich, E. W., and Davis, M. M. (1992). Molecular components of T-cell recognition. *Annu. Rev. Immunol.*, **10**, 835.

2 Kappler, J. W., Roehm, N., and Marrack, P. (1987). T cell tolerance by clonal elimination in the thymus. *Cell*, **49**, 273.

3 Kappler, J. W., Wade, T., White, J., Kushnir, W., Blackman, M., Bill, J., *et al.* (1987). A T cell receptor Vβ segment that imparts reactivity to a class II major histocompatibility complex product. *Cell*, **49**, 263.

4 Kappler, J. W., Staerz, U. D., White, J., and Marrack, P. C. (1988). Self-tolerance eliminates T cells specific for Mls-modified products of the major histocompatibility complex. *Nature*, **332**, 35.

5 MacDonald, H. R., Schneider, R., Lees, R. L., Howe, R. K., Acha-Orbea, H., Festenstein, H., *et al.* (1988). T-cell receptor Vβ use predicts reactivity and tolerance to Mls^a-encoded antigens. *Nature*, **332**, 40.

6 Pullen, A. M., Marrack, P., and Kappler, J. W. (1988). The T-cell repertoire is heavily influenced by tolerance to polymorphic self-antigens. *Nature*, **335**, 796.
7 Abe, R., Vacchio, M. S., Fox, B., and Hodes, R. (1988). Preferential expression of the T-cell receptor Vβ3 gene by Mls[c] reactive T cells. *Nature*, **335**, 827.
8 Fry, A. M. and Matis, L. A. (1988). Self-tolerance alters T-cell receptor expression in an antigen-specific MHC-restricted immune response. *Nature*, **335**, 830.
9 Kanagawa, O., Palmer, E., and Bill, J. (1989). T cell receptor Vβ6 domain imparts reactivity to the Mls-1[a] antigen. *Cell. Immunol.*, **119**, 412.
10 White, J., Herman, A., Pullen, A. M., Kubo, R., Kappler, J. W., and Marrack, P. (1989). The Vβ-specific superantigen staphylococcal enterotoxin B: Stimulation of mature T cells and clonal deletion in neonatal mice. *Cell*, **56**, 27.
11 Candéias, S., Waltzinger, C., Benoist, C., and Mathis, D. (1991). The Vβ17$^+$ T cell repertoire: skewed Jβ usage after thymic selection; dissimilar CDR3's in CD$^+$ versus CD8$^+$ cells. *J. Exp. Med.*, **174**, 989.
12 Smith, H. P., Le, P., Woodland, D. L., and Blackman, M. A. (1992). T cell receptor α-chain influences reactivity to Mls-1 in Vβ8.1 transgenic mice. *J. Immunol.*, **149**, 887.
13 Vacchio, M. S., Kanagawa, O., Tomonari, K., and Hodes, R. J. (1992). Influence of T cell receptor Vα expression on Mls[a] superantigen-specific T cells. *J. Exp. Med.*, **175**, 1405.
14 Waanders, G. A., Lussow, A. R., and MacDonald, H. R. (1993). Skewed T cell receptor Vα repertoire among superantigen reactive murine T cells. *Int. Immunol.*, **5**, 55.
15 Abe, R. and Hodes, R. (1989). T cell recognition of minor lymphocyte stimulating (Mls) gene products. *Annu. Rev. Immunol.* **7**, 683.
16 Acha-Orbea, H., Held, W., Waanders, G. A., Shakhov, A. N., Scarpellino, L., Lees, R., *et al.* (1993). Exogenous and endogenous mouse mammary tumor virus superantigens. *Immunol. Rev.*, **131**, 5.
17 Acha-Orbea, H. and Palmer E. (1991). Mls—a retrovirus exploits the immune system. *Immunol. Today*, **12**, 356.
18 Marrack, P. and Kappler, J. (1990). The staphylococcal enterotoxins and their relatives. *Science*, **248**, 705.
19 Herman, A., Kappler, J. W., Marrack, P., and Pullen, A. M. (1991). Superantigens: Mechanism of T-cell stimulation and role in immune responses. *Annu. Rev. Immunol.*, **9**, 745.
20 Herrmann, T., Maryanski, J. L., Romero, P., Fleischer, B., and MacDonald, H. R. (1990). Activation of MHC class I-restricted CD8$^+$ CTL by microbial T cell mitogens. Dependence upon MHC class II expression of the target cells and Vβ usage of the responder cells. *J. Immunol.*, **144**, 1181.
21 Ildstad, S. T., Vacchio, M. S., Markus, P. M., Hronakes, M. L., Wren, S. M., and Hodes, R. J. (1992). Cross-species transplantation tolerance: rat bone marrow-derived cells can contribute to the ligand for negative selection of mouse T cell receptor Vβ in chimeras tolerant to xenogeneic antigens (Mouse+rat into mouse). *J. Exp. Med.*, **175**, 147.
22 Miller, R. A. and Stutman, O. (1982). Enumeration of IL2-secreting helper T cells by limiting dilution analysis, and demonstration of unexpectedly high levels of IL2 production per responding cell. *J. Immunol.*, **128**, 2258.
23 Fleischer, B. and Mittrücker, H.-W. (1991). Evidence for T cell receptor-HLA class II molecule interaction in the response to superantigenic bacterial toxins. *Eur. J. Immunol.*, **21**, 1331.

24 Rust, C. J. J. and Koning, F. (1993). $\gamma\delta$ T cell reactivity towards bacterial superantigens. *Semin. Immunol.*, **5**, 41.
25 Peck, A. B., Janeway, C. A., and Wigzell, H. (1977). T lymphocyte responses to Mls locus antigens involve recognition of H-2 I region products. *Nature*, **266**, 840.
26 Wall, K. A., Lorber, M. I., Loken, M. R., McClatchey, S., and Fitch, F. W. (1983). Inhibition of proliferation of Mls- and Ia-reactive cloned T cells by a monoclonal antibody against a determinant shared between I-A and I-E. *J. Immunol.*, **131**, 1056.
27 Janeway, C. A. (1990). Self-superantigens? *Cell*, **63**, 659.
28 Held, W., Shakhov, A. N., Waanders, G., Scarpellino, L., Luethy, R., Kraehenbuhl, J.-P., *et al.* (1992). An exogenous mouse mammary tumor virus with properties of Mls-1^a (*Mtv-7*). *J. Exp. Med.*, **175**, 1623.
29 Webb, S., Morris, C., and Sprent, J. (1990). Extrathymic tolerance of mature T cells: Clonal elimination as a consequence of immunity. *Cell*, **63**, 1249.
30 Waanders, G. A., Shakhov, A. N., Held, W., Karapetian, O., Acha-Orbea, H., and MacDonald, H. R. (1993). Peripheral T cell activation and deletion induced by transfer of lymphocyte subsets expressing endogenous or exogenous mouse mammary tumor virus. *J. Exp. Med.*, **177**, 1359.
31 Kawabe, Y. and Ochi, A. (1991). Programmed cell death and extrathymic reduction of Vβ8$^+$ CD4$^+$ T cells in mice tolerant to *Staphylococcus aureus* enterotoxin B. *Nature*, **349**, 245.
32 Festenstein, H. (1973). Immunogenic and biological aspects of *in vitro* allotransformation (MLR) in the mouse. *Transplant. Rev.*, **15**, 62.
33 Abe, R., Kanagawa, O., Sheard, M. A., Malissen, B., and Foo-Phillips, M. (1991). Characterization of a new minor lymphocyte stimulatory system. I. Cluster of self antigens recognized by 'I-E-reactive' Vβs, Vβ5, Vβ11, and Vβ12 T cell receptors for antigen. *J. Immunol.*, **147**, 739.
34 Fleischer, B. and Schrezenmeier, H. (1988). T cell stimulation by staphylococcal enterotoxins. Clonally variable response and requirements for major histocompatibility complex class II molecules on accessory or target cells. *J. Exp. Med.*, **167**, 1697.
35 Lilliehöök, B., Jacobsson, H., and Blomgren, H. (1975). Specifically decreased MLC response of lymphocytes from DBA mice injected with cells from H-2 compatible, M antigen incompatible strain C3H. No such effect after injection of H-2 disparate C3H-hybrid cells. *Scand. J. Immunol.*, **4**, 209.
36 Matossian-Rogers, A. and Festenstein, H. (1976). Modification of murine T cell cytotoxicity by preimmunisation with M locus and H-2 incompatibilities. *J Exp. Med.*, **143**, 456.
37 Pinto, M., Torten, M., and Birnbaum, S. C. (1978). Suppression of the *in vivo* humoral and cellular immune response by staphylococcal enterotoxin B. *Transplantation*, **25**, 320.
38 Halle-Pannenko, O., Pritchard, L., Festenstein, H., and Berumen, L. (1986). Abrogation of lethal graft-versus-host reaction directed against non-H-2 antigens: role of Mls^a and K/I region antigens in the induction of suppressor cells by alloimmunization. *J. Immunogenet.*, **37**, 437.
39 Rammensee, H.-G., Kroschewsky, R., and Frangoulis, B. (1989). Clonal anergy induced in mature Vβ6 T lymphocytes on immunizing Mls-1^b mice with Mls-1^a expressing cells. *Nature*, **339**, 541.
40 Rellahan, B. L., Jones, L. A., Kruisbeek, A. M., Fry, A. M., and Matis, L. A. (1990). *In vivo* induction of anergy in peripheral Vβ8$^+$ T cells by staphylococcal enterotoxin B. *J. Exp. Med.*, **172**, 1091.

41 Kawabe, Y. and Ochi, A. (1990). Selective anergy of Vβ8⁺ CD4⁺ T cells in staphylococcus enterotoxin B-primed mice. *J. Exp. Med.*, **172**, 1065.
42 MacDonald, H. R., Baschieri, S., and Lees, R. K. (1991). Clonal expansion precedes anergy and death of Vβ8⁺ peripheral T cells responding to staphylococcal enterotoxin B *in vivo*. *Eur. J. Immunol.*, **21**, 1963.
43 McCormack, J. E., Callahan, J. E., Kappler, J., and Marrack, P. C. (1993). Profound deletion of mature T cells *in vivo* by chronic exposure to exogenous superantigen. *J. Immunol.*, **150**, 3785.
44 Leonardo, M. (1991). Interleukin-2 programs mouse αβ T lymphocytes for apoptosis. *Nature*, **353**, 858.
45 Kang, S.-M., Beverly, B., Tran, A.-C., Brorson, K., Schwartz, R., and Leonardo, M. (1992). Transactivation by AP-1 is a molecular target of T cell clonal anergy. *Science*, **257**, 1134.
46 Go, C. and Miller, J. (1992). Differential induction of transcription factors that regulate the interleukin-2 gene during anergy induction and restimulation. *J. Exp. Med.*, **175**, 1327.
47 Bergdoll, M. S. (1979). *Staphylococcal intoxications*. Academic Press, New York.
48 Peavy, D. L., Adler, W. H., and Smith, R. T. (1970). The mitogenic effects of endotoxin and staphylococcal enterotoxin B on mouse spleen cells and human peripheral lymphocytes. *J. Immunol.*, **105**, 1453.
49 Langford, M. P., Stanton, G. J., and Johnson, H. M. (1978). Biological effects of staphylococcal enterotoxin A on human peripheral lymphocytes. *Infect. Immunol.*, **22**, 62.
50 Janeway, C. A. J., Yagi, J., Conrad, P. J., Katz, M. E., Jones, B., Vroegop, S., *et al.* (1989). T-cell responses to Mls and to bacterial proteins that mimic its behavior. *Immunol. Rev.*, **107**, 61.
51 Tomai, M., Kolb, M., Majumdar, G., and Beachey, E. H. (1990). Superantigenicity of streptococcal M protein. *J. Exp. Med.*, **172**, 359.
52 Tomai, M. A., Aelion, J. A., Dockter, M. E., *et al.* (1991). T cell receptor V gene usage by human T cells stimulated with the superantigen streptococcal M protein. *J. Exp. Med.*, **174**, 285.
53 Imanishi, K., Igarashi, H., and Uchiyama, T. (1990). Activation of murine T cells by streptococcal pyrogenic exotoxin type A. *J. Immunol.*, **145**, 3170.
54 Nelson, K., Schlievert, P. M., Selander, R. K., and Musser, J. M. (1991). Characterization and clonal distribution of speA gene encoding pyrogenic exotoxin A. *J. Exp. Med.*, **174**, 1271.
55 Braun, M. A., Gerlach, D., Hartwig, U. F., Ozegowski, J.-H., Romagné, F., Carrel, S., *et al.* (1993). Stimulation of human T cells by streptococcal 'superantigen' erythrogenic toxins (Scarlet fever toxins). *J. Immunol.*, **150**, 2457.
56 Tumang, J. R., Posnett, D. N., Cole, B. C., Crow, M. K., and Friedman, S. M. (1990). Helper T cell-dependent human B cell differentiation mediated by a mycoplasmal superantigen bridge. *J. Exp. Med.*, **171**, 2153.
57 Cole, B. C., Kartchner, D. R., and Wells, D. J. (1989). Stimulation of mouse lymphocytes by a mitogen derived from *Mycoplasma arthritides*. VII. Responsiveness is associated with expression of a product(s) of the Vβ8 gene family present on the T cell receptor α/β for antigen. *J. Immunol.*, **142**, 4131.
58 Cole, B. C., Kartchner, D. R., and Wells, D. J. (1990). Stimulation of mouse lymphocytes by a mitogen derived from *Mycoplasma arthritides* (MAM) VIII. Selective activation of T cells expressing distinct Vβ T cell receptors from various strains of mice by the 'superantigen' MAM. *J. Immunol.*, **144**, 425.

59 Stuart, P. M. and Woodward, J. G. (1992). *Yersinia enterolytica* produces superantigen activity. *J. Immunol.*, **148**, 225.
60 Marrack, P., Blackman, M., Kushnir, E., and Kappler, J. (1990). The toxicity of staphylococcal enterotoxin B in mice is mediated by T cells. *J. Exp. Med.*, **171**, 455.
61 Lionetti, P., Spencer, J., Breese, E. J., Murch, S. H., Taylor, J., and MacDonald, T. T. (1993). Activation of mucosal Vβ3+ T cells and tissue damage in human small intestine by the bacterial superantigen *Staphylococcus aureus* enterotoxin B. *Eur. J. Immunol.*, **23**, 664.
62 Miethke, T., Wahl, C., Heeg, K., Echtenacher, B., Krammer, P. H., and Wagner, H. (1992). T cell-mediated lethal shock triggered in mice by the superantigen staphylococcal enterotoxin B: Critical role of tumor necrosis factor. *J. Exp. Med.*, **175**, 91.
63 Carlsson, R. and Sjogren, O. (1985). Kinetics of IL-2 and interferon-γ production, expression of IL-2 receptors, and cell proliferation in human mononuclear cells exposed to staphylococcal enterotoxin A. *Cell. Immunol.*, **96**, 175.
64 Jupin, C., Anderson, S., Damais, C., Alouf, J. E., and Parant, M. (1988). Toxic shock syndrome toxin 1 as an inducer of human tumor necrosis factors and γ interferon. *J. Exp. Med.*, **167**, 752.
65 Fischer, H., Dohlstein, M., Andersson, A., Hedlund, G., Ericsson, P., Hansson, J., *et al.* (1990). Production of TNF-α and TNF-β by staphylococcal enterotoxin A activated human T cells. *J. Immunol.*, **144**, 4663.
66 Herrmann, T., Waanders, G., Chvatchko, Y., and MacDonald, H. R. (1992). The viral superantigen Mls-1^a induces interferon-γ secretion by specifically primed $CD8^+$ cells but fails to trigger cytotoxicity. *Eur. J. Immunol.*, **22**, 2789.
67 Patarca, R., Wei, F.-Y., Iregui, M. V., and Cantor, H. (1991). Differential induction of interferon γ gene expression after activation of T cells by conventional antigen and Mls superantigen. *Proc. Natl. Acad. Sci. USA*, **88**, 2736.
68 Gaugler, B., Langlet, C., Martin, J.-M., Schmitt-Verhulst, A.-M., and Guimezanes, A. (1991). Evidence for quantitative and qualitative differences in functional activation of Mls-reactive T cell clones and hybridomas by antigen or TcR/CD3 antibodies. *Eur. J. Immunol.*, **21**, 2581.
69 Scholl, P. R., Trede, N., Chatila, T. A., and Geha, R. S. (1992). Role of protein tyrosine phosphorylation in monokine induction by staphylococcal superantigen toxic shock syndrome toxin-1. *J. Immunol.*, **148**, 2237.
70 Cardell, S., Höidén, I., and Möller, G. (1993). Manipulation of the superantigen-induced lymphokine response. Selective induction of interleukin-10 or interferon-γ synthesis in small resting $CD4^+$ T cells. *Eur. J. Immunol.*, **23**, 523.
71 Micusan, V. V. and Thibodeau, J. (1993). Superantigens of microbial origins. *Semin. Immunol.*, **5**, 3.
72 Carlsson, R., Fischer, H., and Sjögren, H. O. (1988). Binding of staphylococcal enterotoxin to accessory cells is a requirement for its ability to activate human T cells. *J. Immunol.*, **140**, 2484.
73 Fleischer, B., Schmidt, K.-H., Gerlack, D., and Köhler, W. (1992). Separation of T-cell stimulating activity from Streptococcal M protein. *Infect. Immunol.*, **60**, 1767.
74 Beteley, M. J. and Mekalanos, J. J. (1985). Staphylococcal enterotoxin A is encoded by phage. *Science*, **229**, 185.
75 Fraser, J. D. (1989). High-affinity binding of staphylococcal enterotoxins A and B to HLA-DR. *Nature*, **339**, 221.
76 Fischer, H., Dohlstein, M., Lindwall, M., Sjøgren, H.-O., and Carlsson, R.

(1989). Binding of staphylococcal enterotoxin A to HLA-DR on B cell lines. *J. Immunol.*, **142**, 3151.

77 Mollick, J. A., Cook, R. G., and Rich, R. R. (1989). Class II MHC molecules are specific receptors for staphylococcal enterotoxins. *Science*, **244**, 817.

78 Herrmann, T., Accolla, R. S., and MacDonald, H. R. (1989). Different staphylococcal enterotoxins bind preferentially to distinct major histocompatibility complex class II isotypes. *Eur. J. Immunol.*, **19**, 2171.

79 Dellabonna, P., Peccaud, J., Kappler, J., Marrack, P., Benoist, C., and Mathis, D. (1990). Superantigens interact with MHC class II molecules outside of the antigen groove. *Cell*, **62**, 1115.

80 Hedlund, G., Dohlstein, M., Herrmann, T., Buell, G., Lando, P. A., Segreen, S., *et al.* (1991). A recombinant C-terminal fragment of Staphylococcal enterotoxin A binds to human MHC class II products but does not activate T cells. *J. Immunol.*, **147**, 4082.

81 Karp, D. R. and Long, E. O. (1992). Identification of HLA-DR1 β chain residues critical for binding staphylococcal enterotoxins A and E. *J. Exp. Med.*, **175**, 415.

82 Labrecque, N., Thibodeau, J., and Sékaly, R.-P. (1993). Interactions between staphylococcal superantigens and MHC class II molecules. *Semin. Immunol.*, **5**, 23.

83 Herrmann, T., Baschieri, S., Lees, R. K., and MacDonald, H. R. (1992). *In vivo* responses of $CD4^+$ and $CD8^+$ cells to bacterial superantigens. *Eur. J. Immunol.*, **22**, 1935.

84 Woodland, D. L., Lund, F. E., Happ, M. P., Blackman, M. A., Palmer, E., and Corley, R. B. (1991). Endogenous superantigen expression is controlled by mouse mammary tumor proviral loci. *J. Exp. Med.*, **174**, 1255.

85 Pullen, A. M., Chaoi, Y., Kushnir, E., Kappler, J., and Marrack, P. (1992). The open reading frames in the 3′ long terminal repeats of several mouse mammary tumor virus integrants encode Vβ3-specific superantigens. *J. Exp. Med.*, **175**, 41.

86 Choi, Y., Marack, P., and Kappler, J. (1992). Structural analysis of a mouse mammary tumor virus superantigen. *J. Exp. Med.*, **175**, 847.

87 Acha-Orbea, H., Shakhov, A. N., Scarpellino, L., Kolb, E., Müller, V., Vessaz-Shaw, A., *et al.* (1991). Clonal deletion of Vβ14 positive T cells in mammary tumor virus transgenic mice. *Nature*, **350**, 207.

88 Choi, Y., Kappler, J. W, and Marrack, P. (1991). A superantigen encoded in the open reading frame of the 3′ long terminal repeat of mouse mammary tumor virus. *Nature*, **350**, 203.

89 Kozak, C., Peters, G., Pauley, R., Morris, V., Michaelides, R., Dudley, J., *et al.* H (1987). A standardized nomenclature for endogenous mouse mammary tumor viruses. *J. Virol.*, **61**, 1651.

90 Kuo, W.-L., Vilander, L. R., Huang, M., and Peterson, D. O. (1988). A transcriptionally defective long terminal repeat within an endogenous copy of mouse mammary tumor virus proviral DNA. *J. Virol.*, **62**, 2394.

91 van Klaveren, P. and Bentvelzen, P. (1988). Transactivating potential of the 3′ open reading frame of murine mammary tumor virus. *J. Virol.*, **62**, 4410.

92 Salmons, B., Erfle, V., Brem, G., and Günzburg, W. H. (1990). *naf*, a *trans*-regulating negative-acting factor encoded within the mouse mammary tumor virus open reading frame region. *J. Virol.*, **64**, 6355.

93 Brandt-Carlson, C. and Butel, J. S. (1991). Detection and characterization of a glycoprotein encoded by the mouse mammary tumor virus long terminal repeat gene. *J. Virol.*, **65**, 6051.

94 Winslow, G. M., Scherer, M. T., Kappler, J. W., and Marrack, P. (1992). Detec-

tion and biochemical characterization of the mouse mammary tumor virus 7 superantigen (Mls-1[a]). *Cell*, **71**, 719.
95 Brandt-Carlson, C., Butel, J. S., and Wheeler, D. (1993). Phylogenetic and structural analyses of MMTV LTR sequences of exogenous and endogenous origins. *Virology*, **193**, 171.
96 Held, W., Shakhov, A. N., Izui, S., Waanders, G. A., Scarpellino, L., MacDonald, H. R., *et al.* (1993). Superantigen-reactive $CD4^+$ T cells are required to stimulate B cells after infection with mouse mammary tumor virus. *J. Exp. Med.*, **177**, 359.
97 Jouvin-Marche, E., Cazenave, P.-A., Voegtle, D., and Marche, P. (1992). Vβ17 T-cell deletion by endogenous mammary tumor virus in wild-type-derived mouse strains. *Proc. Natl. Acad. Sci. USA*, **89**, 3232.
98 Jouvin-Marche, E., Marche, P. N., Six, A., Liebe-Gris, C., Voegtle, D., and Cazenave, P.-A. (1993). Identification of an endogenous mammary tumor virus involved in the clonal deletion of Vβ2 cells. *Eur. J. Immunol.*, **23**, 2758.
99 Shakhov, A. N., Wang, H., Acha-Orbea, H., Pauley, R. J., and Wei, W.-Z. (1993). A new infectious mammary tumor virus in the milk of mice implanted with C4 hyperplastic alveolar nodules. *Eur. J. Immunol.*, **23**, 2715.
100 Gollub, K. J. and Palmer, E. (1992). Divergent viral superantigens delete $V\beta 5^+$ T lymphocytes. *Proc. Natl. Acad. Sci. USA*, **89**, 5138.
101 Tomonari, K., Fairchild, S., and Rosenwasser, O. A. (1993). Influence of viral superantigens on Vβ and Vα-specific positive and negative selection. *Immunol. Rev.*, **137**, 131.
102 Korman, A. J., Bourgarel, P., Meo, T., and Rieckhof, G. E. (1992). The mouse mammary tumor virus long terminal repeat encodes a type II transmembrane glycoprotein. *EMBO J.*, **11**, 1901.
103 Knight, A. M., Harrison, G. B., Pease, R. J., Robinson, P. J., Dyson, P. J. (1992). Biochemical evidence of the mouse mammary tumor virus long terminal repeat product. Evidence for the molecular structure of an endogenous superantigen. *Eur. J. Immunol.*, **22**, 879.
104 Tsubura, A., Inaba, M., Imai, S., Murakami, A., Oyaizu, N., Yasumizu, R., *et al.* (1988). Intervention of T-cells in transportation of mouse mammary tumor virus (milk factor) to mammary gland cells *in vivo*. *Cancer Res.*, **48**, 6555.
105 Held, W., Waanders, G. A., Shakhov, A. N., Scarpellino, L., Acha-Orbea, H., and MacDonald, H. R. (1993). Superantigen-induced immune stimulation amplifies mouse mammary tumor virus infection and allows virus transmission. *Cell*, **74**, 529.
106 Golovkina, T. V., Chervonsky, A., Dudley, J. P., and Ross, S. R. (1992). Transgenic mouse mammary tumor virus superantigen expression prevents viral infection. *Cell*, **69**, 637.
107 Acha-Orbea, H. and MacDonald, H. R. (1993). Subversion of host immune responses by viral superantigens. *Trends Microbiol.*, **1**, 32.
108 Peters, G., Brookes, S., Smith, R., Placzek, M., and Dickson, C. (1989). The mouse homolog of the *hst/k-FGF* gene is adjacent to *int-2* and is activated by proviral insertion in some virally induced mammary tumors. *Proc. Natl. Acad. Sci. USA*, **86**, 5678.
109 Nusse, R. and Varmus, H. E. (1982). Many tumors induced by the mouse mammary tumor virus contain a provirus integrated in the same region of the host genome. *Cell*, **31**, 99.
110 Dickson, C., Smith, R., Brookes, S., and Peters, G. (1984). Tumorigenesis by mouse mammary tumor virus: Proviral activation of a cellular gene in the common integration region *int-2*. *Cell*, **37**, 529.
111 Peters, G., Brookes, S., Smith, R., and Dickson, C. (1983). Tumorigenesis by

mouse mammary tumor virus: Evidence for a common region for provirus integration in mammary tumors. *Cell*, **33**, 369.

112 Bentvelzen, P. and Hilgers, J. (1980). Murine mammary tumor virus. In *Viral oncology* (ed. G. Klein), p. 311. Raven Press, New York.

113 Etkind, P. R. (1989). Expression of the *int-1* and *int-2* loci in endogenous mouse mammary tumor virus-induced mammary tumorigenesis in the C3Hf mouse. *J. Virol.*, **63**, 4972.

114 Hügin, A. W., Vacchio, M. S., and Morse, H. C., III. (1991). A virus-encoded 'superantigen' in a retrovirus-induced immunodeficiency syndrome of mice. *Science*, **252**, 424.

115 Imberti, L., Sottini, A., Bettinardi, A., Puoti, M., and Primi, D. (1992). Selective depletion in HIV infection of T cells that bear specific T cell receptor Vβ sequences. *Science*, **254**, 860.

116 Laurence, J., Hodtsev, A. S., and Posnett, D. N. (1992). Superantigen implicated in dependence of HIV-1 replication in T cells on TCR Vβ expression. *Nature*, **358**, 255.

117 Lafon, M., Lafage, M., Martinez-Arends, A., Ramirez, R., Vuiller, F., Charron, D., *et al.* (1992). Evidence for a viral superantigen in humans. *Nature*, **358**, 507.

118 Berumen, L., Festenstein, H., and Halle-Pannenko, O. (1984). Soluble Mls^a antigens: Stimulatory effects *in vitro* versus suppressive effect *in vivo*. *Immunogenetics*, **20**, 33.

119 Platsoucas, C. D., Oleszak, E. L., and Good, R. A. (1986). Immunomodulation of human leukocytes by staphylococcal enterotoxin A: augmentation of natural killer cells and induction of suppressor cells. *Cell. Immunol.*, **97**, 371.

120*a* Brocke, S., Gaur, A., Piercy, C., Gautam, A., Fathman, C. G., *et al.* (1993). Induction of relapsing paralysis in experimental allergic encephalomyelitis by bacterial superantigen. *Nature*, **365**, 642.

120*b* Schiffenbauer, J., Johnson, H. M., Butfilkovski, E. J., Wegrzyn, L., and Soos, J. M. (1993). Staphylococcal enterotoxins can reactivate experimental allergic encephalomyelitis. *Proc. Natl. Acad. Sci. USA*, **90**, 8543.

121 Kotzin, B. L., Babcock, S. K. and Herron, L. R. (1988). Deletion of potentially self-reactive T cell receptor specificities in $L3T4^-$, $Lyt2^-$ T cells of *lpr* mice. *J. Exp. Med.*, **168**, 2221.

122 Singer, P. A., Baldderas, R. S., McEvilly, R. J., Bobardt, M., and Theofilopoulos, A. N. (1989). Tolerance-related Vβ clonal deletions in normal $CD4^-8^-$, $TCR\text{-}\alpha\beta^+$ and abnormal *lpr* and *gld* cell populations. *J. Exp. Med.*, **170**, 1869.

123 Abe, J., Kotzin, B. L., Jujo, K., Melish, M. E., Glode, M. P., Kohsaka, T., *et al.* (1992). Selective expansion of T cells expressing T-cell receptor variable regions Vβ2 and Vβ8 in Kawasaki disease. *Proc. Natl. Acad. Sci. USA*, **89**, 4066.

124 Abe, J., Kotzin, B. L., Meissner, C., Melish, M. E., Takahashi, M., Fulton, D., *et al.* (1993). Characterization of T cell repertoire changes in acute Kawasaki disease. *J. Exp. Med.*, **177**, 791.

125 Paliard, X., West, S. G., Lafferty, J. A., Clements, J. R., Kappler, J. W., Marrack, P., *et al.* (1991). Evidence for the effects of a superantigen in rheumatoid arthritis. *Science*, **253**, 325.

126 Kim, C., Siminovitch, K. A., and Ochi, A. (1991). Reduction of lupus nephritis in MRL/*lpr* mice by a bacterial superantigen treatment. *J. Exp. Med.*, **174**, 1431.

127 Rott, O., Wekerle, H., and Fleischer, B. (1992). Protection from experimental allergic encephalomyelitis by application of a bacterial superantigen. *Int. Immunol.*, **4**, 347.

128 Gascoigne, N. R. J. (1993). Interaction of the T cell receptor with bacterial superantigens. *Semin. Immunol.*, **5**, 13.
129 Fraser, J. D., Urban, R. G., Strominger, J. L., and Robinson, H. (1992). Zinc regulates the function of two superantigens. *Proc. Natl. Acad. Sci. USA*, **89**, 5507.
130 Jones, L. A., Chin, L. T., Longo, D. L., and Kruisbeek, A. M. (1990). Peripheral clonal elimination of functional T cells. *Science*, **250**, 1726.
131 Torres, B. A., Griggs, N. D., and Johnson, H. M. (1993). Bacterial and retroviral superantigens share a common binding region on class II MHC antigens. *Nature*, **364**, 152.
132 Swaminathan, S., Furey, W., Pletcher, J., and Sax, M. (1992). Crystal structure of staphylococcal enterotoxin B, a superantigen. *Nature*, **359**, 801.
133 Kappler, J., Herman, A., Clements, J., and Marrack, P. (1992). Mutations defining functional regions of the superantigen enterotoxin B. *J. Exp. Med.*, **175**, 387.
134 Hudson, K. R., Robinson, H., and Fraser, J. D. (1993). Two adjacent residues in staphylococcal enterotoxins A and E determine T cell receptor Vβ specificity. *J. Exp. Med.*, **177**, 175.
135 Irwin, M. J., Hudson, K. R., Fraser, J. D., and Gascoigne, N. R. J. (1992). Enterotoxin residues determining T-cell receptor Vβ binding specificity. *Nature*, **359**, 841.
136 Gascoigne, N. R. J. and Ames, K. T. (1991). Direct binding of secreted T-cell receptor β chain to superantigen associated with major histocompatibility complex protein. *Proc. Natl. Acad. Sci. USA*, **88**, 613.
137 Cazenave, P. A., Marche, P., Jouvin-Marche, E., Voegtle, D., Bonhomme, F., Bandeira, A., *et al.* (1990). Vβ17-gene polymorphism in wild-derived mouse strains: two amino acid substitutions in the Vβ17 region alter drastically T cell receptor specificity. *Cell*, **63**, 717.
138 Pullen, A., Wade, T., Marrack, P., and Kappler, J. (1990). Identification of the region of the T cell receptor β chain that interacts with the self-superantigen Mls-1^a. *Cell*, **61**, 1365.
139 Choi, Y., Herman, A., DiGusto, D., Wade, T., Marrack, P., and Kappler, J. (1990). Regions of the variable region of the T-cell receptor β-chain that interact with *S. aureus* toxin superantigens. *Nature*, **346**, 471.
140 Pullen, A. M., Bill, J., Kubo, R., Marrack, P., and Kappler, J. W. (1991). Analysis of the interaction site for the self superantigen Mls-1^a on T cell receptor Vβ. *J. Exp. Med.*, **173**, 1183.
141 Pontzer, C. H., Irwin, M. J., Gascoigne, N. R. J., and Johnson, H. M. (1992). T-cell antigen receptor binding sites for the microbial superantigen staphylococcal enterotoxin A. *Proc. Natl. Acad. Sci. USA*, **88**, 7727.
142 Cantor, H., Crump, A. L., Raman, V. K., Liu, H., Markowitz, J. S., Grusby, M. J., *et al.* (1993). Immunoregulatory effects of superantigens: Interactions of staphylococcal enterotoxins with host MHC and non-MHC products. *Immunol. Rev.*, **131**, 27.
143 Labrecque, N., McGrath, H., Subramanyam, M., Huber, B. T., and Sékaly, R.-P. (1993). Human T cells respond to mouse mammary tumor virus-encoded superantigen: Vβ, C. restriction. *J. Exp. Med.*, **177**, 1735.
144 Subramanyam, M., Mohan, N., Mottershead, D., Beutner, U., McLellan, B., Kraus, E., *et al.* (1993). Mls-1 superantigen: Molecular characterization and functional analysis. *Immunol. Rev.*, **131**, 117.
145 Acha-Orbea, H., Held, W., Scarpellino, L., and Shakhov, A. N. (1992). Mls: A link between immunology and retrovirology. *Int. Rev. Immunol.*, **8**, 327.

146 Larsson-Sciard, E. L., Spetz-Hatgberg, A. L., Casrouge, A., and Kourilisky, P. (1990). Analysis of T cell receptor Vβ gene usage in primary mixed lymphocyte reactions: evidence for directive usage by different antigen-presenting cells and Mls-like determinants on T cell blasts. *Eur. J. Immunol.*, **20**, 1223.

147 Gao, E.-K., Kanagawa, O., and Sprent, J. (1989). *J. Exp. Med.*, **170**, 1947.

148 Webb, S. R. and Sprent, J. (1990). Response of mature unprimed $CD8^+$ T cells to Mls^a determinants. *J. Exp. Med.*, **171**, 953.

149 MacDonald, H. R., Lees, R. K., and Chvatchko, Y. (1990). $CD8^+$ T cells respond clonally to Mls-1^a-encoded determinants. *J. Exp. Med.*, **171**, 1381.

150 Chvatchko, Y. and MacDonald, H. R. (1991). $CD8^+$ T cell response to Mls-1^a determinants involves major histocompatibility complex class II molecules. *J. Exp. Med.*, **173**, 779.

151 Kanagawa, O. and Maki, R. (1989). Inhibition of MHC class II-restricted T cell response by Lyt-2 antigen. *J. Exp. Med.* **170**, 901.

152 Waanders, G. A. and MacDonald, H. R. (1992). Hierarchy of responsiveness *in vivo* and *in vitro* among T cells expressing distinct Mls-1^a-reactive Vβ domains. *Eur. J. Immunol.*, **22**, 291.

153 Webb, S. R. and Sprent, J. (1990). Induction of neonatal tolerance to Mls^a antigens by $CD8^+$ T cells. *Science*, **248**, 1643.

154 Sprent, J. and Webb, S. R. (1987). Function and specificity of T cell subsets in the mouse. *Adv. Immunol.*, **41**, 39.

155 Fowlkes, B. J. and Pardoll, D. M. (1989). Molecular and cellular events in T cell development. *Adv. Immunol.*, **44**, 207.

156 Kruisbeek, A. M., Fultz, M. J., Sharrow, S. O., and Singer, A. (1983). Early development of the T cell repertoire: *in vivo* treatment of neonatal mice with antila antibodies interferes with differentiation of I-restricted T cells but not K/D-restricted T cells. *J. Exp. Med.*, **157**, 1932.

157 McDuffie, M., Born, W., Marrack, P., and Kappler, J. (1986). The role of T-cell receptor in thymocyte maturation: Effects *in vivo* of anti-receptor antibody. *Proc. Natl. Acad. Sci. USA*, **83**, 8728.

158 Marusic-Galesic, S., Stephany, D. A., Longo, D. L., and Kruisbeek, A. M. (1988). Development of CD4−8+ cytotoxic T cells requires interactions with class I MHC determinants. *Nature*, **333**, 180.

159 MacDonald, H. R., Hengartner, H., and Pedrazzini, T. (1988). Intrathymic deletion of self-reactive cells prevented by neonatal anti-CD4 antibody treatment. *Nature*, **335**, 174.

160 Fowlkes, B. J., Schwartz, P. H., and Pardoll, D. M. (1988). Deletion of self-reactive thymocytes occurs at a $CD4^+8^+$ precursor stage. *Nature*, **334**, 620.

161 Blackman, M., Kappler, J., and Marrack, P. (1990). The role of the T cell receptor in positive and negative selection of developing T cells. *Science*, **248**, 1335.

162 Ramsdell, F. and Fowlkes, B. J. (1990). Clonal deletion versus clonal anergy: The role of the thymus in inducing self tolerance. *Science*, **248**, 1342–8.

163 Sprent, J., Gao, E.-K., and Webb, S. R. (1990). T cell reactivity to MHC molecules: Immunity versus tolerance. *Science*, **248**, 1357.

164 von Boehmer, H. and Kisielow, P. (1990). Self-nonself discrimination by T cells. *Science*, **248**, 1369.

165 Lorenz, R. G. and Allen, P. M. (1989). Thymic cortical epithelial cells lack full capacity for antigen presentation. *Nature*, **340**, 557.

166 Marrack, P., McCormack, J., and Kappler, J. (1989). Presentation of antigen, foreign major histocompatibility complex proteins and self by thymus cortical epithelium. *Nature*, **338**, 503.

167 Roehm, N., Herron, L., Cambier, J., DiGuisto, D., Haskins, K., Kappler, J., *et al.* (1984). The major histocompatibility complex-restricted antigen receptor on T cells: distribution on thymus and peripheral T cells. *Cell*, **38**, 577.
168 Jenkinson, E., Kingston, R., and Owen, J. J. (1990). Newly generated thymocytes are not refractory to deletion when the α/β component of the T cell receptor is engaged by the superantigen staphylococcal enterotoxin B. *Eur. J. Immunol.*, **20**, 2517.
169 Lin, Y.-S., Lei, H.-Y., Low, T. L. K., Sheen, C. L., Chou, L. J., and Jan, M.-S. (1992). *In vivo* induction of apoptosis in immature thymocytes by staphylococcal enterotoxin B. *J. Immunol.*, **149**, 1156.
170 Pircher, H., Brduscha, K., Steinhoff, U., Kasai, M., Mizouchi, T., Zinkernagel, R. M., *et al.* (1993). Tolerance induction by clonal deletion of $CD4^+8^+$ thymocytes *in vitro* does not require dedicated antigen-presenting cells. *Eur. J. Immunol.*, **23**, 669.
171 Hengartner, H., Odermatt, B., Schneider, R., Schreyer, M., Walle, G., MacDonald, H. R., *et al.* (1988). Deletion of self-reactive T cells prior to entering the thymus medulla. *Nature*, **336**, 388.
172 Pircher, H. P. B. K., Lang, R., Hengartner, H., and Zinkernagel, R. M. (1989). Tolerance induction in double specific T cell receptor transgenic mice varies with antigen. *Nature*, **342**, 559.
173 Berg, L. J., de St. Groth, B. F., Pullen, A. M., and Davis, M. M. (1989). Phenotypic differences between αβ versus β T-cell receptor transgenic mice undergoing negative selection. *Nature*, **340**, 559.
174 Finkel, T. H., Cambier, J. C., Kubo, R. T., Born, W. K., Marrack, P., and Kappler, J. (1989). The thymus has two functionally distinct populations of immature $\alpha\beta^+$ T cells: One population is deleted by ligation of αβ TCR. *Cell*, **58**, 1047.
175 Molina, I. J., Cannon, N. A., Hyman, R., and Huber, B. (1989). Macrophages and T cells do not express $Mls\text{-}1^a$ determinants. *J. Immunol.*, **143**, 39.
176 Webb, S. R., Okamoto, A., Ron, Y., and Sprent, J. (1989). Restricted tissue distribution of Mls^a determinants. *J. Exp. Med.*, **169**, 1.
177 Webb, S. R. and Sprent, J. (1989). T-cell responses and tolerance to Mls^a determinants. *Immunol. Rev.*, **107**, 141.
178 Gollub, K. J. and Palmer, E. (1991). Physiologic expression of two superantigens in the BDF1 mouse. *J. Immunol.*, **747**, 2447.
179 Gollub, K. and Palmer, E. (1993). Aberrant induction of T cell tolerance in B cell suppressed mice. *J. Immunol.*, **150**, 3705.
180 Simpson, E., Dyson, P. J., Knight, A. M., Robinson, P. J., Elliott, J. I., and Altman, D. M. (1993). T-cell receptor selection by mouse mammary tumor viruses and MHC molecules. *Immunol. Rev.*, **131**, 93.
181 MacDonald, H. R. and Lees, R. K. (1990). Programmed death of autoreactive thymocytes. *Nature*, **343**, 642.
182 Mugata, K. and Ochi, A. (1993). The fate of anergic T cells *in vivo*. *J. Immunol.*, **150**, 763.
183 Yagi, J. and Janeway, C. A. (1990). Ligand thresholds at different stages of T cell development. *Int. Immunol.*, **2**, 83.
184 Speiser, D. E., Lees, R. K., Hengartner, H., Zinkernagel, R. M., and MacDonald, H. R. (1989). Positive and negative selection of T cell receptor Vβ domains is controlled by distinct cell populations in the thymus. *J. Exp. Med.*, **170**, 2165.
185 Speiser, D. E., Schneider, R., Hengartner, H., MacDonald, H. R., and Zinkernagel, R. M. (1989). Clonal deletion of self-reactive T cells in irradiation bone

marrow chimeras and neonatally tolerant mice. Evidence for intercellular transfer of Mls-1[a]. *J. Exp. Med.*, **170**, 595.
186 Ramsdell, F., Lantz, T., and Fowlkes, B. J. (1989). A nondeletional mechanism of thymic self-tolerance. *Science*, **246**, 1038.
187 Yuuki, H., Yoshikai, Y., Kishihara, K., Iwasaki, A., Matsuzaki, G., Ogimoto, M., *et al.* (1990). Deletion of self-reactive T cells in nude mice grafted with neonatal allogeneic thymus. *J. Immunol.*, **144**, 474.
188 Smith, H., Chen, I.-M., Kubo, R., and Tung, K. S. I. (1989). Neonatal thymectomy results in a repertoire enriched in T cells deleted in adult thymus. *Science*, **245**, 749.
189 Jones, L. A., Chin, L. T., Merriam, G. R., Nelson, L. M., and Kruisbeck, A. M. (1990). Failure of clonal deletion in neonatally thymectomized mice: Tolerance is preserved through anergy. *J. Exp. Med.*, **172**, 1277.
190 Vanier, L. E. and Prudhomme, G. J. (1992). Cyclosporin A markedly enhances superantigen-induced peripheral T cell deletion and inhibits anergy induction. *J. Exp. Med.*, **176**, 37.
191 Lussow, A. R., Crompton, T., Karapetian, O., and MacDonald, H. R. (1993). Peripheral clonal deletion of superantigen-reactive T cells is enhanced by cortisone. *Eur. J. Immunol.*, **23**, 578.
192 MacDonald, H. R., Lees, R. K., Schneider, R., Zinkernagel, R. M., and Hengartner, H. (1988). Positive selection of $CD4^+$ thymocytes controlled by MHC class II gene products. *Nature*, **336**, 471.
193 Kappler, J. W., Kushnir, E., and Marrack, P. (1989). Analysis of Vβ17a expression in new mouse strains bearing the $V\beta^a$ haplotype. *J. Exp. Med.*, **169**, 1533.
194 Liao, N.-S., Maltzman, J., and Raulet, D. H. (1989). Positive selection determines T cell receptor Vβ14 gene usage by $CD8^+$ T cells. *J. Exp. Med.*, **170**, 135.
195 Benoist, C. and Mathis, D. (1989). Positive selection of the T cell repertoire: where does it occur? *Cell*, **58**, 1027.
196 Berg, L. J., Pullen, A. M., Fazekas de St. Groth, B., Mathis, D., Benoist, C., and Davis, M. M. (1989). Antigen/MHC-specific T cells are preferentially exported from the thymus in the presence of their MHC ligand. *Cell*, **58**, 1035.
197 Bill, J. and Palmer, E. (1989). Positive selection of $CD4^+$ T cells mediated by MHC class II-bearing stromal cell in the thymic cortex. *Nature*, **341**, 649.
198 Lamb, J. R., Skidmore, B. J., Green, N., Chiller, J. M., and Feldman, M. (1983). Induction of tolerance in influenza virus-immune T lymphocyte clones with synthetic peptides of influenza hemagglutinin. *J. Exp. Med.*, **157**, 1434.
199 Jenkins, M. K. and Schwarz, R. H. (1987). Antigen presentation by chemically modified splenocytes induces antigen-specific T cell unresponsiveness *in vitro* and *in vivo*. *J. Exp. Med.*, **165**, 302.
200 Gaspari, A. A., Jenkins, M. K., and Katz, S. I. (1988). Class II MHC-bearing keratinocytes induce antigen-specific unresponsiveness in hapten-specific TH1 clones. *J. Immunol.*, **141**, 2216.
201 Burkly, L. C., Lo, D., Kanagawa, O., Brinster, R. L., and Flavell, R. A. (1989). T cell tolerance by clonal anergy in transgenic mice with nonlymphoid expression of MHC class II I-E. *Nature*, **342**, 564.
202 Schwartz, R. H. (1990). A cell culture model for T lymphocyte clonal anergy. *Science*, **248**, 1349.
203 Kabelitz, D. and Wesselborg, S. (1992). Life and death of a superantigen-reactive human $CD4^+$ T cell clone: staphylococcal enterotoxins induce death by apoptosis

but simultaneously trigger a proliferative response in the presence of HLA-DR$^+$ antigen-presenting cells. *Int. Immunol.*, **4**, 1381.

204 Markmann, J., Lo, D., Naji, A., Palmiter, R. D., Brinster, R. L., and Heber-Katz, E. (1988). Antigen presenting function of class II MHC expressing pancreatic beta cells. *Nature*, **336**, 476.

205 Moskophidis, D., Lechner, F., Pircher, H., and Zinkernagel, R. M. (1993). Virus persistence in acutely infected immunocompetent mice by exhaustion of antiviral cytotoxic effector cells. *Nature*, **362**, 758.

206 Matis, L. A., Glimcher, L. H., Paul, W. E., and Schwartz, R. H. (1983). Magnitude of response of histocompatibility-restricted T-cell clones is a function of the product of the concentrations of antigen and I-A molecules. *Proc. Natl. Acad. Sci. USA*, **80**, 6019.

207 Suzuki, G., Kawase, Y., Koyasu, S., Yahara, I., Kobayashi, Y., and Schwartz, R. H. (1988). Antigen-induced suppression of the proliferative response of T cell clones. *J. Immunol.*, **140**, 1359.

208 Essery, G., Feldman, M., and Lamb, J. R. (1988). Interleukin-2 can prevent and reverse antigen-induced unresponsiveness in cloned human T lymphocytes. *Immunology*, **64**, 413.

209 Ceredig, R. and Corradin, G. (1986). High antigen concentration inhibits T cell proliferation but not interleukin 2 production: examination of limiting dilution microcultures and T cell clones. *Eur. J. Immunol.*, **16**, 30.

210 Blackman, M. R., Gerhardt-Burgert, H., Woodland, D. L., Palmer, E., Kappler, J., and Marrack, P. (1990). A role for clonal inactivation in T cell tolerance to Mls-1^a. *Nature*, **345**, 540.

211 Arnold, B., Schönrich, G., and Hämmerling, G. J. (1993). Multiple levels of peripheral tolerance. *Immunol. Today*, **14**, 12.

212 Bandeira, A., Mengel, J., Burlen-Defranoux, O., and Coutinho, A. (1991). Proliferative T cell anergy to Mls-1^a does not correlate with *in vivo* tolerance. *Int. Immunol.*, **3**, 923.

213 von Boehmer, H. and Sprent, J. (1974). Expression of M locus differences by B cells but not T cells. *Nature*, **249**, 363.

214 Eynon, E. E. and Parker, D. C. (1992). Small B cells as antigen-presenting cells in the induction of tolerance to soluble protein antigens. *J. Exp. Med.*, **175**, 131.

215 Parker, D. C. (1993). T cell-dependent B-cell activation. *Annu. Rev. Immunol.*, **11**, 331.

216 Fuchs, E. J. and Matzinger, P. (1992). B cells turn off virgin but not memory T cells. *Science*, **258**, 1156.

217 Ahmed, A., Scher, I., Smith, A. H., and Sell, K. W. (1977). Studies on non-H-2-linked lymphocyte stimulating determinants. *J. Immunogenet.*, **4**, 201.

218 Faro, J. M., Marcos, A. R., Andreu, J. L., Martinez A. C., and Coutinho, A. (1990). Inside the thymus, Mls antigen is exclusively presented by B lymphocytes. *Res. Immunol.*, **141**, 723.

219 Mazada, O., Watanabe, Y., Gyotoku, J.-I. and Katsura, Y. (1991). Requirement of dendritic cells and B cells in the clonal deletion of Mls-reactive T cells in the thymus. *J. Exp. Med.*, **173**, 539.

220 Inaba, M., Inaba, K., Hosono, M., Kumamoto, T., Ishidia, T., Muramatsu, S., *et al.* (1991). Distinct mechanisms of neonatal tolerance induced by dendritic cells and thymic B cells. *J. Exp. Med.*, **173**, 549.

221 Rocha, B. and von Boehmer, H. (1991). Peripheral selection of the T cell repertoire. *Science*, **251**, 1225.

222 Mamalaki, C., Norton, T., Tanaka, Y., Townsend, A. R., Chandler, P., Simpson, E., *et al.* (1992). Thymic depletion and peripheral activation of class I major histocompatibility complex restricted T cells by soluble peptide in T-cell receptor transgenic mice. *Proc. Natl. Acad. Sci. USA*, **89**, 11342.
223 Qin, S., Cobbold, S. P., Pope, H., Elliott, J., Kioussis, D., Davies, J., *et al.* (1993). 'Infectious' transplantation tolerance. *Science*, **259**, 974.
224 Baschieri, S., Lees, R. K., and MacDonald, H. R. (1993). Clonal anergy to Staphylococcal enterotoxin B *in vivo*. *Eur.J.Immunol.*, **23**, 2661.
225 Acha-Orbea, H. (1993). Bacterial and viral superantigens: roles in autoimmunity? *Ann. Rheumat. Dis.*, **52**, S6.
226 Posnett, D. N. (1993). Do superantigens play a role in autoimmunity? *Semin. Immunol.*, **5**, 65.

Part III

T cell genes

12 Organization of the human TCRB gene complex

M. A. ROBINSON AND P. CONCANNON

1 Introduction

T lymphocytes play a central role in the generation of both humoral and cell-mediated immune responses. Although T cells mediate diverse effector functions, most recognize antigenic peptides presented by molecules encoded within the major histocompatibility complex (MHC). The fine specificity of a T cell is determined by the receptor for antigen (TCR) displayed on the cell surface which is a heterodimer composed of an alpha and a beta chain. Genes for the alpha and beta chains of the TCR are encoded in germline DNA as discontinuous gene segments that rearrange specifically in T cells during development. The variable regions of each of these chains are responsible for antigen recognition and are encoded by juxtaposed variable (V), diversity (D) (for beta chains), and joining (J) gene segments (reviewed in Toyonaga and Mak 1987; Wilson *et al.* 1988; Davis and Bjorkman 1988). TCR V gene segments encode two of the three complementarity determining regions (CDR) which comprise the putative antigen–MHC binding site (Chothia *et al.* 1988). Therefore, a detailed accounting of both the extent and variability of the germline repertoire of V gene segments is critical to understanding the diversity of potential T cell specificities.

We have previously estimated the size of the human TCRBV gene segment repertoire by counting numbers of novel cDNA sequences and bands observed on Southern blots using TCRBV probes (Concannon *et al.* 1986, 1987; Robinson 1991). Unfortunately, these methods may lead to both over- and underestimations of TCRBV gene numbers. Gene segments expressed at low levels may not be detected by cDNA cloning (Robinson 1992). Counting V gene segments by comparison of sequences can lead to overestimation of the number of TCRBV gene segments due to germline polymorphism or to errors such as those introduced by *Taq* polymerase. For example, the number of distinct cDNA sequences reported for some TCRBV families vastly exceeds the number of bands corresponding to those families that can be detected on Southern blots. In contrast, detection of TCRBV gene segments by Southern blotting may fail to detect V gene segments clustered on common fragments, or located on distinct fragments that migrate at the same position. Furthermore, this approach will not discriminate between functional genes and pseudogenes, and will detect only members of families for which probes have been previously isolated.

Genomic mapping represents a useful alternative approach toward establishing the extent of the human TCRB germline repertoire. Genomic mapping allows assignment of distinct TCRBV sequences to specific sites within the TCRB locus making it possible to unambiguously discriminate between the products of alleles of the same gene segment as opposed to different gene segments. This approach has allowed expansion of the estimated TCRB repertoire by identification of new TCRBV genes and alleles. Furthermore, the extent of the estimated TCRBV repertoire has been limited by the identification of pseudogenes and null alleles of TCRBV genes.

2 Mapping TCRBV genes to chromosomes 7 and 9

2.1 Distribution of TCR β genes on *Sfi*I fragments

In order to develop a complete map of the TCRB gene complex, pulsed-field gels of *Sfi*I digested DNA were hybridized with probes that corresponded to 24 of the TCRBV families. The restriction enzyme *Sfi*I is known to reveal two insertion/deletion-related polymorphism (IDRP) within the TCRB gene complex (Seboun *et al.* 1989*b*), thus, DNA donors for these analyses were selected on the basis of genotype such that homozygotes and heterozygotes for the TCRB IDRP were represented by the three individuals selected for study. All TCRBV probe hybridization could be accounted for on six *Sfi*I fragments. Figure 12.1 lists these *Sfi*I fragments and indicates which TCRB genes hybridize with each fragment. Figure 12.2 shows representative Southern blots hybridized with both TCRBV and TCRBC probes. The blot shown in the left-hand panel was hybridized with both TCRBC and TCRBV1 probes. The TCRBC probe hybridized with bands of 145 kb and 125 kb which correspond to the C region IDRP. The TCRBV1 probe hybridized with bands

Sfi fragments	TCRB genes
40kb	V4
60kb	V2,V10,V11,V15,V19
90 kb	V3,V4,V11,V14
110 kb	V2,V8,V10,V15,V16,V17, V18,V19,V21,V24,V25
125/145 kb[a]	V20, C1, C2
280/310 kb[a]	V1,V5,V6,V7,V8,V9,V12,V13, V21,V22,V23

Fig. 12.1 Fragments of *Sfi*I digested genomic DNA that hybridize with TCRB probes. *Sfi*I fragments are listed by size along with the TCRBV probes that hybridize with each fragment. [a]Fragment size depends upon the IDRP type of the DNA donor.

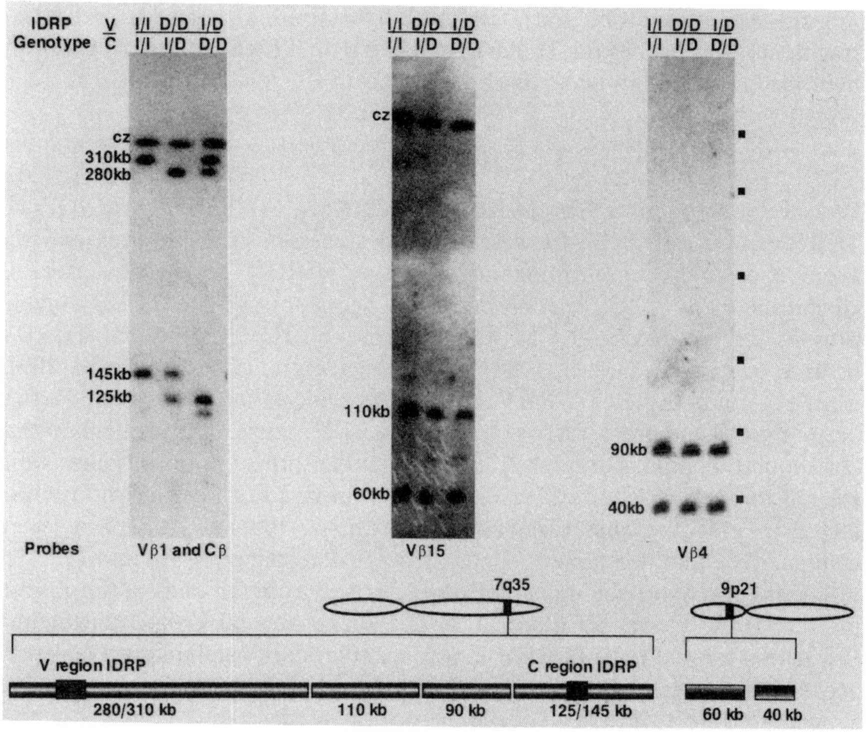

Fig. 12.2 *Top*. Southern blots of DNA samples digested with *Sfi*I, resolved by pulsed-field gel electrophoresis, and hybridized with TCRB probes. The TCRBV and TCRBC region IDRP genotypes of the donors are shown above the lanes on each duplicate panel, and the probes used to hybridize the blot are shown below. The sizes of the observed bands are shown to the *left* of each blot; bands of 310/280 kb correspond to the V region IDRP and bands of 145/125 kb correspond to the C region IDRP. Squares to the *right* of last panel correspond to lambda concatamer size markers of ~ 300, 250, 200, 150, 100, and 50 kb respectively. The compression zone, where DNA fragments do not resolve by size, is marked with cz. *Bottom*. The map locations on chromosomes 7 and 9 are shown for the six *Sfi*I fragments that hybridize with TCRB probes. Sizes are listed below each fragment. The fragments in the TCRB gene complex on chromosome 7 are ordered with the 280/310 kb *Sfi*I fragment containing the V region IDRP at the 5′ end and the 125/145 kb *Sfi*I fragment containing the C region IDRP at the 3′ end of the gene complex. The two *Sfi*I fragments on chromosome 9 map to a similar location by genetic studies but have not been physically linked.

of 310 kb and 280 kb, fragments corresponding to the V region IDRP. No variations in size of fragments outside of the C and V IDRP previously described were observed. Other TCRBV probes revealed additional bands including fragments of 40 kb, 60 kb, 90 kb, and 110 kb (centre and right panels, Figure 12.2). Certain probes hybridized with more than one *Sfi*I fragment. For example, the TCRBV4 probe hybridized with 40 kb and 90 kb

*Sfi*I fragments, a probe for TCRBV11 hybridized with 60 kb and 90 kb fragments, and probes for TCRBV2, TCRBV10, TCRBV15, and TCRBV19 hybridized with 60 kb and 110 kb fragments.

2.2 Two clusters of TCRBV genes

Probes corresponding to TCRBV2, TCRBV4, TCRBV10, TCRBV11, TCRBV15, and TCRBV19 were used to screen cosmid libraries and ten positive cosmids representing all of these TCRBV genes were isolated (Robinson *et al.* 1993; Wei *et al.* 1994). Sequences of the TCRBV genes present on these cosmids, as well as those on three additional TCRBV bearing cosmids (Lai *et al.* 1988) and a phage clone (Charmley *et al.* 1991) were determined. The TCRBV genes were divided into two groups based upon their sequences. One group of TCRBV genes had sequences that correspond to published cDNA sequences. The other group of genes were related to the published cDNA sequences but had not been previously detected in cDNA clones. The clones containing the new sequences were characterized and overlapping segments were determined based upon restriction maps and hybridization with probes derived from the ends of the cosmid inserts. These clones all mapped to the 40 kb and 60 kb *Sfi*I fragments. Sequences for the new TCRBV family members are available in Genbank; accession numbers and appropriate new nomenclature for these genes are shown in Table 12.1.

The coding region sequences of TCRBV2 and TCRBV2(O) sequences show 95.5% identity, TCRBV4 and TCRBV4(O) 96.5% identity, TCRBV10 and TCRBV10(O) 94% identity, TCRBV11 and TCRBV11(O) 93.6% identity, TCRBV15 and TCRBV15(O) 94.1% identity, and TCRBV19 and TCRBV19(O) show 94.9% sequence identity. The sequence of TCRBV19(O) was determined from PCR amplified genomic DNA (Wei *et al.* 1994). Deduced protein sequences of TCRBV2(O), TCRBV4(O), TCRBV10(O), and TCRBV19(O) reveal no obvious defects that would preclude translation. TCRBV11(O) and TCRBV15(O) sequences have deletions and/or insertion of nucleotides that would induce frame shifts compared to their counterparts TCRBV11 and TCRBV15. The heptamer/nonamer rearrangement signal sequences located 3′ of all of the TCRBV(O) genes are similar to the rearrangements signals of functional TCRBV and murine Igκ genes. In spite of their apparently functional coding regions and rearrangement sequences, no transcripts of TCRBV(O) genes rearranged with TCRBC were detected by PCR using primers specific for TCRBV2(O), TCRBV4(O), and TCRBV10(O) and a primer within TCRBC.

Oligonucleotides were constructed that were shown to selectively amplify the TCRBV2(O), TCRBV4(O), TCRBV10(O), TCRBV11(O), and TCRBV15(O) genes from genomic DNA by PCR. These gene-specific oligonucleotides were used to screen somatic cell hybrids in order to determine the chromosomal location of the TCRBV(O) genes. The TCRBV gene

Table 12.1 Reference TCRBV sequences

Family	Member	WHO format	Source of sequence[a]	Reported sequence[b]	Reference	Genbank/EMBL accession no.
Vβ1	1.1	BV1S1	H18.1	HBVT73	Kimura *et al.* 1987	M27381
Vβ2	2.1	BV2S1	HVB14.1	Molt4	Tunnacliffe *et al.* 1985	M12886
	2.OR	BV2S2*1 (O)	HVB22.1		Robinson *et al.* 1993	L05149
Vβ3	3.1	BV3S1	HVB19.1	PL4.4	Concannon *et al.* 1986	M13843
Vβ4	4.1	BV4S1	H46.1	DT110	Kimura *et al.* 1986	X04921
	4.OR	BV4S2 (O)	H28.1		Robinson *et al.* 1993	L05150
Vβ5	5.1	BV5S1	Plate stock	HBVP51	Kimura *et al.* 1986	X04927
	5.2	BV5S2	H127	PL2.5	Concannon *et al.* 1986	M13850
	5.3	BV5S3	H145.5	IGRb08	Ferradini *et al.* 1991	X58801
	5.5	BV5S5	H18.1	Clone 9	Li *et al.* 1991	X61439
	5.6	BV5S6	H26.3	HT415	Plaza *et al.* 1991	X57615
	5.7	BV5S7	H139.1		Wei *et al.* 1994	L26226
	5.8	BV5S8		AL62.24	Hurley *et al.* 1993	M97709
Vβ6	6.1	BV6S1	HVB35.2	HBVP50	Kimura *et al.* 1986	X04934
	6.3	BV6S3	H139.1	ATL12.2	Ikuta *et al.* 1985	M11952
	6.4	BV6S4	H127	IGRb11	Ferradini *et al.* 1991	X58806
	6.5	BV6S5	HVB35.4	IGRb10	Ferradini *et al.* 1991	X58805
	6.7	BV6S7	HVB15.2	ph16	Tillinghast *et al.* 1986	M14262
	6.10	BV6S10 (P)	HVB16.1	Clone 11	Li *et al.* 1991	X61444
	6.11	BV6S11	H26.3	PCR	Gomolka *et al.* 1993	L13762
	6.12	BV6S12 (P)	Plate stock	PCR	Gomolka *et al.* 1993	M97503
	6.14	BV6S14	Plate stock	HT147	Plaza *et al.* 1991	X57607
Vβ7	7.1	BV7S1	Plate stock	PL4.9	Concannon *et al.* 1986	M13855
	7.2	BV7S2	Genomic	PL4.19	Concannon *et al.* 1986	M13856
	7.3	BV7S3	3116.9.A	HT267.2	Plaza *et al.* 1991	X57617

Table 12.1 (*continued*)

Family	Member	WHO format	Source of sequence[a]	Reported sequence[b]	Reference	Genbank/EMBL accession no.
Vβ8	8.1	BV8S1	H7.1	YT35	Yanagi *et al.* 1984	K01571
	8.2	BV8S2	H7.1	PL3.3	Concannon *et al.* 1986	M13858
	8.3	BV8S3	Genomic	BR3.5	Siu *et al.* 1986	X07223
	8.4	BV8S4 (P)	H6.1	pBH9.1R3	Siu *et al.* 1986	X07224
	8.5	BV8S5 (P)	H6.1	M18VB8.5	Siu *et al.* 1986	X06936
Vβ9	9.1	BV9S1	Genomic	PL2.6	Concannon *et al.* 1986	M13859
	9.2	BV9S2 (P)	Genomic	HT307.1	Plaza *et al.* 1991	X57608
Vβ10	10.1	BV10S1	HVB10.1	PL3.9	Concannon *et al.* 1986	M13860
	10.R	BV10 (O)	HVB6.2		Robinson *et al.* 1993	L05151
Vβ11	11.1	BV11S1	HVB25.1	PL3.12	Concannon *et al.* 1986	M13861
	11.R	BV11 (O)	HVB36.1		Robinson *et al.* 1993	L05152
Vβ12	12.2	BV12S2	H7.1	ph27	Tillinghast *et al.* 1986	M14268
	12.3	BV12S3	H6.1	HT96	Plaza *et al.* 1991	X57609
	12.4	BV12S4	H18.1		Wei *et al.* 1994	L26230
Vβ13	13.1	BV13S1	H26.3	HBVP34	Kimura *et al.* 1986	X04932
	13.2	BV13S2	Plate stock	CEM-1	Li *et al.* 1991	X61455
	13.3	BV13S3	HVB16.1	IGRb14	Ferradini *et al.* 1991	X58809
	13.4	BV13S4	H139.1	Clone 4-1	Li *et al.* 1991	X61447
	13.5	BV13S5	HVB35.2	IGRb15	Ferradini *et al.* 1991	X58810
	13.6	BV13S6	H145.5	IGRb16	Ferradini *et al.* 1991	X58815

	13.7	**BV13S7**	**H127**		**Wei** *et al.* 1994	L26228
	13.8	**BV13S8**	**H127**		Wei *et al.* 1994	L26227
	13.9	**BV13S9**	**HVB15.2**		Wei *et al.* 1994	L26229
Vβ14	14.1	BV14S1	HVB19.1	PL8.1	Concannon *et al.* 1986	M13865
Vβ15	15.1	BV15S1	HVB17.2	ATL2-7	Ikuta *et al.* 1985	M11951
	15.R	BV15 (O)	HVB13.2		Robinson *et al.* 1993	L05153
Vβ16	16.1	BV16S1	Genomic	HT370	Plaza *et al.* 1991	X57723
Vβ17	17 1	BV17S1	HVB14.1	HBVT02	Kimura *et al.* 1986	M27388
Vβ18	18.1	BV18S1	HVB14.1	HBVT56	Kimura *et al.* 1986	M27389
Vβ19	19.1	BV19S1	HVB17.2	HBVT72	Kimura *et al.* 1986	M27390
	19.OR	BV19 (O)	Genomic		Wei *et al.* 1994	L26225
Vβ20	20.1	BV20S1	HVB1.2	HUT102	Leiden and Strominger 1986	M13554
Vβ21	21.1	BV21S1	H18.1	TCRBV21.1	Wilson *et al.* 1990	M33233
	21.3	BV21S3	II6.1	TCRBV21.3	Wilson *et al.* 1990	M33235
	21.4	BV21S4	H12.18	VB1W6	Hansen *et al.* 1992	X56665
Vβ22	22.1	BV22S1	Genomic	HT2.10	Plaza *et al.* 1991	X57727
Vβ23	23.1	BV23S1	H7.1	IGRb04	Ferradini *et al.* 1991	X58799
Vβ24	24.1	BV24S1	HVB30.A	IGRb05	Ferradini *et al.* 1991	X58800
Vβ25	25.1	BV25S1	HVB30.A		Wei *et al.* 1994	L26231

[a] Nucleotide sequences were derived from cosmid clones from libraries made from a primary fibroblast line ('HVB' clones), from HeLa cells ('H' clones), from PBL ('3116' clones), from individual unamplified primary plate stocks from those libraries by PCR, or from genomic DNA by PCR. Although in many cases sequences were derived from multiple independent cosmid clones only a single example is listed for brevity.

[b] The nucleotide sequences derived from the indicated genomic sources exactly matched these previously reported sequences. Blank spaces correspond to novel sequences derived from the indicated genomic source.

complex has been localized to the long arm of chromosome 7 (7q35) (Isobe *et al.* 1985). When primer pairs corresponding to TCRBV genes that had been reported as cDNA sequences were used, amplified fragments were detected in DNA samples from all of the somatic cell hybrids that contained human chromosome 7. In contrast, none of the primers specific for any of the five TCRBV(O) genes amplified DNA from these chromosome 7 positive hybrids. All sets of TCRBV(O) primers amplified DNA from every hybrid cell line that contained chromosome 9. These data suggest that TCRBV2(O), TCRBV4(O), TCRBV10(O), TCRBV11(O), TCRBV15(O), and TCRBV19(O) do not map to the TCR β gene complex encoded on chromosome 7 but rather map to chromosome 9. Similar to the designation 'orphon gene' that has been assigned to immunoglobulin V genes encoded outside of the Ig gene complexes TCRBV2(O), TCRBV4(O), TCRBV10(O), TCRBV11(O), TCRBV15(O), and TCRBV19(O) may be considered orphon TCRBV genes.

2.3 Localization of TCRB orphon genes on chromosome 9

In situ hybridization studies were performed in order to more precisely locate the TCRBV(O) genes. Cosmid clones that mapped to the 40 kb *Sfi*I fragment and to the 60 kb *Sfi*I fragment were used as probes for *in situ* hybridization. Both probes showed similar hybridization patterns localizing the orphon Vβ genes on both the 40 kb and the 60 kb *Sfi*I fragments to 9p proximal to the centromere (Robinson *et al.* 1993).

A coding region polymorphism within the TCRBV2(O) sequence that could be detected by PCR was used to localize the TCRBV orphon gene cluster to 9p21 by linkage analysis in the CEPH family collection (Charmley *et al.* 1993*b*). Similar results were obtained with a polymorphism located 3′ of the TCRBV4(O) gene and no recombination was observed between this and the TCRBV2(O) markers (C. E. Day and M. A. Robinson unpublished results). Therefore, the TCRBV(O) genes present on the 40 kb and 60 kb *Sfi*I fragments map to similar locations by *in situ* hybridization and genetic linkage analyses, although the two fragments have not been ordered or physically linked to one another (Figure 12.2). Interestingly, the 9p21–22 region is non-randomly involved in deletions and translocations in leukaemia patients (Diaz *et al.* 1988). Such chromosomal aberrations might involve the rearrangement signals present at the 3′ ends of the TCRBV(O) gene segments.

3 Origin of TCRBV orphon genes

The origin of the orphon genes is unknown. However, several mechanisms which may explain how they arose have been proposed for the generation of VH and Vκ orphon genes (Huber *et al.* 1990; Borden *et al.* 1990; Matsuda *et al.* 1990). The series of six TCRBV orphon genes presumably arose by duplication and intrachromosomal translocation of TCRBV2, TCRBV4,

TCRBV10, TCRBV11, TCRBV15, and TCRBV19 which are encoded on linked *Sfi*I fragments within the TCRB gene complex on chromosome 7. The high degree of sequence identity between the TCRBV orphon genes and their functional counterparts suggest they resulted from one or more relatively recent duplication events. Because all the TCRBV(O) orphon genes map to the same region on the short arm of chromosome 9, it is likely that they were transferred in a single event. However, the cluster of TCRBV(O) genes does not represent a precise duplication of the corresponding portion of the TCRB locus on chromosome 7. TCRBV4 and TCRBV11 are located at opposite ends of the 90 kb *Sfi*I fragment with TCRBV3 and TCRBV14 encoded between them (Figure 12.3); however, homologues of TCRBV3 and TCRBV14 are not present on chromosome 9. TCRBV4(O) is the only orphon gene that cannot be physically linked to the cluster on chromosome 9; therefore, it appears that during the duplication or translocation process, TCRBV3 and TCRBV14 were deleted, and TCRBV4 was separated from the other orphon genes.

3.1 Composite map of the TCRB gene complex

A composite map of the TCRB gene complex linking four of the six *Sfi*I fragments was developed by analysis of Southern blots of conventional and pulsed-field gels (Figure 12.3). This map spans a sequence of approximately 600 kb to 650 kb depending upon the IDRP genotype of the donor. The 280/310 kb fragments containing the V region IDRP is the 5′ most fragment, followed by the 110 kb fragment, the 90 kb fragment, and the 125/145 kb fragments containing the C region IDRP is located at the 3′ end.

The C region IDRP was detected with the restriction enzyme *Cla*I and fragments of 125/145 kb were observed. Probes corresponding to TCRBV3, TCRBV4, TCRBV14, and TCRBV20 hybridized also with these *Cla*I fragments. Since TCRBV3, TCRBV4, and TCRBV14 are present on the 90 kb *Sfi*I fragment, the 90 kb *Sfi*I fragment maps adjacent to and 5′ of the 125/145 kb *Sfi*I C region fragment.

Blots of DNA digested with *Sma*I allow localization of the C region IDRP to the region 5′ of TCRBC. Fragments of 25/45 kb and 10 kb are observed with a TCRBC probe and the 10 kb fragments hybridizes with a probe specific for TCRBC2 (M. A. Robinson unpublished results). Thus, the C region IDRP resides on the same *Sma*I fragments as TCRBC1. The TCRBV20 gene is present on a 17 kb *Sma*I fragment that is located 3′ of TCRBC (Charmley *et al.* 1993*a*).

Identification of the Vβ genes near the borders of *Sfi*I fragments was useful in linking the 90 kb, 110 kb, and 280/310 kb fragments. TCRBV genes located near *Sfi*I sites were identified by Southern blots of single and double enzyme digestions of DNA samples performed with *Bam*HI alone, and *Bam*HI plus *Sfi*I. In this way, the 110 kb *Sfi*I fragment was linked to the 90 kb *Sfi*I fragment based upon a polymorphic *Bam*HI restriction fragment that hybridized with

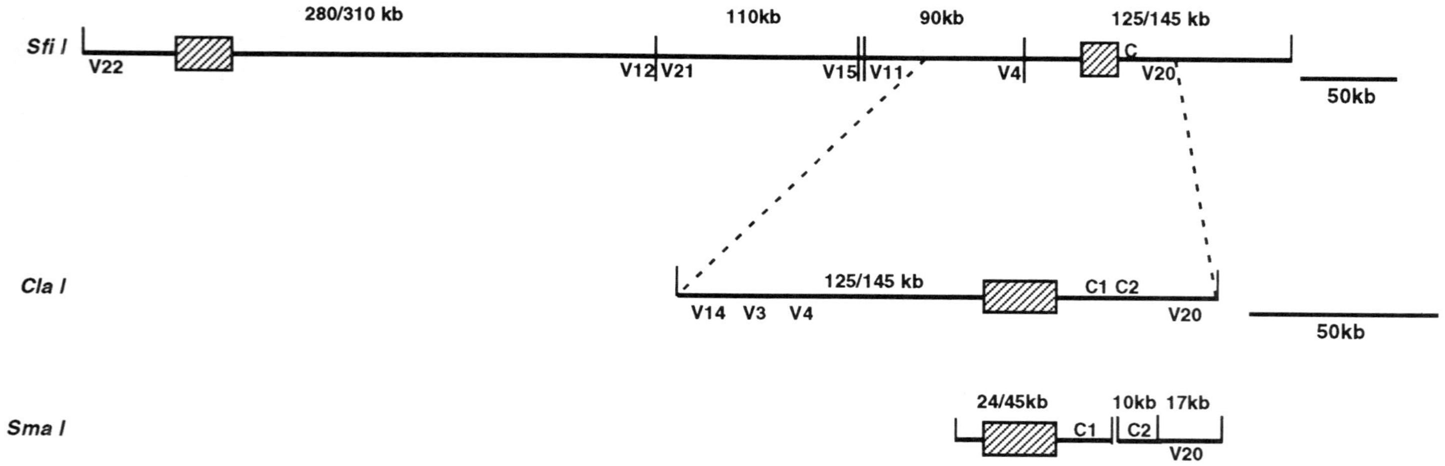

Fig. 12.3 A composite map of the TCRB gene complex. Four *Sfi*I fragments hybridizing with TCRB probes were linked and their order was determined. Vertical lines indicate the location of restriction sites for the enzymes shown at the *left*. Sizes of restriction fragments observed on Southern blots of pulsed-field gels are shown above the horizontal line. TCRBV genes that map to the borders of *Sfi*I fragments and those that hybridize with *Cla*I and *Sma*I fragments are shown beneath, and TCRBC genes are shown above the horizontal lines representing the appropriate fragments. IDRP are represented by hatched boxes which correspond to regions of 'insertion' that are missing in 'deleted' haplotypes.

TCRBV11, TCRBV15, and TCRBV19; TCRBV11 is located on the end of the 90 kb *Sfi*I fragment and TCRBV15 and TCRBV19 are located on the 110 kb *Sfi*I fragment. The 280/310 kb *Sfi*I fragment was linked to the other end of the 110 kb *Sfi*I fragment by a second *Bam*HI RFLP hybridizing with TCRBV12, TCRBV21, and TCRBV8. The *Sfi*I site is located between the TCRBV12 gene on the 280/310 kb fragment and the TCRBV21 and TCRBV8 genes located on the 110 kb *Sfi*I fragment. These results are consistent with our restriction mapping of a previously described cosmid clone (H7.1) spanning this region (Lai *et al.* 1988).

3.2 Duplication events within the TCRB gene complex

Duplication events appear to have played an important role in the evolution of both TCR and Ig complexes. TCR and Ig complexes contain families of V genes that share more than 75% nucleotide identity which presumably arose by duplication events and subsequent mutation. For example, mapping studies of TCRBV6 and TCRBV13 genes have revealed evidence of tandem duplications of clusters of these TCRBV genes. Li *et al.* (1991) isolated four sets of linked TCRBV6 and TCRBV13 loci on phage clones. The genes present in each set represented different family members yet the four sets were found to have related molecular maps. The genes were in the same orientation and the TCRBV6 and TCRBV13 genes were separated from one another by similar distances. Additional mapping studies utilizing larger capacity cloning vectors (cosmids) have revealed the existence of additional clusters of TCRBV6 and TCRBV13 genes as well as indicating that TCRBV5 genes are part of the basic duplication unit (Lai *et al.* 1988; Wei *et al.* 1994).

Another set of TCRBV genes that appear to have duplicated as a cluster include the TCRBV8, TCRBV12, and TCRBV21 gene families. There are five TCRBV8 genes, two of which are pseudogenes and three copies of both TCRBV12, and TCRBV21 gene families. Separate cosmids containing one or two clusters of TCRBV12, TCRBV21, and TCRBV8 have been isolated (Lai *et al.* 1988; Wei *et al.* 1994).

A second type of duplication event with greater evolutionary significance resulted in the creation of the TCRBV orphon gene cluster on chromosome 9. In this case duplication was accompanied by translocation. Although the cluster of TCRBV orphon genes is a family of non-functional genes, such translocation events always have the possibility of including the constant portion of the locus, thereby generating a novel locus for rearranging gene segments. Presumably this is the underlying evolutionary process by which new receptor gene families have developed and diversified.

3.3 Genomic cloning to determine the extent of the TCRBV repertoire

In order to draw a more direct correspondence between the mapping data and previously published TCRBV nucleotide sequences derived from cDNA cloning, cosmid clones containing TCRBV genes were isolated and characterized by both Southern blotting and PCR amplification with subsequent nucleotide sequencing. These studies confirmed the existence of 64 unique human TCRBV gene segments grouped into 25 families. All of the genes were detected by hybridization and by PCR and their nucleotide sequences were determined. Six of the TCRBV genes segments characterized were novel sequences. Fifty-one of the 63 were isolated on one or more cosmid clones. Certain TCRBV gene sequences had been previously reported but were not isolated on any cosmid cloned derived from three different libraries comprising different IDRP haplotypes. PCR was used to detect these sequences in primary unamplified cosmid library plate stocks and in some cases genomic DNA samples. Table 12.1 contains a list of the TCRBV genes identified along with the names of representative cosmids from which sequences were derived. Where applicable, the names and literature references for previously defined nucleotide sequences for the identified TCRBV gene segments are provided.

Table 12.2 contains a listing of the number of TCRBV gene segments by family along with a tally of the number of hybridizing bands observed on genomic Southern blots with the appropriate probes. Almost half of the human TCRBV families contain only a single member. Most of the multi-membered families result from one of the three groups of tandemly duplicated genes described above, the cluster of TCRBV5, TCRBV6, and TCRBV13 genes, the cluster of TCRBV8, TCRBV12, and TCRBV21 genes, or the TCRBV orphon genes and their functional homologues. When the number of defective gene segments is considered the functional repertoire is 53 TCRBV genes. Additional variability in the basic repertoire derives from polymorphism of germline genes.

3.4 Polymorphism in TCRBV gene segments

Polymorphism in TCRBV gene segments may serve to expand or contract the TCR repertoire. A survey of TCRBV sequences reveals that for many TCRBV families, the number of sequences reported significantly exceeds the number of genes in the family. One explanation for more sequences than genes is allelic polymorphism in TCRBV genes segments. Knowledge of the map of the TCRB gene complex and of genomic sequences for TCRBV genes has made it possible to determine the location of a gene having a particular sequence. Such information along with the demonstration that sequence variations segregate as alleles within human families would establish whether sequences correspond to allelic or alternatively, isotypic genes.

Nucleotide substitutions in coding regions have been reported for a number of TCRBV genes; those resulting in replacement substitutions within the TCRBV coding region are listed in Figure 12.4. The locations of the substitutions are shown with respect to regions thought to correspond to functional

Table 12.2 Esimated size of human germline TCRBV repertoire

TCRBV family	Bands on Southern blots	Gene number	Number of functional genes
BV1	1	1	1
BV2	2	2	1
BV3	1	1	1
BV4	2	2	1
BV5	5–8	7	7
BV6	8–10	9	7[a]
BV7	3	3	3
BV8	5	5	3
BV9	2	2	1
BV10	2	2	1
BV11	2	2	1
BV12	3	3	3
BV13	6–9	9	9
BV14	1	1	1
BV15	2	2	1
BV16	1	1	1
BV17	1	1	1
BV18	1	1	1
BV19	2	2	1
BV20	1	1	1[a]
BV21	3	3	3
BV22	1	1	1
BV23	1	1	1
BV24	1	1	1
BV25	1	1	1
Total	58–66	64	53

[a] Common null alleles exist for a TCRBV gene in these families (Charmley *et al.* 1993*a*; Barron and Robinson 1994).

regions of the molecule. Structural models of TCR based upon sequence homology with Ig and the known crystal structure of Ig molecules suggest that the TCR antigen–MHC binding site localizes to three areas (Chothia *et al.* 1988). Two of these regions are encoded within the V gene segment and the third is generated by combinatorial and junctional mechanisms during the assembly of V, D, and J gene segments. A fourth region of hypervariability is found in TCRBV chains and is thought to contain residues important for interactions between superantigens and TCR molecules. Replacement substitutions occurring in the putative binding site may alter TCR function by changing the antigen–MHC binding site; substitutions that occur in framework regions may have an impact upon conformation of the molecule and affect expression by changing alpha/beta chain pairing patterns. Identification

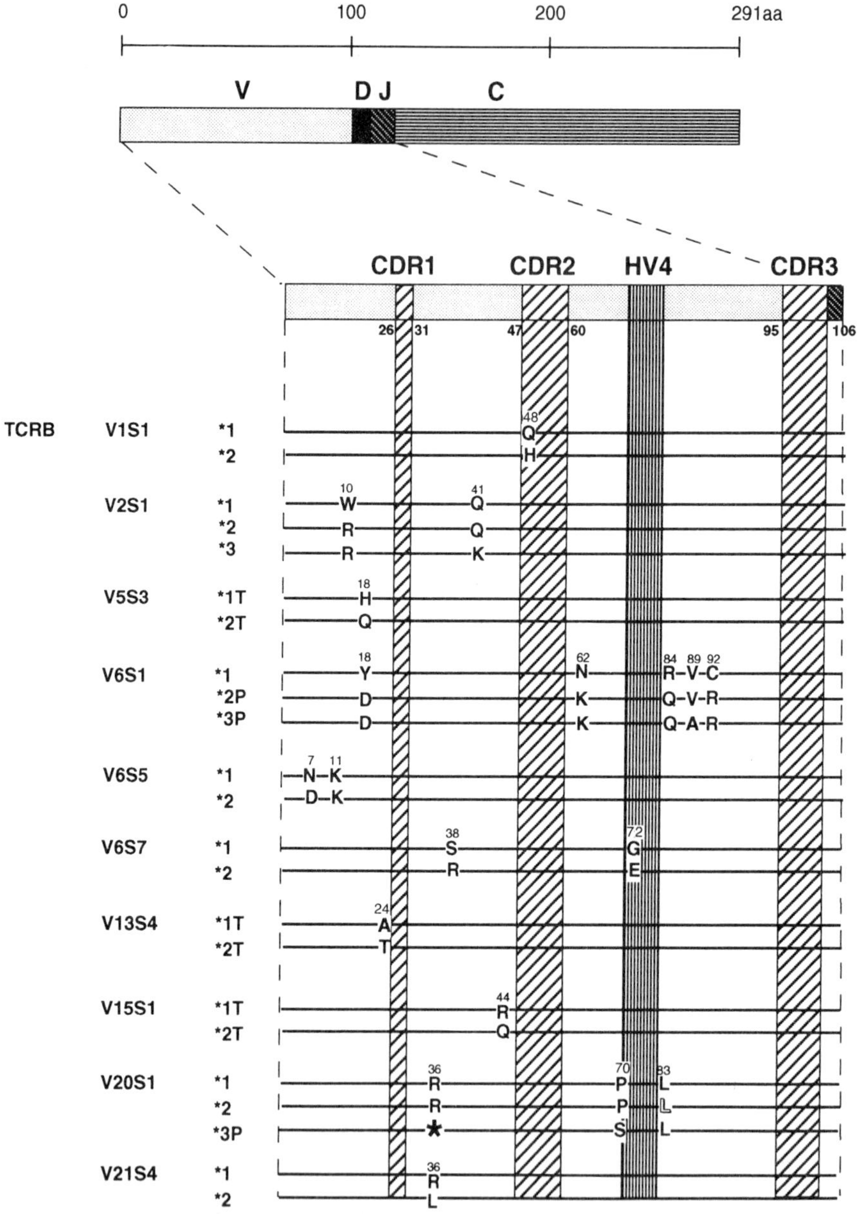

of alleles makes it possible to examine the impact substitutions may have on specific immune responses. Knowledge of the position of the substitutions may reveal functional regions of the molecules.

The listing of TCRBV alleles in Figure 12.4 includes two genes, TCRBV6S1 and TCRBV20, that have null alleles. The null allele of TCRBV20 contains

Fig. 12.4 Listing of TCRBV alleles. A TCRBV chain is schematically represented showing V, D, J, and C regions. Regions of hypervariability are highlighted and residue numbers are given for CDR1, CDR2, and CDR3. Alleles of TCRBV genes are listed showing the location of polymorphic residues. An asterisk (*) represents a stop codon and outline type (𝕃) represents a silent substitution. Sources for sequences are as follows: TCRBV1S1 (Robinson 1989); TCRBV2S1 (Cornelis *et al.* 1993); TCRBV5S3 (Plaza *et al.* 1991); TCRBV6S1 (Barron and Robinson 1994); TCRBV6S5 (Hansen *et al.* 1992); TCRBV6S7 (Li *et al.* 1990); TCRBV13S4 (Plaza *et al.* 1991); TCRBV15S1 (Baccala *et al.* 1991); TCRBV20S1 (Charmley *et al.* 1993*a*); and TCRBV21S4 (Hansen *et al.* 1992).

---

an in-frame stop codon. The two null alleles of TCRBV6S1 contain an arginine replacing the conserved cysteine at position 92 (Lurynik *et al.* 1993; Barron *et al.* 1994). Certain sequences apparently correspond to alleles but since their segregation in families has not been reported these are given tentative allelic designations. Substitutions reported for TCRBV7S2, TCRBV8S2, TCRBV12S2, and TCRBV16 (Plaza *et al.* 1991; Day *et al.* 1992) are silent and thus do not result in protein sequence variation for allelic forms of these genes.

4 Conclusion

Knowledge of the extent of the germline TCR repertoire has been extended by development of an extended map of the human TCRB gene complex. By using pulsed-field gel electrophoresis, the TCRB complex encoded at 7q35 (Isobe *et al.* 1985) was determined to span a distance of ~ 655 kb or ~ 605 kb depending upon the TCR IDRP genotypes of the donor. Initial estimates of the human TCRBV repertoire based on isolation of cDNA clones placed the number of TCRBV genes segments at about 70 (Toyonaga and Mak 1987). In recent studies (Wei *et al.* 1994), the number of TCRBV genes segments was revised to 64 by taking into account sequence information from new TCRBV families, Southern blotting data, and sequences of TCRBV genes obtained from cosmid clones. However, approximately 10% of the known TCRBV gene segments are located outside of the TCRB gene complex. Orphon TCRBV genes encoded on chromosome 9 are not involved in productive rearrangement events with other gene segments within the TCRB gene complex and thus do not contribute to TCR diversity. The orphon status of these genes, which was not readily detected by cDNA cloning or Southern blots, was confirmed by mapping in somatic cell hybrids, by *in situ* hybridization, and by genetic linkage analysis. Genomic sequence analysis would not have revealed that at least three of the orphon genes were pseudogenes because there were no obvious defects in their sequence to preclude expression. When the orphon status of these genes and the number of TCRBV genes within the TCRB gene complex that are pseudogenes is considered the number of functional TCRBV gene segments can be reduced to 53.

Previous comparisons of the organization of the human and murine TCRB loci have led to the proposal that there is extensive synteny in the gene complexes between the two species (Lai *et al.* 1988). Modifications of the map through the inclusion of members of 11 additional TCRBV families and the realignment of some gene segments do not challenge this conclusion. The present data eliminate a number of the discrepancies between the two maps mainly by removal of the orphon gene segments. Present data indicate that a major difference between human and murine TCRB loci is that the human gene complex is larger due to repeated tandem duplication of members of the TCRBV5, TCRBV6, and TCRBV13 families.

The involvement of T lymphocytes in the pathogenesis of a variety of autoimmune diseases has prompted numerous studies to determine the role of TCR genes in disease (Posnett *et al.* 1988; Concannon 1992; Nepom and Concannon 1992; Robinson and Kindt 1992). The majority of these studies have used RFLPs as markers of TCRBV genes. Population-based association studies (Beall *et al.* 1989) and linkage studies in families (Seboun *et al.* 1989*a*) suggest that germline polymorphism of TCRBV gene segments is functionally relevant to disease susceptibility. Knowledge of the genomic organization and polymorphism of human TCRBV gene segments will be necessary in order to further localize TCRBV gene segments that have an impact upon disease susceptibility.

Note added in proof

A report characterizing the TCRBV region IDRP has been published (Zhao *et al.* 1994). The TCRBV region IDRP contains three TCRBV gene segments that are missing from deleted TCRBV haplotypes. These genes are BV7S3, BV9S2 (a pseudogene), and a second identical copy of BV13S2. Estimates for numbers of genes present in the human TCRBV repertoire shown in Table 12.2 were determined for inserted TCRBV haplotypes. The sequence designated 13.9 in Table 12.1 is an allele of BV13S2 present in deleted TCRBV haplotypes.

References

Baccala, R., Kono, D. H., Walker, S., Balderas, R. S., and Theofilopoulos, A. N. (1991). Genomically imposed and somatically modified human thymocyte Vβ gene repertoires. *Proc. Natl. Acad. Sci. USA*, **88**, 2908–12.

Barron, K. S. and Robinson, M. A. (1994). The human T cell receptor variable gene segment, TCRBV6S1 has two null alleles. *Hum. Immunol.*, **40**, 17–19.

Beall, S. S., Concannon, P., Charmley, P., McFarland, H. F., Gatti, R. A., Hood, L. E., *et al.* (1989). The germline repertoire of T cell receptor beta-chain genes in patients with chronic progressive multiple sclerosis. *J. Neuroimmunol.*, **21**, 59–66.

Boitel, B., Ermonval, M., Panina-Bordignon, P., Mariuzza, R. A., Lanzavecchia, A., and Acuto, O. (1992). Preferential Vβ gene usage and lack of junctional sequence conservation among human T cell receptors specific for a tetanus toxin-derived peptide: evidence for a dominant role of a germline-encoded V region in antigen/major histocompatibility complex recognition. *J. Exp. Med.*, **175**, 765–77.

Borden, P., Jaenichen, R., and Zachau, H. G. (1990). Structural features of transposed human V*K* genes and implications for the mechanism of their transpositions. *Nucleic Acids Res.*, **18**, 2101–7.

Charmley, P., Beall, S. S., Concannon, P., Hood, L., and Gatti, R. A. (1991). Further localization of a multiple sclerosis susceptibility gene on chromosome 7q using a new T cell receptor beta-chain DNA polymorphism. *J. Neuroimmunol.*, **32**, 231–40.

Charmley, P., Wang, K., Hood, L., and Nickerson, D. A. (1993*a*). Identification and physical mapping of a polymorphic human T cell receptor V beta gene with a frequent null allele. *J. Exp. Med.*, **177**, 135–43.

Charmley, P., Wei, S., and Concannon, P. (1993*b*). Polymorphisms in the Tcrb-V2 gene segments localize the Tcrb orphon genes to human chromosome 9p21. *Immunogenetics*, **38**, 283–6.

Chothia, C., Boswell, D. R., and Lesk, R. M. (1988). The outline structure of the T-cell alpha beta receptor. *EMBO J.*, **7**, 3745–55.

Concannon, P. (1992). T-cell receptor gene polymorphism: Applications in the study of autoimmune disease. In *Manual of clinical laboratory immunology* (ed. Rose, DeMacario, Fahey, Friedman, and Penn), 4th edn., pp. 885–9.

Concannon, P., Pickering, L. A., Kung, P., and Hood, L. (1986). Diversity and structure of human T-cell receptor beta-chain variable region genes. *Proc. Natl. Acad. Sci. USA*, **83**, 6598–602.

Concannon, P., Gatti, R. A., and Hood, L. E. (1987). Human T cell receptor Vβ gene polymorphism. *J. Exp. Med.*, **165**, 1130–40.

Cornelis, F., Pile, K., Loveridge, J., Moss, P., Harding, R., Julier, C., *et al.* (1993). Systematic study of human alpha beta T cell receptor V segments shows allelic variations resulting in a large number of distinct T cell receptor haplotypes. *Eur. J. Immunol.*, **23**, 1277–83.

Davis, M. M. and Bjorkman, P. J. (1988). T-cell antigen receptor genes and T-cell recognition. *Nature*, **334**, 395–402.

Day, C. E., Zhao, T., and Robinson, M. A. (1992). Silent allelic variants of a T-cell receptor Vβ12 gene are present in diverse human populations. *Hum. Immunol.*, **34**, 196–202.

Diaz, M. O., Ziemin, S., Le Beau, M. M., Pitha, P., Smith, S. D., Chilcote, R. R., *et al.* (1988). Homozygous deletion of the alpha- and beta 1-interferon genes in human leukemia and derived cell lines. *Proc. Natl. Acad. Sci. USA*, **85**, 5259–63.

Duby, A. D. and Seidman, J. G. (1986). Abnormal recombination products result from aberrant DNA rearrangement of the human T-cell antigen receptor β-chain gene. *Proc. Natl. Acad. Sci. USA*, **83**, 4890–4.

Ferradini, L., Roman-Roman, S., Azocar, J., Michalaki, H., Trievel, F., and Hercend, T. (1991). Studies on the human T cell receptor αβ variable region genes. II. Identification of four additional Vβ subfamilies. *Eur. J. Immunol.*, **21**, 935–42.

Gomolka, M., Epplen, C., Buitkamp, J., and Epplen, J. T. (1993). Novel members and germline polymorphisms in the human T-cell receptor Vβ6 family. *Immunogenetics*, **37**, 257–65.

Hansen, T., Ronningen, K. S., Ploski, R., Kimura, A., and Thorsby, E. (1992).

Coding region polymorphisms of human T-cell receptor V beta 6.9 and V beta 21.4. *Scand. J. Immunol.*, **36**, 285–90.

Huber, C., Thiebe, R., Hameister, H., Smola, H., Lötscher, E., and Zachau, H. G. (1990). A human immunoglobulin kappa orphon without sequence defects may be the product of a pericentric inversion. *Nucleic Acids Res.*, **18**, 3475–8.

Hurley, C. K., Steiner, N., Wagner, A., Geiger, M. J. Eckels D. D., and Rosen-Bronson, S. (1993). Nonrandom T cell receptor usage in the allorecognition of HLA-DR1 microvariation. *J Immunol.*, **150**, 1314–24.

Ikuta, K., Ogura, T., Shimizu, A., and Honjo, T. (1985). Low frequency of somatic mutation in beta-chain variable region genes of human T-cell receptors. *Proc. Natl. Acad. Sci. USA*, **82**, 7701–5.

Isobe, M., Erikson, J., Emanuel, B., Nowell, P., and Croce, C. (1985). Location of gene for beta subunit of human T-cell receptor at band 7q35, a region prone to rearrangements in T cells. *Science*, **228**, 580–3.

Kimura, N., Toyonaga, B., Yoshikai, Y., Triebel, F., Debre, P., Minden, M. D., *et al.* (1986). Sequences and diversity of human T-cell receptor β chain variable region genes. *J. Exp. Med.*, **164**, 739–50.

Kimura, N., Toyonaga, B., Yoshikai, Y., Du, R.-P., and Mak, T. W. (1987). Sequences and repertoire of the human T-cell receptor α and β chain variable region genes in thymocytes. *Eur. J. Immunol.*, **17**, 375–83.

Lai, E., Concannon, P., and Hood, L. (1988). Conserved organization of the human and murine T-cell receptor beta-gene families. *Nature*, **331**, 543–6.

Leiden, J. M. and Strominger, J. L. (1986). Generation of diversity of the β chain of the human T-lymphocyte receptor for antigen. *Proc. Natl. Acad. Sci. USA*, **83**, 4456–60.

Li, Y., Szabo, P., Robinson, M. A., Dong, B., and Posnett, D. N. (1990). Allelic variations in the human T cell receptor V beta 6.7 gene products. *J. Exp. Med.*, **171**, 221–30.

Li, Y., Szabo, P., and Posnett, D. N. (1991). The genomic structure of human Vβ 6 T cell antigen receptor genes. *J. Exp. Med.*, **174**, 1537–47.

Luyrink, L., Gabriel, C. A., Thompson, S. D., Grom, A. A., Maksymowych, W. P., Choi, E., *et al.* (1993). Reduced expression of a human Vβ6. 1 T-cell receptor allele. *Proc. Natl. Acad. Sci. USA*, **90**, 4369–73.

Malhotra, U., Spielman, R., and Concannon, P. (1992). Variability in T cell receptor Vβ gene usage in human peripheral blood lymphocytes. Studies of identical twins, siblings, and insulin-dependent diabetes mellitus patients. *J. Immunol.*, **149**, 1802–8.

Matsuda, F., Shin, E. K., Hirabayashi, Y., Nagaoka, H., Yoshida, M. C., Zong, S. Q., *et al.* (1990). Organization of variable region segments of the human immunoglobulin heavy chain: duplication of the D5 cluster within the locus and interchromosomal translocation of variable region segments. *EMBO J.*, **9**, 2501–6.

Nepom, G. T. and Concannon, P. (1992). Molecular genetics of autoimmunity. In *The autoimmune diseases* (ed. N. R. Rose and I. R. Mackay), pp. 127–52. Academic Press, Inc., NY.

Plaza, A., Kono, D. H., and Theofilopoulos, A. N. (1991). New human Vβ genes and polymorphic variants. *J. Immunol.*, **147**, 4360–5.

Posnett, D. N., Gottlieb, A., Bussel, J. B., Friedman, S. M., Chiorazzi, N., Li, Y., *et al.* (1988). T cell antigen receptors in autoimmunity. *J. Immunol.*, **141**, 1963–9.

Robinson, M. A. (1989) Allelic sequence variations in the hypervariable region of a T-cell receptor β chain: correlation with restriction fragment length polymorphism in human families and populations. *Proc. Natl. Acad. Sci. USA*, **86**, 9422–6.

Robinson, M. A. (1991). The human T cell receptor beta-chain gene complex contains at least 57 variable gene segments. Identification of six V beta genes in four new gene families. *J. Immunol.*, **146**, 4392–7.

Robinson, M. A. (1992). Usage of human T-cell receptor Vβ, Jβ, Cβ, and Vα gene segments is not proportional to gene number. *Hum. Immunol.*, **35**, 60–7.

Robinson, M. A. and Kindt, T. J. (1985). Segregation of polymorphic T-cell receptor genes in human families. *Proc. Natl. Acad. Sci. USA*, **82**, 3804–8.

Robinson, M. A. and Kindt, T. J. (1992). Linkage between T cell receptor genes and susceptibility to multiple sclerosis: a complex issue. *Reg. Immunol.*, **4**, 274–83.

Robinson, M. A., Mitchell, M. P., Wei, S., Day, C. E., Zhao, T. M., and Concannon, P. (1993). Organization of human T-cell receptor beta-chain genes: clusters of V beta genes are present on chromosomes 7 and 9. *Proc. Natl. Acad. Sci. USA*, **90**, 2433–7.

Seboun, E., Robinson, M. A., Doolittle, T. H., Ciulla, T. A., Kindt, T. J., and Hauser, S. L. (1989*a*). A susceptibility locus for multiple sclerosis is linked to the T cell receptor beta chain complex. *Cell*, **57**, 1095–100.

Seboun, E., Robinson, M. A., Kindt, T. J., and Hauser, S. L. (1989*b*). Insertion/deletion-related polymorphisms in the human T cell receptor beta gene complex. *J. Exp. Med.*, **170**, 1263–70.

Siu, G., Strauss, E. C., Lai, E., and Hood, L. E. (1986). Analysis of a human Vβ gene subfamily. *J. Exp. Med.*, **164**, 1600–14.

Smith, L. R., Plaza, A., Singer, P. A., and Theofilopoulos, A. N. (1990). Coding sequence polymorphisms among V beta T cell receptor genes. *J. Immunol.*, **144**, 3234–7.

Tillinghast, J. P., Behlke, M. A., and Loh, D. Y. (1986). Structure and diversity of the human T-cell receptor β chain variable region genes. *Science*, **233**, 879–83.

Toyonaga, B. and Mak, T. W. (1987). Genes of the T-cell antigen receptor in normal and malignant T cells. *Annu. Rev. Immunol.*, **5**, 585–620.

Tunnacliffe, A., Kefford, R., Milstein, C., Forster, A., and Rabbitts, T. H. (1985). Sequence and evolution of the human T-cell antigen receptor β-chain genes. *Proc. Natl. Acad. Sci. USA*, **82**, 5068–72.

Wei, S., Charmley, P., Robinson, M. A., and Concannon, P. (1994). The extent of the human germline T cell receptor Vβ gene segment repertoire determined by genomic cloning. *Immunogenetics*, **40**, 27–36.

Wilson, R. K., Lai, E., Concannon, P., Barth, R. K., and Hood, L. E. (1988). Structure, organization and polymorphism of murine and human T-cell receptor alpha and beta chain gene families. *Immunol. Rev.*, **101**, 149–204.

Wilson, R. K., Lai, E., Kim, L. D., and Hood, L. E. (1990). Sequence and expression of a novel human T-cell receptor β-chain variable gene segment subfamily. *Immunogenetics*, **32**, 406–12.

Yanagi, Y., Yoshikai, Y., Leggett, K., Clark, S. P., Aleksander, I., and Mak, T. W. (1984). A human T cell specific cDNA clone encodes a protein having extensive homology to immunoglobulin chains. *Nature*, **308**, 145–9.

Zhao, T. M., Whitaker, S., and Robinson, M. A. (1994). Insertion/deletion of variable gene segments in human TCRB haplotypes, *J. Exp. Med.*, **180**, 1405–14.

13 Genomic organization of T cell receptor genes in the mouse

NICHOLAS R. J. GASCOIGNE

1 Introduction

T cells, like B cells, have a clonally variable receptor through which they can be activated by antigens. The T cell receptor (TCR) comes in two major forms, which define the first branches of the T cell family tree. Thus T cells are either αβ cells or γδ cells. Those bearing a TCR consisting of α and β chains are in the majority among mature T cells, and their functions in antigen-specific responses are well understood. The γδ T cells have a much more obscure role. They mainly inhabit epithelial tissues and their TCR sequences suggest that they frequently have tissue-specific recognition specificity.

The specificity of each of the TCRs is determined by the rearrangement of genetic elements during T cell ontogeny. For example, to produce a functional TCR β chain (TCRB) gene, three elements encoded in the germline DNA must be rearranged. These are the variable (TCRBV),[1] diversity (TCRBD), and joining (TCRBJ) gene elements that become fused together to make a V region exon. The D element is found only in the TCRB and δ chain (TCRD) genes, not in the α (TCRA) or γ chain (TCRG) genes (Davis and Bjorkman 1988).

The TCR genes are arranged in three loci on separate chromosomes. The TCRB locus is on chromosome 6 (Lee *et al.* 1984), TCRG is on chromosome 13 (Kranz *et al.* 1985), and the TCRA and TCRD loci are both encoded on chromosome 14 (Kranz *et al.* 1985; Chien *et al.* 1987*a*). The organization of the α and δ loci is particularly interesting because the δ locus is within the α locus. Therefore, a chromosome that has undergone α chain rearrangement will have deleted the TCRD genes. This review deals with the structure and organization of the TCR genes, and is intended to provide background for the other chapters in this section dealing with recombination, allelic exclusion, and polymorphism.

2 TCRB locus

The β chain was the first of the TCR genes to be cloned (Hedrick *et al.* 1984*a,b*), and the first for which the genomic organization was determined.

[1] Nomenclature follows WHO-IUIS nomenclature sub-committee on TCR designation (1993).

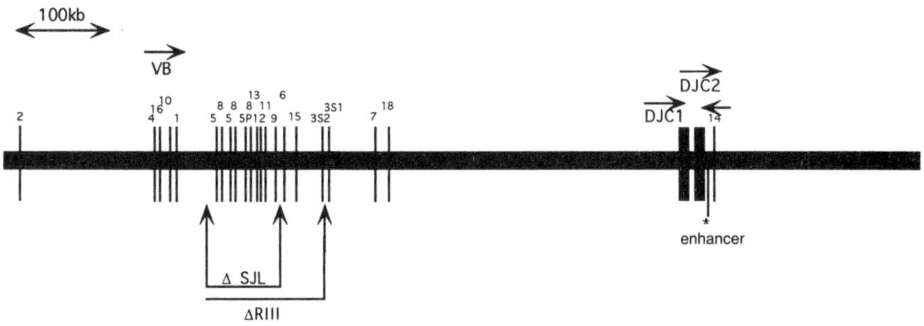

Mouse TCRB locus

Fig. 13.1 Map of the mouse TCRB locus from the 1 Vβ[b] haplotype. The transcription orientation of the gene segments is shown by *arrows* and the V regions are numbered. ΔSJL and ΔRIII refer to the deletions in these two mouse strains. SJL is the Vβ[a] haplotype and RIII is the TCRBV*3 (Vβ[c]) haplotype. This information is mainly from Chou *et al.* (1987*b*), Lindsten *et al.* (1987*b*), and Lee and Davis (1988). See text for other references.

The overall organization is shown in Figure 13.1. The V genes are almost all located upstream of two clusters, each containing a D segment, six J segments, and the C region gene. Rearrangement of any V element can be to the 5′ or the 3′ DJC cluster.

2.1 TCRBV elements

There are approximately 25 TCRBV genes in the mouse. These extend over ~ 450 kb of DNA located ~ 300 kb upstream of the two clusters of D, J, and C elements (Chou *et al.* 1987*b*; Lindsten *et al.* 1987*b*). The one exception is V14, which is downstream of the TCRBC2 gene, and in the opposite transcriptional orientation (Malissen *et al.* 1986). The order of the main array of TCRBV elements has been determined by field-inversion gel electrophoresis, cosmid mapping, and studies of TCRBV deletion in T cell lines (Chou *et al.* 1987*a,b*; Lai *et al.* 1987; Lindsten *et al.* 1987*b*; Lee and Davis 1988). The majority of TCRBV elements are single copy genes that are not closely related to one another (Patten *et al.* 1984; Barth *et al.* 1985; Behlke *et al.* 1985). The main exceptions are the V5 and V8 families. These each have three members and are tandemly arranged, probably resulting from gene duplication events (Chou *et al.* 1987*a*) (Figure 13.1). The downstream V5 element is a pseudogene (Chou *et al.* 1987*a*). The closeness of these genes appears to allow unusual RNA splicing. The V5S1 signal peptide exon has been observed to splice to the V8S2 V exon, forming a functional gene product (Chou *et al.* 1987*a*). It also splices to its expected V5S1 V exon. The

V3 gene family contains two members, V3S1 (Vβ3 or 3.1 in the old nomenclature) and V3S2 (previously Vβ17a or Vβ3.2) (Kappler *et al.* 1987).

There are four known haplotypes of the TCRB locus. $V\beta^b$ is the haplotype of the vast majority of laboratory mouse strains. $V\beta^a$ is present in the strains SJL, SWR, C57/L, and C57/BR, and differs from $V\beta^b$ mainly in having a deletion of ten V elements (Figure 13.1). The following V regions are deleted: 5, 8, 9, 11, 12, and 13 (Behlke *et al.* 1986; Lindsten *et al.* 1987*b*; Chou *et al.* 1987*b*; Lee and Davis 1988). There are also point mutations between $V\beta^b$ and $V\beta^a$ regions, including a threonine to alanine change in V15 (Behlke *et al.* 1986; Gascoigne *et al.* 1986), and a valine to phenylalanine change in V3 (Smith *et al.* 1990). This latter change has a marked effect on T cell recognition of antigen, so that the repertoire of T cells from $V\beta^a$ strains of mice responding to pigeon cytochrome *c* is very different from the repertoire in $V\beta^a$ strains (Gahm *et al.* 1991). Another point mutation with a large effect is in V3S2. The $V\beta^a$ V3S2 gene is active, whereas the $V\beta^a$ V3S2 gene is a pseudogene due to a chain termination codon (Wade *et al.* 1988). The $V\beta^c$ haplotype is known only from the strain RIIIS/J. It has a similar but larger deletion to $V\beta^a$. In addition to the V elements lost in $V\beta^a$, it has also deleted V6, 15, and 3S2 (Haqqi *et al.* 1989) (Figure 13.1). A fourth TCRB haplotype is present in NZW (Kotzin *et al.* 1985). This strain has a deletion in the DJC clusters (see below), having lost TCRBC1, TCRBD2, and TCRBJ2 genes (Kotzin *et al.* 1985).

2.2 TCRBD, J, and C genes

Figure 12.2 shows the organization of the D, J, and C loci. They form two very similar clusters. The two D elements and the twelve J regions are all unique sequences, and contribute significantly to the diversity of the T cell repertoire since they encode most of the third complementarity determining region (CDR3) of the β chain protein (Davis and Bjorkman 1988). The two C region genes are very strongly conserved, particularly in the exon encoding the extracellular region of the protein (Gascoigne *et al.* 1984). Each of the DJC clusters is arranged with a D element ~ 600 bp upstream of the most 5′ J region (J1S1 or J2S1) (Kavaler *et al.* 1984; Siu *et al.* 1984). The six functional J elements of each cluster are located within 1.2 kb and 1.8 kb (Gascoigne *et al.* 1984; Chien *et al.* 1984; Malissen *et al.* 1984). Both J region clusters include one clear pseudo-J element. In the 5′ (J1) cluster, this is downstream of J1S6 (Gascoigne *et al.* 1984; Malissen *et al.* 1984), and in the 3′ (J2) cluster it is between S5 and S6 (Chien *et al.* 1984). About 2.5 kb downstream of the most 3′ J region (J1S6 or J2S6) is the first exon of the TCRBC gene (Gascoigne *et al.* 1984; Malissen *et al.* 1984). The V14 gene is ~ 7 kb downstream of the TCRBC2 gene, in the opposite transcriptional orientation to the rest of the TCRB locus gene elements (Malissen *et al.* 1986). The TCRB enhancer has been identified within a 550 bp fragment located ~ 5 kb downstream of TCRBC2 (Krimpenfort *et al.* 1988).

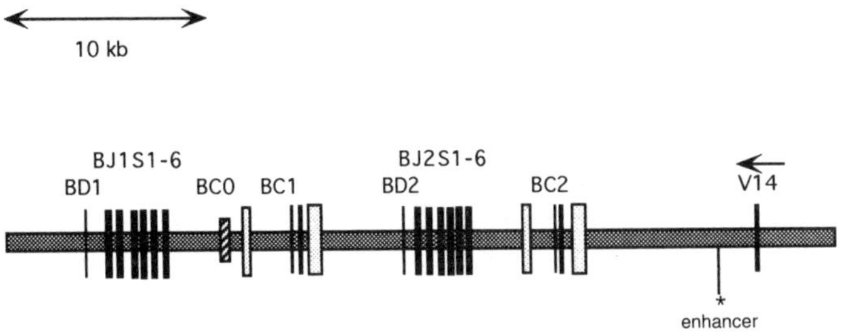

TCRB DJC loci

Fig. 13.2 The 3′ end of the TCRB locus. This shows the two TCRBDJC clusters, each arranged with one D, followed by six J regions, and the four C region exons. The transcription orientation of all these elements is left to right. The TCRBV14 element downstream is in the opposite orientation. The alternate, but probably useless, C0 exon, sometimes spliced between J and C, is also shown. Derived mainly from Chien *et al.* (1984), Gascoigne *et al.* (1984), Malissen *et al.* (1984, 1986), Kavaler *et al.* (1984), Siu *et al.* (1984), and Behlke and Loh (1986).

The TCRBC genes are both arranged in four exons, closely corresponding to the domains of the protein (Gascoigne *et al.* 1984; Malissen *et al.* 1984). The first exon encodes the extracellular globular part of the constant region of the β chain. The second exon is the smallest, encoding only six amino acids, including the cysteine residue for the interchain disulfide bond to the α chain. The third exon encodes mainly the transmembrane segment of the protein, leaving the fourth exon for the positive charged six residue cytoplasmic tail plus 3′ untranslated region. There are no sequence differences between the proteins encoded by exons 1 and 2 of the two C genes, and there is no evidence for any functional differences between the proteins encoded by the 5′ or 3′ DJC clusters.

An extra exon (C0) of the TCRB chain has been found in some β chain cDNA clones. It is spliced in-frame between the J and C elements, and would result in a protein 24 amino acids longer than the normal mature protein. The exon is located less than 1 kb upstream of TCRBC1 (Behlke and Loh 1986). It is used in ~ 1% of β chain mRNA in peripheral T cells, and in about 5–10% of β chain mRNA from adult or fetal thymocytes (Dent *et al.* 1989). There is no evidence for any functional significance of these messages, nor for a protein product. In NZW mice, where TCRBC1 is deleted, ~ 18% of cDNA clones from a thymus library contained this exon (Behlke and Loh 1986). The corresponding exon in rat has mutations in the splice sites stopping it from being used as part of the β chain, and no evidence for a homologous region was found in human DNA (Dent *et al.* 1989).

3 TCRA/TCRD locus

The genes encoding the α and δ chains of TCR are remarkable for being intermingled (Chien *et al.* 1987*a*). As can be seen from Figure 13.3A, the TCRDD, J, and C elements are in between the V and J segments of the TCRA locus. Early in T cell ontogeny, rearrangements within the TCRA locus were noted that did not correspond to expression of α chain (Lindsten *et al.* 1987*a*). These were due to rearrangements of the TCRD locus, which, like the γ chain, is rearranged and expressed earlier than the TCRB and TCRA genes (Chien *et al.* 1987*b*; Raulet 1989; Allison and Havran 1991).

3.1 TCRA and TCRD V elements

The V elements used by the α and δ chains include some that have only been found in one chain or the other, but numerous V genes can be used by both α and δ chains (Elliott *et al.* 1988; Bluestone *et al.* 1988; Jameson *et al.* 1991; Seto *et al.* 1994).[2] The TCRDV105 element is like BV14 (above) in that it is 2.5 kb 3′ of the C gene, in the opposite transcription orientation (Iwashima *et al.* 1988). There are probably ~ 100 V gene elements in the TCRA/D locus (Arden *et al.* 1985; Becker *et al.* 1985; Davis and Bjorkman 1988), encompassing more than 370 kb of DNA (Jouvin-Marche *et al.* 1990). The majority of these V genes fall into families of several members (Arden *et al.* 1985; Becker *et al.* 1985; Singer *et al.* 1988; Klotz *et al.* 1989; Jouvin-Marche *et al.* 1989). Southern blot experiments demonstrated that most members of TCRAV families are not closely linked. Instead, they are interspersed with members of other families (Singer *et al.* 1988; Klotz *et al.* 1989; Jouvin-Marche *et al.* 1989). It is now clear from V region deletion and cosmid mapping that the families have arisen by several rounds of gene duplication. The linkage map of the $V\alpha^a$ locus from BALB/c is shown in Figure 13.4 (Jouvin-Marche *et al.* 1990). In the genome of this mouse strain, there are five large clusters of V genes, each of which contains a member of eight TCRAV families (Jouvin-Marche *et al.* 1990). These clusters are intermingled with other smaller clusters containing members of two or three different families. The V gene clusters of BALB/c fall into three types. Cluster I is repeated at least four times and contains one member each of TCRAV1, 2, 3, 4, 7, 8, 10, and 11. A likely fifth repetition of cluster I is found at the 3′ end of the locus, but this does not seem to have all of the V elements present in the other clusters (Jouvin-Marche *et al.* 1990). For the case where it was determined, the transcriptional orientation of the V segments in each cluster is the same (Jouvin-Marche *et al.* 1990). The organization of the TCRAV2 family,

[2] α chain and δ chain V regions are denoted by the same numbering system, e.g. Vα11 would be TCRAV11 or TCRDV11. The V regions used exclusively in δ chains are numbered from 101, e.g. Vδ5 becomes TCRDV105.

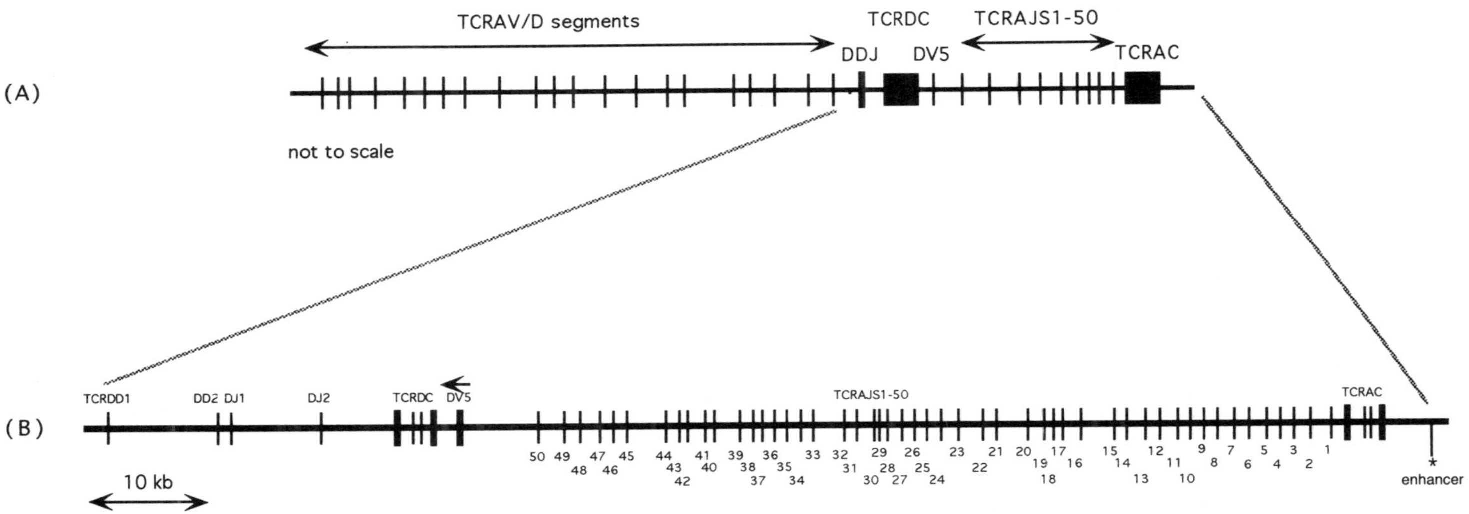

Mouse TCRA/D locus

Fig. 13.3 Map of the mouse TCRA/TCRD locus. (A) is a schematic diagram showing the relationship of the α and δ chain V elements to the D, J, and C of the TCRD gene, and to the TCRAJ and TCRAC genes. See text for references. (B) is a scale map of the gene segment containing all of the 50 TCRAJ regions, the TCRD elements upstream, and the TCRAC elements downstream. This derives from Iwashima *et al.* (1988), Wilson *et al.* (1992), and Koop *et al.* (1992).

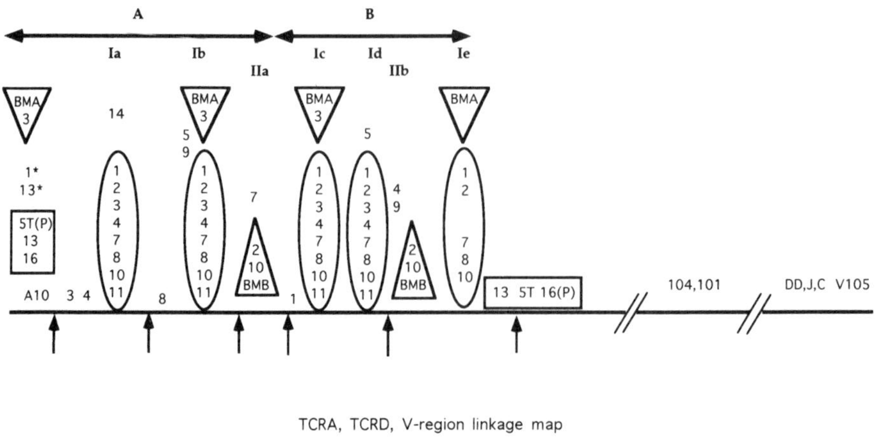

Fig. 13.4 Schematic map of the BALB/c TCRAV elements (from Jouvin-Marche *et al.* 1990; Marche *et al.* 1993, and P. Marche, personal communication). There are four types of clusters of contiguous V gene segments, based on repetitive patterns in the V locus. Where the relative order of V segments has not been determined, they are shown in a vertical line. Where there are several members of the same family in the same region, it is denoted by an asterisk. Type I clusters are shown as elongated ovals, type II clusters are shown as triangles, type III clusters as inverted triangles, and type IV clusters as rectangles. The relative order of V genes in type I clusters (ovals) has not been fully determined (but see text). The arrows show the points of rearrangement in a series of T cell lines that were used in these mapping studies (Jouvin-Marche *et al.* 1990; Marche *et al.* 1993). Note: TCRDV segments that have only been found as part of δ chains are denoted as V101, 102, etc. (previously Vδ1, Vδ2, etc.). Vδ3 is identical to Vα6 and is therefore denoted as V6.

comprising nine members (TCRAV2S1–9), has been analysed by cosmid mapping. In each case, V2 is downstream of a TCRAV10 element and, in at least five cases, upstream of a TCRAV7 element (Jouvin-Marche *et al.* 1990). The V7 elements are present in type I clusters, but not in type II clusters. The longest of the cosmid maps representing a type I cluster contains gene segments in the order V1, 8, 10, 2, 7, where each V region is separated by ~ 4–9 kb. A flanking sequence probe for the cluster II V2 genes distinguishes them from V2 elements in cluster I. Thus V2 and V10 are found in both cluster types (Jouvin-Marche *et al.* 1990). Cluster II is repeated twice and includes one each of TCRAV2, 10, and V (BM.B) segments. Cluster III includes a V3 and a V (BM.A) element and is repeated at least three times.

The few TCRAV or TCRDV genes that have one or two members are found at the extreme 5′ or 3′ flanks of the V locus, suggesting that the duplication events occurred within the V region locus, not affecting the outlying V genes (Jouvin-Marche *et al.* 1990). The 3′ end of the TCRAV region of BALB/c has recently been investigated using cosmid mapping and sequence analysis (Seto *et al.* 1994). These data show TCRAV2S6, V8S4,

V5S3, V10S6, V16S1(P), V6 (same as Vδ3), and V102 linked on three cosmids spanning ~ 80 kb. These are positioned an undetermined distance upstream of TCRDV104 and V101, upstream of the δ chain D, J, and C segments (Seto *et al.* 1994). Comparison of these data with the mapping studies of Jouvin-Marche *et al.* (1990) suggests that these cosmids may overlap with the most 3′ of the type I clusters, although the precise order of the V elements is ambiguous.

There is significant polymorphism in the TCRAV locus between mouse strains. Restriction fragment length polymorphism analysis has identified five haplotypes (Singer *et al.* 1988; Klotz *et al.* 1989; Jouvin-Marche *et al.* 1989). These are identified as $V\alpha^{a-e}$. The $V\alpha^{a}$ haplotype is found in A/J, AKR, BALB, C3H, and CBA, amongst common laboratory strains. The C57-related strains have $V\alpha^{b\prime}$ NZB, SJL, SWR, PL, and NOD have $V\alpha^{c}$. $V\alpha^{d}$ is found in the DBA strains and in NZW. The $V\alpha^{e}$ haplotype is known from RIII/Rd, and appears to have been formed from a recombination between $V\alpha^{a}$ and $V\alpha^{b}$ (Jouvin-Marche *et al.* 1989). The different TCRAV haplotypes have been subjected to varying numbers of duplication events, or other genetic transpositions, resulting in different numbers of V gene clusters. The related genes in the different clusters have undergone point mutations to produce the large degree of V region polymorphism found in sequence studies.

Polymorphism between members of the same TCRAV family has been studied for the V2, V8, and V11 families. In the two latter families, much of the polymorphism is focused on the CDR1 or CDR2 regions (Chou *et al.* 1986; Jameson *et al.* 1991). This is less clear in the V2 family, but out of the ten residues that are variable between at least two of seven members, seven are in CDR1 or 2 (Smith *et al.* 1992; Tang *et al.* 1994).

3.2 TCRAJ and C genes

There are 50 α chain J region gene segments in the mouse. This has recently been determined by the complete sequencing of ~ 95 kb of DNA representing the region from TCRDJS1, through TCRDC, TCRDV105, the TCRAJ segments, TCRAC, and the α chain enhancer (Wilson *et al.* 1992; Koop *et al.* 1992) (see Figure 13.3). Knowledge of the genomic order of the J regions has allowed Jα numbering to be rationalized (Koop *et al.* 1992). The J regions are found within a 60 kb region, the most 3′ segment (TCRAJ1) being ~ 1 kb 5′ of the first Cα exon (Wilson *et al.* 1992; Koop *et al.* 1992). The TCRAC gene is composed of four exons (Hayday *et al.* 1985*a*; Winoto *et al.* 1985). As with the TCRBC gene, exon one encodes the extracellular, immunoglobulin-like domain and exon two the 'hinge' region with the cysteine residue for αβ chain pairing. Unlike the gene for the β chain, exon three encodes both the transmembrane region and the cytoplasmic tail, and a fourth exon encodes only 3′ untranslated region with the polyadenylation signals. The α chain enhancer has been localized about 3 kb 3′ of the TCRAC gene (Winoto and

Baltimore 1989*b*). In addition, two silencer elements that switch off the α chain enhancer in γδ expressing cells, have been mapped between TCRAC and the enhancer (Winoto and Baltimore 1989*a*).

3.3 TCRDD, J, and C genes

The organization of these gene segments is as follows: D1, D2, J1, J2, C (followed by V5) (Iwashima *et al.* 1988; Koop *et al.* 1992) (see Figure 13.3). The two D elements, which can be used individually or together, and the N regions between V–D, D–D, and D–J segments, give the δ chain an enormous potential CDR3 repertoire (Elliott *et al.* 1988; Davis and Bjorkman 1988). The TCRDC exons are arranged in the same way as the α chain, with three coding exons followed by one non-coding exon (Iwashima *et al.* 1988). The δ chain enhancer has been located in the mouse locus 4.2 kb 3′ of TCRDC (~ 1.4 kb 3′ of TCRDV105) on the basis of sequence comparison to the human δ chain enhancer (Koop *et al.* 1992).

4 TCRG locus

This is now the most completely mapped of any of the TCR loci (Hayday *et al.* 1985*b*; Garman *et al.* 1986; Iwamoto *et al.* 1986; Traunecker *et al.* 1986; Pelkonen *et al.* 1987; Woolf *et al.* 1988; Vernooij *et al.* 1993). The whole TCRG locus from BALB/c is shown in Figure 13.5. It encompasses four VJC clusters arranged over 205 kb (Raulet 1989; Vernooij *et al.* 1993). This arrangement is unique amongst the TCR genes, and is similar to the arrangement of the immunoglobulin λ chain genes (Eisen and Reilly 1985). Rearrangement of the γ-chain V genes is mostly limited to the J regions within the same cluster. The TCRG1 cluster contains four V segments TCRGV2, 3, 4, and 5, ~ 17 kb upstream of TCRGJ1 and C1. These V genes are little related to each other or to the V genes of the other clusters. This provides the TCRG1 cluster with the highest potential diversity of any of the γ genes. However, V3 and V4 are used in dendritic epithelial γδ cells of skin (V3), tongue, and female reproductive tract (V4), and they express no diversity in junctional (CDR3) sequences. V2 and V5 are present in lung γδ cells (V2, plus V4) and intestinal intraepithelial γδ cells (V5), and these γ chain sequences are very diverse (Allison and Havran 1991). The enhancer element is 2.5 kb downstream of the C region gene (Vernooij *et al.* 1993).

The TCRG3 cluster contains one V, one J, and one C region. Both TCRGC3(P) and J3(P) are pseudogenes (hence 'P'), and this region is deleted from some strains of mice (Hayday *et al.* 1985*b*; Iwamoto *et al.* 1986; Traunecker *et al.* 1986). There is a probable enhancer element 3.5 kb 3′ of the C gene (Vernooij *et al.* 1993). The TCRG2 cluster is in the opposite transcriptional orientation from the rest of the TCRG locus. It is arranged head to head with the TCRG4 cluster, so that V1S2 is only ~ 3.5 kb 5′ of

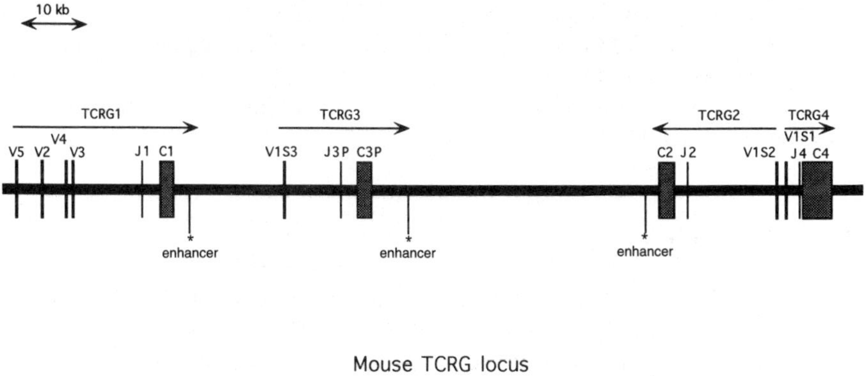

Fig. 13.5 Map of the TCRG locus showing the four VJC gene clusters with their relative orientations. P refers to pseudogene segments. Adapted from Vernooij *et al.* (1993), from data in numerous publications (see text).

V1S1 (in the TCRG4 cluster) (Hayday *et al.* 1985*b*). Its probable enhancer is located ~ 4 kb downstream of the C exons (Vernooij *et al.* 1993). TCRG4 is the smallest of the γ chain clusters, being only about 13 kb long (Hayday *et al.* 1985*b*; Traunecker *et al.* 1986; Vernooij *et al.* 1993). The distance between V1S1 and J4 is only ~ 2 kb. TCRGV5 has been found rearranged to this J region in one instance (Pelkonen *et al.* 1987). No enhancer has been found in this region, either functionally, by cross-hybridization, or by sequence searches (Vernooij *et al.* 1993).

The TCRG3, 2, and 4 clusters have very closely related sequences in V region genes (V1S1-3). Of the J regions, J1 and J2 are identical in amino acid sequence. The only other functional J region is J4, and this differs significantly from the others, particularly in the N-terminal end which will form part of CDR3 (Hayday *et al.* 1985*b*; Iwamoto *et al.* 1986; Traunecker *et al.* 1986).

4.1 TCRGC exon structure

The sequences of TCRGC1, 2, and 3(P) are very close. The major difference between C1 and C2 is in a five amino acid insertion in C1 relative to C2. The TCRGC1, C2, and C3(P) genes are comprised of three exons; the extracellular domain, followed by a hinge exon (as in all the other TCR C region genes). The third exon encodes the transmembrane region and cytoplasmic tail as well as the 3′ untranslated regions (Hayday *et al.* 1985*b*). The insertion in C1 is just upstream of the cysteine residue in the hinge. The C4 gene is significantly different from the others in sequence, having only 66% identity at the amino acid level. It has a longer hinge region than C1 or C2: 23 amino acids longer than C2. As with the insertion in C1, this insertion is in the region immediately N-terminal to the hinge cysteine residue (Iwamoto *et al.* 1986). This longer hinge region is encoded by two exons, thus there are a total of four exons for TCRGC4 (Vernooij *et al.* 1993). There are therefore significant differences in the three γ-chains in the length of the segment connecting the variable domain to the transmembrane region. This type of polymorphism has also been found in the human γ chains, but its functional significance is uncertain (Raulet 1989).

5 Conclusion

Each of the TCR genes has structural features promoting or constraining the variability of the protein products, or regulating the temporal expression of the genes. The organization of the genes has given us numerous clues as to the functional regions of the TCR proteins and to the generation of T cell diversity. The enormous mapping and sequencing projects that have recently been published, and those that will doubtless soon follow, will close out this chapter in the understanding of the TCR.

Acknowledgements

Thanks to Dr Bee-Cheng Sim for critically reviewing the manuscript. This work was supported by NIH grants GM46134 and GM48002. N.R.J.G. is a Scholar of the Leukemia Society of America. This is manuscript 8599-IMM from The Scripps Research Institute.

References

Allison, J. P. and Havran, W. L. (1991). The immunobiology of T cells with invariant γδ antigen receptors. *Annu. Rev. Immunol.*, **9**, 679–705.

Arden, B., Klotz, J. L., Siu, G., and Hood, L. E. (1985). Diversity and structure of genes of the α family of mouse T-cell antigen receptor. *Nature*, **316**, 783–7.

Barth, R. K., Kim, B. S., Lan, N. C., Hunkapiller, T., Sobieck, N., Winoto, A., *et al.* (1985). The murine T-cell receptor uses a limited repertoire of expressed Vβ gene segments. *Nature*, **316**, 517–23.

Becker, D. M., Patten, P., Chien, Y., Yokota, T., Eshhar, Z., Giedlin, M., *et al.* (1985). Variability and repertoire size of T-cell receptor Vα gene segments. *Nature*, **317**, 430–4.

Behlke, M. A. and Loh, D. Y. (1986). Alternative splicing of T cell receptor beta chain transcripts. *Nature*, **322**, 379–82.

Behlke, M. A., Spinella, D. G., Chou, H. S., Sha, W., Hartl, D. L., and Loh, D. Y. (1985). T-cell receptor β-chain expression: dependence on relatively few variable region genes. *Science*, **229**, 566–70.

Behlke, M. A., Chou, H. S., Huppi, K., and Loh, D. Y. (1986). Murine T-cell receptor mutants with deletions of β-chain variable region genes. *Proc. Natl. Acad. Sci. USA*, **83**, 767–71.

Bluestone, J. A., Cron, R. Q., Cotterman, M., Houlden, B. A., and Matis, L. A.

(1988). Structure and specificity of T cell receptor γδ on major histocompatibility complex antigen-specific CD3$^+$, CD4$^-$, CD8$^-$ T lymphocytes. *J. Exp. Med.*, **168**, 1899–916.

Chien, Y., Gascoigne, N. R. J., Kavaler, J., Lee, N. E., and Davis, M. M. (1984). Somatic recombination in a murine T-cell receptor gene. *Nature*, **309**, 322–6.

Chien, Y., Iwashima, M., Kaplan, K. B., Elliott, J. F., and Davis, M. M. (1987*a*). A new T-cell receptor gene located within the alpha locus and expressed early in T-cell differentiation. *Nature*, **327**, 677–82.

Chien, Y., Iwashima, M., Wettstein, D. A., Kaplan, K. B., Elliott, J. F., Born, W., *et al.* (1987*b*). T-cell receptor δ gene rearrangements in early thymocytes. *Nature*, **330**, 722–7.

Chou, H. S., Behlke, M. A., Godambe, S. A., Russell, J. H., Brooks, C. G., and Loh, D. Y. (1986). T cell receptor genes in an alloreactive CTL clone: implications for rearrangement and germline diversity of variable gene segments. *EMBO J.*, **5**, 2149–55.

Chou, H. S., Anderson, S. J., Louie, M. C., Godambe, S. A., Pozzi, M. R., Behlke, M. A., *et al.* (1987*a*). Tandem linkage and unusual RNA splicing of the T-cell receptor β-chain variable-region genes. *Proc. Natl. Acad. Sci. USA*, **84**, 1992–6.

Chou, H. S., Nelson, C. A., Godambe, S. A., Chaplin, D. D., and Loh, D. Y. (1987*b*). Germline organization of the murine T cell receptor β-chain genes. *Science*, **238**, 545–8.

Davis, M. M. and Bjorkman, P. J. (1988). T-cell antigen receptor genes and T-cell recognition. *Nature*, **334**, 395–402.

Dent, A. L., Fink, P. J., and Hedrick, S. M. (1989). Characterization of an alternative exon of the murine T cell receptor β-chain. *J. Immunol.*, **143**, 322–8.

Eisen, H. N. and Reilly, E. B. (1985). Lambda chains and genes in inbred mice. *Annu. Rev. Immunol.*, **3**, 337–65.

Elliott, J. F., Rock, E. P., Patten, P. A., Davis, M. M., and Chien, Y. (1988). The adult T-cell receptor α-chain is diverse and distinct from that of fetal thymocytes. *Nature*, **331**, 627–31.

Gahm, S.-J., Fowlkes, B. J., Jameson, S. C., Gascoigne, N. R. J., Cotterman, M. M., Kanagawa, O., *et al.* (1991). Profound alteration in an αβ T-cell antigen receptor repertoire due to polymorphism in the first complementarity-determining region of the β chain. *Proc. Natl. Acad. Sci. USA*, **88**, 10267–71.

Garman, R. D., Doherty, P. J., and Raulet, D. H. (1986). Diversity, rearrangement and expression of murine T cell gamma genes. *Cell*, **45**, 733–42.

Gascoigne, N. R. J., Chien, Y., Becker, D. M., Kavaler, J., and Davis, M. M. (1984). Genomic organization and sequence of T-cell receptor β-chain constant- and joining-region genes. *Nature*, **310**, 387–91.

Gascoigne, N. R. J., Waters, S., Elliott, J. F., Victor-Kobrin, C., Goodnow, C., Davis, M. M., *et al.* (1986). Expression of T cell receptor genes in an antigen-specific hybridoma and radiation-induced variants. *J. Exp. Med.*, **164**, 113–30.

Haqqi, T. M., Banerjee, S., Anderson, G. D., and David, C. S. (1989). RIII S/J (H-2^r); an inbred mouse strain with a massive deletion of T cell receptor Vβ genes. *J. Exp. Med.*, **169**, 1903–9.

Hayday, A. C., Diamond, D. J., Tanigawa, G., Heilig, J. S., Folsom, V., Saito, H., *et al.* (1985*a*). Unusual organization and diversity of T-cell receptor α-chain genes. *Nature*, **316**, 828–32.

Hayday, A. C., Saito, H., Gillies, S. D., Kranz, D. M., Tanigawa, G., Eisen, H. N., *et al.* (1985*b*). Structure, organization, and somatic rearrangement of T cell gamma genes. *Cell*, **40**, 259–69.

Hedrick, S. M., Cohen, D. I., Neilsen, E. A., and Davis, M. M. (1984*a*). Isolation of cDNA clones encoding T cell-specific membrane-associated proteins. *Nature*, **308**, 149–53.

Hedrick, S. M., Neilsen, E. A., Kavaler, J., Cohen, D. I., and Davis, M. M. (1984*b*). Sequence relationships between putative T-cell receptor polypeptides and immunoglobulins. *Nature*, **308**, 153–8.

Iwamoto, A., Rupp, F., Ohashi, P. S., Walker, C. L., Pircher, H., Joho, R., *et al.* (1986). T cell-specific gamma genes in C57BL/10 mice. Sequence and expression of new constant and variable regions. *J. Exp. Med.*, **163**, 1203–12.

Iwashima, M., Green, A., Davis, M. M., and Chien, Y. (1988). Variable region (Vδ) gene segment most frequently utilized in adult thymocytes is 3′ of the constant (Cδ) region. *Proc. Natl. Acad. Sci. USA*, **85**, 8161–5.

Jameson, S. C., Nakajima, P. B., Brooks, J. L., Heath, W., Kanagawa, O., and Gascoigne, N. R. J. (1991). The T cell receptor Vα11 gene family. Analysis of allelic sequence polymorphism and demonstration of Jα-region dependent recognition by allele-specific antibodies. *J. Immunol.*, **147**, 3185–93.

Jouvin-Marche, E., Goncalves Morgado, M., Trede, N., Marche, P. N., Couez, D., Hue, I., *et al.* (1989). Complexity, polymorphism, and recombination of mouse T-cell receptor α gene families. *Immunogenetics*, **30**, 99–104.

Jouvin-Marche, E., Hue, I., Marche, P. N., Liebe-Gris, C., Marolleau, J.-P., Malissen, B., *et al.* (1990). Genomic organization of the mouse T cell receptor Vα family. *EMBO J.*, **9**, 2141–50.

Kappler, J. W., Wade, T., White, J., Kushnir, E., Blackman, M., Bill, J., *et al.* (1987). A T cell receptor Vβ segment that imparts reactivity to a class II major histocompatibility complex product. *Cell*, **49**, 263–71.

Kavaler, J., Davis, M. M., and Chien, Y. (1984). Localization of a T-cell receptor diversity-region element. *Nature*, **310**, 421–3.

Klotz, J. L., Barth, R. K., Kiser, G. L., Hood, L. E., and Kronenberg, M. (1989). Restriction fragment length polymorphisms of the mouse T-cell receptor gene families. *Immunogenetics*, **29**, 191–201.

Koop, B. F., Wilson, R. K., Wang, K., Vernooij, B., Zaller, D., Kuo, C. L., *et al.* (1992). Organization, structure and function of 95 kb of DNA spanning the murine T-cell receptor Cα/Cδ region. *Genomics*, **13**, 1209–30.

Kotzin, B. L., Barr, V. L., and Palmer, E. (1985). A large deletion within the T-cell receptor beta-chain gene complex in New Zealand white mice. *Science*, **229**, 167–71.

Kranz, D. M., Saito, H., Disteche, C. M., Swissheim, K., Pravtcheva, D., Ruddle, F. H., *et al.* (1985). Chromosomal locations of the murine T-cell receptor alpha-chain gene and the T-cell gamma gene. *Science*, **277**, 941–5.

Krimpenfort, P., de Jong, R., Uematsu, Y., Dembic, Z., Ryser, S., von Boehmer, H., *et al.* (1988). Transcription of T cell receptor β-chain genes is controlled by a downstream regulatory element. *EMBO J.*, **7**, 745–50.

Lai, E., Barth, R. K., and Hood, L. (1987). Genomic organization of the mouse T-cell receptor β-chain gene family. *Proc. Natl. Acad. Sci. USA*, **84**, 3846–50.

Lee, N. E. and Davis, M. M. (1988). T cell receptor β-chain genes in BW5147 and other AKR tumors. Deletion order of murine Vβ gene segments and possible 5′ regulatory regions. *J. Immunol.*, **140**, 1665–75.

Lee, N. E., D'Eustachio, P., Pravtcheva, D., Ruddle, F. H., Hedrick, S. M., and Davis, M. M. (1984). Murine T cell receptor beta chain is encoded on chromosome 6. *J. Exp. Med.*, **160**, 905–13.

Lindsten, T., Fowlkes, B. J., Samelson, L. E., Davis, M. M., and Chien, Y. (1987*a*).

Plates

```
  mDV6 --------dakttqpdsmestegetvhlpcshatis-gneyiywyrqvplqgpeyvthglqqnt-----tnsmaflaiasdrksstlilthvslrdaavyhcilrv
  hAV4 ---------kttq-ptsmdcaegraanlpcnhstis-gneyvywyrqihsqgpqyiihglknne-----tnemasliitedrksstlilphatlrdtavyycivr
 mAV12 -------lakttq-ppsmeayegqevnvscshtnia-tseyiywyrqvphqgpqfiiqgykdyv-----vnevaslfisadrklstlslpwvslrdaavyycivtd
 hAV20 ---------kttq-pismdsyegqevnitcshnnia-tndyitwyqqfpsqgprfiiqgykt-----kvtnevaslfipadrksstlslprvslsdtavyycl
mADV17 --------qkvtqtqtsisvmekttvtmdcvyetrd-ssyflfwykqtasgeivflirqdsykke--nateghyslnfqkpkssigliitatqiedsavyfcamre
 hAV12 --------qkvtqaqteisvvekedvtldcvyetrd-ttyylfwykqppsgelvflirrnsfdeq--neisgryswnfqkstssfnftitasqvvdsavyfca
mDV104 --------qtvsqpqkkksvqvaesatldctydtsd-tnyllfwykqqg-gqvtlvilqeaykqy--natlnrfsvnfqkaaksfsleisdsqlgdaatyfcalme
 hAV14 --------qtvtqsqpemsvqeaetvtlsctydtse-sdyylfwykqppsrqmilvirqeaykqq--natenrfsvnfqkaaksfslkisdsqlgdaamyfca
 hADV6 --------qkitqtqpgmfvqekeavtldctydtsd-psyglfwykqpssgemifliyqgsydqq--nategrysl nfqkarksanlvisasqlgdsamyfcamre
hDV101 --------qkvtqaqssvsmpvrkavtlnclyetsw-wsyyifwykqlpskemiflirqgs-deq--naksgrysvnfkkaaksvaltisalqledsakyfcalge
 mADV7 --------ekviqvwstasrqegeeltldcsyetsq-vlyhlfwykhllsgemvflirqtsssta--kersgrysvvfqkslksisliisalqpddsgkyfcalwel
 hAV28 -------edkvvqspqslvvhegdtvtlncsyemt--nfrslqwykqe-kkaptflfmltssgie---kksgrlssildkkelfsilnitatqtgdsavylca
 hAV18 -------ilnveqgpqslhvqegdstnftcsfpss--nfyalhwyrwetaktpealfvmtlngde---kkkgrisatlntkegysylyikgsqpedsatylca
 hAV30 -------edqvtqspealrlqegesssl ncsytvs--glrglfwyrqdpgkgpefl ftlysagee---kekerlkatltkke--sflhitapkpedsatylca
 mAV18 --------qlaeenlwalsvhegesvtvncsykt---sitalqwyrqksgegpaqlilirsnere---krngrlratldtssqsssl sitatrcedtavyfcat
  hAV3 --------qqgeedpqalsiqegenatmncsykt---sinnlqwyrqnsgrglvhlilirsnere---khsgrlrvtldtskksssllitasraadtasyfca
  hAV5 --------qkieqnsealniqegktatltcnytny--spaylqwyrqdpgrgpvflllireneke---krkerlkvtfdttlkqslfhitasqpadsatylca
 hAV23 -------kqevtqipaalsvpegenlvlncsftds--aiynlqwfrqdpgkgltsllliqssqre---ptsgrlnasldkssgrstlyiaasqpgdsatylca
 mAV10 --------qqvqqspaslvlqegenaelqcnfss---tatrlqwfyqhpggrlvslfynpsgt-----khtgrltsttvtnerrgslhisssqttdsgtyfcaid
 hAV13 --------iqveqsppdlilqeganstlrcnfsd---svnnlqwfhqnpwgqlinlfyipsgt-----kqngrlsattvaterysllyisssqttdsgvyfcav
 mAV11 --------dqveqspsalslhegtgsalrcnftt---tmravqwfrknsrgslinlfylasgt-----kengrlksafdskerystlhirdaqledsgtyfcaae
 hAV27 -------elkveqnplflsmqegknytiycnystt--sdr-lywyrqdpgksleslfvllsngav---kqegrlmasldtkarlrtlhitaavhdlsatyfca
 mAV14 -------ktqveqspqslvvrqgencvlqcnysvt--pdnhlrwfkqdtgkglvsltvlvdqkd---ktsngrysatldkdakhstlhitatlldd tatyicvv
 hAV24 -------knqveqspqsliilegknctlqcnytvs--pfsnlrwykqdtgrgpvsltimtfsent---ksngrytatldadtkqsslhitasqlsdsasyicv
 hAV19 -------kneveqspqnltaqegefitincsysv---gisalhwlqqhpgggivslfmlssg-----kkkhgrliatiniqekhsslhitashprdsavyica
mDV102 --------qmlhqspqsltiqegdevtmscnlsts---lyallwyrqgddgslvslvtlqkggde---kskdkitakldkkmqqsslqiqasqpshsgtylcggk
 hAV26 --------qeleqspqslivqegknltincttssk---tlyglywykqkygeglifl mmlqkggee---kshekitakldekkqqsslhitasqpshagiylcg
 hAV29 -------qqpv-qspqavilregedavincsssk---alysvhwyrqkhgeapvflmillkggeq---kghekisasfnekkrqsslyltasqlsysgtyfcg
 hAV10 --------qlleqspqflsiqegenltvycnsss---vfsslqwyrqepgegpvllvtvvtggev---kklkrltfqfgdarkdsslhitaaqtgdtglylca
 hAV32 --------qqvmqipqyqhvqegedfttycnsst---tlsniqwykqrpgghpvfliqlvksgev---kkqkrltfqfgeakknsslhitatqttdvgtyfcagd
 hAV25 --------qqlnqspqsmfiqegedvsmnctsss---ifntwlwykqdpgegpvllialykagel---tsngrltaqfgitrkdsflnisasipsdvgiyfca
  mAV1 -------qqkvqqspeslivpegamtslnctfsds--asqyfawyrqhsgkapkalmsifsnge----keegrftihlnkaslhfslhirdsqpsdsalylcavs
  hAV2 ------qqkeveqdpgplsvpegaivslnctysns--afqyfmwyrqysrkgpellmytyssgn----kedgrftaqvdksskyislfirdsqpsdsatylcams
  mAV2 qqkeksdqqqvrqspqsltvwegettilncsyeds--tfdyfpwyrqfpgkspalliaislvsn---kkedgrftiffnkrekklslhitdsqpgdsatyfcaas
hADV17 qqkeksdqqqvkqspqslivqkggipiincayent--afdyfpwyqqfpgkgpalliairpdvse---kkegrftisfnksakqfslhimdsqpgdsatyfcaa
 mAV20 --------qqvkqsspsltvqeggililncdyend--mfdyfawykkypdnsptllisvrsnvd---kredgrftvflnksgkhfslhitasqpedtavylcaag
hADV21 qqk--nddqqvkqnspslsvqegrisilncdytns--mfdyflwykkypaegptflisissikd---knedgrftvflnksakhlslhivpsqpgdsavyfcaas
  mAV5 --------eqveqrpphlsvregdsaviictytdp--nsyyffwykqepgaglqllmkvfsste---inegqgftvllnkkdkqlslnltaahpgdsavyfcav
 hAV15 --------edveqsl-flsvregdssvinctytds--sstylywykqepgaglqlltyifsnmdm---kqdqrltvllnkkdkhlslriadtqtgdsaiyfcae
```

```
mAV15 ---------ekveqhestlsvregdsavinctytdt--assyfpwykqeagkglhfvidirsnvd---rkqsqrlivlldkkakrfslhitatqpedsaiyfcaaa
 hAV8 ---------enveqhpstlsvqegdsavikctysds--asnyfpwykqelgkrpqliidirsnvge---kkdqriavtlnktakhfslhitetqpedsavyfcaa
 mAV9 ---------eqveqlpsilrvqegssasinctyens--asnyfpwykqepgenpkliidirsnme---rkqtqglivlldkkakrfslhitdtqpgdsamyfcaas
mAV19 ---------qgveqp-aklmsvegtfarvnctysts--gfnglswyqqregqapvflsyvvldgl----kdsghfstflsrsngysylllltelqikdsasylcavr
 hAV7 ---------qsleq-psevtavegaivqinctyqts--gfyglswyqqhdggaptflsynaldgle----etgrfssflsrsdsygylllqelqmkdsasyfcavr
 mAV4 ---------dsviqmqgqvtfsendslfinctystt--gyptlfwyvqysgegpqlllqvttann---kgssrgfeatydkgttsfhlqktsvqeidsavyycais
hAV22 ---------nsvtqmegpvtlseeafltinctytat--gypslfwyvqypgeglqlllkatkadd---kgsnkgfeatyrkettsfhlekgsvqvsdsavyfcal
hAV31 ---------nsvkqt-gqitvsegasvtmnctytst--gyptlfwyveypskplqllqretm-------ensknfgggnik-dknspivkysvqvsdsavyycllg
 mAV8 ---------dsvtqteglvtltkglpvmlnctyqtty-s-pflfwyvqhlneapkllllksstdnk---rtehqgfyatlhkssssfhlqkssvqlsdsalyfcals
mAV13 ---------qsvaqpdahvtlsegaslqlrcsysys--aapylfwyvqypgqslqfllkyitgdtv--vkgtkgfeaefrksnssfnlkkspahwsdsakyfcal
 mAV3 ---------qsvtqpdarvtvsegaslqlrckysys--atpylfwyvqyprqglqlllkyysgdpv--vqgvngfeaefsksnssfhlrkasvhwsdsavyfcavs
 hAV1 ---------qsvtqlgshvsvsegalvllrcnysss--vppylfwyvqypnqglqlllkytsaatl--vkgingfeaefkksetsfhltkpsahmsdaaeyfcavs
hAV16 ---------qsvaqpedqvnvaegnpltvkctysvs--gnpylfwyvqypnrglqfllkyitgdnl--vkgsygfeaefnksqtsfhlkkpsalvsdsalyfcavr
 hAV9 ---------qrvtqpekllsvfkgapvelkcnysys--gspelfwyvqysrqrlqlllrhis------resikgftadlnkgetsfhlkkpfaqeedsamyyca
mDV105 ---------ltltqsstdqtvasgtevtllctynads-pnpdlfwyrkrpdrsfqfilyrddtsshdadfvqgrfsvkhskanrtfhlvispvsledsatyycasgy
hDV103 ---------dkvtqsspdqtvasgsevvllctydtvy-snpdlfwyrirpdysfqfvfygdnsrsegadftqgrfsvkhiltqkafhlvispvrtedsatyycaf
 hAV11 --------kdqvfq-pstvassegavveifcnhsvs--naynffwylhfpgcaprllvkgskp------sqqgrynmtyerf--sssllilqvreadaavyycav
hDV102 ---------ielvpehqtvpvsigvpatlrcsmkgeaignyyinwyrktqgntitfiyrekdiygp---gfkdnfqgdidiaknlavlkilapserdegsyycacdt
mDV101 -------ldvylepvaktftvvagdpasfyctvtggdmknyhmswykkngtnalflvyklnsnstdg-gksnlkgkiniskn--qfildiqkatmkdagtyycgsd
   No. ........1........10........20......--.30........40........50....--...60........70........80........90.....
Indel. HGFEDCBA1..................27AB.......................55AB..........................................
```

Plate 1 Alignment of the human (h) and mouse (m) Vα/δ peptide sequences. Only one representative member of each subfamily is included. For sources of sequences and references, see Tables 13.1 and 13.4. Amino acid residues are colour-coded to highlight functionally conserved features. Hydrophobic residues are blue; tyrosine (y) and histidine (h) are light blue in a hydrophobic preference; asparagine, glutamine, serine, and threonine (n, q, s, and t) are green; aspartate and glutamate (d and e) are purple; arginine and lysine (r and k) are red; cysteine (c) is pink. More than 50% occurrence of a property results in colouring. S and T are also coloured in a hydrophobic preference. All glycines (g, orange) and prolines (p, yellow) are coloured. The Genetic Data Environment (GDE)-editor from Steven Smith of Harvard University and the colour mask from Julie Thompson of EMBL were used, together with Clustal W, a greatly improved multiple sequence alignment program (Thompson *et al.* 1994). For example, residue specific gap penalties and locally reduced gap penalties in hydrophilic regions encourage new gaps in potential loop regions rather than regular secondary structure. By means of this program, the eight β-strands consisting of alternating hydrophobic residues were clearly aligned, and the gaps (dashes) were mainly placed into the loops that correspond to the first and second complementarity determining region (CDR1 and CDR2) of immunoglobulins. For numbering system, see Arden and co-workers (1995*a,b*). Indel, insertions or deletions.

```
mBV18  -llyqkpnrdicqsgtslkiqcvad-sqvvsmfwyqqfqeqslmlmatanegseatyesgftkdkfpisrpnltfstltvnnarpgdssiyfcssr
 hBV4  -visqkpsrdicqrgtsltiqcqvd-sqvtmmfwyrqqpgqsltliatanqgseatyesgfvidkfpisrpnltfstltvsnmspedssiylcsve
 hBV2  -vvsqhpswvicksgtsvkiecrsldfqattmfwyrqfpkqslmlmatsnegskatyeqgvekdkflinhasltlstltvtsahpedssfyicsar
mBV15  -lvyqyprrticksgtsmrmecqavgfqatsvawyrqspqktfelialstvnsaikyeqnftqekfpishpnlsfssmtvlnayledrglylcgar
 mBV6  -iitqtpkfligqegqkltlkcqqn-fnhdtmywyrqdsgkglrliyysi-tendlqkgdls-egydasrekkssfsltvtsaqknemavflcass
hBV17  -gitqspkylfrkegqnvtlsceqn-lnhdamywyrqdpgqglrliyysq-ivndfqkgdia-egysvsrekkesfpltvtsaqknptafylcass
hBV13  -gvtqtpkfqvlktgqsmtlqcaqd-mnheymswyrqdpgmglrlihysv-gagitdqgevp-ngynvsrsttedfplrllsaapsqtsvyfcassy
hBV12  -gitqsprhkvtetgtpvtlrchqt-enhrymywyrqdpghglrlihysy-gvkdtdkgevs-dgysvsrsktedflltlesatssqtsvyfcaise
 mBV8  -avtqsprskvavtggkvtlschqt-nnhdymywyrqdtghglrlihysy-vadstekgdip-dgykasrpsqenfslilelaslsqtavyfcassd
hBV11  ------------gtgkkitlecsqt-mghdkmywyqqdpgmelhlihysy-gvnstekgdls-sestvsrirtehfpltlesarplhtsqylcass
 mBV7  -kvtqmsrylikrmgenvllecgqd-mshetmywyrqdpglglqliyisy-dvdsnsegdip-kgyrvsrkkrehfslildsaktnqtsvyfc
 hBV3  -kvtqssrylvkrtgekvflecvqd-mdhenmfwyrqdpglglrliyfsy-dvkmkekgdip-egysvsrekkerfslilesastnqtsmylcass
hBV14  -qvtqnprylitvtgkkltvtcsqn-mnheymswyrqdpglglrqiyysm-nvevtdkgdvp-egykvsrkekrnfplilespspnqtslyfcas
hBV15  -dvtqtprnritktgkrimlecsqt-kghdrmywyrqdpglglrliyysf-dvkdinkgeis-dgysvsrqaqakfslslesaipnqtalyfcatsdl
 mBV1  -kitqsprylilgrank-sleceqh-lghnamywykqsaekppelmflyn-lkqlirnetvp-srfipecpdssklllhisavdpedsavyfcass
 hBV7  -evtqtpkhlvmgmtnkkslkceqh-mghramywykqkakkppelmfvys-yeklsinesvp-srfspecpnssllnlhlalqpedsalylcass
 mBV4  -kiiqkpkylvavtgsekiliceqy-lghnamywyrqsakkplefmfsys-yqklmdnqtas-srfqpqsskknhldlqitalkpddsatyfcass
mBV10  -avfqtpnyhvtqvgnevsfnckqt-lghdtmywykqdskkllkimfsyn-nkqlivnetvp-rrfspqssdkahlnlriksvepedsavylcassz
 hBV9  -avsqtpkylvtqmgndksikceqn-lghdtmywykqdskkflkimfsyn-nkeliinetvp-nrfspkspdkahlnlhinslelgdsavyfcass
 mBV5  -gvvqspryiikgkgersilkcipi-sghlsvawyqqtqgqelkffiqhy-dkmerdkgnlp-srfsvqqfddyhsemnmsaleledsavyfcassl
hBV23  -gviqsprhlikekretatlkcypi-prhdtvywyqqgpgqdpqflisfy-ekmqsdkgsip-drfsaqqfsdyhselnmsslelgdsalyfcassl
 hBV1  -gvtqtpkhlitatgqrvtlrcspr-sgdlsvywyqqsldqglqfliqyy-ngeerakgnil-erfsaqqfpdlhselnlsslelgdsalyfcas
 hBV5  -gvtqtprylіktrgqqvtlscspi-sghrsvswyqqtpgqglqflfeyf-setqrnkgnfp-grfsgrqfsnsrsemnvstlelgdsalylcassl
 mBV3  -kviqtprylvkgqgqkakmrcipe-kghpvvfwyqqnknnefkflinfq-nqevlqqidmtekrfsaecpsnspcsleiqsseagdsalylcassl
hBV19  -kvtqtpghlvkgkgqktkmdctpe-kghtfvywyqqnqnkefmllisfq-neqvlqetemhkkrfssqcpknapcslailssepgdtalylcass
mBV17  -gvtqtprylvkekgqkahmscspe-kghtafywyqqnqkqeltflisfr-neeimeqtdlvkkrfsakcssnsrcileilsseeddsalylcassl
mBV20  -avtqkprylikmkgqeaemkcipe-kghtavfwyqqkqskelkfliyfq-nqqpldqidmvkerfsavcpssslcslgirtceaedsalylcsss
mBV19  -gvvqnprhlvkgkeqkakmdctpi-nghsyvvwyykkpgeelkflvyfq-nediidkidmigknisakcpakkpctieiqssksltdsavyfcassq
hBV10  -kvtqrprllvkaseqkakmdcvpi-kahsyvvywyrkkleeelkflvyfq-neeliqkaeiinerflaqcsknssctleiqstesgdtalyfcass
hBV25  eevaqtpkhlvrgegqkaklycapi-kghsyvfwyqqvlknefkflisfq-nenvfdetgmpkerfsakclpnspcsleiqatkledsavyfcassq
hBV18  -gvmqnprhlvrrrgqearlrcspm-kghshvywyrqlpeeglkfmvylq-keniidesgmpkerfsaefpkegpsilriqqvvrgdsaayfcassp
mBV16  -kvlqipshqiidmgqmvtlncdpv-snhlyfywykqilgqqmeflvnfy-nvkvmekskljfkdqfsverpdgsyftlkiqptaledsavyfcassl
hBV22  -evtqtpshqvtqmgqevilhcvpi-snhlyfywyrqilgqkveflvsfy-nneisekseifddqfsverpdgsnftlkirstkledsamyfcass
mBV11  -gviqtprhkvtgkgqeatlwcepi-sghsavfwyrqtivqgleflttyfr-nqapiddsgmpkerfsaqmpnqshstlkiqstqpqdsavylcassl
 hBV8  -gviqsprhevtemgqevtlrckpi-sghnslfwyrqtmmrglelliyfn-nnvpiddsgmpedrfsakmpnasfstlkiqpseprdsavyfcassl
mBV12  -----------------------------------------------yfr-skslmedggafkdrfkaemlnssfstlkiqptepkdsavylcas
```

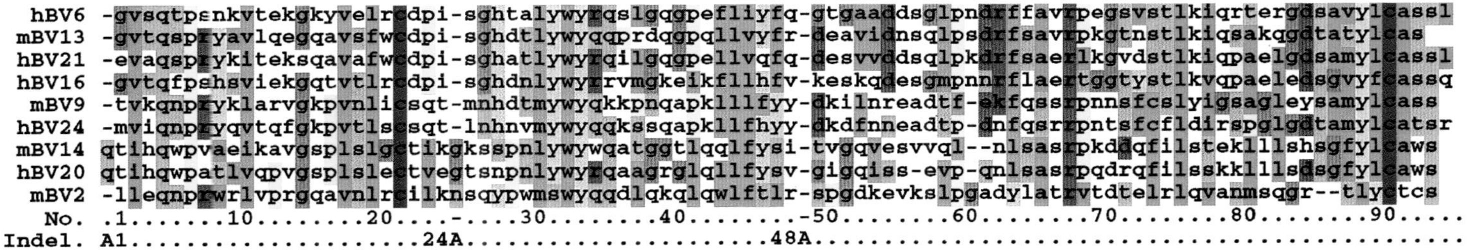

Plate 2 Deduced Vβ peptide sequence alignment. The murine and human sequences used for this alignment are listed in Tables 13.2 and 13.5, respectively. A stop codon at the 3′ end of the germline mBV10 gene segment is indicated by z. See Plate 1 for an explanation.

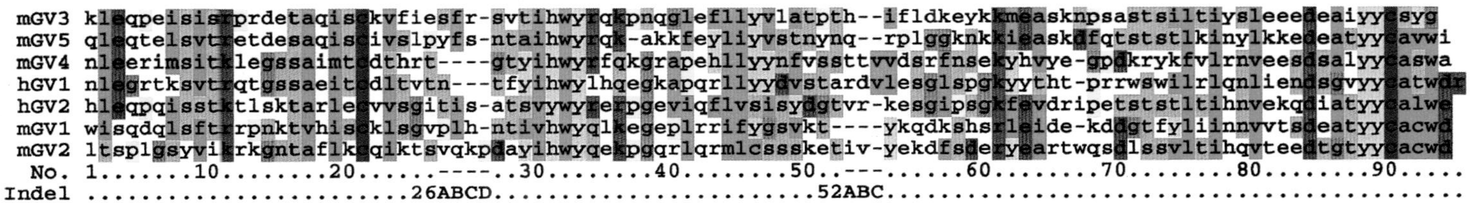

Plate 3 Deduced Vγ peptide sequence alignment. The murine and human sequences used for this alignment are listed in Tables 13.3 and 13.6, respectively. See Plate 1 for an explanation.

```
AV4    ---------kttq-ptsmdcaegraanlpcnhstis-gneyvywyrqihsqgpqyiihglknne--------tnemaslliitedrksstlilphatlrdtavyycivr
AV20   ---------kttq-pismdsyegqevnitcshnnia-tndyitwyqqfpsqgprfiiqgyktk--------vtnevaslfipadrksstlslprvslsdtavyycl
AV12   --------qkvtqaqteisvvekedvtldcvyetrd-ttyylfwykqppsgelvflirrnsfdeq----neisgryswnfqkstssfnftitasqvvdsavyfca
AV14   --------qtvtqsqpemsvqeaetvtlsctydtse-sdyylfwykqppsrqmilvirqeaykqq----natenrfsvnfqkaaksfslkisdsqlgdaamyfca
ADV6   --------qkitqtqpgmfvqekeavtldctydtsd-psyglfwykqpssgemiflliyqgsydqq----nategrysInfqkarksanlvisasqlgdsamyfcamre
DV101  --------qkvtqaqssvsmpvrkavtlnclyetsw-wsyyifwykqlpskemiflirqgs-deq----naksgrysvnfkkaaksvaltisalqledsakyfcalge
AV28   -------edkvvqspqslvvhegdtvtlncsyemt--nfrslqwykqe-kkaptflfmltssgie-----kksgrlssildkkelfsilnitatqtgdsavylca
AV18   -------ilnveqgpqslhvqegdstnftcsfpss--nfyalhwyrwetaktpealfvmtlngde-----kkkgrisatlntkegysylyikgsqpedsatylca
AV30   -------edqvtqspealrlqegesssl ncsytvs--glrglfwyrqdpgkgpeflftlysagee-----kekerlkatltkke--sflhitapkpedsatylca
AV3    --------qqgeedpqalsiqegenatmncsykts---innlqwyrqnsgrglvhlilirsnere-----khsgrlrvtldtskksssllitasraadtasyfca
AV5    --------qkieqnsealniqegktatltcnytny--spaylqwyrqdpgrgpvflllireneke-----krkerlkvtfdttlkqslfhitasqpadsatylca
AV23   -------kqevtqipaalsvpegenlvlncsftds--aiynlqwfrqdpgkgltsllliqssqre-----ptsgrlnasldkssgrstlyiaasqpgdsatylca
AV13   --------iqveqsppdlilqeganstlrcnfsds---vnnlqwfhqnpwgqlinlfyipsgt-------kqngrlsattvaterysllyisssqttdsgvyfcav
AV27   -------elkveqnplflsmqegknytiycnystt---sdrlywyrqdpgkslesltvllsngav-----kqegrlmasldtkarlrtlhitaavhdlsatyfca
AV24   -------knqveqspqsliilegknctlqcnytvs---pfsnlrwykqdtgrgpvsltimtfsent-----ksngrytatldadtkqsslhitasqlsdsasyicv
AV19   -------kneveqspqnltaqegefitincsysvg---isalhwlqqhpgggivslfmlssgk-------kkhgrliatiniqekhsslhitashprdsavyica
AV26   --------qeleqspqslivqegknltincts skt---lyglywykqkygeglifImmlqkggee-----kshekitakldekkqqsslhitasqpshagiylcg
AV29   -------qqpv-qspqavilregedavincssska---lysvhwyrqkhgeapvflmillkggeq------kghekisasfnekkrqsslyltasqlsysgtyfcg
AV10   --------qlleqspqflsiqegenltvycnsssv---fsslqwyrqepgegpvllvtvvtggev-----kklkrltfqfgdarkdsslhitaaqtgdtglylca
AV32   --------qqvmqipqyqhvqegedfttycnsstt---lsniqwykqrpgghpvfliqlvksgev-----kkqkrltfqfgeakknsslhitatqttdvgtyfcagd
AV25   --------qqlnqspqsmfiqegedvsmnctsssi---fntwlwykqdpgegpvllialykagel-----tsngrltaqfgitrkdsflnisasipsdvgiyfca
AV2    ------qqkeveqdpgplsvpegaivslnctysns--afqyfmwyrqysrkgpellmytyssgn-------kedgrftaqvdksskyislfirdsqpsdsatylcams
ADV17  qqkeksdqqqvkqspqslivqkggipiincayent--afdyfpwyqqfpgkgpalliairpdvse-----kkegrftisfnksakqfslhimdsqpgdsatyfcaa
ADV21  qqk--nddqqvkqnspslsvqegrisilncdytns--mfdyflwykkypaegptflisissikdk-----nedgrftvflnksakhlslhivpsqpgdsavyfcaas
AV15   --------edveqsl-flsvregdssvinctytds--sstylywykqepgaglqlltyifsnmdm-----kqdqrltvllnkkdkhlslriadtqtgdsaiyfcae
AV8    --------enveqhpstlsvqegdsavikctysds--asnyfpwykqelgkrpqliidirsnvge-----kkdqriavtlnktakhfslhitetqpedsavyfcaa
AV7    --------qsleq-psevtavegaivqinctyqts--gfyglswyqqhdggaptflsynaldgle-------etgrfssflsrsdsygyllqelqmkdsasyfcavr
AV22   --------nsvtqmegpvtlseeafltinctytat--gypslfwyvqypgeglqlllkatkaddk-----gsnkgfeatyrkettsfhlekgsvqvsdsavyfcal
AV31   --------nsvkqt-gqitvsegasvtmnctytst--gyptlfwyveypskplqllqretm-----------ensknfgggnikdkns-pivkysvqvsdsavyyclg
AV1    --------qsvtqlgshvsvsegalvllrcnysss--vppylfwyvqypnqglqlllkytsaatl-----vkgingfeaefkksetsfhltkpsahmsdaaeyfcavs
AV16   --------qsvaqpedqvnvaegnpltvkctysvs--gnpylfwyvqypnrglqfllkyitgdnl-----vkgsygfeaefnksqtsfhlkkpsalvsdsalyfcavr
AV9    --------qrvtqpekllsvfkgapvelkcnysys--gspelfwyvqysrqrlqlllrhis----------resikgftadlnkgetsfhlkkpfaqeedsamyyca
DV103  --------dkvtqsspdqtvasgsevvllctydtvy-snpdlfwyrirpdysfqfvfygdnsrsega---dftqgrfsvkhiltqkafhlvispvrtedsatyycaf
AV11   -------kdqvfq-pstvassegavveifcnhsvs--naynffwylhfpgcaprllvkgskp--------sqqgrynmtyerf--ssslliIqvreadaavyycav
DV102  --------ielvpehqtvpvsigvpatlrcsmkgeaignyyinwyrktqgntitfiyrekdiygp-----gfkdnfqgdidiaknlavlkilapserdegsyycacdt
BV4    ---------visqkpsrdicqrgtsltiqcqvds---qvtmmfwyrqqpgqsltliatanqgseat-yesgfvidkfpisrpnltfs-tltvsnmspedssiylcsve
BV2    ---------vvsqhpswvicksgtsvkiecrsldf--qattmfwyrqfpkqslmlmatsnegskat-yeqgvekdkflinhasltls-tltvtsahpedssfyicsar
BV17   ---------gitqspkylfrkegqnvtlsceqnl---nhdamywyrqdpgqglrliyysqivndf--qkgdia-egysvsrekkesf-pltvtsaqknptafylcass
BV13   ---------gvtqtpkfqvlktgqsmtlqcaqdm---nheymswyrqdpgmglrlihysvgagit--dqgevp-ngynvsrsttedf-plrllsaapsqtsvyfcassy
BV12   ---------gitqsprhkvtetgtpvtlrchqte---nhrymywyrqdpghglrlihysygvkdt--dkgevs-dgysvsrsktedf-lltlesatssqtsvyfcaise
```

```
BV11 --------------------gtgkkitlecsqtm---ghdkmywyqqdpgmelhlihysygvnst--ekgdls-sestvsrirtehf-pltlesarplhtsqylcass
 BV3 ---------kvtqssrylvkrtgekvflecvqdm---dhenmfwyrqdpglglrliyfsydvkmk--ekgdip-egysvsrekkerf-slilesastnqtsmylcass
BV14 ---------qvtqnprylitvtgkkltvtcsqnm---nheymswyrqdpglglrqiyysmnvevt--dkgdvp-egykvsrkekrnf-plilespspnqtslyfcas
BV15 ---------dvtqtprnritktgkrimlecsqtk---ghdrmywyrqdpglglrliyysfdvkdi--nkgeis-dgysvsrqaqakf-slslesaipnqtalyfcatsdl
 BV7 ---------evtqtpkhlvmgmtnkkslkceqhm---ghramywykqkakkppelmfvysyekls--inesvp-srfspecpnssll-nlhlhalqpedsalylcass
 BV9 ---------avsqtpkylvtqmgndksikceqnl---ghdtmywykqdskkflkimfsynnkeli--inetvp-nrfspkspdkahl-nlhinslelgdsavyfcass
BV23 ---------gviqsprhlikekretatlkcypip---rhdtvywyqqgpgqdpqflisfyekmqs--dkgsip-drfsaqqfsdyhs-elnmsslelgdsalyfcassl
 BV1 ---------gvtqtpkhlitatgqrvtlrcsprs---gdlsvywyqqsldqglqfliqyyngeer--akgnil-erfsaqqfpdlhs-elnlsslelgdsalyfcas
 BV5 ---------gvtqtprylikt rgqqvtlscspis---ghrsvswyqqtpgqglqflfeyfsetqr--nkgnfp-grfsgrqfsnsrs-emnvstlelgdsalylcassl
BV19 ---------kvtqtpghlvkgkgqktkmdctpek---ghtfvywyqqnqnkefmllisfqneqvl--qetemhkkrfssqcpknapc-slailssepgdtalylcass
BV10 ---------kvtqrprllvkaseqkakmdcvpik---ahsyvywyrkkleeelkflvyfqneeli--qkaeiinerflaqcsknssc-tleiqstesgdtalyfcass
BV25 --------eevaqtpkhlvrgegqkaklycapik---ghsyvfwyqqvlknefkflisfqnenvf--detgmpkerfsakclpnspc-sleiqatkledsavyfcassq
BV18 ---------gvmqnprhlvrrrgqearlrcspmk---ghshvywyrqlpeeglkfmvylqkenii--desgmpkerfsaefpkegps-ilriqqvvrgdsaayfcassp
BV22 ---------evtqtpshqvtqmgqevilhcvpis---nhlyfywyrqilgqkveflvsfynneis--ekseifddqfsverpdgsnf-tlkirstkledsamyfcass
 BV8 ---------gviqsprhevtemgqevtlrckpis---ghnslfwyrqtmmrglelliyfnnnvpi--ddsgmpedrfsakmpnasfs-tlkiqpseprdsavyfcassl
 BV6 ---------gvsqtpsnkvtekgkyvelrcdpis---ghtalywyrqslgqgpefliyfqgtgaa--ddsglpndrffavrpegsvs-tlkiqrtergdsavylcassl
BV21 ---------evaqsprykiteksqavafwcdpis---ghatlywyrqilgqgpellvqfqdesvv--ddsqlpkdrfsaerlkgvds-tlkiqpaelgdsamylcassl
BV16 ---------gvtqfpshsviekgqtvtlrcdpis---ghdnlywyrrvmgkeikfllhfvkeskq--desgmpnnrflaertggtys-tlkvqpaeledsgvyfcassq
BV24 ---------mviqnpryqvtqfgkpvtlscsqtl---nhnvmywyqqkssqapkllfhyydkdfn--neadtp-dnfqsrrpntsfc-fldirspglgdtamylcatsr
BV20 --------qtihqwpatlvqpvgsplslectvegt--snpnlywyrqaagrglqllfysvgigqi---ssevp-qnlsasrpqdrqf-ilsskkllsdsgfylcaws
 GV1 ---------nlegrtksvtrqtgssaeitcdltvt--ntfyihwylhqegkapqrllyydvstardvlesglspgkyytht-prrwswilrlqnliendsgvyycatwdr
 GV2 ---------hleqpqisstktlsktarlecvvsgitisatsvywyrerpgeviqflvsisydgtv-rkesgipsgkfevdripetststltihnvekqdiatyycalwe
```

Plate 4 Alignment of peptide sequences from each of the different human TCRV families. Vα/δ, Vβ, and Vγ sequences are listed in Tables 13.4, 13.5, and 13.6, respectively. For explanation, see Plate 1; for discussion, see text.

Transient rearrangements of the T cell antigen receptor α locus in early thymocytes. *J. Exp. Med.*, **166**, 761–5.

Lindsten, T., Lee, N. E., and Davis, M. M. (1987*b*). Organization of the T-cell antigen receptor β-chain locus in mice. *Proc. Natl. Acad. Sci. USA*, **84**, 7639–43.

Malissen, M., Minard, K., Mjolsness, S., Kronenberg, M., Goverman, J., Hunkapiller, T., *et al.* (1984). Mouse T cell antigen receptor: structure and organization of constant and joining gene segments encoding the β polypeptide. *Cell*, **37**, 1101–10.

Malissen, M., McCoy, C., Blanc, D., Trucy, J., Devaux, C., Schmitt-Verhulst, A.-M., *et al.* (1986). Direct evidence for chromosomal inversion during T-cell receptor β gene rearrangements. *Nature*, **319**, 28–33.

Marche, P. N., Six, A., Gahéry, H., Gris-Liebe, C., Cazenave, P.-A., and Jouvin-Marche, E. (1993). T cell receptor Vα gene segment with alternate splicing in the junctional region. *J. Immunol.*, **151**, 5319–27.

Patten, P., Yokota, T., Rothbard, J., Chien, Y., Arai, K., and Davis, M. M. (1984). Structure, expression and divergence of T-cell receptor β-chain variable regions. *Nature*, **312**, 40–6.

Pelkonen, J., Traunecker, A., and Karjalainen, K. (1987). A new mouse TCR Vγ gene that shows remarkable evolutionary conservation. *EMBO J.*, **6**, 1941–4.

Raulet, D. H. (1989). The structure, function, and molecular genetics of the γ/δ T cell receptor. *Annu. Rev. Immunol.*, **7**, 175–207.

Seto, D., Koop, B. F., Deshpande, P., Howard, S., Seto, J., Wilk, E., *et al.* (1994). Organization, sequence, and function of 34.5 kb of genomic DNA encompassing several murine T-cell receptor α/δ variable gene segments. *Genomics*, **20**, 258–66.

Singer, P. A., McEvilly, R. J., Balderas, R. S., Dixon, F. J., and Theofilopoulos, A. N. (1988). T-cell receptor α-chain variable-region haplotypes of normal and autoimmune laboratory mouse strains. *Proc. Natl. Acad. Sci. USA*, **85**, 7729–33.

Siu, G., Kronenberg, M., Strauss, E., Haars, R., Mak, T. W., and Hood, L. (1984). The structure, rearrangement and expression of Dβ gene segments of the murine T-cell antigen receptor. *Nature*, **311**, 344–50.

Smith, L. R., Plaza, A., Singer, P. A., and Theophilopoulos, A. N. (1990). Coding sequence polymorphisms among Vβ T cell receptor genes. *J. Immunol.*, **144**, 3234–7.

Smith, H. P., Nguyen, P., Woodland, D. L., and Blackman, M. A. (1992). T cell receptor α-chain influences reactivity to Mls-1 in Vβ8.1 transgenic mice. *J. Immunol.*, **149**, 887–96.

Tang, X. X., Ikegaki, N., and Heber-Katz, E. (1994). Nucleotide sequences of three new members of the mouse Vα2 gene family. *Mol. Immunol.*, **31**, 79–82.

Traunecker, A., Oliveri, F., Allen, N., and Karjalainen, K. (1986). Normal T cell development is possible without functional γ chain. *EMBO J.*, **5**, 1589–93.

Vernooij, B. T. M., Lenstra, J. A., Wang, K., and Hood, L. (1993). Organization of the murine T-cell receptor γ locus. *Genomics*, **17**, 566–74.

Wade, T., Bill, J., Marrack, P. C., Palmer, E., and Kappler, J. W. (1988). Molecular basis for the nonexpression of Vβ17 in some strains of mice. *J. Immunol.*, **141**, 2165–7.

WHO IUIS nomenclature sub-committee on TCR designation. (1993). Nomenclature for T-cell receptor (TCR) gene segments of the immune system. *WHO Bull.*, **71**, 113–15.

Wilson, R. K., Koop, B. F., Chen, C., Halloran, N., Sciammis, R., and Hood, L. (1992). Nucleotide sequence analysis of 95 kb near the 3′ end of the murine T-cell receptor α/δ chain locus: strategy and methodology. *Genomics*, **13**, 1198–208.

Winoto, A. and Baltimore, D. (1989*a*). αβ Lineage-specific expression of the α T cell receptor gene by nearby silencers. *Cell*, **59**, 649–55.
Winoto, A. and Baltimore, D. (1989*b*). A novel, inducible and T cell-specific enhancer located at the 3′ end of the T cell receptor α locus. *EMBO J.*, **8**, 729–33.
Winoto, A., Mjolsness, S., and Hood, L. (1985). Genomic organization of the genes encoding mouse T-cell receptor α-chain. *Nature*, **316**, 832–6.
Woolf, T., Lai, E., Kronenberg, M., and Hood, L. (1988). Mapping genomic organization by field inversion and two-dimensional gel electrophoresis: application to the murine T cell receptor γ gene family. *Nucleic Acids Res.*, **16**, 3863–75.

14 Nature of variation among variable genes

BERNHARD ARDEN, STEPHEN P. CLARK, DIETER KABELITZ, AND TAK W. MAK

1 Introduction

T lymphocytes recognize their antigenic peptides through the action of the heterodimeric T cell receptor (TCR) which is composed of an α and β chain for most mature lymphocytes, although a small proportion of cells use a γδ heterodimer instead (Ferrick *et al.* 1989). Like the immunoglobulins, the T cell receptor proteins are encoded in the genome as variable gene segments (V), diversity segments (D; except in the case of α and γ chains), joining segments (J) and constant regions (C). The random assortment of the various V, D, and J elements, as well as junctional diversity that occurs during recombination, provides an essentially limitless repertoire for antigen recognition.

The V gene segments have been named according to their family (α, β, γ, or δ) and how closely related they are to other members of the family. Traditionally, 75% identity at the nucleotide level has been used as the cut-off between subfamilies (Barth *et al.* 1985; Arden *et al.* 1985; Toyonaga and Mak 1987). The TCRV gene segments are composed of two exons separated by an intron of 100 to 400 base pairs, with the first exon encoding a hydrophobic leader sequence of about 50 base pairs that was removed in all pair-wise comparisons. The mature V region starts near the end of the first exon, and the second exon encodes the remainder of the V region. The large number of gene segments has resulted in a certain amount of confusion in the literature because there are many instances where different groups have given the same gene different names, or have given different genes the same name. Furthermore, it has been unclear how to name certain V segments which can be linked to both α and δ J and C regions. In order to resolve these problems, an international group has been formed to assign consistent names to all the variable gene segments (Williams *et al.* 1993). As part of that effort, we have collected all the available mouse (Arden *et al.* 1995*a*) and human (Arden *et al.* 1995*b*) T cell receptor variable gene sequences and determined how they are related to each other. We then have compared the V genes between these species (Clark *et al.* 1995).

Table 14.1 Murine TCRA/D-V gene segments

Official designation[a]	Previous designation	Clone name	Strain	Reference	GENBank/EMBL accession number
AV1S3		5H	B10.A(5R)	Winoto *et al.* 1985	X02833
AV2S5		8I	A/J	Lai *et al.* 1988*b*	—
AV3S1		λ2.2	BALB/c	Hayday *et al.* 1985*a*	X02857
AV4S6		DA.33.C2	C57BL/6	Spinella *et al.* 1987	M16675
AV5S1	Vα5	TA72	C57BL/Ka	Arden *et al.* 1985	X02933
DV6S2	AV6		BALB/c	Seto *et al.* 1994	M94080
ADV7S1	Vδ7.1	RL6.14	C57BL/6	Arden *et al.*, unpublished	—
AV8S3		F3.4	C57BL/6	Chou *et al.* 1986	X06306
AV9S1	Vα13.1		BALB/c	Yague *et al.* 1988	M38681
AV10S6		p109s	Std:ddY	Toda *et al.* 1988	X17178
AV11S1	Vα11.1	B10	B10.A	Fink *et al.* 1986	X03860
AV12S1		BDFLI	B6 × DBA/2	Dembic *et al.* 1986	X03668
AV13S1		BM1037	B6.H-2bm10	Costlow *et al.*, unpublished	—
AV14S1	Vα14.1		C57BL/6	Koseki *et al.* 1989	D90229
AV15S1		SJL-HE-1.1	SJL	Sutherland *et al.* 1991	X57397
ADV17S2		5.3.18α	B6.H-2bm12	Sherman *et al.* 1987	M16119
AV18S2		BM.A	CBA/J	Couez *et al.* 1991	—
AV19S1		A10	B10.A	Malissen *et al.* 1988	M22604
AV20S1		5T		Marche *et al.*, unpublished	—
DV101S1		7-17.1	AKR/J	Asarnow *et al.* 1988	M23545
DV102S1	DV2		BALB/c	Seto *et al.* 1994	M94080
DV104S1	Vδ4	Z10	BALB/c × 129	Elliott *et al.* 1988	M37280
DV105S1	Vδ5	Z72	B10.D2dm1	Koop *et al.* 1992	M64239
		glVδ5	BALB/c	Iwashima *et al.* 1988	M23382
		glδ7.3	B10.D2dm2	Korman *et al.* 1989	M23095

[a] Only representative subfamily members are listed, whose peptide sequences are shown in the alignment in Plate 1. Pseudogenes were excluded from this analysis. For complete listing and nomenclature, see Arden *et al.* (1995*a*). S, subfamily member.

2 Murine variable genes

In the mouse, 87 α chain V gene segments (TCRAV) have been identified, but only 23 for the β chain (TCRBV), 7 for the γ chain (TCRGV) and 16 for the δ chain (TCRDV). The TCRDV gene segments are interspersed with the TCRAV gene segments. Traditionally, gene segments being 75% or more homologous with one another at the DNA level were considered members of a subfamily (Barth *et al.* 1985; Arden *et al.* 1985), a level of similarity required for cross-hybridization on genomic Southern blots. We show that the mouse TCRV gene segments have evolved as subfamilies of highly homologous genes. Using the cut-off of 75%, different subfamilies are clearly separated from each other, rather than a broad continuum of more or less related genes. There are only a few exceptions of V gene segments that are equally distant from either of two subfamilies.

We grouped the TCRAV gene segments into 20 different subfamilies (Table 14.1) using the subfamily designation originally suggested by Arden and co-workers (1985) for AV1–AV10. A closer look revealed that only very few subfamilies contain members that share less than 80% similarity at the DNA level: AV4 (76%) and AV7 (78%). Accordingly, very few members from different subfamilies share above 65% similarity: AV3 with AV13 (71–77%), AV5 with AV9 (69–72%) and with AV15 (67–71%), and AV9 with AV15 (76%) (Arden *et al.* 1995*a*). With these exceptions, it is evident that TCRAV gene segments do not share a continuum from low to high similarity, rather they can be grouped into significantly distinct subfamilies of highly homologous gene segments. An alignment of the Vα peptide sequences representing each TCRAV subfamily is shown in Plate 1. The TCRAV subfamilies range in size from 1 to 15 members (including different putative alleles, see below). Eight subfamilies have greater than six members, twelve subfamilies contain less than three members.

TCRAV and TCRDV gene segments are included in the same alignment, because they are interspersed on chromosome 14 and many of them are quite closely related: of the 16 TCRDV gene segments identified to date, 12 display greater than 75% similarity to one of the TCRA subfamilies. Originally, the TCRDV gene segments were numbered as a separate family (Becker *et al.* 1985). In order to be as consistent as possible with the original TCRD subfamily names, we preserved the numbers of those TCRDV gene segments that lack homology to TCRAV genes by starting above 100 (e.g. DV101 for Vδ1), leaving a gap, where TCRDV gene segments (Vδ3) could be assigned to homologous TCRA subfamilies. Vα6 and Vδ3 become AV6S1 and DV6S2 respectively. These probably represent two different alleles of the same locus, because Southern blots from several strains indicate that ADV6 is a single member subfamily (Arden *et al.* 1985; Elliott *et al.* 1988). Both, AV6S1 and DV6S2 are functionally transcribed (Arden *et al.* 1985; Elliott *et al.* 1988), and they are about equally frequently expressed in α chains or δ chains

Table 14.2 Murine TCRB V gene segments

Official designation[a]	Previous designation	Clone name	Strain	Reference	GENBank/EMBL accession number
BV1S1A1		BW14	AKR	Lee *et al.* 1988	M20177
BV2S1	Vβ2	AR1	C57L	Barth *et al.* 1985	X02780
BV3S1A1		3H.25	C57BL/6	Goverman *et al.* 1985	M12415
BV4S1	Vβ4	TB3	C57BL/Ka	Barth *et al.* 1985	X02781
BV5S1	Vβ5.1		C57BL/6	Chou *et al.* 1987	M15613
BV6S1A1		C9	C57BL/6	Iwamoto *et al.* 1987	X05738
BV7S1		pHDS11	BALB/B	Saito *et al.* 1984	X00696
BV8S1	Vβ8.1		C57BL/6	Chou *et al.* 1987	M15616
BV9S1	Vβ2		BALB/c	Behlke *et al.* 1985	M13677
BV10S1A1		Vβ10-8	C57BL/6	Hirama *et al.* 1991	X56725
BV11S1		LVAK	C57BL/6	Epplen *et al.* 1986	N00046
BV12S1T		NZW8	NZW	Behlke and Loh 1986	M30880
BV13S1		BVI/5.β11	CBA/J	Heuer *et al.* 1991	M31648
BV14S1	Vβ14	VB14GL	B10.WR7	Malissen *et al.* 1986	X03277
BV15S1A1		FN1-18	C57BL/6	Gascoigne *et al.* 1986	X04047
BV16S1A1		4.C3	B10.A	Fink *et al.* 1986	X03865
BV17S1A3	Vβ17a2		PWK	Cazenave *et al.* 1990	M61184
BV18S1	Vβ18		BALB/c	Louie *et al.* 1989	X16695
BV19S1	Vβ19a		SJL	Louie *et al.* 1989	—
BV20S1	Vβ20	K9	BALB/c	Six *et al.* 1991	X59150

[a] Only those gene segments are listed which were used for the alignment in Plate 2. For complete listing and detailed explanation of nomenclature, see Arden *et al.* (1995*a*). S, subfamily member. A, allele. T, tentative designation, if not proven to be a true allele.

respectively, in neonatal thymocytes (Happ and Palmer 1989). Among the homologous TCRAV and TCRDV genes, there are three V gene segments that each were rearranged to form both an α chain and a δ chain message (ADV7S1, ADV7S2, and ADV17S2). In the new nomenclature V gene segments that are rearranged to both, TCRAJ and TCRDJ gene segments are designated ADV to refer to the germline gene segment (Williams *et al.* 1993). We are currently investigating whether these V gene segments can also be functionally expressed both in αβ and in γδ receptors.

For the TCRB family, 20 different subfamilies have been defined, 18 of which are single member subfamilies (Table 14.2). We followed the numbering system (BV1–BV8) originally proposed by Barth and co-workers (1985), and extended (BV9–BV13) by Behlke and co-workers (1985), and (BV14–BV16) by Behlke and co-workers (1986). Subsequently, newly discovered TCRBV subfamilies were consistently numbered by the authors. We named the five pseudogene subfamilies described by Louie and co-workers BV21–BV25. It is concluded that most, if not all, TCRBV subfamilies have been identified (Louie *et al.* 1989). An alignment of the Vβ peptide sequences is shown in Plate 2. The overall pairwise similarity of the TCRB sequences is below 55% at the DNA level. As is the case for the TCRA family, very few of the different TCRB subfamilies share greater than 65% of their nucleotide sequences: BV3 with BV17 (73–74%), BV3 with BV20 (72–73%), and BV17 with BV20 (74–75%).

For the seven murine TCRGV genes, three different nomenclatures have been used by the laboratories of Tonegawa (Heilig and Tonegawa 1986), Raulet (Garman *et al.* 1986), and Karjalainen (Traunecker *et al.* 1986; Pelkonen *et al.* 1987). For cross-referencing, see Raulet (1989). In agreement with the laboratories that originally named the TCRGV genes and with the TCR nomenclature sub-committee, we suggest a new consensus nomenclature (Arden *et al.* 1995*a*). GV1S1 to GV4S1 are named in the order of genomic localization upstream of TCRGC1 from 3′ to 5′. This is the one exception. All other TCRV genes are named according to sequence similarity. GV1S1 to GV4S1 are single member subfamilies that share less than 42% similarity at the nucleotide level. They are all rearranged to TCRGJ1. GV5S1 to GV5S3 are members of one subfamily with greater than 94% similarity with one another. They are rearranged to TCRGJ4, J2, and J3 respectively (Table 14.3; for the genomic organization of the TCRGV gene segments, see Lai *et al.* 1989). The peptide alignment of the functional TCRGV gene segments is shown in Plate 3.

3 Human variable genes

In the human, 86 different TCRA and TCRDV gene segments have been published up to EMBL release 38, March 1994 (Stoehr and Cameron 1991). These were grouped into 32 different TCRAV and 3 TCRDV subfamilies

Table 14.3 Murine TCRGV gene segments

Official designation[a]	Previous designation	Clone name	Strain	Reference	GENBank/EMBL accession number
GV1S1	Vγ3	TC-13	BALB/c	Garman *et al.* 1986	M13337
GV2S1	Vγ4	TC-11	BALB/c	Garman *et al.* 1986	M13338
GV3S1A1	Vγ2	TC-17	BALB/c	Garman *et al.* 1986	M13336
GV4S1A1	Vγ4.4	BW3.8.1	AKR	Pelkonen *et al.* 1987	X05501
GV5S1A1	Vγ1	V10.8B	BALB/c	Hayday *et al.* 1985*b*	M12832

[a] See Tables 14.1 and 14.2 for an explanation.

(Table 14.4). There are two larger subfamilies, TCRAV1 and TCRAV2, containing at least 5 and 3 members, respectively. Four subfamilies contain only two members and 29 are single member subfamilies. The subfamilies are overall quite distinct from each other, sharing about 40% of their nucleotide sequence. However, there are borderline cases: The AV3 and AV5 subfamilies have 75–80% similarity and the AV8 and AV15 subfamilies are 65–80% similar. Apart from these ambiguities, there are a few subfamilies with pairwise similarities just below 75%: AV6 compared with AV12 and AV14, AV17 with AV21, and AV26 with AV29. Conversely, one member of the AV1 subfamily shares less than 75% with some members of the same subfamily (Arden *et al.* 1995*b*).

TCRA and TCRD gene sequences are grouped together, because they are interspersed on chromosome 14. Some TCRA and TCRDV gene segments are quited closely related. A TCRD gene segment that was previously called Vδ4 is identical with AV6S1 (Guglielmi *et al.* 1988) and may thus be named DV6S1. Recently, DV6S1 was shown to be expressed also in α chain message (Moss *et al.* 1993) and the germline gene segment may therefore be referred to as ADV6S1. Similarly, Vδ5, isolated by Takihara and co-workers (1989), is identical to AV21S1 (Leiden *et al.* 1986). Therefore, we call the germline V gene segment ADV21S1, to indicate that this V gene segment can be found in both α and δ chain transcripts. There remain only three TCRDV gene segments that are not homologous to TCRAV subfamilies. These are called DV101, DV102, and DV103 according to the rules of the WHO-IUIS nomenclature subcommittee on TCR designation (Williams *et al.* 1993). A protein alignment containing one member of each subfamily is shown in Plate 1.

The size of the human TCRBV gene repertoire is 138 identified V gene segments up to EMBL release 38. These may be grouped into 26 subfamilies (Table 14.5). The TCRBV5, 6, 8, and 13 subfamilies contain five or more members. Ten subfamilies contain 2–3 members and 12 are single-member subfamilies. All in all, the TCRBV gene sequences are more similar to each other than the TCRAV sequences. Members from different subfamilies

Table 14.4 Human TCRA/DV gene segments

Official designation[a]	Previous designation	Clone name	Reference	GENBank/EMBL accession number
AV1S2A1T	Vα1.2	pY14	Yanagi *et al.* 1985	M12423
AV2S2A1T	Vα2.2	AG110	Klein *et al.* 1987	M17655
AV3S1	Vα3.1	HAP05	Yoshikai *et al.* 1986	X04948/M13726
AV4S2A1T	Vα4.2	HAVT01	Kimura *et al.* 1987	M27372
AV5S1		IGRa10	Roman-Roman *et al.* 1991	X58747
ADV6S1A1		GERM V	Guglielmi *et al.* 1988	M21626
AV7S1A1			Ikeda, unpublished	D21847
AV8S1A1	Vα8.1	HAP41	Yoshikai *et al.* 1986	X04954/M13733
AV9S1	Vα9.1	HAP36	Yoshikai *et al.* 1986	X04942/M13737
AV10S1A2	AV10.1a	WADM22B	Obata *et al.* 1993	D13075
AV11S1A1T	Vα11.1	HAP02	Yoshikai *et al.* 1986	X04936/M13739
AV12S1		HTA73	Plaza, unpublished	X70310
AV13S1	Vα13.1	S26.2	Hurley *et al.* 1993	M97722
AV14S1	Vα14.1	HAVT20	Kimura *et al.* 1987	M27375
AV15S1			Devaux *et al.* 1991	S60795
AV16S1A1T	Vα1.6	AG21	Klein *et al.* 1987	M17651
ADV17S1A1T	Vα13.1	AB11	Klein *et al.* 1987	M17660
AV18S1	Vα14.1	AB21	Klein *et al.* 1987	M17661
AV19S1	Vα15.1	AC24	Klein *et al.* 1987	M17662
AV20S1	Vα16.1	AE212	Klein *et al.* 1987	M17663
ADV21S1A1	Vα21b		Wright *et al.* 1991	–
AV22S1A2T	AV22.1a	WADM13D	Obata *et al.* 1993	D13072
AV23S1	Vαw23	IGRa01	Roman-Roman *et al.* 1991	X58736
AV24S1	Vαw24	IGRa02	Roman-Roman *et al.* 1991	X58737
AV25S1	Vαw25	IGRa03	Roman-Roman *et al.* 1991	X58738
AV26S1	Vαw26	IGRa04	Roman-Roman *et al.* 1991	X58739
AV27S1	Vαw27	IGRa05	Roman-Roman *et al.* 1991	X58740
AV28S1A2T		IGRa15	Roman-Roman *et al.* 1991	X61070
AV29S1A2T	Vαw29.n		Santamaria *et al.* 1993	L06883
AV30S1A1T	Vα30		Moss *et al.* 1993	X68696/S59345
AV31S1			Bernard *et al.* 1993	X73521
AV32S1	Vα30	KT2	Boitel *et al.* 1992	M64350
DV101S1	Vδ1	K15A	Satyanarayana *et al.* 1988	M22198
DV102S1A1T	TRDV2	λLY67Vδ2	Dariavach *et al.* 1989	X15207
DV103S1A1T	Vδ2	λPl1	Loh *et al.* 1988	–

[a] Only representative members of each subfamily are listed as they were used for the alignment in Plate 1. For complete listing and detailed explanation of nomenclature, see Arden *et al.* (1995*b*). Following the letter S referring to different subfamily members, alleles are denoted by the letter A. In those cases where they are not proven to be alleles, T is used for tentative designation. Pseudogenes were excluded from this analysis.

Table 14.5 Human TCRBV gene segments

Official designation[a]	Previous designation	Clone name	Reference	GENBank/EMBL accession number
BV1S1A1	Vβ1.2	HBVT73	Kimura *et al.* 1987	M27381
BV2S1A1		pUCM4-4	Tunnacliffe *et al.* 1985	M12886
BV3S1		HT12	Plaza *et al.* 1991	X57610
BV4S1A1T			Kalams *et al.* 1994	Z29580
BV5S1A1T	Vβ5.1	HBP51	Kimura *et al.* 1986	X04927
BV6S1A1	Vβ6.1	HBP50	Kimura *et al.* 1986	X04934
BV7S1A1		IGRb19	Ferradini *et al.* 1991	X58813
BV8S1	Vβ8.1	M18H7.1B7	Siu *et al.* 1986	X07192
BV9S1A1T		DE5	Wedderburn *et al.* 1993	Z23044
BV10S1	Vβ10.1	PL3.9	Concannon *et al.* 1986	M13860/M16309
BV11S1A2T	Vβ11.3	1.3	Jores *et al.* 1993	X74845
BV12S1A1	BV12S2		Slightom *et al.* 1994	U03115
BV13S1	Vβ13.1	PL4.24	Concannon *et al.* 1986	M13863
BV14S1		ph21	Tillinghast *et al.* 1986	M14267
BV15S1	Vβ15.1	ATL2-1G	Ikuta *et al.* 1985	M11951
BV16S1A1			Smith *et al.* 1987	X06154
BV17S1A1T	Vβ17.1	HBVT02	Kimura *et al.* 1987	M27388
BV18S1	Vβ18.1	HBVT56	Kimura *et al.* 1987	M27389
BV19S1	Vβ19.1	HBVT72	Kimura *et al.* 1987	M27390
BV20S1A1	Vβ18A	H29	Charmley *et al.* 1993*a*	Z13967
BV21S1	BV21.1		Wilson *et al.* 1990	M33233
BV22S1A1T	Vβ23		Robinson 1991	M62379
BV23S1A2T	BV23S1		Slightom *et al.* 1994	U03115
BV24S1A3T	BV24S1		Slightom *et al.* 1994	U03115
BV25S1A1T	BV25S1	HVB30.A	Wei *et al.* 1994	L26231

[a] Only those gene segments are listed that were used for the alignment in Plate 2. See Table 14.4 for an explanation.

usually have a pairwise similarity of 52% as opposed to 40% for the TCRAV gene segments. However, the TCRBV subfamily definition, as is the case for the human TCRAV family, has same ambiguities. The BV1 and BV5 subfamilies are closely related: BV1 displays 75–80% similarity to some members of the BV5 subfamily. Likewise, the BV12 and 13 subfamilies are closely related. In contrast, the BV7 subfamily contains members sharing only 70–75% of their nucleotide sequence. The relatedness between the peptide sequences of the TCRB family is shown in the alignment in Plate 2.

There are six functional human TCRGV gene segments (Strauss *et al.* 1987; Huck *et al.* 1988). Five of them belong to the GV1 subfamily, being greater than 80% similar to each other. The other functional gene is the single member of the GV2 subfamily and shares less than 50% of its nucleotide sequence with those of the GV1 subfamily (Table 14.6). The peptide sequences are shown in Plate 3. GV3 and GV4 were recently shown to be pseudogenes (Zhang *et al.* 1994).

Table 14.6 Human TCRGV gene segments

Official designation[a]	Previous designation	Clone name	Reference	GENBank/EMBL accession number
GV1S3A1T	Vγ3	λSH4	Lefranc *et al.* 1986	M13430
GV2S1A2	Vγ2.2	λA6	Forster *et al.* 1987	X08086

[a] See Table 14.4 for an explanation.

4 Allelic polymorphism

The identification of allelic polymorphism in TCRV gene segments has been hampered by the difficulty in distinguishing allelic polymorphism from variation between subfamily members. In large TCRV gene subfamilies, in particular in most murine TCRAV genes, the presence of alleles can not be inferred from the number of bands detected in genomic Southern hybridization. In the AV11 subfamily, some effort was made to study allelic polymorphism (Jameson *et al.* 1991). We feel that distinguishing between new subfamily members and alleles requires each variant to be mapped to its locus. Due to the difficulty of mapping minor variants, genomic sequencing may be needed.

Since the murine TCRB family mostly consists of single-member subfamilies and the TCRB locus has been mapped in several strains (Lai *et al.* 1988*a*), it has been possible to define true alleles at the nucleotide level. Alleles were designated by the letter A followed by a number indicating the order of their discovery. A refers to asterisk or allele. It is not recommended to use an asterisk (or any other special characters) in computerized symbols. For example, the symbol BV17S1A1 stands for the complete name TCRB-V17S1*1 (Lyon 1989). There are two major haplotypes of the TCRB complex (Klotz *et al.* 1989; Smith *et al.* 1990) which are distinguished by small letter superscripts *a* and *b*. However, individual TCRBV elements occasionally vary, even when comparing different strains bearing the same major TCRB haplotype. Therefore, we chose to use numbers to designate alleles.

In the human, a high frequency polymorphism has recently been found for the AV8S1 gene segment (Cornélis *et al.* 1993). Using denaturing gradient gel electrophoresis (DGGE) on PCR amplified DNA pooled from many individuals, Charmley and co-workers (1994) confirmed the polymorphism of AV8S1 and found two additional coding region mutations in the AV2 and AV7 subfamilies, respectively. These mutations occur at frequencies from 31–69% in the population. The single coding region mutation T to C in the alleles AV2S1A1 and AV2S1A2, respectively, results in a phenylalanine to serine change. In AV7S1A1, a C to G or alanine to glycine substitution has occurred leading to its allelic form AV7S1A2. Using V-locus specific PCR, Reyburn and co-workers (1993) have isolated two alleles of the AV6S1A1 gene segment. Both differ from AV6S1A1 by the same two nucleotide changes, which result in two amino acid changes. One differs by an additional silent substitution from the other two alleles. They also found a productive nucleotide polymorphism in the AV10 subfamily. Wright and co-workers (1991) selected a single member subfamily, AV21 and corrected the sequence of AV21S1A1N1 so that ADV21S1A1N2 is most likely to represent its allelic form. (N stands for non-productive substitution. See Arden *et al.* (1995*b*) for detailed explanation of nomenclature.)

In the TCRB family, a frequent null allele of the BV20 gene results from a nucleotide substitution that creates a stop codon (Charmley *et al.* 1993*a*). The two segments BV2S1A1 and BV2S1A3 have been demonstrated to be alleles (Plaza *et al.* 1991). They differ from each other by two amino acid substitutions, and can be detected by *Alu*I digestion of the PCR products (Charmley *et al.* 1993*b*). A third allele at this locus, BV2S1A2, has recently been identified (Cornélis *et al.* 1993). Day and co-workers (1992) examined coding region variation of BV12S2 and identified three common silent allelic variants. These polymorphisms segregated codominantly in family studies. Using a monoclonal antibody, the two segments BV6S5A1 and BV6S5A2 were shown to be allelic products of the same locus. These phenotypes are stable over time and inherited in families. One of the two non-conservative changes is a point mutation within a *Bam*H1 site. This *Bam*H1 site, which is present in BV6S5A1 and absent in BV6S5A2, allows us to distinguish the two alleles by restriction fragment length polymorphism (RFLP). This is an exception to the difficulty of correlating between RFLP and sequence variation in coding regions. The substitution of one amino acid residue between two allelic forms of BV1S1 showed a close correlation with RFLP (Robinson 1989). Because this correlation is not absolute, the use of specific oligonucleotide probes to characterize allelic forms of TCRV genes will provide important tools for studies of immune responsiveness and disease susceptibility. An absence of tight linkage was found between the above three polymorphisms, indicating a lack of linkage disequilibrium across the TCRBV locus (Cornélis *et al.* 1993). Since a large number of TCR haplotypes exists resulting from different combinations of polymorphic V gene segments, we suggest that alleles be named numerically irrespective of particular haplotypes.

5 The α/δ locus: how different are TCRAV and DV genes?

Currently, 13 different TCRDV genes have been identified in the mouse, if one does not include sequences that may be derived from putative alleles. DV101 to DV105 are single member subfamilies, and the others are single TCRDV gene segments homologous to different TCRAV subfamilies, whereas the ADV7 subfamily contains at least three TCRDV gene segments. One may expect to find more TCRDV gene segments, since Northern-hybridizations have revealed that TCRDV genes are members in almost any of the TCRAV subfamilies (Happ and Palmer 1989). In the human, five different TCRDV genes have been counted. DV101 to DV103 are single member subfamilies. Recently, an α-chain mRNA was isolated containing an allele of AV6S1 that was identical to the Vδ4 gene segment (Moss *et al.* 1993). Vδ4 may therefore be termed DV6S1. Vδ5 is identical to AV21 and may be termed DV21. Thus, both in mice and in humans, certain members of different subfamilies can be used interchangeably as part of αβ and γδ receptors, but the α and δ repertoires are largely mutually exclusive. Although it has been demonstrated, using TCRDV- and TCRAC-specific PCR primers, that all TCRDV gene segments (with the exception of mouse TCRDV101) can rearrange to TCRAJ gene segments (Sottini *et al.* 1991; Spieß *et al.* 1992), in peripheral blood lymphocytes only 1% of the productive TCRA transcripts contain a TCRDV gene segment (Moss *et al.* 1993). There is accumulating evidence that the TCRA and TCRD repertoires are largely separate due to both a highly controlled rearrangement process and selection for the expression of functional receptors.

This raises the question of whether AV and DV gene segments have different regulatory elements and/or whether α and δ variable region peptides have distinct structural features. The latter could influence pairing with β and γ chains, respectively, or binding to ligands involved in repertoire selection. The α and δ peptides are aligned in Plate 1. Residues conserved in Vα peptides are also found in Vδ peptides and all secondary structural features appear to be conserved between α and δ. However, one residue can be found that distinguishes Vα and Vδ. Glutamic acid at position 14 is conserved in all Vα peptides (with three exceptions) but absent in 8 out of 11 Vδ peptides. This residue is located in a loop in immunoglobulins (Ig) pointing away from the antigen-binding site. It is probably exposed to the solvent and could be involved in binding to a ligand that interacts with only one of the two chains, α or δ. Alternatively, residue 14 could be involved in specific interaction with the constant region discriminating between Cα and Cδ. In the Ig heavy chain, a residue of similar location in V_H, Leucine or Valine at position 11, contacts several residues of C_H1 (Saul *et al.* 1978). Another residue, glutamine at position 37, although a less convincing case, is frequently absent in most Vδ peptides. This residue forms part of the interface between Ig V_H and V_L. Interestingly, this residue is also replaced in most Vγ peptides. It is tempting to speculate that polar residues of opposite charge, lysine-37 in hDV102 and mDV101 may be responsible for restricted pairing with hGV2 and mGV2, respectively, both carrying glutamic acid-35.

Given the extreme homology between Vα and Vδ peptides, one would expect that, conversely, Vβ and Vγ peptides should be structurally related. We have compared the four human TCRV families in one alignment (Plate 4). In fact, Vγ peptides are more closely related to Vβ as opposed to Vα/δ peptides in that the length of their second complementarity determining region (CDR2) equals that of the Vβ peptides. The CDR2 length of Vβ peptides exceeds that of Vα peptides by 4 residues on the average (Plate 4).

About 30 sites are responsible for the conformation of the Ig V domain framework. Similar residues are found at the homologous sites in the TCR Vα and Vβ peptides (Chothia *et al.* 1988). We found these also conserved in Vγ peptides. Moreover, two residues present in V_L and V_H but absent in Vα and Vβ are also highly conserved in Vγ, alanine-19 and isoleucine-30. The latter are only subtle changes, and it remains to be determined whether γδ receptors have a more Ig like structure as opposed to αβ receptors.

6 Comparison of human and murine variable gene segment subfamilies

In the mouse, eight of the 20 TCRAV subfamilies consist of 6–15 members, including putative alleles. Only five are single member subfamilies. In contrast, 18 of 20 TCRBV subfamilies are single member subfamilies. Surprisingly, the opposite can be observed in the human. Here, the TCRB subfamilies are on the average much larger than the TCRA subfamilies. Only two of the 32 human TCRA subfamilies, AV1 and AV2 contain more than two members, whereas 7 of the 26 different human TCRB subfamilies contain from 3 to 9 members. This raises the question as to whether the TCRA and TCRB repertoires have evolved differently in human and mouse.

To estimate the relatedness between the mouse and human TCRV gene repertoires, the TCR AD, B, and GV gene families, respectively, were compared between the two species. One V gene segment from each subfamily was included in this analysis. The result was that a member of a certain mouse subfamily was closer to a human V gene segment than to a member of another mouse subfamily. Surprisingly, this was true of almost every subfamily of the TCRAD and B families. The comparison between the mouse and human TCRAD families is shown in Figure 14.1. For example, mouse AV19 is the homologue of human AV7, and mouse DV101 appears to be homologous to human DV102. Although this dendrogram does not represent a phylogenetic tree, there seems to be no question that members of different subfamilies from a species are often highly diverged, and pairs of V gene segments are often shared by the two species. In the major histocompatibility complex (MHC), pairs of polymorphic alleles have been shown to persist through speciation events, and were therefore termed trans-specific (Arden and Klein 1982; Klein 1987). Our general observation that pairs of TCRV gene segments are frequently conserved between mouse and human strongly supports the hypothesis of trans-species evolution, i.e. retention of ancestral V gene segments through speciation. Trans-species polymorphism of MHC alleles and trans-specific evolution of TCRV gene segments may indicate co-evolution of the two gene families.

The interspecies comparison between the TCRBV gene segments is shown in Figure 14.2. A comparison of the TCRBV segments between subfamilies within a species shows identities below 40% at the amino acid level (Arden

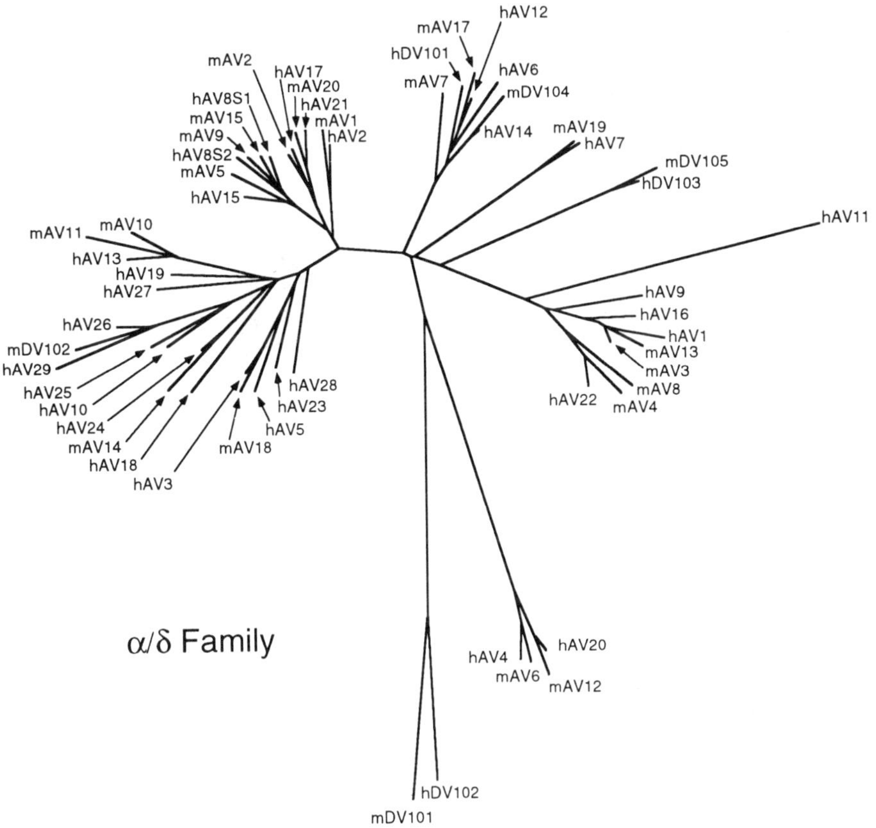

Fig. 14.1 Dendrogram showing the relationship between the human and murine Vα/δ subfamilies. The longer the line that joins two sequences, the greater the divergence. Leader peptide sequences were removed before comparison of the peptide sequences.

et al. 1995*a*; Arden *et al.* 1995*b*). However, when the mouse and human TCRBV segments are compared across species, 14 of the 20 mouse TCRBV subfamilies are greater than 60% similar to their human homologues at the protein level. In the physical maps of the human and murine TCRBV families, the relative order of the TCRBV homologues is highly conserved, apart from internal duplications in the human TCRBV locus (Lai *et al.* 1988*a*). Similarly, V segments from 14 of the 20 mouse TCRAV subfamilies have human counterparts sharing greater than 60% of their amino acid sequence. Despite this high conservation between man and mouse, homologous subfamilies have evolved very differently in the two species. Murine AV4 and AV8 have reached a size of about 10 members each, whereas their only human counterpart AV22 is a single member subfamily. Some subfamilies completely lack homologous genes in the other species. We have, therefore, not attempted

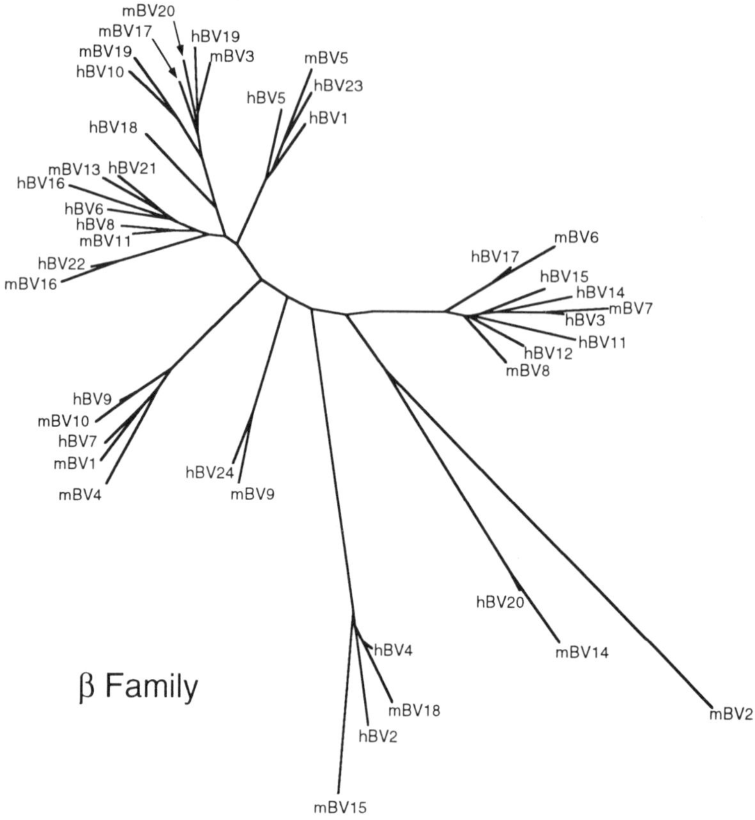

Fig. 14.2 Dendrogram of the human and murine Vβ subfamilies. For explanation, see Figure 14.1.

to design a uniform nomenclature for both human and mouse V segments. Members from individual subfamilies are dispersed throughout the TCRAV locus (Wang *et al.* 1994). Even when their genomic order will be known, there may not be an obvious system for a uniform human/mouse nomenclature.

Out of all the human TCRADV gene segments, DV102 shows the greatest homology to mouse DV101. Even though both are only 34% similar at the amino acid and 60% similar at the nucleotide level, they are obviously homologues in the dendrogram in Figure 14.1. Both V genes are the first to be expressed in ontogeny, in the early fetal wave of γδ-receptor expression in mouse and human (Havran and Allison 1988; Krangel *et al.* 1990; van der Stoep *et al.* 1990). In addition, both V gene segments are localized most proximal to the mouse and human TCRDD gene segments (Hata *et al.* 1989; Malissen *et al.* 1992). In this pair of V genes, homology at the nucleotide level therefore correlates with similar expression patterns as well as genomic localization. The fetal wave of human DV102 expression is followed by dominant expression of DV101 after birth. A mouse TCRDV gene relatively closely related to human DV101 is ADV7 (Figure 14.1; 54% amino acid and 71% nucleotide sequence similarity). Murine DV7 is predominantly found in neonatal thymus (Happ and Palmer 1989). In the human, at least one TCRAV gene segment (AV21) maps between DV101 and DV102 (Hata *et al.* 1989). Similarly, the mouse DV7 gene segments and the TCRAV gene segments are interspersed, the 3′-most DV7 gene segment mapping close to mouse DV101 (Jouvin-Marche *et al.* 1990). Both human DV101 and mouse DV7 gene segments also rearrange to TCRAJ gene segments (Miossec *et al.* 1991; Arden unpublished). In contrast to human DV101, its mouse homologue may have duplicated several times to reach a subfamily size of at least three members. The striking conservation of structure, genomic localization, and expression pattern between mouse and human suggest that these TCRDV gene segments may serve similar functions in these species. This conservation extends to human TCRDV103, which is homologous to murine TCRDV105 (Figure 14.1) and, like murine TCRDV105, is located 3′ to the TCRDC segment, in an inverted orientation. Both share 68% of their amino acid sequence and 74% of their nucleotide sequence. TCRDV105 predominates in adult thymus and spleen (Takagaki *et al.* 1989; Ezquerra *et al.* 1990) and also TCRDV103 only appears later in ontogeny (Krangel *et al.* 1990).

Surprisingly, human TCRGV2, which preferentially pairs with TCRDV102 in early fetal ontogeny (Casorati *et al.* 1989; Krangel *et al.* 1990), does not display significant homology to TCRGV1, its murine counterpart (Figure 14.3),

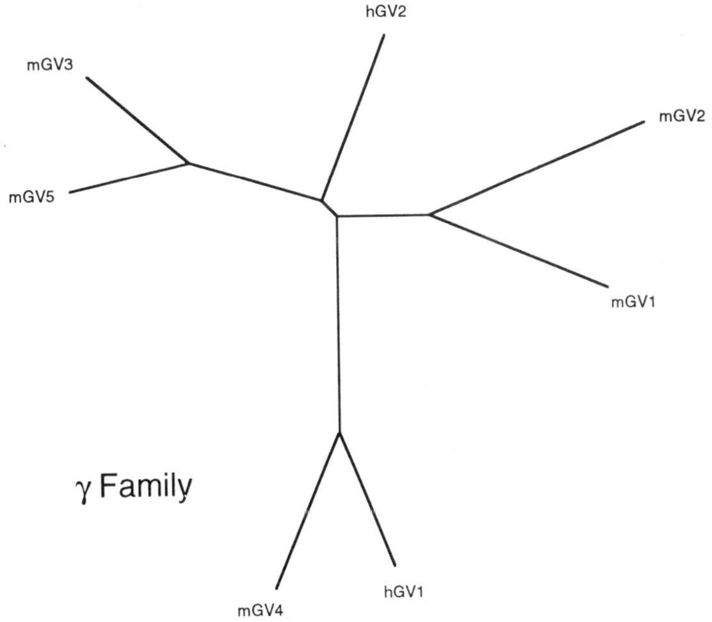

Fig. 14.3 Dendrogram of the human and murine Vγ subfamilies. For explanation, see Figure 13.1.

which is expressed in the first fetal wave together with TCRDV101. There is very little homology between the human and mouse TCRG subfamilies. There are only two values above 40% at the protein level: human GV1 with mouse GV4 at 42% identity, and human GV3 with mouse GV3 at 45%, human GV3, however, being a pseudogene (Zhang *et al.* 1994). The evolutionary conservation of the relative order of TCRV gene segments on the physical maps may arise from the need to express sets of TCRV gene segments at distinct developmental stages, as for example the overrepresentation of the 3′-most V gene segments during fetal and neonatal ontogeny.

The homologues compared above were analysed for codon usage as well as the use of silent nucleotides in individual codons. In many silent site positions, the same nucleotide is shared between V gene segments in both species despite amino acid differences caused by replacement substitutions in the same codon. Leucine is encoded by two codons differing by one as opposed to two silent changes, with one exception (*Leu* at position 75, TTA → CTT in human DV101 → mouse DV7S1). At this position, two silent changes also occur between different subfamilies in the mouse. Similarly, it occurs only in one case that a serine residue is encoded by entirely different codons in different species (*Ser* at site 11, TCC → AGC in human DV101 → mouse DV7S1). At the same site, an entirely different codon (TCT) is also used in the closely related AV17 subfamily in the mouse. The alignments of mouse and human V gene segment sequences facilitate a systematic analysis of codon usage and silent mutations. In addition to the above comparisons, a more systematic study confirmed that convergent evolution does not play a major part, and further supports the hypothesis of trans-species evolution.

Acknowledgements

We are grateful to Constanze Taylor for assistance in preparing this manuscript, and Toby J. Gibson of EMBL, Heidelberg for advice in constructing the peptide alignments.

References

Arden, B. and Klein, J. (1982). Biochemical comparison of major histocompatibility complex molecules from different subspecies of *Mus musculus*: evidence for *trans*-specific evolution of alleles. *Proc. Natl. Acad. Sci. USA*, **79**, 2342–6.

Arden, B., Klotz, J. L., Siu, G., and Hood, L. E. (1985). Diversity and structure of genes of the α family of mouse T-cell antigen receptor. *Nature*, **316**, 783–7.

Arden, B., Clark, S. P., and Mak, T. W. (1995*a*). Mouse T-cell receptor variable gene segment families. *Immunogenetics*, in press

Arden, B., Clark, S. P., and Mak, T. W. (1995*b*). Human T-cell receptor variable gene segment families. *Immunogenetics*, in press.

Asarnow, D. M., Kuziel, W. A., Bonyhadi, M., Tigelaar, R. E., Tucker, P. W., and Allison, J. P. (1988). Limited diversity of γδ antigen receptor genes of Thy-1^+ dendritic epidermal cells. *Cell*, **55**, 837–47.

Barth, R. K., Kim, B. S., Lan, N. C., Hunkapiller, T., Sobieck, N., Winoto, A., *et al.* (1985). The murine T-cell receptor uses a limited repertoire of expressed Vβ gene segments. *Nature*, **316**, 517–23.

Becker, D. M., Patten, P., Chien, Y. H., Yokota, T., Eshhar, Z., Giedlin, M., *et al.* (1985). Variability and repertoire size of T-cell receptor Vα gene segments. *Nature*, **317**, 430–4.

Behlke, M. A. and Loh, D. Y. (1986). Alternative splicing of murine T-cell receptor β-chain transcripts. *Nature*, **322**, 379–82.

Behlke, M. A., Spinella, D. G., Chou, H. S., Sha, W., Hartl, D. L., and Loh, D. Y. (1985). T-cell receptor β-chain expression: dependence on relatively few variable region genes. *Science*, **229**, 566–70.

Behlke, M. A., Chou, H. S., Huppi, K., and Loh, D. Y. (1986). Murine T-cell receptor mutants with deletions of β-chain variable region genes. *Proc. Natl. Acad. Sci. USA*, **83**, 761–71.

Bernard, O., Groettrup, M., Mugneret, F., Berger, R., and Azogui, O. (1993). Molecular analysis of T-cell receptor transcripts in a human T-cell leukemia bearing a t(1;14) and an inv(7); cell surface expression of a TCR β chain in the absence of α chain. *Leukemia* **7**, 1645–53.

Boitel, B., Ermonval, M., Panina-Bordignon, P., Mariuzza, R. A., Lanzavecchia, A., and Acuto, O. (1992). Preferential Vβ gene usage and lack of junctional sequence conservation among human T cell receptors specific for a tetanus toxin-derived peptide: evidence for a dominant role of a germline-encoded V region in antigen/major histocompatability complex recognition. *J. Exp. Med.*, **175**, 765–77.

Casorati, G., de Libero, G., Lanzavecchia, A., and Migone, N. (1989). Molecular analysis of human γ/δ^+ clones from thymus and peripheral blood. *J. Exp. Med.*, **170**, 1521–35.

Cazenave, P. A., Marche, P. N., Jouvin-Marche, E., Voegtlé, D., Bonhomme, F., Bandeira, A., and Coutinho, A. (1990). Vβ17 gene polymorphism in wild-derived mouse strains: two amino acid substitutions in the Vβ17 region greatly alter T cell receptor specificity. *Cell*, **63**, 717–28.

Charmley, P., Wang, K., Hood, L., and Nickerson, D. A. (1993*a*). Identification and physical mapping of a polymorphic human T cell receptor Vβ gene with a frequent null allele. *J. Exp. Med.*, **177**, 135–43.

Charmley, P., Wei, S., and Concannon, P. (1993*b*). Polymorphisms in the Tcrb-V2 gene segments localize the Tcrb orphon genes to human chromosome 9p21. *Immunogenetics*, **38**, 283–6.

Charmley, P., Nickerson, D. A., and Hood, L. (1994). Polymorphism detection and sequence analysis of human T-cell receptor V alpha-chain-encoding gene segments. *Immunogenetics*, **39**, 138–45.

Chothia, C., Boswell, D. R., and Lesk, A. M. (1988). The outline structure of the T-cell αβ receptor. *EMBO J.*, **7**, 3745–55.

Chou, H. S., Behlke, M. A., Godambe, S. A., Russell, J. H., Brooks, C. G., and Loh, D. Y. (1986). T cell receptor genes in an alloreactive CTL clone: implications for rearrangement and germline diversity of variable gene segments. *EMBO J.*, **5**, 2149–55.

Chou, H. S., Anderson, S. J., Louie, M. C., Godambe, S. A., Pozzi, M. R., Behlke, M. A., *et al.* (1987). Tandem linkage and unusual RNA splicing of the T-cell receptor β-chain variable-region genes. *Proc. Natl. Acad. Sci. USA*, **84**, 1992–6.

Clark, S. P., Arden, B., and Mak, T. W. (1995). Comparison of human and murine T-cell receptor variable gene segment subfamilies. *Immunogenetics*, in press.

Concannon, P., Pickering, L. A., Kung, P., and Hood, L. (1986). Diversity and structure of human T-cell receptor β-chain variable region genes. *Proc. Natl. Acad. Sci. USA*, **83**, 6598–602.

Cornélis, F., Pile, K., Loveridge, J., Moss, P., Harding, R., Julier, C., and Bell, J. (1993). Systematic study of human αβ T cell receptor V segments shows allelic variations resulting in a large number of distinct T cell receptor haplotypes. *Eur. J. Immunol.*, **23**, 1277–83.

Couez, D., Malissen, M., Buferne, M., Schmitt-Verhulst, A. M., and Malissen, B. (1991). Each of the two productive T cell receptor α-gene rearrangements found in both the A10 and BM 3.3 T cell clones give rise to an α chain which can contribute to the constitution of a surface-expressed αβ dimer. *Int. Immunol.*, **3**, 719–29.

Dariavach, P. and Lefranc, M.-P. (1989). First genomic sequence of the human T-cell receptor δ2 gene (TRDV2). *Nucleic Acids Res.*, **17**, 4880.

Day, C. E., Zhao, T., and Robinson, M. A. (1992). Silent allelic variants of a T-cell receptor Vβ12 gene are present in diverse human populations. *Hum. Immunol.*, **34**, 196–202.

Dembic, Z., Haas, W., Weiss, S., McCubrey, J., Kiefer, H., von Boehmer, H., *et al.* (1986). Transfer of specificity by murine α and β T-cell receptor genes. *Nature*, **320**, 232–8.

Devaux, B., Bjorkman, P. J., Stevenson, C., Greif, W., Elliott, J. F., Sagerström, C., *et al.* (1991). Generation of monoclonal antibodies against soluble human T cell receptor polypeptides. *Eur. J. Immunol.*, **21**, 2111–19.

Elliott, J. F., Rock, E. P., Patten, P. A., Davis, M. M., and Chien, Y. H. (1988). The adult T-cell receptor δ-chain is diverse and distinct from that of fetal thymocytes. *Nature*, **331**, 627–31.

Epplen, J. T., Bartels, F., Becker, A., Nerz, G., Prester, M., Rinaldy, A., *et al.* (1986). Change in antigen specificity of cytotoxic T-lymphocytes is associated with the rearrangement and expression of a T-cell receptor β-chain gene. *Proc. Natl. Acad. Sci. USA*, **83**, 4441–5.

Ezquerra, A., Cron, R. Q., McConnell, T. J., Valas, R. B., Bluestone, J. A., and Coligan, J. E. (1990). T cell receptor δ-gene expression and diversity in the mouse spleen. *J. Immunol.*, **145**, 1311–17.

Ferradini, L., Roman-Roman, S., Azocar, J., Michalaki, H., Triebel, F., and Hercend, T. (1991). Studies on the human T cell receptor α/β variable region genes. II. Identification of four additional Vβ subfamilies. *Eur. J. Immunol.*, **21**, 935–42.

Ferrick, D. A., Ohashi, P. S., Wallace, V., Schilham, M., and Mak, T. W. (1989). Thymic ontogeny and selection of αβ and γδ T cells. *Immunology Today*, **10**, 403–7.

Fink, P. J., Matis, L. A., McElligott, D. L., Bookman, M., and Hedrick, S. M. (1986). Correlations between T-cell specificity and the structure of the antigen receptor. *Nature*, **321**, 219–26.

Forster, A., Huck, S., Ghanem, N., Lefranc, M.-P., and Rabbitts, T. H. (1987). New subgroups in the human T cell rearranging Vγ gene locus. *EMBO J.*, **6**, 1945–50.

Garman, R. D., Doherty, P. J., and Raulet, D. H. (1986). Diversity, rearrangement, and expression of murine T cell γ genes. *Cell*, **45**, 733–42.

Gascoigne, N. R. J., Waters, S., Elliott, J. F., Victor-Kobrin, C., Goodnow, C., Davis, M. M., *et al.* (1986). Expression of T cell receptor genes in an antigen-specific hybridoma and radiation-induced variants. *J. Exp. Med.*, **164**, 113–30.

Goverman, J., Minard, K., Shastri, N., Hunkapiller, T., Hansburg, D., Sercarz, E., *et al.* (1985). Rearranged β T cell receptor genes in a helper T cell clone specific for lysozyme: no correlation between Vβ and MHC restriction. *Cell*, **40**, 859–67.

Guglielmi, P., Davi, F., d'Auriol, L., Bories, J.-C., Dausset, J., and Bensussan, A. (1988). Use of a variable α region to create a functional T-cell receptor δ chain. *Proc. Natl. Acad. Sci. USA*, **85**, 5634–8.

Happ, M. P. and Palmer, E. (1989). Thymocyte development: an analysis of T cell receptor gene expression in 519 newborn thymocyte hybridomas. *Eur. J. Immunol.*, **19**, 1317–25.

Hata, S., Clabby, M., Devlin, P., Spits, H., de Vries, J. E., and Krangel, M. S. (1989). Diversity and organization of human T cell receptor δ variable gene segments. *J. Exp. Med.*, **169**, 41–57.

Havran, W. L. and Allison, J. P. (1988). Developmentally ordered appearance of thymocytes expressing different T-cell antigen receptors. *Nature*, **335**, 443–5.

Hayday, A. C., Diamond, D. J., Tanigawa, G., Heilig, J. S., Folsom, V., Saito, H., *et al.* (1985*a*). Unusual organization and diversity of T-cell receptor α-chain genes. *Nature*, **316**, 828–32.

Hayday, A. C., Saito, H., Gillies, S. D., Kranz, D. M., Tanigawa, G., Eisen, H. N. *et al.* (1985*b*). Structure, organization, and somatic rearrangement of T cell γ genes. *Cell*, **40**, 259–69.

Heilig, J. S. and Tonegawa, S. (1986). Diversity of murine γ genes and expression in fetal and adult T lymphocytes. *Nature*, **322**, 836–40.

Heuer, J., Degwert, J., Pauels, H. G., and Koelsch, E. (1991). T cell receptor α and β gene expression in a murine antigen-specific T suppressor lymphocyte clone with cytolytic potential. *J. Immunol.*, **146**, 775–82.

Hirama, T., Takeshita, S., Matsubayashi, Y., Iwashiro, M., Masuda, T., Kuribayashi, K., *et al.* (1991). Conserved V(D)J junctional sequence of cross-reactive cytotoxic T-cell receptor idiotype and the effect of a single amino acid substitution. *Eur. J. Immunol.*, **21**, 483–8.

Huck, S., Dariavach, P., and Lefranc, M.-P. (1988). Variable region genes in the human T-cell rearranging gamma (TRG) locus: V-J junction and homology with the mouse genes. *EMBO J.*, **7**, 719–26.

Hurley, C. K., Steiner, N., Wagner, A., Geiger, M. J., Eckels, D. D., and Rosen-Bronson, S. (1993). Nonrandom T cell receptor usage in the allorecognition of HLA-DR1 microvariation. *J. Immunol.*, **150**, 1314–24.

Ikuta, K., Ogura, T., Shimizu, A., and Honjo, T. (1985). Low frequency of somatic mutation in β-chain variable region genes of human T-cell receptors. *Proc. Natl. Acad. Sci. USA*, **82**, 7701–5.

Iwamoto, A., Ohashi, P. S., Pircher, H., Walker, C. L., Michalopoulos, E. E., Rupp, F., *et al.* (1987). T cell receptor variable gene usage in a specific cytotoxic T cell response. Primary structure of the antigen-MHC receptor of four hapten-specific cyotoxic T cell clones. *J. Exp. Med.*, **165**, 591–600.

Iwashima, M., Green, A., Davis, M. M., and Chien, Y. H. (1988). Variable region (Vδ) gene segment most frequently utilized in adult thymocytes is 3′ of the constant (Cδ) region. *Proc. Natl. Acad. Sci. USA*, **85**, 8161–5.

Jameson, S. C., Nakajima, P. B., Brooks, J. L., Heath, W., Kanagawa, O., and Gascoigne, N. R. J. (1991). The T cell receptor Vα11 gene family: analysis of allelic sequence polymorphism and demonstration of Jα region-dependent recognition by allele-specific antibodies. *J. Immunol.*, **147**, 3185–93.

Jores, R. and Meo, T. (1993). Few V gene segments dominate the T cell receptor β-chain repertoire of the human thymus. *J. Immunol.*, **151**, 6110–22.

Jouvin-Marche, E., Hue, I., Marche, P. N., Liebe-Gris, C., Marolleau, J.-P.,

Malissen, B. *et al.* (1990). Genomic organization of the mouse T cell receptor Vα family. *EMBO J.*, **9**, 2141–50.

Kalams, S. A., Johnson, R. P., Trocha, A. K., Dynan, M. J., Ngo, H. S., D'Aquila, R. T. *et al.* (1994). Longitudinal analysis of T cell receptor (TCR) gene usage by human immunodeficiency virus 1 envelope-specific cytotoxic T lymphocyte clones reveals a limited TCR repertoire. *J. Exp. Med.*, **179**, 1261–71.

Kimura, N., Toyonaga, B., Yoshikai, Y., Triebel, F., Debre, P., Minden, M. D., *et al.* (1986). Sequences and diversity of human T-cell receptor β-chain variable region genes. *J. Exp. Med.*, **164**, 739–50.

Kimura, N., Toyonaga, B., Yoshikai, Y., Du, R.-P., and Mak, T. W. (1987). Sequences and repertoire of the human T cell receptor α and β chain variable region genes in thymocytes. *Eur. J. Immunol.*, **17**, 375–83.

Klein, J. (1987). Origin of major histocompatibility complex polymorphism: the trans-species hypothesis. *Hum. Immunol.*, **19**, 155–62.

Klein, M. H., Concannon, P., Everett, M., Kim, L. D. H., Hunkapiller, T., and Hood, L. (1987). Diversity and structure of human T-cell receptor α-chain variable region genes. *Proc. Natl. Acad. Sci. USA*, **84**, 6884–8.

Klotz, J. L., Barth, R. K., Kiser, G. L., Hood, L. E., and Kronenberg, M. (1989). Restriction fragment length polymorphisms of the mouse T-cell receptor gene families [published erratum appears in *Immunogenetics* (1989) **30**, 235]. *Immunogenetics*, **29**, 191–201.

Koop, B. F., Wilson, R. K., Wang, K., Vernooij, B., Zaller, D., Kuo, C. L., *et al.* (1992). Organization, structure, and function of 95 kb of DNA spanning the murine T-cell receptor Cα/Cδ region. *Genomics*, **13**, 1209–30.

Korman, A. J., Maruyama, J., and Raulet, D. H. (1989). Rearrangement by inversion of a T-cell receptor δ variable region gene located 3′ of the δ constant region gene. *Proc. Natl. Acad. Sci. USA*, **86**, 267–71.

Koseki, H., Imai, K., Ichikawa, T., Hayata, I., and Taniguchi, M. (1989). Predominant use of a particular α-chain in suppressor T cell hybridomas specific for keyhole limpet hemocyanin. *Int. Immunol.*, **1**, 557–64.

Krangel, M. S., Yssel, H., Brocklehurst, C., and Spits, H. (1990). A distinct wave of human T cell receptor γ/δ lymphocytes in the early fetal thymus: evidence for controlled gene rearrangement and cytokine production. *J. Exp. Med.*, **172**, 847–59.

Lai, E., Concannon, P., and Hood, L. (1988*a*). Conserved organization of the human and murine T-cell receptor β-gene families. *Nature*, **331**, 543–6.

Lai, M. Z., Huang, S. Y., Briner, T. J., Guillet, J. G., Smith, J. A., and Gefter, M. L. (1988*b*). T cell receptor gene usage in the response to λ repressor cI protein: an apparent bias in the usage of a Vα gene element. *J. Exp. Med.*, **168**, 1081–97.

Lai, E., Wilson, R. K., and Hood, L. E. (1989). Physical maps of the mouse and human immunoglobulin-like loci. *Adv. Immunol.*, **46**, 1–59.

Lee, N. E. and Davis, M. M. (1988). T cell receptor β-chain genes in BW5147 and other AKR tumors: deletion order of murine Vβ gene segments and possible 5′ regulatory regions. *J. Immunol.*, **140**, 1665–75.

Lefranc, M.-P., Forster, A., Baer, R., Stinson, M. A., and Rabbitts, T. H. (1986). Diversity and rearrangement of the human T cell rearranging γ genes: nine germ-line variable genes belonging to two subgroups. *Cell* **45**, 237–46.

Leiden, J. M., Fraser, J. D., and Strominger, J. L. (1986). The complete primary structure of the T-cell receptor genes from an alloreactive cytotoxic human T-lymphocyte clone. *Immunogenetics*, **24**, 17–23.

Loh, E. Y., Cwirla, S., Serafini, A. T., Phillips, J. H., and Lanier, L. L. (1988).

Human T-cell-receptor δ chain: genomic organization, diversity, and expression in populations of cells. *Proc. Natl. Acad. Sci. USA*, **85**, 9714–18.

Louie, M. C., Nelson, C. A., and Loh, D. Y. (1989). Identification and characterization of new murine T cell receptor β-chain variable region (Vβ) genes. *J. Exp. Med.*, **170**, 1987–98.

Lyon, M. F. (1989). Rules and guidelines for gene nomenclature. In *Genetic Variants and Strains of the Laboratory Mouse*, 2nd edn. (ed. M. F. Lyon and A. G. Searle), pp. 1–11. Oxford University Press, Oxford.

Malissen, M., McCoy, C., Blanc, D., Trucy, J., Devaux, C., Schmitt-Verhulst, A. M., *et al.* (1986). Direct evidence for chromosomal inversion during T-cell receptor β-gene rearrangements. *Nature*, **319**, 28–33.

Malissen, M., Trucy, J., Letourneur, F., Rebaï, N., Dunn, D. E., Fitch, F. W., *et al.* (1988). A T cell clone expresses two T cell receptor α genes but uses one αβ heterodimer for allorecognition and self MHC-restricted antigen recognition. *Cell*, **55**, 49–59.

Malissen, M., Trucy, J., Jouvin-Marche, E., Cazenave, P.-A., Scollay, R., and Malissen, B. (1992). Regulation of TCR α and β gene allelic exclusion during T-cell development. *Immunol. Today*, **13**, 315–22.

Miossec, C., Caignard, A., Ferradini, L., Roman-Roman, S., Faure, F., Michalaki, H., *et al.* (1991). Molecular characterization of human T cell receptor α chains including a Vδ1-encoded variable segment. *Eur. J. Immunol.*, **21**, 1061–4.

Moss, P. A. H., Rosenberg, W. M. C., Zintzaras, E., and Bell, J. I. (1993). Characterization of the human T cell receptor α-chain repertoire and demonstration of a genetic influence on Vα usage. *Eur. J. Immunol.*, **23**, 1153–9.

Obata, F., Tsunoda, M., Kaneko, T., Ito, K., Ito, I., Masewicz, S., *et al.* (1993). Human T-cell receptor TCRAV, TCRBV, and TCRAJ sequences newly found in T-cell clones reactive with allogeneic HLA class II antigens. *Immunogenetics*, **38**, 67–70.

Pelkonen, J., Traunecker, A., and Karjalainen, K. (1987). A new mouse TCR Vγ gene that shows remarkable evolutionary conservation. *EMBO J.*, **6**, 1941–4.

Plaza, A., Kono, D. H., and Theofilopoulos, A. N. (1991). New human Vβ genes and polymorphic variants. *J. Immunol.*, **147**, 4360–5.

Raulet, D. H. (1989). The structure, function, and molecular genetics of the γ/δ T cell receptor. *Annu. Rev. Immunol.*, **7**, 175–207.

Reyburn, H., Cornélis, F., Russell, V., Harding, R., Moss, P., and Bell, J. (1993). Allelic polymorphism of human T-cell receptor V alpha gene segments. *Immunogenetics*, **38**, 287–91.

Robinson, M. A. (1989). Allelic sequence variations in the hypervariable region of a T-cell receptor β chain: correlation with restriction fragment length polymorphism in human families and populations. *Proc. Natl. Acad. Sci. USA*, **86**, 9422–6.

Robinson, M. A. (1991). The human T cell receptor β-chain gene complex contains at least 57 variable gene segments. Identification of six Vβ genes in four new gene families. *J. Immunol.*, **146**, 4392–7.

Roman-Roman, S., Ferradini, L., Azocar, J., Genevée, C., Hercend, T., and Triebel, F. (1991). Studies on the human T cell receptor α/β variable region genes. I. Identification of 7 additional Vα subfamilies and 14 Jα gene segments. *Eur. J. Immunol.*, **21**, 927–33.

Saito, H., Kranz, D. M., Takagaki, Y., Hayday, A. C., Eisen, H. N., and Tonegawa, S. (1984). Complete primary structure of a heterodimeric T-cell receptor deduced from cDNA sequences. *Nature*, **309**, 757–62.

Santamaria, P., Lewis, C., and Barbosa, J. J. (1993). Amino acid sequences of seven Vβ, eight Vα, and thirteen Jα novel human TCR genes. *Immunogenetics*, **38**, 163.

Satyanarayana, K., Hata, S., Devlin, P., Roncarolo, M. G., De Vries, J. E., Spits, H., *et al.* (1988). Genomic organization of the human T-cell antigen-receptor α/β locus. *Proc. Natl. Acad. Sci. USA*, **85**, 8166–70.

Saul, F. A., Amzel, M. L., and Poljak, R. J. (1978). Preliminary refinement and structural analysis of the Fab fragment from human immunoglobulin New at 2.0 Å resolution. *J. Biol. Chem.*, **253**, 585–97.

Seto, D., Koop, B. F., Deshpande, P., Howard, S., Seto, J., Wilk, E., *et al.* (1994). Organization, sequence, and function of 34.5 kb of genomic DNA encompassing several murine T-cell receptor α/δ variable gene segments. *Genomics*, **20**, 258–66.

Sherman, D. H., Hochman, P. S., Dick, R., Tizard, R., Ramachandran, K. L., Flavell, R. A., *et al.* (1987). Molecular analysis of antigen recognition by insulin-specific T-cell hybridomas from B6 wild-type and bm12 mutant mice. *Mol. Cell. Biol.*, **7**, 1865–72.

Siu, G., Strauss, E. C., Lai, E., and Hood, L. E. (1986). Analysis of a human Vβ gene subfamily. *J. Exp. Med.*, **164**, 1600–14.

Six, A., Jouvin-Marche, E., Loh, D. Y., Cazenave, P. A., and Marche, P. N. (1991). Identification of a T cell receptor β chain variable region, Vβ20, that is differentially expressed in various strains of mice. *J. Exp. Med.*, **174**, 1263–6.

Slightom, J. L., Siemieniak, D. R., Sieu, L. C., Koop, B. F., and Hood, L. (1994). Nucleotide sequence analysis of 77.7 kb of the human Vβ T-cell receptor gene locus: direct primer-walking using cosmid template DNAs. *Genomics*, **20**, 149–68.

Smith, L. R., Plaza, A., Singer, P. A., and Theofilopoulos, A. N. (1990). Coding sequence polymorphisms among Vβ T cell receptor genes. *J. Immunol.*, **144**, 3234–7.

Smith, W. J., Tunnacliffe, A., and Rabbits, T. H. (1987). Germline sequence of two human T-cell receptor Vβ genes: Vβ8.1 is transcribed from a TATA-box promoter. *Nucleic Acids Res.*, **15**, 4991.

Sottini, A., Imberti, L., Fiordalisi, G., and Primi, D. (1991). Use of variable human Vδ genes to create functional T cell receptor α chain transcripts. *Eur. J. Immunol.*, **21**, 2455–9.

Spieß, S., Kuhröber, A., Schirmbeck, R., and Reimann, J. (1992). Bone marrow cells of athymic nude mice express functional T cell receptor α chain transcripts rearranged to Vδ2, 3, 4, 5, 6 genes. *Eur. J. Immunol.*, **22**, 1939–42.

Spinella, D. G., Hansen, T. H., Walsh, W. D., Behlke, M. A., Tillinghast, J. P., Chou, H. S., *et al.* (1987). Receptor diversity of insulin-specific T cell lines from C57BL (H-2^b) mice. *J. Immunol.*, **138**, 3991–5.

Stoehr, P. J. and Cameron, G. N. (1991). The EMBL data library. *Nucleic Acids Res.*, **19** (Suppl.), 2227–30.

Strauss, W. M., Quertermous, T., and Seidman, J. G. (1987). Measuring the human T cell receptor γ-chain. *Science*, **237**, 1217–19.

Sutherland, R. M., Paterson, Y., Scherle, P. A., Gerhard, W., and Caton, A. J. (1991). A new mouse T-cell receptor α chain variable region family. *Immunogenetics*, **34**, 372–5.

Takagaki, Y., Nakanishi, N., Ishida, I., Kanagawa, O., and Tonegawa, S. (1989). T cell receptor-γ and -δ genes preferentially utilized by adult thymocytes for the surface expression. *J. Immunol.*, **142**, 2112–21.

Takihara, Y., Reimann, J., Michalopoulos, E., Ciccone, E., Moretta, L., and Mak, T. W. (1989). Diversity and structure of human T cell receptor δ chain genes in peripheral blood γ/δ bearing T lymphocytes. *J. Exp. Med.*, **169**, 393–405.

Thompson, J. D., Higgins, D. G., and Gibson, T. J. (1994). CLUSTAL W: improving the sensitivity of progressive multiple sequence alignment through sequence weight-

ing, position specific gap penalties and weight matrix choice. *Nucleic Acids Res.* **22**, 4673–80.
Tillinghast, J. P., Behlke, M. A., and Loh, D. Y. (1986). Structure and diversity of the human T-cell receptor β-chain variable region genes. *Science*, **233**, 879–83.
Toda, M., Fujimoto, S., Iwasato, T., Takeshita, S., Tezuka, K., Ohbayashi, T., *et al.* (1988). Structure of extrachromosomal circular DNAs excised from T-cell antigen receptor α and δ-chain loci. *J. Mol. Biol.*, **202**, 219–31.
Toyonaga, B. and Mak, T. W. (1987). Genes of the T-cell antigen receptor in normal and malignant T cells. *Ann. Rev. Immunol.*, **5**, 585–620.
Traunecker, A., Oliveri, F., Allen, N., and Karjalainen, K. (1986). Normal T cell development is possible without 'functional' γ chain genes. *EMBO J.*, **5**, 1589–93.
Tunnacliffe, A., Kefford, R., Milstein, C., Forster, A., and Rabbitts, T. H. (1985). Sequence and evolution of the human T-cell antigen receptor β-chain genes. *Proc. Natl. Acad. Sci. USA*, **82**, 5068–72.
van der Stoep, N., de Krijger, R., Bruining, J., Koning, F., and van den Elsen, P. (1990). Analysis of early fetal T-cell receptor δ chain in humans. *Immunogenetics*, **32**, 331–6.
Wang, K., Klotz, J. L., Kiser, G., Bristol, G., Hays, E., Lai, E., *et al.* (1994). Organization of the V gene segments in mouse T-cell antigen receptor α/δ locus. *Genomics*, **20**, 419–28.
Wedderburn, L. R., O'Hehir, R. E., Hewitt, C. R. A., Lamb, J. R., and Owen, M. J. (1993). In vivo clonal dominance and limited T-cell receptor usage in human $CD4^+$ T-cell recognition of house dust mite allergens. *Proc. Natl. Acad. Sci. USA*, **90**, 8214–18.
Wei, S., Charmley, P., Robinson, M. A., and Concannon, P. (1994). The extent of the human germline T-cell receptor V beta gene segment repertoire. *Immunogenetics*, **40**, 27–36.
Williams, A. F., Strominger, J. L., Bell, J., Mak, T. W., Kappler, J., Marrack, P., *et al.* (1993). Nomenclature for T-cell receptor (TCR) gene segments of the immune system. *WHO Bull.*, **71**, 113–15.
Wilson, R. K., Lai, E., Kim, L. D. H., and Hood, L. E. (1990). Sequence and expression of a novel human T-cell receptor β-chain variable gene segment subfamily. *Immunogenetics*, **32**, 406–12.
Winoto, A., Mjolsness, S., and Hood, L. (1985). Genomic organization of the genes encoding mouse T-cell receptor α-chain. *Nature*, **316**, 832–6.
Wright, J. A., Hood, L., and Concannon, P. (1991). Human T-cell receptor Vα gene polymorphism. *Hum. Immunol.*, **32**, 277–83.
Yague, J., Blackman, M., Born, W., Marrack, P., Kappler, J., and Palmer, E. (1988). The structure of Vα and Jα segments in the mouse. *Nucleic Acids Res.*, **16**, 11355–64.
Yanagi, Y., Chan, A., Chin, B., Minden, M., and Mak, T. W. (1985). Analysis of cDNA clones specific for human T cells and the α and β chains of the T-cell receptor heterodimer from a human T-cell line. *Proc. Natl. Acad. Sci. USA*, **82**, 3430–4.
Yoshikai, Y., Kimura, N., Toyonaga, B., and Mak, T. W. (1986). Sequences and repertoire of human T-cell receptor α chain variable region genes in mature T lymphocytes. *J. Exp. Med.*, **164**, 90–103.
Zhang, X.-M., Tonnelle, C., Lefranc, M.-P., and Huck, S. (1994). T cell receptor γ cDNA in human fetal liver and thymus: variable regions of γ chains are restricted to VγI or V9, due to the absence of splicing of the V10 and V11 leader intron. *Eur. J. Immunol.*, **24**, 571–8.

15 T cell receptor V(D)J recombination: mechanisms and developmental regulation

HAYDN M. PROSSER AND SUSUMU TONEGAWA

1 Introduction

The vertebrate immune system has evolved highly variable antigen receptor molecules in order to identify and neutralize the enormous range of invasive organisms which may be encountered (Tonegawa 1983; Blackwell and Alt 1988; Davis 1988). The variability of the two basic forms of receptors, T cell receptor (TCR) and the immunoglobulins (Ig), is dependent upon a common process of site-specific recombination within the encoding DNA. This V(D)J recombination activity enables a specific combination of the multiple variable (V), diversity (D) (in certain cases), and joining (J) region gene segments to fuse into a single genetic unit encoding the variable region of the antigen receptors. Additional variability is accomplished by the imprecise joining of the V, D, and J gene segments due to the loss of and/or addition of nucleotides at the join.

Since its original description V(D)J recombination has remained a subject of intensive investigation. However, the essence of how the mechanism operates remains a mystery. There is no solid evidence as to the identity of the enzymatic component(s) of the V(D)J recombinase; although some promising leads have emerged. Prime among these leads is the cloning of the recombination activating genes, RAG-1 and RAG-2, whose products are capable of making cells competent for V(D)J recombination which are normally lacking in this activity (Schatz *et al.* 1989; Oettinger *et al.* 1990). In the absence of the ability to directly study the biochemistry of the V(D)J recombinase much of our knowledge of the system stems from investigations of the substrates and products of the process.

This chapter discusses what is known about V(D)J recombination within thymocytes which gives rise to the two forms of TCR, αβ, and γδ. However, due to the common nature of V(D)J recombination within both the T and B branches of lymphocyte development much evidence will be drawn from studies on pre-B cells and the rearrangement of Ig genes. In addition to the discussion of the mechanism of recombination there will be a summary of the role of recombination in T cell development with particular reference to evidence from mutant mice with TCRs disrupted by homologous recombination.

2 V(D)J recombination

2.1 Recombination substrates

The T cell receptor (and Ig) V, (D), and J gene segments are flanked by conserved recombination signal sequences (RSS) (Figure 15.1a) (Tonegawa 1983; Hesse *et al.* 1989). The RSSs act as recognition sites for the V(D)J recombinase activity and delineate the points at which somatic recombination occurs. Each RSS is comprised of two consensus sequences, a heptamer and an A/T rich nonamer, which are separated by a non-conserved spacer of either 12(+/− 1 bp) or 23(+/− 1 bp) nucleotides in length. Recombination is governed by the '12–23' rule whereby RSSs with a 12 nucleotide spacer may only recombine with RSSs with a 23 nucleotide spacer and vice versa. The arrangement of RSSs determines which combinations of TCR gene rearrangements are possible (Figure 15.1b). Thus V gene segments, with a 23 bp spacer RSS 3′ of the segment, can recombine with D gene segments or directly with J gene segments, both with a 12 bp spacer RSS 5′ of the segment. The D gene segments of TCR β and δ are flanked by a 5′ 23 bp spacer RSS

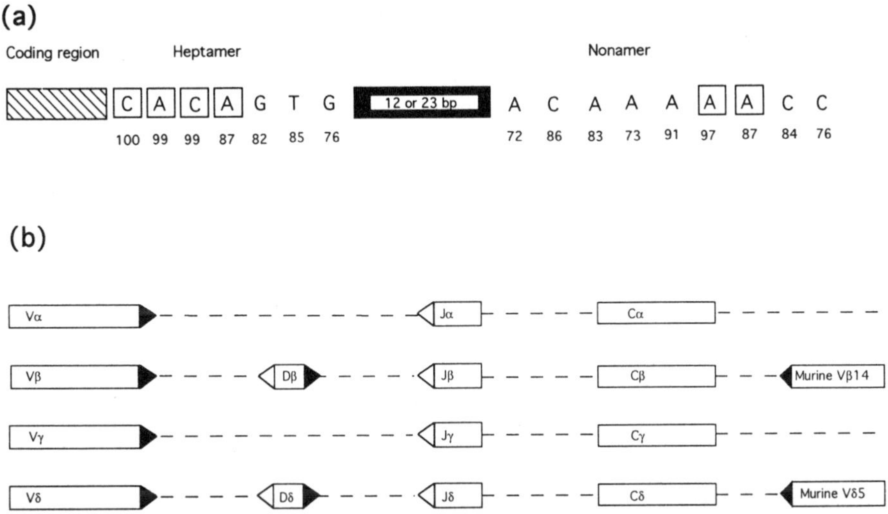

Fig. 15.1 (a) The consensus sequences of the heptamer and nonamer components of the RSS (Hesse *et al.* 1989). The numbers below each nucleotide represent the percentage with which nucleotides are found the RSSs of a large number of antigen receptor loci. The boxed nucleotides are the most important components of the RSS based on mutational analysis. (b) Schematic representation of the arrangement of 12 bp spacer RSSs (clear triangles) and 23 bp spacer RSSs (filled triangles) in relation to the TCR coding sequences (rectangles). The illustration is not to scale and does not represent the TCR loci accurately. The Vβ14 gene found 3′ of the constant region in the murine TCR β locus is indicated.

and a 3′ 12 bp spacer RSS so that they can recombine with both V and J gene segments, and with other D gene segments.

V(D)J recombination can be studied experimentally by the introduction of artificial recombination substrates into appropriate cell types (Akira *et al.* 1987; Hesse *et al.* 1987). Artificial recombination substrates are bacterial plasmid constructs containing RSSs divided by intervening DNA. The development of a quantitative assay for V(D)J recombination activity using artificial recombination substrates has allowed the importance of each nucleotide of a heptamer or nonamer sequence to be assessed by mutagenesis (Hesse *et al.* 1989) (Figure 15.1b). The fact that RSSs taken out of the context of antigen receptor genes are still functional as recombination substrates signifies that they are the only DNA motifs necessary to direct V(D)J recombination. The conservation of the RSSs between TCR and Ig loci and the observation that the same RSS are functional in both B and T cells are strong arguments in favour of a single V(D)J recombinase being shared by all lymphocytes (Yancopoulos *et al.* 1986; Lieber *et al.* 1988*a*).

2.2 Recombination products

V(D)J recombination of the TCR loci most frequently occurs by deletion of the intervening chromosomal DNA separating the oppositely oriented RSSs of the V, D, or J fusion partners (Figure 15.2a). Two products arise from this process, each associated with a particular joint. A ‘coding joint’ is formed between the coding regions of the rearranged genes within the chromosome. The coding joints exhibit considerable variability due to the loss and/or addition of nucleotides. The second product is a microcircle comprised of deleted intervening DNA including the two RSSs. The two RSSs of the microcircle fuse without junctional variation at their respective heptamer sequences in a head to head fashion to form the ‘signal joint’. The existence of microcircle DNA by-products was demonstrated by their isolation and characterization in thymocytes undergoing V(D)J recombination (Fujimoto and Yamaguchi 1987; Okazaki *et al.* 1987; Okazaki and Sakano 1988; Winoto and Baltimore 1989*a*). An alternative form of rearrangement is encountered in situations where the RSS of a V gene segment is oriented in the same direction as the RSS of the J or D gene segment fusion partners (Figure 15.2b). This is true in the case of the murine Vβ4 (Malissen *et al.* 1986) and Vδ5 gene segments (Korman *et al.* 1989; Iwashima *et al.* 1989) which are located 3′ of the D, J, and constant region gene segments. Such V gene segments rearrange by a mechanism of DNA inversion instead of deletion.

Artificial recombination substrates can undergo deletion and inversion depending on the relative orientation of the RSSs. The plasmids can be recovered after exposure to the cells V(D)J recombinase apparatus to facilitate examination of the recombination products by DNA sequencing. Through this approach two aberrant forms of joint were identified at high frequency in the inversion configuration of artificial substrates. ‘Hybrid joints’ are

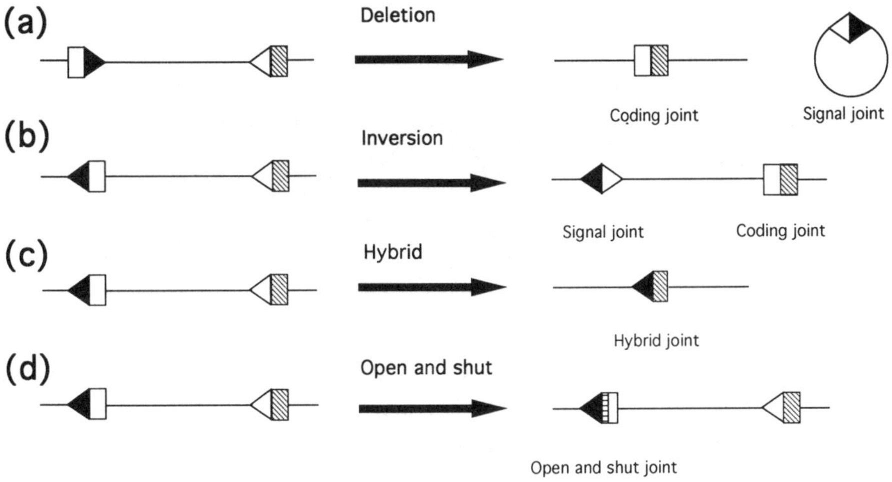

Fig. 15.2 The different forms of V(D)J recombination. The two most common forms of V(D)J recombination at antigen receptor loci are (a) deletion and (b) inversion. Aberrant forms of recombination occur at inversion-type recombination substrates. These are represented by (c) hybrid joints, formed between signal and coding ends, and (d) open and shut joints where horizontal bars indicate the loss and addition of nucleotides at the coding end. Filled triangles represent 23 bp spacer RSSs and clear triangles represent 12 bp spacer RSSs. Coding regions are indicated as rectangles.

formed when a RSS joins to the coding region of its fusion partner (Lewis *et al.* 1988; Moryzycka-Wroblewska *et al.* 1988) (Figure 15.2c). 'Open and shut joints' result when no inversion has taken place so that RSS and coding regions remain in the original configuration (Lewis *et al.* 1988) (Figure 15.2d). However, bases have been lost or added between the coding region and its RSS indicating that this joint must have opened and closed. Artificial recombination substrates have been utilized to develop a quantitative assay for V(D)J recombination (Hesse *et al.* 1987) (Figure 15.3). This assay was useful for purposes such as defining the RSSs by mutagenesis (Hesse *et al.* 1989) and studying the activity of the recombination activating genes RAG-1 and RAG-2 (Oettinger *et al.* 1990).

Aberrant V(D)J recombination events occur in mice possessing the severe combined immunodeficiency (*scid*) mutation (Bosma *et al.* 1983). *scid* is a spontaneous autosomal recessive mutation identified in a laboratory mouse colony. The defect leads to massive reduction in the numbers of mature B and T cells. This is due to the much reduced frequency with which *scid* mice form coding joints (Schuler *et al.* 1986; Lieber *et al.* 1988*b*; Blackwell *et al.* 1989). In contrast the frequency of signal joint formation remains normal; although the joint is less precise than usual. The use of *scid* mice as a tool in dissecting the V(D)J recombination mechanism will be discussed further.

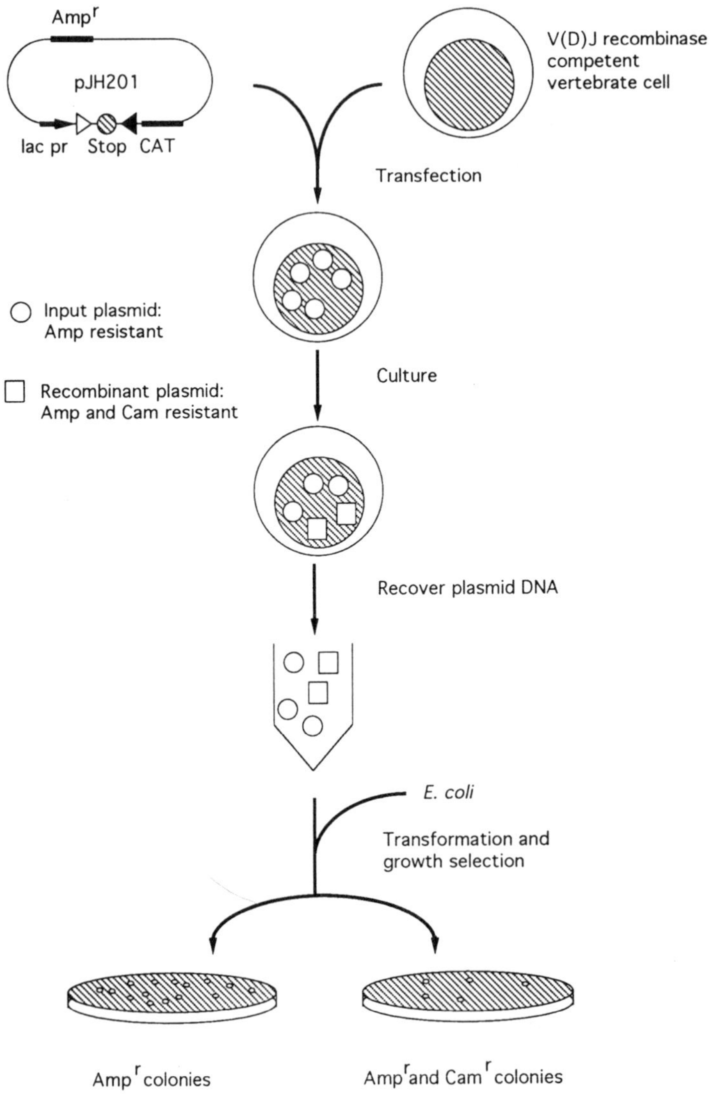

2.3 Junctional diversity and coding joint formation

The loss and/or gain of nucleotides at the coding joint of TCR genes greatly increases the variability of the TCR. The imprecision of the coding joint means that only one-third of all V(D)J recombination events will produce a TCR gene in the correct reading frame. The imprecise joining encountered at the coding joints provides some interesting insights into the V(D)J recombination process, particularly with regard to intermediate DNA structures which are formed. The loss of nucleotides can be explained by exonucleolytic

Fig. 15.3 The essential features of the quantitative assay for V(D)J recombination developed by Hesse *et al.* (1987) are outlined. A bacterial plasmid (pJH201) contains two selectable markers for bacterial growth; the β-lactamase (Amp^r) gene, providing ampicillin resistance, and the chloramphenicol acetyltransferase (CAT) gene, providing chloramphenicol (Cam) resistance. Transcription of the CAT gene from the lac promoter is blocked by the presence of a prokaryotic transcription terminator (circle) which is flanked by a 12 bp RSS (clear triangle) and a 23 bp RSS (filled triangle). On transfection of JH201 into V(D)J recombinase competent cells a fraction of the plasmids will recombine (proportional to the V(D)J recombination rate), deleting the transcription terminator and enabling the plasmid to confer Cam resistence in *E. coli.* After culturing, the plasmid is recovered and used to transform *E. coli* which are grown on medium containing ampicillin alone and on medium containing ampicillin and chloramphenicol. The ratio of the number of *E. coli* colonies on ampicillin and chloramphenicol medium against the number of colonies on ampicillin medium alone is a measurement of the rate of V(D)J recombination.

---

activity. The gain of nucleotides at the coding joint is due to both a non-templated and a templated mechanism. The non-templated addition is accomplished by terminal deoxynucloetide transferase (TdT) activity within early lymphoid cells (Landau *et al.* 1987; Kallenbach *et al.* 1992; Gilfillan *et al.* 1993; Komori *et al.* 1993). The nucleotides, termed N nucleotides, are added in a random pattern to the free ends of the coding strands prior to joining.

The templated form of nucleotide addition was originally identified in the TCR loci by analysis of the junctional sequences in fetal thymic TCR γδ T cells (Lafaille *et al.* 1989). The identification was aided by the fact that fetal thymic TCR γδ T cell populations are very homogeneous in TCR junctional sequence, in contrast to the case in adult TCR γδ T cells. One or two additional nucleotides, termed P nucleotides, were found at the coding joints which were specific to the junctional borders of particular TCR gene segments. The P nucleotide(s) were complementary to the last base(s) of the coding end thus forming one-half of a palindromic sequence. A model of templated nucleotide addition was proposed in which transfer of nucleotides from one strand of a double-stranded DNA break to the other leads to the creation of the palindromic sequence. Evidence of P nucleotide addition has been found in other antigen receptor loci at coding ends where no nucleotide loss has occurred. P nucleotide addition appears to be an early event in processing of the coding joint as, in the majority of cases, subsequent nucleotide loss masks the existence of this process.

The existence of recombination substrates at an intermediate stage of V(D)J recombination has been documented *in vivo* by analysis of the TCR δ locus of thymocytes (Figure 15.4) (Roth *et al.* 1992*a*). Southern analysis performed on thymus DNA using probes specific for the Dδ2 and Jδ1 regions of the TCR δ locus revealed that 10–20% of the restriction fragments hybridizing to these probes were broken at points corresponding to the RSSs. The double-stranded breaks resulted in truncation of the expected full-length

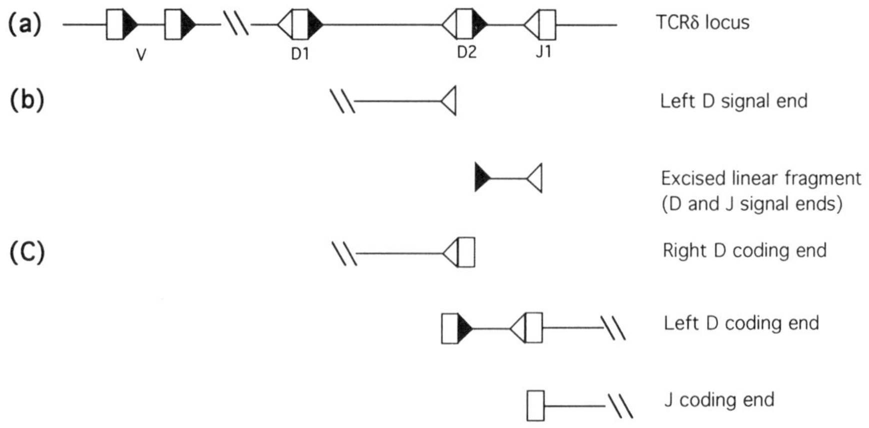

Fig. 15.4 (a) Representation of the TCR δ locus used in the study of recombination intermediate structures (Roth *et al.* 1992*a*). The distances are not to scale and only two V regions are shown. (b) The broken DNA fragments found at the thymic Dδ2 and Jδ1 loci of wild-type mice. All fragments end in RSSs. (c) In *scid* mice additional broken DNA fragments are found in thymic DNA with coding sequences at their ends. Filled triangles represent 23 bp spacer RSSs and clear triangles represent 12 bp spacer RSSs. Coding regions are indicated as rectangles.

hybridizing band. The abundance of these linear DNA molecules suggests that the double-stranded ends are very stable, perhaps as a result of protection by binding of components of the V(D)J recombinase. In wild-type mice the only intermediaries identified were the RSS end 5′ of Dδ2 and the excised linear DNA between Dδ2 and Jδ1 with a RSS at each extremity. Thus, only signal ends were identified and coding end intermediates were not observed. This anomaly was ascribed to the rate of coding end ligation being higher than that of RSS ends so that the abundance of DNA fragments with coding ends was below the level of detection. As previously discussed, mice with the *scid* mutation are defective in coding joint formation. Repetition of the Southern analysis on thymus DNA from *scid* mice enabled the coding joints to be detected. It therefore appears that the *scid* mutation imposes a block on the processing of the intermediate structures which form the coding joint resulting in their accumulation within the thymocyte.

Further analysis of the TCR δ coding joints produced in the *scid* thymus yielded more information on the structure of the intermediate products and on how the coding joint may be processed (Roth *et al.* 1992*b*). Thymic genomic DNA was subjected to two-dimensional electrophoreses and Southern blotted. The second electrophoresis proceeded under denaturing conditions which separated the two strands of a DNA fragment. Different sizes of DNA fragments with double-stranded breaks would fall on a diagonal under these conditions. However, DNA fragments with *scid* TCR δ coding joints exhibited reduced electrophoretic mobility in the second dimension indicating a doubling of size. This result was explained by the two DNA strands being joined at the coding end in a hairpin loop structure.

The existence of a hairpin loop at the coding end DNA intermediates provides a highly plausible explanation for the creation of P nucleotides (Figure 15.5). A symmetrical nicking of the hairpin loop by some component of V(D)J recombinase complex followed by filling-in of the single-stranded product would produce P nucleotide addition to the coding end. The *scid* defect might therefore represent an inability to correctly nick the hairpin loop in order to create a product suitable for ligation. In wild-type mice one or two P nucleotide additions are seen after formation of the coding joint. This may indicate that nicking of the hairpin loop in normal mice is only marginally offset from the apex. However, it is also possible that P nucleotide addition greater than two nucleotides length normally occur but are deleted by exonucleolytic activity. The very rare coding joints found in *scid* mice can possess long P nucleotide additions of 12 to 15 nucleotides in length (Ferrier *et al.* 1990*a*; Kienker *et al.* 1991; Schuler *et al.* 1991). This occurrence may represent an aberrant nicking of the hairpin loop due to the *scid* mutation or alternatively a fortuitous nicking in the vicinity of the hairpin loop due to DNA damage unrelated to V(D)J recombination, leading to an intermediate which is then processed as normal.

2.4 Factors responsible for V(D)J recombination

The evidence reviewed thus far from analysis of recombination substrates, intermediate structures, and recombination products provides an outline of the probable activities involved in V(D)J recombination. The RSSs are recognized by components of the V(D)J recombinase producing double-stranded breaks in the DNA at the border of the RSS. The free coding ends form an intermediate hairpin loop structure which is later nicked, frequently off-centre, and the single-stranded product is filled-in with P nucleotides. Nucleotides are lost from or added to the coding ends before ligation to form the variable coding joint. In contrast the signal joints normally lack variability being formed by the precise fusion of two heptamer sequences.

Little is known about the protein factors responsible for most of these activities. The best characterized factor which plays a role in V(D)J recombination is the enzyme terminal deoxynucleotide transferase (TdT). TdT was proposed to be responsible for N nucleotide addition to the coding joints (Landau *et al.* 1987; Kallenbach *et al.* 1992). The function of TdT was unequivocally demonstrated by the creating of mutant mice lacking TdT which were deficient in N region diversity (Gilfillan *et al.* 1993; Komori *et al.* 1993). Though TdT is important for the variability, and so functioning, of the TCR it should be noted that it is not essential for V(D)J recombination.

As previously mentioned, the factor rendered defective by the *scid* mutation is vital to the processing of the coding junction intermediates (Schuler *et al.* 1986; Lieber *et al.* 1988*b*; Blackwell *et al.* 1989). In addition to the

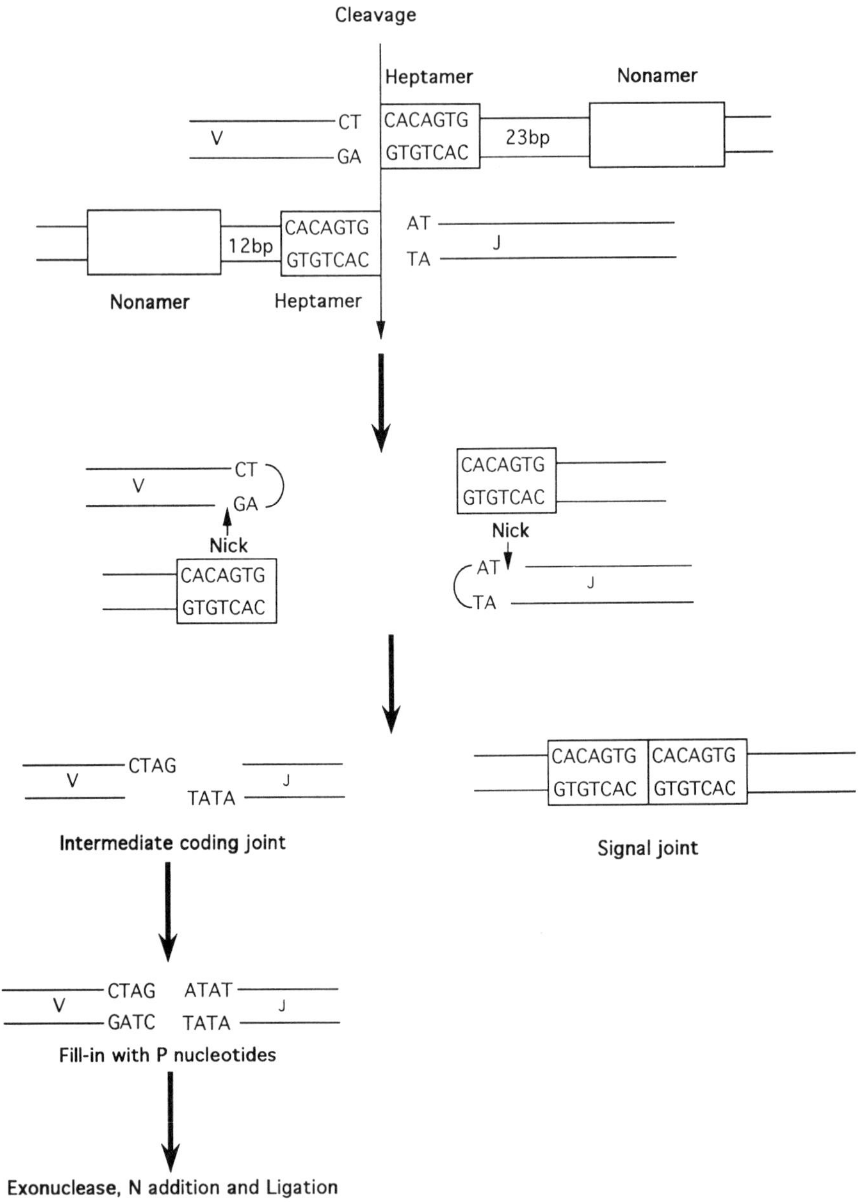

Fig. 15.5 A model for the events during antigen receptor recombination. Hairpin loop structural intermediates are formed at the coding ends which are later nicked and filled-in to create templated, or P, nucleotides. A two nucleotide addition is indicated, although longer additions are possible. Subsequently the coding ends are subjected to exonuclease digestion and N nucleotide addition prior to ligation to form the coding joint. Recombination between V and J coding regions is represented.

specific defect in lymphoid cells, non-lymphoid cells from *scid* mice also exhibit a mutant phenotype, being more sensitive to X- and γ-irradiation than normal cells (Fulop and Phillips 1990; Biederman *et al.* 1991; Hendrickson *et al.* 1991). The *scid* factor was recently identified as the catalytic subunit of the DNA dependent protein kinase (DNA-PK) (Blunt *et al.* 1995). The nature of the *scid* factor was indicated by the observation that mutant Chinese hamster ovary cell lines with defects in DNA double-strand break repair (X-ray sensitivity) also lacked the ability to direct V(D)J recombination when co-transfected with the recombinase activating genes (RAG-1 and RAG-2) and artificial recombination substrates (Alt *et al.* 1992; Taccioli *et al.* 1993, 1994*a*). One of the DNA double-strand break repair mutants, V3, showed a preferential impairment in the joining of coding joints, as in the *scid* mutation, whereas a further two of these lines (xrs-6 and XR-1) were deficient in the formation of coding and signal joints (Alt *et al.* 1992; Pergola *et al.* 1993; Taccioli *et al.* 1993, 1994*a*). All three mutant cell lines fell into different complementation groups (Taccioli *et al.* 1994*a*). Additionally the DNA repair defect of V3 and *scid* cells, and the V(D)J recombination defect of *scid* cells are complemented by human chromosome 8 (Itoh *et al.* 1993; Kirchgessner *et al.* 1993; Komatsu *et al.* 1993; Banga *et al.* 1994). Several lines of evidence indicate that the V3 and *scid* mutations are mutations of the same gene and that this gene is the catalytic subunit of DNA-PK (Blunt *et al.* 1995). Both V3 and *scid* cells lack DNA-PK activity and, in extracts of these cells, the activity can be restored by the addition of purified DNA-PK. Yeast artificial chromosome containing the DNA-PK catalytic subunit gene restored DNA-PK activity to V3 cells and complemented the radiosensitivity and V(D)J recombination competence of V3 cells.

The DNA-PK catalytic subunit associates with two nuclear DNA end binding subunits of approximately 70 kDa and 80 kDa size (Ku70 and Ku80 respectively). The xrs-6 cell line is defective in functional Ku (Getts and Stamato 1994; Rathmell and Chu 1994; Taccioli *et al.* 1994*b*), lacks detectable DNA-PK activity (Finnie *et al.* 1995) and is complemented by expression of Ku80 cDNA (Smider *et al.* 1994; Taccioli *et al.* 1994*b*). Thus both *scid*/V3 and XR-6 cells possess mutations in components of DNA-PK. The precise role that DNA-PK plays in DNA double-strand break repair and V(D)J recombination remains to be determined.

As the heptamer and nonamer sequences of the RSS represent probable DNA binding sites for components of the V(D)J recombinase efforts have been made to identify and clone factors which bind these sequences. Biochemical evidence for heptamer binding (Aguilera *et al.* 1987; Hamaguchi *et al.* 1989) and nonamer binding (Halligan and Desiderio 1987; Li *et al.* 1989) activities has been found. Furthermore, a heptamer binding protein (RBP-Jκ) has been purified and cloned (Hamaguchi *et al.* 1989; Matsunami *et al.* 1989). The predicted amino acid sequence of RBP-Jκ contains a 40 amino acid region of homology with the conserved motif of integrase proteins responsible for site-specific recombination in yeast, bacteria, and bacteriophage. RBP-Jκ is therefore a strong candidate for a factor recognizing the heptamer in V(D)J recombination.

It is apparent from the evidence presented thus far on the make-up of the V(D)J recombinase that there is a gap between our knowledge of activities and our knowledge of the factors responsible. Thus, in some cases V(D)J recombinase activities have been identified in the absence of any candidate factors. In other cases factors have been identified but conclusive proof of their involvement in V(D)J recombination is lacking due to the absence of a suitable experimental system to study their proposed role. However, there are two factors, the RAG-1 and RAG-2 proteins with an indisputable role in V(D)J recombination, though their exact function remains unknown.

3 The recombination activating genes, RAG-1 and RAG-2

The proposition that a single genetic locus might be capable of activating V(D)J recombinase activity stemmed from the surprising observation that transfection of NIH3T3 cells with sheared human or murine genomic DNA could induce occasional V(D)J recombination (Schatz and Baltimore 1988). Although the recombination events were rare (about 0.1% of transfectants) it was possible to identify them because V(D)J recombination led to activation of a drug resistance gene contained within an artificial substrate integrated into chromosomal DNA. The region encoding the V(D)J recombinase activity was narrowed down by serial genomic transfection of oligonucleotide-tagged genomic DNA. The RAG-1 gene was finally identified as a probe which detected a single 6.6 kb mRNA in transfectants, as well as pre-B and pre-T cells (Schatz *et al.* 1989). However, transfection of the RAG-1 gene alone, as a 18 kb genomic clone (clone 12C.2), or as independently isolated cDNAs, activated recombination in NIH3T3 cells at a rate 1000-fold lower than that expected if RAG-1 were the only gene responsible. This observation led to speculation that a second recombination activating gene (RAG-2), located close to RAG-1 but not fully contained within clone 12C.2 was necessary for full V(D)J recombination activity. Indeed, a probe isolated from 12C.2 did detect a predominant 2.2 kb mRNA species with the appropriate tissue specificity (Oettinger *et al.* 1990).

Thus, two genes in close proximity, RAG-1 and RAG-2, were identified which, when co-expressed, act synergistically to activate V(D)J recombination in NIH3T3 cells. RAG-1 and RAG-2 have since been shown to activate V(D)J recombination in other cells where this activity is not normally present, including CHO cells (Kallenbach *et al.* 1992) and neuronally differentiated p19 embryonal carcinoma cells (Schatz *et al.* 1992). The products of recombination within RAG-1 and RAG-2 expressing NIH3T3 cells are normal; an imprecise coding joint with frequent base loss and an invariable signal joint. The absolute necessity of RAG-1 and RAG-2 for V(D)J recombination is underscored by gene disruption in mice by homologous recombination

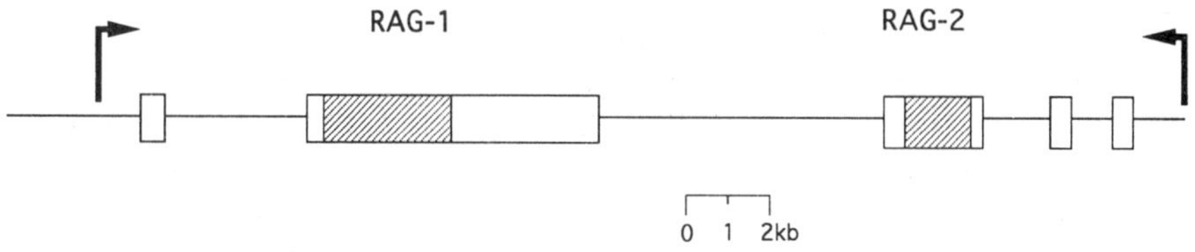

Fig. 15.6 The genomic locus containing RAG-1 and RAG-2. Coding regions of the RAG-1 and RAG-2 genes are hatched and untranslated regions are clear. RAG-2 has multiple 5′ untranslated exons of which only two are represented. *Arrows* represent the direction of transcription.

(Mombaerts *et al.* 1992*b*; Shinkai *et al.* 1992). Inactivation of either RAG-1 or RAG-2 by this method prevents V(D)J recombination at any TCR or Ig locus. The low frequency of V(D)J recombination seen in NIH3T3 cells on transfection of RAG-1 alone was therefore ascribed to low level expression of the endogenous RAG-2 gene.

The RAG-1/RAG-2 locus has been mapped to chromosome 2 in the mouse and the syntenic chromosome 11p in humans (Oettinger *et al.* 1992). The locus has an unusual structure (Figure 15.6). RAG-1 and RAG-2 are positioned within 8 kb of each other and are transcribed in opposite orientations (Oettinger *et al.* 1990). The entire coding and 3′ untranslated regions of RAG-1 and RAG-2 are each contained within one exon. RAG-1 possesses one exon containing 5′ untranslated sequence whereas RAG-2 contains multiple undetermined 5′ untranslated exons (Schatz *et al.* 1992). The structure of the RAG-1/RAG-2 locus is conserved in species as diverse as chicken (Carlson *et al.* 1991) and *Xenopus* (Greenhalgh *et al.* 1993) suggesting that selective pressure might act to maintain this arrangement. Such selective pressure might act if, for example, RAG-1 and RAG-2 shared some *cis*-acting transcriptional regulatory components. It is unlikely that the RAG-1 and RAG-2 genomic structure originated as a gene duplication event as the two genes are completely unrelated in DNA sequence. The unusually compact nature of the RAG-1/RAG-2 locus has led to speculation that it may have originated as a part of a viral or fungal recombination mechanism system which integrated into the ancestral genome and evolved to perform its present function in V(D)J recombination (Oettinger *et al.* 1990).

In addition to the structure of the genetic locus, the predicted amino acid sequences of RAG-1 and RAG-2 are also highly conserved between species. Upon low stringency hybridization the coding sequences of the RAG-1 and RAG-2 genes cross-hybridize with all vertebrate species tested (Schatz *et al.* 1989; Oettinger *et al.* 1990). The amino acid sequences of mouse and human RAG-1 and RAG-2 are 90% identical. The identity between mouse and chicken is 75% for RAG-1 and 70% for RAG-2 (Carlson *et al.* 1991). RAG-1 possesses no known DNA binding motifs. RAG-1 does, however, contain a 50 amino acid cysteine-rich sequence, termed the RING finger, with homology to motifs in a number of otherwise unrelated proteins of widely varying function including *S. cerevisiae* RAD-18 gene, human ret transforming gene, the human RING-1 gene, and the human rpt-1 interleukin-2 receptor regulator gene (Freemont *et al.* 1991; Lovering *et al.* 1993). This sequence is related to known zinc finger motifs and may represent a DNA binding domain (Lovering *et al.* 1993). Mutational analysis of RAG-1 has demonstrated that the RING finger motif is dispensible for V(D)J recombination of extracellular substrates in fibroblast cells (Sadofsky *et al.* 1993; Silver *et al.*, 1993). A particularly interesting homology is found between the C-terminal half of RAG-1 and the *S. cerevisiae* gene HPR1 (Wang *et al.* 1990). The HPR1 gene functions in yeast to suppress excision recombination. HPR1 itself has homology with yeast topoisomerase, although there is no evidence that HPR1 or RAG-1 function is related to topoisomerase activity (Aguilera and Klein 1990). Deletion of the HPR1 homologous region leads to inactivation of RAG-1 V(D)J recombinase activity (Sadofsky *et al.* 1993; Silver *et al.* 1993). However, point mutation of a potential topoisomerase active site (tyrosine 998) did not drastically effect recombinase activity (Kallenbach *et al.* 1993; Sadofsky *et al.* 1993; Silver *et al.* 1993). The RAG-2 gene and protein have no homologies with any known sequence.

Both RAG-1 and RAG-2 proteins localize to the nuclei of thymocytes and of transfected fibroblasts (Lin and Desiderio, 1993; Sadofsky *et al.* 1993; Silver *et al.* 1993). Within the nucleus expression of the RAG-2 protein, and thus recombinase activity, is regulated in the cell cycle dependent manner. RAG-2 protein accumulates in the G0/G1 phase, is rapidly degraded before entry into S phase and remains at a low level through the S, G2, and M phases (Lin and Desiderio 1994). The pattern of expression of the RAG-2 protein during the cell cycle corresponds to the appearance of broken ended signal sequences (Schlissel *et al.* 1993). A possible mechanism for this regulation is the destabilization of the RAG-2 protein by a cyclin dependent kinase. Several lines of evidence support this view. RAG-2 is phosphorylated *in vivo* at position threonine 490 leading to destabilization of the protein (Lin and Desiderio, 1993). Mutation of residue threonine 490 extends the half life of RAG-2 *in vivo*. Furthermore threonine 490 of RAG-2 is phosphorylated *in vitro* by $p34^{cdc2}$. Restriction of V(D)J recombinase activity to the G0/G1 phase of the cell cycle may be important to avoid the deleterious effects of DNA double-strand breaks occurring during mitosis or S phase. Indeed, universal mechanisms exist to prevent a delay passage through the cell cycle in response to DNA double-strand breaks (Kastan *et al.* 1992; Murray 1992).

Despite their obvious central roles in V(D)J recombination the function of RAG-1 and RAG-2 remains unknown. There is no evidence to suggest that RAG-1 and RAG-2 interact with each other to achieve their activity. Similarly there is no evidence that RAG-1 or RAG-2 bind DNA. Using a genetic screen a protein named Rch1 has recently been shown to interact with the RAG-1 protein and thus may potentially be a component of a recombinase complex (Cuomo *et al.* 1994). Rch1 bears similarity to the yeast nuclear envelope protein SRP1. There are two possible ways of envisaging RAG-1 and RAG-2 action. The first possibility is that RAG-1 and RAG-2 encode the lineage-specific components of the V(D)J recombinase enzyme. The alternative view is that RAG-1 and RAG-2 may be regulatory proteins which function to activate the V(D)J recombinase either by post-translational modification or by controlling gene expression. What limited circumstantial evidence exists has been proposed to support the former assertion and is detailed below (Oettinger 1992; Schatz *et al.* 1992).

1. An essential component of the V(D)J recombinase mechanism present throughout vertebrate species would be expected to display high evolutionary conservation. RAG-1 and RAG-2 are highly conserved throughout their length as might be expected if they have to interact with other components of the V(D)J recombinase and possibly DNA. A regulatory protein might have more leeway for variation in primary structure while retaining its function.
2. Following transfection of NIH3T3 cells with RAG-1 and RAG-2 the only discernible phenotype gained is V(D)J recombination activity (Oettinger *et al.* 1990). If RAG-1 and RAG-2 were regulatory proteins it might be expected that other lymphoid markers such as TdT, Oct-2, or B220 might be expressed.
3. The primary phenotype detected in RAG-1 or RAG-2 knock-out mice is the loss of V(D)J recombination (Mombaerts *et al.* 1992*b*; Shinkai *et al.* 1992). If RAG-1 and RAG-2 were regulators, other developmental abnormalities might be apparent in these mutant mice.
4. RAG-1 has homology to a *S. cerevisiae* gene, HPR1, which is known to function in a recombination process, albeit in the suppression of this recombination (Wang *et al.* 1990).
5. The expression of RAG-2 has been correlated with a second vertebrate recombination mechanism, that of Ig gene conversion in the bursa of Fabricus in chickens (Carlson *et al.* 1991). However, it does not appear to be essential for this process as gene disruption of RAG-2 in cell lines does not abolish this process (Takeda *et al.* 1992).

Whatever the function of RAG-1 and RAG-2 proves to be, their co-ordinated expression in pre-T and pre-B cells is essential for V(D)J activity. There is no data available on the *cis*-elements responsible for regulating expression. However, recent studies have yielded useful data on the control of RAG-1 and RAG-2 expression in T and B lineages; data from both lineages will be discussed because of probable general applicability of control mechanisms.

4 Regulation of TCR V(D)J recombination in thymocyte development

The onset of V(D)J recombination is strictly controlled in different T cell lineages. During fetal thymic development at the TCR β locus Dβ to Jβ rearrangement precedes Vβ to Dβ rearrangement which in turn precedes TCR α rearrangement. Both TCR γ and TCR δ rearrangements occur at about the same time as TCR β rearrangement leading to expression of TCR γδ on a distinct set of T cells. No full V(D)J rearrangement of TCR genes has been detected in B cells. However, there are examples of incomplete Dβ to Jβ rearrangements in B cells and conversely Ig D_H to J_H rearrangements in T cells (Forster *et al.* 1980; Kurosawa *et al.* 1981; Cook and Balaton 1987). As all examples of V(D)J recombination share the same mechanism but not all TCR and Ig loci rearrange concurrently the control of onset of rearrangement cannot be at the level of the recombinase. Control of V(D)J recombination must therefore depend on the particular loci. An accessibility model has been proposed to explain how different loci are selectively presented as substrates for the V(D)J recombinase (Alt *et al.* 1987; Blackwell and Alt 1988). In this model a locus is made receptive to the V(D)J recombinase by transcription from the unrearranged gene segments. The mode of operation of transcription is unknown but may involve a relaxation of chromatin structure. The first evidence in favour of the accessibility model came from the study of pre-B cells where germline transcripts were found in loci undergoing V(D)J recombination (van Ness *et al.* 1981; Reth and Alt 1984; Yancopoulos and Alt 1985; Lennon and Perry 1985, 1990; Schlissel *et al.* 1991*a*). Germline transcripts from unrearranged TCR genes are also found in thymocytes (Calman and Peterlin 1986; Pardoll *et al.* 1987). More recently, the introduction of heat shock inducible RAG-1 and RAG-2 genes into a recombinationally inert B cell line has demonstrated the importance of transcriptional enhancers in promoting the accessibility of V(D)J segments for recombination (Oltz *et al.* 1993). Heat shock of the cell line lead to expression of the recombinase activating genes and rapid rearrangement of chromasomally integrated V(D)J recombination substrates, but only when these substrate constructs contained transcriptional enhancers.

The accessibility model therefore implicates *cis*- and *trans*-acting components of the transcriptional apparatus of the antigen receptor genes as the controlling mechanism for initiation of V(D)J recombination. Indeed, induction of transcription at the Igκ locus in pre-B cells by treatment with LPS and at the Igμ locus of a pre-T cell line by transfection of a transcription factor binding the Ig heavy chain gene enhancer do increase the rate of recombination at these loci (Schlissel and Baltimore 1989; Schlissel *et al.* 1991*b*). Various *cis*-acting elements of the TCR genes have been identified. These include transcriptional enhancers 3′ of the TCR α (Winoto and Baltimore 1989*b*; Ho *et al.* 1989), TCR β (Krimpenfort *et al.* 1988; McDougall *et al.* 1988; Gottschalk and Leiden 1990; Takeda *et al.* 1990; Prosser *et al.* 1991), and TCR γ (Kappes *et al.* 1991; Spencer *et al.* 1991) constant regions, and in the J to constant region intron of the TCR δ gene (Bories *et al.* 1990; Redondo *et al.* 1990), and promoter elements of the murine Vβ genes (Anderson *et al.* 1988, 1989). Interestingly, a deletion of the TCR β locus by homologous recombination abolished all forms of TCR β recombination although the TCR β enhancer remained intact (Mombaerts *et al.* 1992*b*). This finding suggests that the TCR β enhancer is not in isolation responsible for directing TCR β recombination. The TCR β deletion does, however, remove the Jβ2 to Cβ2 intron where a potential regulatory element of unknown function has been identified as a DNaseI hypersensitive site (Hashimoto *et al.* 1990). Transcriptional silencer activities have been described for the TCR α (Winoto and Baltimore 1989*c*), TCR β (Krimpenfort *et al.* 1988; Takeda *et al.* 1990; Prosser *et al.* 1991), and TCR γ (Bonneville *et al.* 1990; Ishida *et al.* 1990) loci. However, the role of these activities in controlling TCR rearrangement and/or transcription remains unproven.

The importance of the specificity of transcriptional activation in the regulation of tissue-specific recombination has been revealed by the generation of transgenic mice using chimeric minigene constructs (Ferrier *et al.* 1990*b*). These constructs contained unrearranged TCR β V, D, and J gene segments in association with the Cμ gene in the presence and absence of the Igμ enhancer. The presence of the Igμ enhancer was necessary for recombination to occur and allowed incomplete D to J recombination in both B cells and T cells. However, complete V to DJ recombination was restricted to T cells implying that tissue-specific transcription elements associated with the Vβ gene were responsible for directing the tissue-specific rearrangement. Incorporation of the TCRβ and TCRα enhancers into this minigene construct in place of the Igμ enhancer underlined the importance of the enhancer in directing appropriate V(D)J recombination (Capone *et al.* 1993). In this case all recombination, including D to J recombination was predominantly T cell specific. Also recombination occurred stage specifically. Thus, the TCRα and TCRβ enhancer containing minigene constructs rearranged respectively when the endogenous TCRα and TCRβ gene segments rearranged. The intronic immunoglobulin heavy chain enhancer plays a similar role in directing endogenous heavy chain rearrangement (Chen *et al.* 1993). Thus, replacement of the enhancer region with the Neo^r gene by gene targeting caused a *cis*-acting block of endogenous J_H rearrangement.

A possible scenario for control of the ordered rearrangement of TCR genes is that recombination of one TCR gene is necessary for the induction of the recombination of its partner TCR gene. By analogy evidence has been presented suggesting that Ig heavy chain recombination induces Igκ chain rearrangement (Reth and Alt 1984), although there is also evidence to the contrary (Blackwell *et al.* 1989). However, in TCR β knock-out mice rearrangement of the TCR α locus proceeded as normal (Mombaerts *et al.* 1992*a*). Therefore TCR α rearrangement is not dependent on prior TCR β rearrangement.

Inactivation of rearrangement is important both to prevent multiple, and unselected, specificities in lymphocytes, and to avoid the possibility of aberrant chromosomal rearrangements leading to immortalization of the cell. It has been shown that RAG-1 and RAG-2 mRNA expression is actively regulated during thymic development thus restricting recombinase activity to specific developmental stages. Expression of RAG-1 and RAG-2 peaks in two waves *in vivo*. The first occurs at the double negative ($CD4^-CD8^-$) thymocyte stage corresponding to the appearance of TCRβ, γ, and δ gene transcripts, and the second occurs at the double positive ($CD4^+CD8^+$) thymocyte stage when full-length TCRa transcripts appear (Wilson *et al.* 1994). Between these two stages RAG-1 and RAG-2 expression is down-regulated. Down-regulation at the double-negative stage can be mimicked by phorbol ester and calcium ionophore treatment of thymocytes *in vitro*. At the $CD4^-CD8^{lo}$ thymocyte stage (the immediate progenitor to the double-negative stage) RAG-1 and RAG-2 messenger RNA was reduced by a post-transcriptional mechanism upon either cross-linking the TCR or treatment with phorbol esters *in vitro* (Takahama and Singer 1992). Down-regulation of RAG-1 and RAG-2 messenger RNA at the double positive stage of thymocyte development may be induced either by (1) binding of antigen and MHC to the TCR during the process of positive selection (Borgulya *et al.* 1992; Brandle *et al.* 1992) or (2) by non-specific stimuli such as cross-linking TCR with antibodies and treatment with phorbol esters and calcium ionophore (Turka *et al.* 1991).

However, other mechanisms must operate to restrict the availability of loci for V(D)J recombination. For example, productive rearrangement of one TCRβ gene prevents rearrangement of the allelic gene. Additionally, mice made transgenic for a rearranged TCR β gene allelically excluded endogenous TCR β rearrangement although the recombination apparatus remained functional, as testified by the normal rearrangement of the TCR α loci (Uematsu *et al.* 1988; von Boehmer 1990). Thus, allelic exclusion operates at the level of the TCR β gene itself. The mechanism by which allelic exclusion operates to shut down recombination is unknown. Transgenesis using frame shifted and/or truncated TCR β genes indicate that translation of the TCR β constant region gene segment is required for allelic exclusion of the endogenous TCR β genes (Krimpenfort *et al.* 1989). It is not yet established whether TCR β must be expressed at the cell surface to achieve allelic exclusion. The T cell-specific protein-tyrosine kinase $p56^{lck}$, which is physically associated with the cytoplasmic tails of the CD4 or CD8 molecules and is activated upon antigenic stimulation of the TCR, is likely to be involved in allelic exclusion of TCR β (Anderson *et al.* 1992). Over-expression of *lck* in transgenic mice reduced Vβ to Dβ segment joining while other TCR genes rearranged as normal. This suggests that a signal transduced via $p56^{lck}$ may cause TCR β allelic exclusion.

4.1 Thymocyte development in mice with disrupted TCR genes

Gene knock-out experiments have revealed the important role of TCR rearrangements in directing the development of thymocytes. It is well established that transition from the TCR $\alpha\beta^+$ double-positive ($CD4^+CD8^+$) to

TCR $\alpha\beta^+$ single-positive ($CD4^+$ or $CD8^+$) stage of thymocyte development is dependent on binding of TCR $\alpha\beta$ to thymic MHC molecules and is associated with the process of positive selection. The earlier transition from double-negative ($CD4^-CD8^-$) to double-positive ($CD4^+CD8^+$) thymocytes has been shown to be dependent on productive rearrangement of the TCR β gene. Disruption of the TCR β gene by homologous recombination prevents this transition, which can be reconstituted by introduction of a TCR β transgene (Mombaerts *et al.* 1992*a*). Analogous disruption of the TCR α locus has no effect on the double-negative to double-positive transition (Mombaerts *et al.* 1992*a*). Disruption of the *lck* gene in mice also prevents the transition of thymocytes from the double-negative to double-positive stage (Molina *et al.* 1992). This result suggests $p56^{lck}$ is important in transducing the signal responsible for the maturation of thymocytes to the double-positive stage.

Developmental mechanisms in murine fetal TCR γ and TCR δ gene rearrangement have been studied using homologous recombination (Itohara *et al.* 1993) and transgenic techniques (Asarnow *et al.* 1993). Murine fetal TCR γ and TCR δ genes possess several distinguishing characteristics. They are divided into subsets each with a specific combination of V, (D), and J genes segments. Each subset develops at a distinctive time and is expressed in specific tissues. Fetal TCR $\gamma\delta$ T cells are also very limited in their TCR V(D)J junctional sequences in contrast to adult TCR γ and TCR δ genes. There are two interpretations of how these features are developed. Either they are intrinsic features of the fetal TCR $\gamma\delta$ T cells which are programmed intracellularly to develop this way, or only these types of TCR $\gamma\delta$ T cell populations are selected to proliferate by some extracellular system akin to thymic selection. Despite disruption of the TCR δ locus by homologous recombination fetal TCR γ transcripts showed the normal features of VJ combinations, tissue specificity, and limited junctional diversity (Itohara *et al.* 1993). As no extracellular mechanism of selection can operate on T cells with a disrupted TCR δ gene, because there is no cell surface expression of the TCR, the features associated with the recombination of fetal TCR $\gamma\delta$ T cells must be programmed intracellularly. The limited junctional diversity of fetal TCR γ genes was also studied using a transgene comprised of a minilocus of unrearranged $V\gamma$ and $J\gamma$ genes in which the $V\gamma$ genes were mutated by the generation of a frame shift in the DNA sequence (Asarnow *et al.* 1993). Although no TCR γ protein was available for selection from outside the cell, the coding junctions formed upon rearrangement of the transgene in newborn mice were characteristically lacking in diversity. Hence, this study also supports the notion of an intracellular mechanism controlling murine fetal TCR $\gamma\delta$ sequences.

5 Summary

Much has been learnt about the activities involved in V(D)J recombination by studying recombination substrates, intermediates, and products. In contrast progress in identifying and elucidating the function of the essential components of the V(D)J recombinase itself has been limited. Intuitively, it seems likely that the RAG-1 and RAG-2 proteins represent the essential tissue-specific components of the V(D)J recombinase, although the evidence in favour of this proposition is very weak at present. In the absence of any other leads, the most productive route to understanding the V(D)J recombinase mechanism most probably lies in studying the roles of RAG-1 and RAG-2. Progress in this endeavour has been slow, due in a large part to technical difficulties involved in working with RAG-1 and RAG-2. No doubt studies addressing questions such as whether RAG-1 and RAG-2 proteins interact with each other, with other components of the V(D)J recombinase, or with DNA substrates, are likely to warrant research for some time to come. The ultimate goal of such research would be to establish a cell-free system for V(D)J recombination in order to study the biochemical functioning of the V(D)J recombinase component(s).

Acknowledgements

We thank David Gerber, Asa Abeliovich, and Juan Lafaille for their suggestions on the manuscript. H. M. P. is supported by a Damon Runyon-Walter Winchell Cancer Research Fund Fellowship, DRG-1132.

References

Aguilera, R. J., Akira, S., Okazaki, K., and Sakano, H. (1987). A pre-B cell nuclear protein which specifically interacts with the immunoglobulin V-J recombination sequences. *Cell*, **51**, 909–17.

Aguilera, A. and Klein, H. L. (1990). HPR1, a novel yeast gene that prevents intrachromosomal excision recombination shows carboxy-terminal homology to the *Saccharomyces cerevisiae* TOP1 gene. *Mol. Cell. Biol.*, **10**, 1439–51.

Akira, S., Okazaki, K., and Sakano, H. (1987). Two pairs of recombination signals are sufficient to cause immunoglobulin V-(D)-J joining. *Science*, **238**, 1134–8.

Alt, F., Blackwell, T. K, and Yancopoulos, V. (1987). Development of the primary antibody repertoire. *Science*, **238**, 1079–87.

Alt, F. W., Oltz, E. M., Young, F., Gorman, J., Taccioli, G., and Chen, J. (1992). VDJ recombination. *Immunology Today*, **13**, 306–14.

Anderson, S. J., Chou, H. S., and Loh, D. Y. (1988). A conserved sequence in the T-cell receptor β-chain promoter. *Proc. Natl. Acad. Sci. USA*, **85**, 3551–4.

Anderson, S. J., Miyake, S., and Loh, D. Y. (1989). Transcription from a murine T-cell receptor Vβ promoter depends on a conserved decamer motif similar to the cyclic AMP response element. *Mol. Cell. Biol.*, **9**, 4835–45.

Anderson, S. J., Abraham, K. M., Nakayama, T., Singer, A., and Perlmutter, R. M. (1992). Inhibition of TCRβ gene rearrangement by overexpression of the non-receptor protein tyrosine kinase p56lck. *EMBO J.*, **11**, 4877–86.

Asarnow, D. M., Cado, D., and Raulet, D. H. (1993). Selection is not required to produce invariant T-cell receptor γ-gene junctional sequences. *Nature*, **362**, 158–60.

Banga, S. S., Hall, K. T., Sandhu, A. K., Weaver, D. T., and Athwal, R. S. (1994). Complementation of V(D)J recombination defect and X-ray sensitivity of *scid* mouse cells by human chromosome 8. *Mutat. Res.*, **315**, 239–47.

Biedermann, K. A., Sun, J. R., Gaccia, A. J., Totso, L. M., and Brown, J. M. (1991). *scid* mutation in mice confers hypersensitivity to ionizing radiation and a deficiency in DNA double-strand break repair. *Proc. Natl. Acad. Sci. USA*, **88**, 1394–7.

Blackwell, T. K. and Alt, F. W. (1988). Immunoglobulin genes. In *Molecular immunology* (ed. B. D. Hames and D. M. Glover), pp. 1–60. IRL, Washington, DC.

Blackwell, T. K., Moore, M., Yancopoulos, G., Suh, H., Lutzker, S., Selsing, E., *et al.* (1986). Recombination between immunoglobulin variable region gene segments is enhanced by transcription. *Nature*, **324**, 585–9.

Blackwell, T. K., Malynn, B. A., Pollock, R. R., Ferrier, P., Covey, L. R., Fulop, G. M, *et al.* (1989). Isolation of *scid* pre-B cells that rearrange κ light chain genes: formation of normal signal and abnormal coding joins. *EMBO J.*, **8**, 735–42.

Blunt, T., Finnie, N. J., Taccioli, G. E., Smith, G. C. M., Demengeot, Gottlieb, T. M., *et al.* (1995). Defective DNA-dependent protein kinase activity is linked to V(D)J recombination and DNA repair defects associated with the murine *scid* mutation. *Cell*, **80**, 813–23.

Bonneville, M., Ishida, I., Mombaerts, P., Katsuki, M., Verbeek, S., Berns, A., *et al.* (1990). Blockage of αβ T-cell development by TCR γδ transgenes. *Nature*, **342**, 931–4.

Bories, J. C., Loiseau, P., d'Auriol, L., Gontier, C., Bensussan, A., Degos, L., *et al.* (1990). Regulation of transcription of the human T cell antigen receptor δ chain gene. *J. Exp. Med.*, **171**, 75–83.

Borulya, P. Kishi, H., Uematsu, Y., and von Boehmer, H. (1992). Exclusion and inclusion of α and β T cell receptor alleles. *Cell*, **69**, 529–37.

Bosma, G. C., Custer, R. P., and Bosma, M. J. (1983). A severe combined immunodeficiency mutation in the mouse. *Nature*, **301**, 527–30.

Bosma, G. C., Davisson, M. T., Ruetsch, N. R., Sweet, H. O., Schultz, L. D., and Bosma, M. J. (1989). The mouse mutation severe combined immune deficiency (*scid*) is on chromosome 16. *Immunogenetics*, **29**, 54–7.

Brandle, D., Muller, C., Rulicke, T., Hengartner, H., and Pircher, H. (1992). Engagement of the T cell receptor during positive selection in the thymus down-regulates RAG-1 expression. *Proc. Natl. Acad. Sci. USA*, **89**, 19529–33.

Calman, A. F. and Peterlin, M. (1986). Expression of T cell receptor genes in human B cells. *J. Exp. Med.*, **164**, 1940–57.

Capone, M., Watrin, F., Fernex, C., Horvat, B., Krippl, B., Wu, L., Scollay, R., and Ferrier, P. (1993). TCRβ and TCRα gene enhancers confer tissue- and stage specificity on V(D)J recombination events. *EMBO J.*, **12**, 4335–46.

Carlson, L. M., Oettinger, M. A., Schatz, D. G., Masteller, E. L., Hurley, E. A., McCormack, W. T., *et al.* (1991). Selective expression of RAG-2 in chicken B cells undergoing gene conversion. *Cell*, **64**, 201–8.

Chen, J., Young, F., Bottaro, A., Stewart, V., Smith, R. K., and Alt, F. W. (1993). Mutations of the intronic IgH enhancer and its flanking sequences differentially affect accessibility of the IH locus. *EMBO J.*, **12**, 4635–45.

Cook, W. D. and Balaton, A. M. (1987). T-cell receptor and immunoglobulin genes are rearranged together in Abelson virus-transformed pre-B and pre-T cells. *Mol. Cell. Biol.*, **7**, 266–72.

Cuomo, C. A., Kirch, S. A., Gyuris, J., Brent, R., and Oettinger, M. A. (1994). Rch1, a protein that specifically interacts with the RAG-1 recombination-activating protein. *Proc. Natl. Acad. Sci. USA*, **91**, 6156–60.

Davis, M. M. (1988). T cell antigen receptor genes. In *Molecular immunology* (ed. B. D. Hames and D. M. Glover), pp. 61–79. IRL, Washington, DC.

Ferrier, P., Covey, L. R., Li., S. C., Suh, H., Malynn, B. A., Blackwell, T. K., *et al.* (1990a). Normal recombination substrate V_H to DJ_H rearrangements in pre-B cell lines from *scid* mice. *J. Exp. Med.*, **171**, 1909–18.

Ferrier, P., Krippl, B., Blackwell, T. K., Furley, A. J. W., Suh, H., Winoto, *et al.* (1990*b*). Separate elements control DJ and VDJ rearrangement in a transgenic recombination substrate. *EMBO J.*, **9**, 117–25.

Finnie, N. J., Gottlieb, T. M., Blunt, T., Jeggo, P., and Jackson, S. P. (1995). DNA-PK activity is absent in xrs-6 cells; implications for site-specific recombination and DNA double-strand break repair. *Proc. Natl. Acad. Sci. USA*, **92**, 320–4.

Forster, A., Hobart, M., Hengartner, H., and Rabbitts, T. H. (1980). An immunoglobulin heavy chain gene is altered in two T cell clones. *Nature*, **286**, 897–9.

Freemont, P. S., Hanson, I. M., and Trowsdale, J. (1991). A novel cysteine-rich sequence motif. *Cell*, **64**, 483–4.

Fujimoto, S. and Yamaguchi, H. (1987). Isolation of an excision product of the T-cell receptor α-chain gene rearrangement. *Nature*, **338**, 430–2.

Fulop, G. M. and Phillips, R. A. (1990). The *scid* mutation in mice causes a general defect in DNA repair. *Nature,* **347**, 479–82.

Getts, R. C., and Stamato, T. D. (1994). Absence of Ku-like DNA end-binding activity in the xrs double-strand DNA repair deficient mutant. *J. Biol. Chem.*, **269**, 15981–4.

Gilfillan, S., Dierrich, A., Lemeur, M., Benoist, C., and Mathis, D. (1993). Mice lacking TdT: Mature animals with an immature lymphocyte repertoire. *Science*, **261**, 1175–8.

Gottschalk, L. R. and Leiden, J. M. (1990). Identification and functional characterization of the human T-cell receptor β gene transcriptional enhancer: common nuclear proteins interact with the transcriptional regulatory elements of the T-cell receptor α and β genes. *Mol. Cell. Biol.*, **10**, 5486–95.

Greenhalgh, P., Olesen, C. E., and Steiner, L. A. (1993). Characterization and expression of recombination activating genes (RAG-1 and RAG-2) in *Xenopus laevis*. *J. Immunol.*, **151**, 3100–10.

Halligan, B. D. and Desiderio, S. V. (1987). Identification of a DNA binding protein that recognizes the nonamer recombinational signal sequence of immunoglobulin genes. *Proc. Natl. Acad. Sci. USA*, **84**, 7019–23.

Hamaguchi, Y., Matsunami, N., Yamamoto, Y., and Honjo, T. (1989). Purification and characterization of a protein that binds to the recombination signal sequence of the immunoglobulin J kappa segment. *Nucleic Acids Res.*, **17**, 9015–26.

Hashimoto, Y., Maxam, A. M., and Green, M. I. (1990). Identification of tissue specific nuclear proteins: DNA sequence and protein binding regions in the T cell receptor β J-C intron. *Nucleic Acids Res.*, **18**, 3027–33.

Hendrikson, E. A., Qin, X.-Q., Bump, E. A., Schatz, D. G., Oettinger, M. A., and Weaver, D. T. (1991). A link between double stranded break repair and V(D)J recombination: the *scid* mutation. *Proc. Natl. Acad. Sci. USA*, **88**, 1394–7.

Hesse, J. E., Lieber, M. R., Gellert, M., and Mizuuchi, K. (1987). Extrachromosomal DNA substrates in pre-B cells undergo inversion or deletion at immunoglobulin V-(D)-J joining signals. *Cell*, **49**, 775–83.

Hesse, J. E., Lieber, M. R., Mizuuchi, K., and Gellert, M. (1989). V(D)J recombination: a functional definition of the joining signals. *Genes Dev.*, **3**, 1053–61.

Ho, I.-C., Yang, L.-H., Morle, G., and Leiden, J. M. (1989). A T-cell-specific transcriptional enhancer element 3′ of Cα in the human T-cell receptor α locus. *Proc. Natl. Acad. Sci. USA*, **86**, 6714–18.

Ishida, I., Verbeek, S., Bonneville, M., Itohara, S., Berns, A. and Tonegawa, S. (1990). T-cell receptor γδ and γ transgenic mice suggest a role of a γ gene silencer in the generation of αβ T cells. *Proc. Natl. Acad. Sci. USA*, **87**, 3067–71.

Itoh, M., Hamatani, K., Komatsu, K., Arakai, R., Takayama, K., and Abe, M. (1993). Human chromosome 8 (p12-q22) complement radiosensitivity in severe combined immune deficiency (*scid*) mouse. *Radiat. Res.*, **134**, 364–8.

Itohara, S., Mombaerts, P., Lafaille, J., Iacomini, J., Nelson, A., Clarke, A. R., *et al.* (1993). T cell receptor δ gene mutant mice: independent generation of αβ T cells and programmed rearrangements of γδ TCR genes. *Cell*, **72**, 337–48.

Iwashima, M., Green, A., Davis, M. M., and Chien, Y.-H. (1989). Variable region (Vδ) gene segment most frequently utilized in adult thymocytes is 3′ of the constant (Cδ) region. *Proc. Natl. Acad. Sci. USA*, **85**, 8161–5.

Kallenbach, S., Doyen, N., Fanton d'Andon, M., and Rougeon, F. (1992). Three lymphoid-specific factors account for all junctional diversity characteristics of somatic assembly of T cell receptor and immunoglobulin genes. *Proc. Natl. Acad. Sci. USA*, **89**, 2799–803.

Kallenbach, S., Brinkmann, T., and Rougeon, F. (1993). RAG-1: a topoisomerase? *International Immunology*. **5**, 231–232.

Kappes, D. J., Browne, C. P., and Tonegawa, S. (1991). Identification of a T-cell-specific enhancer at the locus encoding T-cell antigen receptor γ chain. *Proc. Natl. Acad. Sci. USA*, **88**, 2204–8.

Kastan, M. B., Zhan, Q., El-Diery, W. S., Carrier, F., Jacks, T., Walsh, W. V., *et al.* (1992). A mammalian cell cycle checkpoint pathway using p53 and GADD45 is defective in ataxia-telangiectasia. *Cell*, **71**, 587–97.

Kienker, L. J., Kuziel, W. A., and Tucker, P. W. (1991). T cell receptor γ and δ gene junctional sequences in SCID mice: excessive P nucleotide insertion. *J. Exp. Med.*, **174**, 769–73.

Kirchgessner, C. U. (1993). Complementation of the radiosensitive phenotype in severe combined immunodeficient mice by human chromosome 8. *Cancer Res.*, **53**, 6011.

Komatsu, K., Ohta, T., Jinno, Y., Nikawa, N., and Okamura, Y. (1993). Functional complementation in mouse *scid* gene to the pericentric region of human chromosome 8. *Hum. Mol. Genet.*, **7**, 1031–4.

Komori, T., Okada, A., Stewart, V., and Alt, F. (1993). Lack of N regions in antigen receptor variable region genes of TdT-deficient lymphocytes. *Science*, **261**, 1171–5.

Korman, A. J., Maruyama, J., and Raulet, D. H. (1989). Rearrangement by inversion of a T-cell receptor δ variable region gene located 3′ of the constant region gene. *Proc. Natl. Acad. Sci. USA*, **86**, 267–71.

Krimpenfort, P., de Jong, R., Uematsu, Y., Dembic, Z., Ryser, S., von Boehmer, H., *et al.* (1988). Transcription of T cell receptor β-chain genes is controlled by a downstream regulatory element. *EMBO J.*, **7**, 745–50.

Krimpenfort, P., Ossendorp, F., Borst, J., Melief, C., and Berns, A. (1989). T cell depletion in transgenic mice carrying a mutation for TCRβ. *Nature*, **34**, 742–6.

Kurosawa, Y., von Boehmer, H., Hass, W., Sakano, H., Traunecker, A., and Tonegawa, S. (1981). Identification of D segments of immunoglobulin heavy chain genes and their rearrangement in T lymphocytes. *Nature*, **290**, 565–70.

Lafaille, J. J., DeCloux, A., Bonneville, M., Takagaki, Y., and Tonegawa, S. (1989). Junctional sequences of T cell receptor γδ genes: implications for γδ cell lineages and for a novel intermediate of V-(D)-J joining. *Cell*, **59**, 859–70.

Landau, N. R., Schatz, D. G., Rosa, M., and Baltimore, D. (1987). Increased frequency of N-regional insertion in a murine pre-B-cell line infected with a terminal deoxynucleotide transferase retroviral expression vector. *Mol. Cell. Biol.*, **7**, 3237–43.

Lennon, G. G. and Perry, R. P. (1985). Cμ-containing transcripts initiate heterogeneously with the IgH enhancer region and contain a novel 5′-nontranslatable exon. *Nature*, **318**, 475–8.

Lennon, G. G. and Perry, R. P. (1990). The temporal order of appearance of transcripts from unrearranged and rearranged Ig genes in murine fetal liver. *J. Immunol.*, **144**, 1983–7.

Lewis, S. M., Hesse, J. E., Mizuuchi, K., and Gellert, M. (1988). Novel strand exchanges in V(D)J recombination. *Cell*, **55**, 1099–107.

Li, M., Morzycka-Wroblewska, E., and Desderio, S. (1989). NBP, a protein that specifically binds an enhancer of gene rearrangement: purification and characterization. *Genes Dev.*, **3**, 1801–13.

Lieber, M. R., Hesse, J. E., Mizuuchi, K., and Gellert, M. (1988*a*). Developmental stage specificity of the lymphoid V(D)J recombination activity. *Genes Dev.*, **1**, 751–61.

Lieber, M. R., Hesse, J. E., Lewis, S., Bosma, G. C., Rosenberg, N., Mizuuchi, K., *et al.* (1988*b*). The defect in murine severe immune deficiency: Joining of signal sequences but not coding segments in V(D)J recombination. *Cell*, **55**, 7–16.

Lin, W-C. and Desiderio, S. (1993). Regulation of V(D)J recombination activator protein RAG-2 by phosphorylation. *Science*, **260**, 953–9.

Lin, W-C. and Desiderio, S. (1994). Cell cycle regulation of V(D)J recombination-activating protein RAG-2. *Proc. Natl. Acad. Sci. USA*, **91**, 2733–7.

Lovering, R., Hansen, I. M., Borden, K. L., Martin, S., O'Reilly, N. J., Evan, *et al.* (1993). Identification and preliminary characterization of a protein motif related to a zinc finger. *Proc. Natl. Acad. Sci. USA*, **90**, 2112–16.

Malissen, M., McCoy, C., Blanc, D., Trucy, J., Devaux, C., Schmitt-Verhulst, A.-M., *et al.* (1986). Direct evidence for chromosomal inversion during T cell receptor β-gene rearrangements. *Nature*, **319**, 28–33.

Matsunami, N., Hamaguchi, Y., Yamamoto, Y., Kuze, K., Kanagawa, K., Matsuo, H., *et al.* (1989). A protein binding to the Jκ recombination sequence of immunoglobulin genes contains a sequence related to the integrase motif. *Nature*, **342**, 934–7.

McDougall, S., Peterson, C. L., and Calame, K. (1988). A transcriptional enhancer 3′ of Cβ2 in the T cell receptor β locus. *Science*, **241**, 205–8.

Molina, T. J., Kishihara, K., Siderovski, D. P., van Ewijk, W., Narendran, A., Timms, E., *et al.* (1992). Profound block in thymocyte development in mice lacking $p56^{lck}$. *Nature*, **357**, 161–4.

Mombaerts, P., Clarke, A. R., Rudnicki, M. A., Iacomini, J., Itohara, S., Lafaille, J. J., *et al.* (1992*a*). Mutations in the T-cell antigen receptor genes α and β block thymocyte development at different stages. *Nature*, **360**, 225–31.

Mombaerts, P., Iacomini, J., Johnson, R. S., Herrup, K., Tonegawa, S., and Papaioannou, V. E. (1992*b*). RAG-1 deficient mice have no mature B and T lymphocytes. *Cell*, **68**, 869–77.

Moryzycka-Wroblewska, E., Lee, F. E. H., and Desderio, S. V. (1988). Unusual immunoglobulin gene rearrangement leads to replacement of recombinational signal sequences. *Science*, **242**, 261–3.

Murray, A. W. (1992). Creative blocks: Cell-cycle checkpoints and feedback controls. *Nature*, **359**, 599–604.

Oettinger, M. A. (1992). Activation of V(D)J recombination by RAG-1 and RAG-2. *Trends Genet.*, **8**, 413–16.

Oettinger, M. A., Schatz, D. G., Gorka, C., and Baltimore, D. (1990). RAG-1 and RAG-2, adjacent genes that synergistically activate V(D)J recombination. *Science*, **248**, 1517–23.

Oettinger, M. A., Stanger, B., Schatz, D. G., Glaser, T., Call, K., Housman, D., *et al.* (1992). The recombination activating genes RAG-1 and RAG-2 are on chromosome 2 in mice. *Immunogenetics*, **35**, 97–101.

Okazaki, K. and Sakano, H. (1988). Thymocyte circular DNA excised from T cell receptor α-δ gene complex. *EMBO J.*, **7**, 1669–74.

Okazaki, K., Davis, D. D., and Sakano, H. (1987). T cell receptor β gene sequences in the circular DNA of thymocyte nuclei: direct evidence for intramolecular DNA deletion in V-D-J joining. *Cell*, **49**, 477–85.

Oltz, E. M., Alt, F. W., Lin, W-C., Chen, J., Taccioli, G., Desiderio, S., and Rathbun, G. (1993). A V(D)J recombinase-inducible B-cell line: Role of transcriptional enhancer elements in directing V(D)J recombination. *Mol. Cell Biol.*, **13**, 6223–30.

Pardoll, D. M., Fowlkes, B. J., Lechler, R. I., Germain, R. N., and Schwartz, R. H. (1987). Early genetic events in T cell development analyzed by *in situ* hybridization. *J. Exp. Med.*, **165**, 1624–38.

Pergola, F., Zdzienicka, M. Z., and Lieber, M. R. (1993). V(D)J recombination in mammalian cell mutants defective in DNA double-strand break repair. *Mol. Cell. Biol.*, **13**, 3464–71.

Prosser, H. M., Lake, R. A., Wotton, D., and Owen, M. J. (1991). Identification and functional analysis of the transcriptional enhancer of the human T cell receptor β gene. *Eur. J. Immunol.*, **21**, 161–6.

Rathmell, W. K. and Chu, G. (1994). Involvement of the Ku autoantigen in the cellular response to DNA double-strand breaks. *Proc. Natl. Acad. Sci. USA,* **91**, 7623–7.

Redondo, J. M., Hata, S., Brocklehurst, C., and Krangel, M. S. (1990). A T cell-specific transcriptional enhancer within the human T cell receptor δ locus. *Science*, **247**, 1225–9.

Reth, M. and Alt, F. (1984). Novel immunoglobulin heavy chains are produced from DJ_H gene segment rearrangements in lymphoid cells. *Nature*, **312**, 418–23.

Reth, M., Petrac, E., Wiese, P., Lobel, L., and Alt, F. (1987). Activation of Vκ gene rearrangement in pre-B cells follows the expression of membrane-bound immunoglobulin heavy chains. *EMBO J.*, **6**, 3299–305.

Roth, D. B., Menetski, J. P., Nakajima, P. B., Bosma, M. J., and Gellert, M. (1992*a*). V(D)J recombination: broken DNA molecules with covalently sealed (Hairpin) coding ends in *scid* mouse thymocytes. *Cell*, **70**, 983–91.

Roth, D. B., Nakajima, P. B., Menetski, J. P., Bosma, M. J., and Gellert, M. (1992*b*). V(D)J recombination in mouse thymocytes: double stranded breaks near T cell receptor δ rearrangement signals. *Cell*, **69**, 41–53.

Sadofsky, M. J., Hesse, J. E., McBlane, J. F., and Gellert, M. (1993). Expression and V(D)J recombination activity of mutated RAG-1 Proteins. *Nucl. Acids Res.*, **21**, 5644–50.

Schatz, D. G., and Baltimore, D. (1988). Stable expression of immunoglobulin gene V(D)J recombinase activity by gene transfer into 3T3 fibroblasts. *Cell*, **53**, 107–15.

Schatz, D. G., Oettinger, M. A., and Baltimore, D. (1989). The V(D)J recombination actvivating gene (RAG-1). *Cell*, **59**, 1035–48.

Schatz, D. G., Oettinger, M. A., and Schlissel, M. S. (1992). V(D)J recombination: molecular biology and regulation. *Annu. Rev. Immunol.*, **10**, 359–83.

Schlissel, M. S. and Baltimore, D. (1989). Activation of immunoglobulin kappa gene rearrangement correlates with induction of germline kappa gene transcription. *Cell*, **58**, 1001–7.

Schlissel, M. S., Concoran, L. M., and Baltimore, D. (1991*a*). Virus-transformed pre-B cells show ordered activation but not inactivation of immunoglobulin gene rearrangement and transcription. *J. Exp. Med.*, **173**, 711–20.

Schlissel, M., Voronova, A., and Baltimore, D. (1991*b*). Helix-loop-helix transcription factor E47 activates germline immunoglobulin heavy-chain gene transcription and rearrangement in a pre-T cell line. *Genes Dev.*, **5**, 1367–76.

Schuler, W., Weiler, I. J., Schuler, A., Phillips, R. A., Rosenberg, N., Mak., T. K., *et al.* (1986). Rearrangement of antigen receptor genes is defective in mice with severe immune deficiency. *Cell*, **46**, 963–72.

Schuler, W., Ruetsch, N. R., Amsler, M., and Bosma, M. J. (1991). Coding joint formation of endogenous T cell receptor genes in lymphoid cells from *scid* mice: unusual P-nucleotide additions in VJ-coding joints. *Eur. J. Immunol.*, **21**, 589–96

Shinkai, Y., Rathbun, G., Lam, K.-P., Oltz, E. M., Stewart, V., Mendelsohn, M., *et al.* (1992). RAG-1 deficient mice lack mature lymphocytes owing to inability to initiate V(D)J rearrangement. *Cell*, **68**, 855–67.

Silver, D. P., Spanopoulou, E., Mulligan, R. C., and Baltimore, D. (1993). Dispensible sequence motifs in the RAG-1 and RAG-2 genes for plasmid V(D)J recombination. *Proc. Natl. Acad. Sci. USA*, **90**, 6100–4.

Smider, V., Rathmell, W. K., Lieber, M. R., and Chu, G. (1994). Restoration of X-ray resistance and V(D)J recombination in mutant cells by Ku cDNA. *Science*, **266**, 288–91.

Spencer, D. M., Hsiang, Y.-H., Goldman, J. P., and Raulet, D. H. (1991). Identification of a T-cell-specific transcriptional enhancer located 3′ of Cγ1 in the murine T-cell receptor γ locus. *Proc. Natl. Acad. Sci. USA*, **88**, 800–4.

Taccioli, G. E., Rathbun, G., Oltz, E., Stamato, T., Jeggo, P. A., and Alt, F. W. (1993). Impairment of V(D)J recombination in double-strand break repair mutants. *Science*, **260**, 207–10.

Taccioli, G. E., Chenge, H.-L., Varghese, A. J., Whitmore, G., and Alt, F. W. (1994*a*). A DNA repair defect in Chinese hamster ovary cells affect V(D)J recombination similarly to the murine *scid* mutation. *J. Biol. Chem.*, **269**, 7439–42.

Taccioli, G. E., Gottlieb, T. M., Blunt, T., Priestley, A., Demengeot, J., Mizuta, *et al.* (1994*b*). Ku80: product of the XRCC5 gene and its role in DNA repair and V(D)J recombination. *Science*, **265**, 1442–5.

Takahama, Y. and Singer, A. (1992). Post-transcriptional regulation of early T cell development by T cell Receptor Signals. *Science*, **258**, 1456–62.

Takeda, J., Cheng, A., Mauxion, F., Nelson, C. A., Newberry, R. D., Sha, W. C., *et al.* (1990). Functional analysis of the murine T-cell receptor β enhancer and characteristics of its DNA-binding proteins. *Mol. Cell. Biol.*, **10**, 5027–35.

Takeda, S., Masteller, E. L., Thompson, C. B., and Buerstedde, J.-M. (1992). RAG-2 expression is not essential for chicken immunoglobulin gene conversion. *Proc. Natl. Acad. Sci. USA*, **89**, 4023–7.

Tonegawa, S. (1983). Somatic generation of antibody diversity. *Nature*, **302**, 575–81.

Turka, L. A., Schatz, D. G., Oettinger, M. A., Chun, J. J. M., Gorka, C., Lee, K., *et al.* (1991). Thymocyte expression of the recombination activating genes RAG-1 and RAG-2 can be terminated by T-cell receptor cross-linking. *Science*, **253**, 778–81.

Uematsu, Y., Ruser, S., Dembic, Z., Borgulya, P., Krimpenfort, P., Berns, A., *et al.* (1988). In transgenic mice the introduced functional T cell receptor β gene prevents expression of endogenous β genes. *Cell*, **52**, 831–41.

van Ness, B., Weigert, M., Coleclough, C., Mather, E., Kelley, D., and Perry, R. (1981). Transcription of the unrearranged mouse Cκ locus: Sequence of the initiation region and comparison of activity with a rearranged Vκ-Cκ gene. *Cell*, **27**, 593–602.

von Boehmer, H. (1990). Developmental biology of T cells in T cell-receptor transgenic mice. *Annu. Rev. Immunol.*, **8**, 1369–73.

Wang, J. C., Caron, P. R., and Kim, R. A. (1990). The role of DNA topoisomerases in recombination and genome stability: a double edged sword? *Cell*, **64**, 403–6.

Wilson, A., Held, W., and MacDonald, H. R. (1994). Two waves of recombinase gene expression in developing thymocytes. *J. Exp. Med.* **179**, 1355–60.

Winoto, A. and Baltimore, D. (1989*a*). Separate lineages of T cells expressing the βα and γδ receptors. *Nature*, **338**, 430–2.

Winoto, A. and Baltimore, D. (1989*b*). A novel, inducible and T cell-specific enhancer located at the 3′ end of the T cell receptor α locus. *EMBO J.*, **8**, 729–33.

Winoto, A. and Baltimore, D. (1989*c*). αβ lineage-specific expression of the α T cell receptor gene by nearby silencers. *Cell*, **59**, 649–55.

Yancopoulos, G. and Alt, F. (1985). Developmentally controlled and tissue-specific expression of unrearranged V_H gene segments. *Cell*, **40**, 271–81.

Yancopoulos, G., Blackwell, T. K., Suh, H., Hood, L., and Alt, F. W. (1986). Introduced T cell receptor variable region gene segments recombine in pre-B cells: evidence that B and T cells use a common recombinase. *Cell*, **44**, 251–9.

16 Allelic exclusion of T cell antigen receptor genes

BERNARD MALISSEN AND MARIE MALISSEN

1 Introduction

T cells can be divided into two subsets based on the structure of their T cell antigen receptors (TCR). In the adult, most T cells express a TCR heterodimer consisting of α and β chains, whereas a minor population expresses an alternative TCR made of γ and δ chains. Each of these four TCR chains includes a clonally variable (V) region. During intrathymic development, the genes encoding the TCR V regions are assembled via a series of site-specific DNA recombinations. The $CD4^{low}$ precursor cells entering the thymus carry their TCR gene loci in germline configuration and may develop along the γδ or αβ lineages (Wu *et al.* 1991). As shown in Figure 16.1, upon commitment to the αβ lineage, immature $CD4^-CD8^-CD3^-$ ('triple-negative') thymocytes differentiate into $CD4^+CD8^+$ ('double-positive') cells, a small percentage of which mature into either $CD4^+CD8^-$ or $CD4^-CD8^+$ ('mature single-positive') cells which correspond to the end stage of the intrathymic differentiation pathway (Shortman *et al.* 1992). The onset of TCR β chain gene rearrangement arises between the $CD44^+CD25^+$ and the $CD44^-CD25^+$ triple-negative stages of thymocyte differentiation (Godfrey *et al.* 1993). TCR α chain genes appear to rearrange at, or at the transition to, the double-positive stage (Pearse *et al.* 1989; Capone *et al.* 1993). β chain V genes are formed through the assembly of single members from each of three discrete libraries of gene segments (or subexons) denoted variable (Vβ), diversity (Dβ), and joining (Jβ). α chain V genes are assembled from two separate libraries of gene segments called variable (Vα) and joining (Jα). During V to J, V to D, or D to J recombination, the coding elements are imprecisely joined to each other; often a few bases are removed from one or both extremities (Gellert 1992). In some instances, variability at the junction is further amplified by the inclusion of extra bases that were not present in either of the precursor sequences. As a consequence, V(D)J joining events could result either in productive rearrangements that maintain an open reading frame throughout the gene, or in out-of-frame, non-productive rearrangements.

Because T lymphocytes are diploid cells, this recombination process could, in principle, generate T cell clones expressing two productive rearrangements at the α and/or β chain loci and therefore more than one TCR αβ chain combination. In the mouse, the expression of a productively rearranged TCR

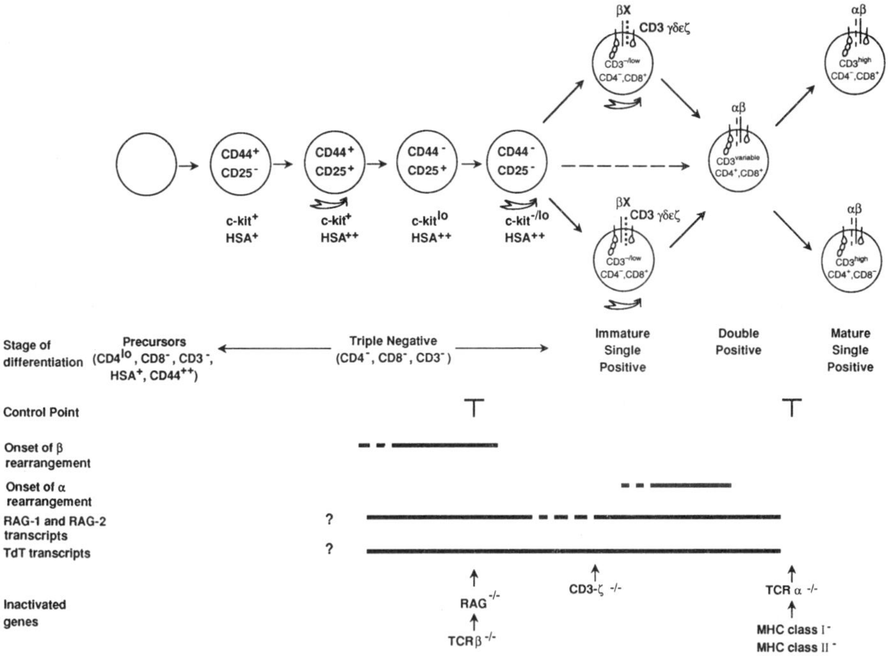

Fig. 16.1 Main stages of αβ T cell intrathymic development. Developmental control points (Shortman 1992) appear to ensure that T cells do not complete their intrathymic differentiation programme in the absence of productive TCR gene rearrangements or if they express TCR αβ combinations with inappropriate specificities (i.e. TCR combinations which are self-reactive, unable to co-operate with the set of co-expressed self-MHC molecules, or displayed with a mismatched CD4/CD8 co-receptor). These 'proof-reading' mechanisms may operate via the triggering of the TCR–CD3 complexes expressed at the surface of developing thymocytes. By activating intracellular effectors, TCR engagement may modulate gene expression pattern and induce the phenotypic shifts outlined in this figure. For instance, productive TCR β chain gene rearrangement appear required for the maturation out of the triple-negative compartment. It is not yet clear whether the role played by the TCR β chain in the modulation of gene expression pattern relies on its cellular display or requires its surface expression in association with some CD3 subunits (these putative α-less TCR–CD3 complexes are denoted βX). At a later stage, both positive selection (as reflected by the transition from the double-positive to the single-positive stage) and negative selection are controlled by the specificity of the clonally variable TCR αβ dimer and its interplay with a set of surface molecules among which CD4, CD8, and MHC molecules play a determining role. Rearrangement at the TCR β locus precedes rearrangement at the TCR α locus. The time windows within which RAG and TdT transcripts have been found are indicated. Also shown are the stages at which development is arrested in mice deficient for RAG-1 or RAG-2 ($RAG^{-/-}$), TCR β (TCR $\beta^{-/-}$), CD3 ζ/η (CD3 $\zeta^{-/-}$), TCR α (TCR $\alpha^{-/-}$), β2-microglobulin (MHC class $I^{-/-}$), and I-A α or β (MHC class $II^{-/-}$). In postnatal thymus most of the cell proliferation (⤳) occurs before the expression of fully assembled TCR αβ dimers.

β transgene has been shown to prevent complete V–DJ rearrangement of endogenous β genes (Uematsu *et al.* 1988), and this has led to the assumption that T cells have developed mechanisms of allelic exclusion that ensure that a single T cell clone makes one, and only one, TCR β chain. However, the analysis of the configuration and expression of TCR α genes in mature T cell lines (Malissen *et al.* 1988) and TCR αβ transgenic mice (von Boehmer 1990) has led to question whether allelic exclusion operates for the products of the TCR α locus. This review summarizes the current state of knowledge concerning the allelic exclusion of TCR α and β gene products.

It should be pointed out at this stage that the term 'allelic exclusion' has been used over the years to denote two distinct phenomena. Its first use corresponds to genotypic allelic exclusion and is intended to mean that the acquisition of a productive V(D)J rearrangement at one allele of a given antigen-receptor locus stops all further rearrangement at the other allele. Its second use corresponds to phenotypic allelic exclusion and indicates that a given clone expresses only one functional combination of antigen-receptor chains at the surface (the one cell–one specificity rule). As illustrated below, several mechanisms, only one of which is genotypic allelic exclusion, can account for phenotypic allelic exclusion.

2 Allelic exclusion in functional $\alpha\beta^+$ T cell clones

2.1 TCR α gene rearrangements

Initial clues as to the mechanisms of TCR α allelic exclusion have been provided by the analysis of the patterns of TCR α gene rearrangements displayed by mature $\alpha\beta^+$ T cells. Given the large size of the Jα cluster and the complexity of the Vα family, this tedious approach has been limited to date to a panel of 31 mouse T cell clones and its conclusion can be summarized as follows (Table 16.1). First, in most T cell clones, both α alleles have undergone V–J joining events. Only two clones have kept a Jα cluster in an unrearranged configuration (subsequently referred to as α°), and in both instances this feature was attributable to the presence of a non-productive VδDδJδ rearrangement. Secondly, 22 of the 31 clones were found to contain one productively rearranged α allele (denoted as α^+) and one non-productively rearranged allele (referred to as α^-). In the seven remaining clones, the two α alleles appeared productively rearranged in that they showed a V–J junction that has maintained the proper translational reading frame. Note that a single productive β chain gene rearrangement was identified in each of these seven clones. From these observations, the possibility arose that approximately 20% of the mature T cell clones express more than one αβ combination at their surface and consequently violate the rule of allelic exclusion. However, when three T cell clones carrying two in-frame VαJα rearrangements were further analysed, a single αβ combination was found to be expressed at their cell surface (reviewed in Malissen *et al.* 1992). Thus,

Table 16.1 Configuration of the TCR α and β alleles in T cell clones expressing functional TCR αβ heterodimers

	Number of clones analysed	Configuration[a]		
		α^+/α°	α^+/α^-	α^+/α^+
TCR α	31	2[b] (6%)	22 (71%)	7 (23%)
		β^+/β^{DJ}	β^+/β^-	β^+/β^+
TCR β	10	5	4	1

$\alpha\beta^+$ T cells have been categorized according to their TCR α or TCR β genotypes; (0): an unrearranged locus, (+): a productively rearranged gene, (−): a non-productively rearranged gene, DJ: a partially rearranged TCR β gene.

[a] The approach used to determine the productive or non-productive status of most of the α alleles compiled in this table is based on partial sequencing of the Vα genes (Casanova *et al.* 1991). Therefore, given the existence of Vα pseudogene segments, the frequency of cytolytic T cell clones with an α^+/α^+ genotype may have been slightly overestimated (see Malissen *et al.* 1992).

[b] In both clones with an α^+/α° genotype, one chromosome shows a non-productive VδDδJδ rearrangement and therefore contributes a Jα cluster in unrearranged configuration.

TCR α allelic exclusion can be achieved in a sizeable proportion of T cells through the non-productive rearrangement of one of the α alleles (for example, in clones with an α^+/α^- genotype, Table 16.1), and, in the absence of genotypic allelic exclusion, via mechanisms operating at the post-translational level and involving either the preferential pairing of one of the two α chains with the single β chain, or the inability of certain α chains to assemble properly with their associated β chain.

Further information on TCR α gene allelic exclusion have been obtained from the analysis of human T cells with a panel of anti-Vα antibodies (Padovan *et al.* 1993). According to this study, up to one-third of peripheral blood T cells express two distinct α chains at the cell surface. Moreover, analysis of independent T cell clones using the same pair of Vα genes indicated that the surface levels of the two Vα varied greatly from one clone to another and were inversely proportional; a feature probably attributable to the differential pairing of the two α chains with the single expressed β chain. Along the same line, it should be noted that a significant fraction of human γδ T cells have been recently found to express on their surface a single δ chain paired with either of two distinct γ chains (Davodeau *et al.* 1993). As will be discussed below, these results contrast significantly with the ones observed at the TCR β locus.

2.2 TCR β gene rearrangements

A limited number of $\alpha\beta^+$ T cell clones have been analysed for all of their TCR β gene rearrangements (reviewed in Malissen *et al.* 1992). As summarized in Table 16.1, half have undergone a single productive VDJ rearrangement (denoted as β^+) and achieved a partial DβJβ rearrangement (referred to as β^{DJ}) on their second β allele (note that the assembly of a complete Vβ gene proceeds in two separately controlled steps involving an initial D–J joining and a subsequent V–DJ rearrangement). In four clones, the productive β allele was found associated with a non-productive VDJ rearrangement (denoted as β^-). Finally, a unique T cell clone, showed two productive VDJ rearrangements and expressed the products of both alleles at the cell surface (Schittek *et al.* 1989). In view of the limited number of TCR β alleles analysed to date (Table 16.1), it is difficult to raise any conclusion as to the frequency of $\alpha\beta^+$ T cells expressing two distinct β chains at their surface.

The existence of a sizeable proportion of mature $\alpha\beta^+$ T cells with a β^+/β^{DJ} genotype (Table 16.1) has been confirmed by the analysis of a panel of 58 short-term human T cell clones (Seboun *et al.* 1992). VDJ rearrangements on one or both alleles were present in 60% and 40% of the clones, respectively. The existence of approximately equal numbers of $\alpha\beta^+$ T cells with either one or two VDJ rearrangements at the TCR β locus contrasts with the situation at the TCR α locus where nearly all of the $\alpha\beta^+$ T cells analysed displayed two VJ rearrangements (Table 16.1). The different frequencies of V–(D) J recombination events observed at these two loci might be accounted for by:

1. The total number of V and (D) J elements that can participate in recombination (the TCR α locus contains 49 Jα and as many as 100 Vα segments, whereas the TCR β locus has far fewer Vβ and Dβ–Jβ elements available for rearrangement).
2. The relative efficiency of their respective recombination signal sequences (Ramdsen and Wu 1991; Gerstein and Lieber 1993*b*).
3. The time elapsed during which the TCR α or TCR β loci are accessible to recombination events (Reynaud *et al.* 1989; McCormack *et al.* 1991).
4. The possible occurrence of a high frequency of secondary TCR α rearrangements (Marolleau *et al.* 1988; Fondell and Marcu 1992).

3 Secondary TCR α rearrangements

Given the large number of Vα and Jα gene segments and the absence of Dδ gene segments, a single α chain locus can experience successive rounds of V–J rearrangements. Indirect evidence for the existence of such secondary TCR α gene rearrangements has been provided by:

1. The presence of productive and non-productive Vα–Jα rearrangements in the circular DNA excised during V–J joinings (Okazaki and Sakano 1988).
2. The determination of the use of Jα gene segments during T cell development (Thompson *et al.* 1990; Roth *et al.* 1991).
3. The analysis of a virally induced mouse T lymphoma (Marolleau *et al.* 1988; Fondell and Marcu 1992).

In the third instance, subcloning of these $CD4^-CD8^-TCR\ \alpha\beta^+$ lymphoma yielded several lines which had undergone secondary rearrangements at the TCR α locus, while maintaining a stable pattern of β gene rearrangement. In each of the secondary rearrangements analysed, a fully assembled VJ join had been deleted via a recombination event involving the joining of an upstream Vα segment to a downstream Jα segment. Both productively and non-productively rearranged α alleles were found to serve as substrates for secondary rearrangements. As originally hypothesized for the immunoglobulin κ genes (Feddersen and Van Ness 1990), these secondary rearrangements may happen at a high frequency during T cell development and serve to 'correct' (replace) either a previous non-productive VJ join, or an α^+ allele expressing an 'inadequate' (non-selectable) α chain product (see below). It should be stressed, however, that most of these speculations have been based either on the study of virally-transformed cell lines, or on indirect evidence. Therefore, they can not be used to ascertain whether secondary TCR α rearrangements occur in normal thymocytes and represent physiologically important events. For instance, the productive Vα–Jα joins found in the DNA circles excised from the TCR α locus have been interpreted as the remnants of primary rearrangement and used to support the view that TCR α gene rearrangement can continue even after the production of a first α chain. However, it is equally possible that these episomal in-frame Vα–Jα joins occurred subsequently to the excision event.

4 Allelic exclusion in TCR transgenic mice

Mice carrying productively rearranged TCR transgenes have greatly contributed to our present views on TCR allelic exclusion. For instance, the analysis of T cell clones derived from TCR β and TCR αβ transgenic mice showed that the rearrangements of the endogenous β genes were almost completely stopped prior to the V–DJ stage (Uematsu *et al.* 1988). As previously suggested for Ig transgenic mice (Nussenzweig *et al.* 1988), this phenotype may have resulted from the expression of the fully assembled TCR β transgene at an abnormally early stage (Borgulya *et al.* 1992; Wilson *et al.* 1992; Nikolic-Zugic *et al.* 1993). Thus, the accelerated expression of the β transgene may have speeded up the transition to the double-positive stage and shortened the developmental window within which V–DJ recombination normally occurs. As a consequence, the probability of endogenous V–DJ rearrangements would have been greatly reduced in these mice.

In contrast to the findings for the TCR β locus, it has been shown that αβ TCR transgenic mice are able to express their endogenous TCR α genes to varying levels (Berg *et al.* 1988; von Boehmer 1990; Brandle *et al.* 1992). This 'leakiness' may have resulted from the deletion of the α transgene in some T cell clones (Bluthmann *et al.* 1988), or from a delay in the onset of expression of the α TCR transgene compared to that of the endogenous TCR α genes.

Alternatively, the expression of a transgenic αβ TCR dimer may have been necessary but not sufficient to shut off endogenous α gene rearrangements.

Particularly relevant to these issues is the analysis of mice expressing α and β TCR transgenes derived from a $CD8^-$ cytolytic T cell clone specific for the H-Y antigen and restricted by the D^b MHC molecule (von Boehmer 1988). The introduction of α and β TCR transgenes into female $H\text{-}2^b$ mice resulted in the accelerated expression of both transgenic chains at the surface of the late $CD4^-CD8^-$ (Borgulya *et al.* 1992; von Boehmer, 1992; Nikolic-Zugic *et al.* 1993). In normal mice significant levels of TCR αβ dimers are only found at the $CD4^+CD8^+$ stage (Figure 16.1). Most of the $CD4^+CD8^+$ and $CD4^-8^+$ thymocytes were found to express both transgenic chains at their surface. In contrast, the $CD4^+8^-$ thymocytes expressed on their surface TCR dimers consisting of transgenic β chains and endogenous α chains, even though they still transcribed the α TCR transgene. The emergence of these $CD4^+8^-$ mature cells may be explained by the fact that α gene rearrangements can still occur in $CD4^+8^+$ thymocytes expressing the transgenic αβ TCR on their surface. Accordingly, before their recruitment into the $CD4^-8^+$ compartment, some $CD4^+8^+$ clones may have generated endogenous α chains capable of a better match with the transgenic β chain. In some cases, the new surface expressed αβ TCR dimers could have been positively selected by the thymic class II MHC molecules, allowing the corresponding clones to be recruited into the $CD4^+8^-$ mature subset. These data, together with the ones obtained in αβ TCR transgenic mice co-expressing MHC class I molecules that did not positively select the transgenic TCR (von Boehmer 1988; Brandle *et al.* 1992), support the hypothesis that the mere expression of an αβ TCR dimer on the surface of $CD4^+8^+$ and late $CD4^-8^-$ thymocytes may not be sufficient to shut down the expression of the recombination activating genes RAG-1 and RAG-2. Further, positive selection via the expressed αβ TCR dimer may be needed to trigger the maturation of the corresponding thymocytes to a differentiation stage devoid of recombinase activity (Malissen *et al.* 1988; Borgulya *et al.* 1992; Brandle *et al.* 1992). Thus, the levels of endogenous TCR α chain gene rearrangements should be inversely proportional to the extent of positive selection. Three observations support the view that the triggering of the TCR present at the surface of developing T cells is necessary to turn off RAG-1 and RAG-2. First, RAG-1 and RAG-2 expression can be readily detected in double-positive TCR^+ thymocytes (Turka *et al.* 1991). Secondly, in *in vitro* experiments, cross-linking of the TCR–CD3 complex expressed at the surface of unfractionated human thymocytes has been found to result in significant down-regulation of both RAG-1 and RAG-2 transcripts (Turka *et al.* 1991). Thirdly, *in vivo* studies have shown that positive selection is associated with the shut off of RAG expression since small $CD4^+8^+$ TCR^{low} cells express 10- to 20-fold higher levels than small $CD4^+8^+$ TCR^{high} cells which probably result from positive selection (Borgulya *et al.* 1992). In line with this last observation, it has been noted that the levels of RAG-1 transcription were reduced in the thymic cortex from transgenic mice expressing TCR α and β chain genes derived from a virus-specific $CD8^+$ T cell clone and undergoing positive selection (Brandle *et al.* 1992). Recent work on T and B lymphocyte selection have provided additional examples of a link between antigen receptor engagement and RAG expression (Ma *et al.* 1992; Gay *et al.* 1993; Tiegs *et al.* 1993; Campbell and Hashimoto 1993).

5 Toward an integrated model of allelic exclusion

It is tempting to adopt for TCR genes a 'regulated' model of allelic exclusion similar to the one originally elaborated for the mouse immunoglobulin genes (reviewed in Alt *et al.* 1992). According to this model, TCR β genes should be rearranged first (only on one allele or sequentially on both alleles if the first rearrangement is non-productive) and tested for the expression of a TCR β chain (steps 1 to 4 in Figure 16.2). By acting intracellularly or via its insertion at the cell surface, this TCR β chain should stop further β gene rearrangements and relay signals that will trigger rearrangement at the α chain locus. Similarly, α gene rearrangements would be attempted sequentially for one or both α alleles (steps 5 and 6 in Figure 16.2) until a resulting TCR αβ dimer can be recognized by some intra- or submembrane molecular sensor and signal the cell to arrest further α gene rearrangements. In this fixed programme, the product of each joining event is 'assayed' before the next joining event is started and, according to its functional or non-functional status should condition the occurrence of the next developmental step.

The existence of 'proof-reading' mechanisms operating at different stages of the intrathymic developmental cascade (Figure 16.1) has been recently

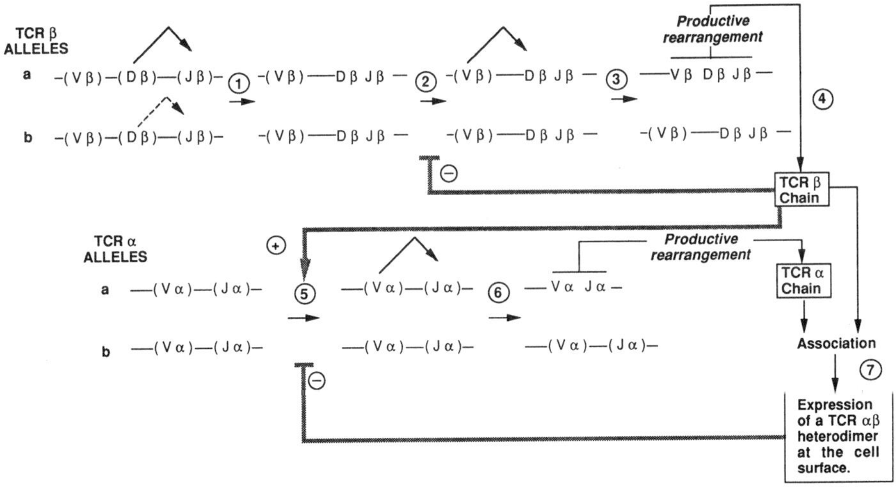

Fig. 16.2 An 'ordered' and 'regulated' model of allelic exclusion of TCR α and β genes (see text).

illustrated by Mallick and co-workers (1993). To minimize the possible effects of positive selection, the authors have analysed TCR α deficient mice and showed that their pool of double-positive cells contained TCR β chain genes which were predominantly productively rearranged. Thus, TCR β chain expression appears to be a prerequisite for the transition to the double-positive compartment. In line with this idea, it should be noted that mice with disrupted TCR β chain genes have profoundly reduced numbers of double-positive cells (Mombaerts *et al.* 1992; Shinkai *et al.* 1993). As discussed above, TCR β chain expression has been hypothesized to be required for the suppression of β chain gene rearrangement as well as the induction of α chain gene rearrangement (Figure 16.2). However, mice with disrupted TCR β genes were still capable of rearranging their TCR α genes (Mombaerts *et al.* 1992). Taken together, these data suggest that rearrangements at the TCR β and TCR α loci occur sequentially but without a causal relationship. As recently hypothesized for the pre-B cell receptor complex (Schatz *et al.* 1992; Ehlich *et al.* 1993; Lassoued *et al.* 1993), the role of the β chain may be limited to the up-regulation of the rate of α chain gene rearrangement.

As depicted in Figure 16.3, in the absence of secondary rearrangements, the ratio of the number of $\alpha\beta^+$ T cells containing a +/0 (or +/DJ) β genotype to that containing a +/− β genotype would be expected to be about 1.5. The observed ratio was about one in the small sample collated in Table 16.1. Note that the human data (Seboun *et al.* 1992) gave a ratio of 1.5 for a sample of 58 clones. These numbers are consistent with the 'regulated' model outlined in Figure 16.2. For those cells with a functional β chain, the same regulated process would then occur at the α locus, giving a final ratio of α^+/α° to α^+/α^- genotypes of 1.5. The observed ratio (about 0.09, Table 16.1) is much lower than predicted, since few mature $\alpha\beta^+$ T cells have maintained one Jα cluster in germline (α°) configuration (Table 16.1). Furthermore, there is a high frequency (23%) of α^+/α^+ cells, which would not be predicted by the process shown in Figure 16.2. This suggests that the 'regulated' model does not apply to the TCR α genes. At least three alternative models could better explain the α chain data. First, VαJα rearrangements could be attempted quasi-simultaneously on both α alleles. Given this assumption, and in the absence of secondary VJ rearrangements, the expected frequency (20%) of functional $\alpha\beta^+$ T cells with an α^+/α^+ genotype approximates to the one determined in Table 16.1 (23%). Secondly, TCR α gene rearrangements at both alleles could still occur in a sequential mode but, once properly expressed at the cell surface of $CD4^+CD8^+$ thymocytes a given Vα/Vβ pair has only a low probability of being positively selected by the intrathymic self-MHC products. Consequently, the $CD4^+CD8^+$ thymocyte clones expressing 'inadequate' (that is, non-selectable) αβ dimers do not progress to a recombinase negative stage and may still undergo rearrangements on the other allele. Two α chains could then be expressed and compete for the single β chain. A possible objection to this model would be that endogenous α chains can rearrange in αβ transgenic mice in which a selectable receptor is already

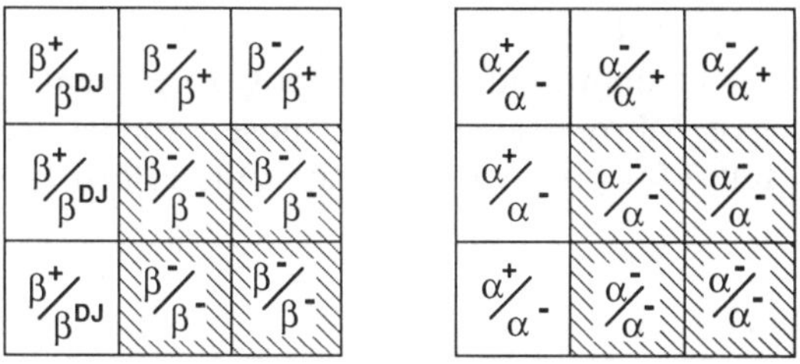

Fig. 16.3 Predicted frequency of TCR α and TCR β genotypes according to the 'regulated' model of allelic exclusion. Since the V – (D)J joining process is essentially a random one, and assuming reading frame errors to be the main source of non-functional V genes, from a cohort of nine thymocytes, three might be expected to produce a functional Vβ gene as a result of their first attempt of rearrangement. The corresponding TCR β chains will stop further β rearrangement and the resulting mature T cells would later contain one functionally rearranged β allele and one partially rearranged (for the majority), or unrearranged β allele. Of the remaining six thymocytes with a non-productive β allele, an additional third might be expected to produce a functional Vβ gene upon subsequent rearrangement of their second allele and later display a +/− β genotype. Accordingly, the ratio of the number of mature $\alpha\beta^+$ T cells containing a +/DJ (or +/o) β genotype to that containing a +/− β genotype would be expected to be about 1.5. For those cells with a functional β chain (see Mallick *et al.* 1993), the same regulated process would then occur at the α locus, giving a secondary loss of 4/9 cells and a final ratio of α^+/α° to α^+/α^- genotypes of 1.5. Finally, two points are worth noticing. First, owing to the fact that each of the two mouse Dβ gene segments can be used in all three reading frames, the determination of the frequency of TCR β genotypes has only to take into consideration the single V–DJ joining event. Secondly, it is generally agreed that the V–(D)J joining process is a random one which results in about one-third of the joins being in a productive configuration. Note that contrary to this expectation, the presence or absence of short regions of homology in the coding ends participating in the recombination may have a guiding role and increase the probability of forming a productive join (Asarnow *et al.* 1993; Gerstein and Lieber 1993*a*; Itohara *et al.* 1993). For instance, the recurrent invariant joins of the TCR γ chains used by the epithelial-associated γδ T cells of the skin and reproductive tract appear to result from the presence at the breakpoints of short homologous regions which are either germline encoded or P nucleotide contributed. DJ: a partially rearranged TCR β locus; +: a productively rearranged α or β gene; −: a non-productively rearranged α or β gene. Cells with a −/− genotype are committed to die within the thymus (hatched squares).

expressed on the cell surface. However, this objection is valid only if positive selection is 100% effective, an unlikely situation since these same αβ transgenic mice do not have a highly increased production of mature thymocytes or T cells (Huesmann *et al.* 1991; Kelly *et al.* 1993).

A third, and not mutually exclusive, model to account for the paucity of cells with an α^+/α° genotype is that the time-span elapsing between the expression of a selectable αβ dimer and its interaction with intrathymic MHC products is sufficiently long to allow an α gene rearrangement on the other allele. Any of these three models could have secondary rearrangements on either allele built into it. However, some special restrictions on the timing of secondary rearrangements would have to apply to prevent the deletion of a previous productive VJ rearrangement whose product was being tested (perhaps successfully) for positive selection at the cell surface.

6 Conclusions

Altogether, the data described in this review illustrate the differences between the mechanisms that achieve allelic exclusion for the TCR α and TCR β genes. Whereas TCR β genes undergo a tight allelic exclusion, the events underlying that of the TCR α genes appear to be less stringent and may depend on secondary signals resulting from positive selection.

The different frequencies of V–(D)J recombination events observed at the TCR α and β loci may be accounted for by the following features. First, once it is accessible to the recombinase, the TCR α locus may, in comparison to the β locus, display a markedly higher probability of rearranging a V gene within a given unit of time. The difference in the kinetics of V–(D)J rearrangement may be due to the dissimilarity in the size of the pool of V and (D)J substrates available for recombination and/or to the relative efficiency of the recombination signal sequences. Secondly, the α-less TCRβ–CD3 complexes hypothesized to be expressed at the surface of immature thymocytes may be efficiently selected by a superantigen-like ligand and be capable of delivering rapid shutdown signals to the β locus. In contrast, the TCR αβ combinations expressed at the surface of $CD4^+CD8^+$ thymocytes may have a low probability of being selected by thymic MHC molecules and the process of TCR αβ selection may follow slow kinetics, allowing more than one VJ rearrangement. Interestingly, the same parameters (e.g. the length of time during which rearrangement are allowed or, the probability of making a productive rearrangement within a given unit of time . . .) have been also used to account for the seemlingly ordered nature of Ig heavy and light chain gene rearrangements (Schatz *et al.* 1992; Ehlich *et al.* 1993).

Acknowledgements

We thank A. Lanzavecchia and J.-L. Casanova for sharing data with us prior to publication, and N. Guglietta and C. Beziers La Fosse for the preparation of the manuscript. Work in the authors' laboratory was supported by CNRS and INSERM.

References

Alt, F. W., Oltz, E. M., Young, F., Gorman, J., Taccioli, G., and Chen, J. (1992). VDJ recombination. *Immunol. Today*, **13**, 306–314.

Asarnow, D. M., Cado, D., and Raulet, D. H. (1993). Selection is not required to produce invariant T-cell receptor γ-gene junctional sequences. *Nature*, **362**, 158–60.
Berg, L., Fazekas, B., and Ivars, F. *et al.* (1988). Expression of T-cell receptor alpha-chain genes in transgenic mice. *Mol. Cell. Biol.*, **8**, 5459–69.
Bluthmann, H., Kisielow, P., Uematsu, Y., *et al.* (1988). T-cell-specific deletion of T-cell receptor transgenes allows functional rearrangement of endogenous α and β genes. *Nature*, **334**, 156–9.
Borgulya, P., Kishi, H., Uematsu, Y., and von Boehmer, H. (1992). Exclusion and inclusion of α and β T cell receptor alleles. *Cell*, **69**, 529–37.
Brändle, D., Miller, C., Rllicke, T., Hengartner, H., and Pircher, H. (1992). Engagement of the T-cell receptor during positive selection in the thymus down-regulates RAG-1 expression. *Proc. Natl. Acad. Sci. USA*, **89**, 9529–33.
Campbell, J. J. and Hashimoto, Y. (1993). Recombinase activating gene expression in thymic subpopulations. A transitional cell type has lost RAG-2 but not RAG-1. *J. Immunol.*, **150**, 1307–13.
Capone, M., Watrin, F., Fernex, C., Horvat, B., Krippl, B., Wu, L., *et al.* (1993). TCR β and TCR α gene enhancers confer tissue- and stage-specificity on V(D)J recombination events. *EMBO J.*, **12**, 4335–46.
Casanova, J. L., Romero, P., Widmann, C., Kourilsky, P., and Maryanski, J. L. (1991). T cell receptor genes in a series of class I major histocompatibility complex-restricted cytotoxic T lymphoyte clones specific for a *Plasmodium berghei* nonapeptide: implications for T cell allelic exclusion and antigen-specific repertoire. *J. Exp. Med.*, **174**, 1371–83.
Davodeau, F., Peyrat, M. A., Houde, I., Hallet, M. M., De Libero, G., Vie, H., *et al.* (1993). Surface expression of two distinct functional antigen receptors on human γδ T cells. *Science*, **260**, 1800–2.
Ehlich, A., Schaal, S., Gu, H., Kitamura, D., Muller, W, and Rajewsky, K. (1993). Immunoglobulin heavy and light chain genes rearrange independently at early stages of B cell development. *Cell*, **72**, 695–704.
Feddersen, R. M. and Van Ness, B. G. (1990). Corrective recombination of mouse immunoglobulin kappa alleles in Abelson murine leukemia virus-transformed pre-B cells. *Mol. Cell. Biol.*, **10**, 569–76.
Fondell, J. D. and Marcu, K. B. (1992). Transcription of germ line Vα segments correlates with ongoing T-cell receptor α-chain rearrangements. *Mol. Cell. Biol.*, **12**, 1480–9.
Fondell, J. D., Marolleau, J. P., Primi, D., and Marcu, K. B. (1990). On the mechanism of non-allelically excluded Vα-Jα T cell receptor secondary rearrangements in a murine T cell lymphoma. *J. Immunol.*, **144**, 1094–103.
Gay, D., Saunders, T., Camper, S., and Weigert, M. (1993). Receptor editing: an approach by autoreactive B cells to escape tolerance. *J. Exp. Med.*, **177**, 999–1008.
Gellert, M. (1992). Molecular analysis of V(D)J recombination. *Annu. Rev. Genet.*, **26**, 425–46.
Gerstein, R. M. and Lieber, M. R. (1993*a*). Extent to which homology can constrain coding exon junctional diversity in V(D)J recombination. *Nature*, **363**, 625–7.
Gerstein, R. M. and Lieber, M. R. (1993*b*). Coding end sequence can markedly affect the initiation of V(D)J recombination. *Genes Dev.*, **7**, 1459–69.
Godfrey, D. I., Kennedy, J., Suda, T., and Zlotnik, A. (1993). A developmental pathway involving four phenotypically and functionally distinct subsets of CD3−CD4−CD8− triple-negative adult mouse thymocytes defined by CD44 and CD25 expression. *J. Immunol.*, **150**, 4244–52.
Huesmann, M., Scott, B., Kisielow, P., and von Boehmer, H. (1991). Kinetics and

efficacy of positive selection in the thymus of normal and T cell receptor transgenic mice. *Cell*, **66**, 533–40.

Itohara, S., Mombaerts, P., Lafaille, J., Iacomini, J., Nelson, A., Clarke, A. R., *et al.* (1993). T cell receptor δ gene mutant mice: independent generation of αβ T cells and programmed rearrangements of γδ TCR genes. *Cell*, **72**, 337–48.

Kelly, K. A., Pircher, H., von Boehmer, H., Davis, M. M., and Scollay, R. (1993). Regulation of T cell production in T cell receptor transgenic mice. *Eur. J. Immunol.*, **23**, 1922–8.

Lassoued, K., Nunez, C. A., Billips, L., Kubagawa, H., Monteiro, R. C., Le Blen, T. W., *et al.* (1993). Expression of surrogate light chain receptors is restricted to a late stage in pre-B cell differentiation. *Cell*, **73**, 73–86.

Ma, A., Fisher, P., Dildrop, R., Oltz, E., Rathbun, G., Achacoso, P., *et al.* (1992). Surface IgM mediated regulation of RAG gene expression in E 5-N-myc B cell lines. *EMBO J.*, **11**, 2727–34.

Malissen, M., Trucy, J., Letourneur, F., Rebai, N., Dunn, D. E., Fitch, F. W., *et al.* (1988). A T cell clone expresses two T cell receptor α genes but uses one αβ heterodimer for allorecognition and self MHC-restricted antigen recognition. *Cell*, **55**, 49–59.

Malissen, M., Trucy, J., Jouvin-Marche, E., Cazenave, P. A., Scollay, R., and Malissen, B. (1992). Regulation of TCR α and β gene allelic exclusion during T-cell development. *Immunol. Today*, **13**, 315–22.

Mallick, C. A., Dudley, E. C., Viney, J. L., Owen, M. J., and Hayday, A. C. (1993). Rearrangement and diversity of T cell receptor β chain genes in thymocytes: a critical role for the β chain in development. *Cell*, **73**, 513–19.

Marolleau, J. P., Fondell, J. D., Malissen, M., Trucy, J., Barbier, E., Marcu, K. B., *et al.* (1988). The joining of germ-line Vα to Jα genes replaces the preexisting Vα-Jα complexes in a T cell receptor α,β positive T cell line. *Cell*, **55**, 294–300.

McCormack, W. T., Tjoelker, L. W., and Thompson, C. B. (1991). Avian B-cell development: generation of immunoglobulin repertoire by gene conversion. *Annu. Rev. Immunol.*, **9**, 219–41.

Mombaerts, P., Clarke, A. R., Rudnicki, M. A., Iacomini, J., Itohara, S., Lafaille, J. J., *et al.* (1992). Mutations in T-cell antigen receptor genes α and β block thymocyte development at different stages. *Nature*, **360**, 225–31.

Nikolic-Zugic, J., Andjelic, S., Teh, H. S., and Jain, N. (1993). The influence of rearranged T cell receptor αβ transgenes on early thymocyte development. *Eur. J. Immunol.*, **23**, 1699–704.

Nussenzweig, M. C., Schmidt, E. V., and Shaw, A. C., *et al.* (1988). A human immunogloblin gene reduces the incidence of lymphomas in c-myc-bearing transgenic mice. *Nature*, **336**, 446–50.

Okasaki, K. and Sakano, H. (1988). Thymocyte circular DNA excised from T cell receptor αβ gene complex. *EMBO J.*, **7**, 1669–74.

Padovan, E., Casorati, G., Dellabona, P., Meyer, S., Brockhaus, M., and Lanzavecchia, A. (1993). Expression of two T cell receptor α chains: dual receptor T cells. *Science*, **262**, 422–4.

Pearse, M., Wu, L., Egerton, M., Wilson, A., Shortman, K., and Scollay, R. (1989). A murine early thymocyte developmental sequence is marked by transient expression of the interleukin 2 receptor. *Proc. Natl. Acad. Sci. USA*, **86**, 1614–18.

Ramsden, D. A. and Wu, G. E. (1991). Mouse κ light-chain recombination signal sequences mediate recombination more frequently than do those of λ light chain. *Proc. Natl. Acad. Sci. USA*, **88**, 10721–5.

Reynaud, C. A., Dahan, A., Anquez, V., and Weill, J. C. (1989). Somatic hyper-conversion diversifies the single Vh gene of the chicken with a high incidence in the D region. *Cell*, **59**, 171–83.

Roth, M. E., Holman, P. O., and Kranz, D. M. (1991). Non random use of Jα gene segments. Influence of Vα and Jα gene location. *J. Immunol.*, **147**, 1075–81.

Schatz, D. G., Oettinger, M. A., and Schlissel, M. S. (1992). V(D)J recombination: molecular biology and regulation. *Annu. Rev. Immunol.*, **10**, 359–83.

Schittek, B., Unkelbach, E., and Rajewsky, K. (1989). Violation of allelic exclusion of the T cell receptor β genes in a helper T cell clone. *Int. Immunol.*, **1**, 273–80.

Seboun, E., Joshi, N., and Hauser, S. L. (1992). Haplotypic origin of β-chain genes expressed by human T-cell clones. *Immunogenetics*, **36**, 363–8.

Shinkai, Y., Koyasu, S., Nakayama, K.-I., Murphy, K. M., Loh, D. Y., Reinherz, E. L., *et al.* (1993). Restoration of T cell development in RAG-2-deficient mice by functional TCR transgenes. *Science*, **259**, 822–5.

Shortman, K. (1992). Cellular aspects of early T-cell development. *Curr. Opinion Immunol.*, **4**, 140–6.

Thompson, S. D., Pelkonen, J., Rytkonen, M., Samaridis, J., and Hurwitz, J. L. (1990). Nonrandom rearrangement of T cell receptor Jα genes in bone marrow T cell differentiation cultures. *J. Immunol.*, **144**, 2829–34.

Tiegs, S. L., Russell, D. M., and Nemazee, D. (1993). Receptor editing in self-reactive bone marrow B cells. *J. Exp. Med.*, **177**, 1009–20.

Turka, L. A., Schatz, D. G., Oettinger, M. A., Chun, J. J., Gorka, C., Lee, K., *et al.* (1991). Thymocyte expression of RAG-1 and RAG-2: termination by T cell receptor cross-linking. *Science*, **253**, 778–81.

Uematsu, Y., Ryser, S., Dembic, Z., Borgulya, P., Krimpenfort, P., Berns, A., *et al.* (1988). In transgenic mice the introduced functional T cell receptor β gene prevents expression of endogenous β genes. *Cell*, **52**, 831–41.

von Boehmer, H. (1988). The developmental biology of T lymphocytes. *Annu. Rev. Immunol.*, **6**, 309–26.

von Boehmer, H. (1990). Developmental biology of T cells in T cell-receptor transgenic mice. *Annu. Rev. Immunol.*, **8**, 531–56.

von Boehmer, H. (1992). Thymic selection: a matter of life and death. *Immunol. Today*, **13**, 454–8.

Wilson, A., Pircher, H., Ohashi, P., and MacDonald, H. R. (1992). Analysis of immature (CD4−CD8−) thymic subsets in T-cell receptor αβ transgenic mice. *Dev. Immunol.*, **2**, 85–94.

Wu, L., Scollay, R., Egerton, M., Pearse, M., Spangrude, G. J., and Shortman, K. (1991). CD4 expressed on earliest T-lineage precursor cells in the adult murine thymus. *Nature*, **349**, 71–4.

Part IV

T cell proteins

17 Plasticity of the TCR–CD3 complex

COX TERHORST, STEVE SIMPSON, BAOPING WANG, JIAN SHE, CRAIG HALL, MANLEY HUANG, TOM WILEMAN, KLAUS EICHMANN, GEORG HOLLÄNDER, CHRISTIAAN LEVELT, AND MARK EXLEY

1 Introduction

The T cell repertoire is characterized by an extremely diverse population of different surface receptors which facilitate highly specific responses to a large number of different antigens. Variable TCR αβ and γδ receptors are associated with the invariant CD3 γ, δ, ε, and ζ proteins, thus forming the TCR–CD3 complex. Within the TCR–CD3 complex a division of labour exists: TCR αβ or γδ recognize antigen and CD3 γ, δ, ε, and ζ control assembly and signal transduction.

The prototype T lymphocyte recognizes antigen via the αβ-TCR–CD3 complex as a specific MHC–peptide antigen combination on the interface of T lymphocytes and antigen-presenting cells. This so-called MHC restriction, develops as a consequence of negative and positive selection of thymocytes during their maturation within the thymus. These selection processes are a necessity, because the variable elements of the TCR β and α genes rearrange (in that order) at random in early thymocytes. A substantial number of these early thymocytes contain non-productively rearranged TCR α and β mRNAs or TCR α and β polypeptide chains, which fail to pair and form a stable TCR–CD3 complex on the cell surface. These thymocytes are thought to be eliminated by *apoptosis* as they are unable to receive the appropriate signals necessary to progress to the next developmental stage. Remaining thymocytes, which successfully express αβ TCR–CD3 complexes on their surface and which carry both the CD4 and the CD8 marker, are either negatively or positively selected by MHC and peptide (Hogquist *et al.* 1994). Thymocytes which have been positively selected are termed *mature thymocytes* ($CD4^+$ or $CD8^+$), express few if any autoreactive TCR, and are exported to the periphery. A second population of peripheral T cells has been found to display an alternative clonotypic structure termed the γδ TCR–CD3 complex. In contrast to αβ T cells, no common restriction element (like MHC) is presently known for γδ T cells and potential selection principles involved in their development are unknown.

The exceptionally complicated interactions of TCR–CD3 complex with

Table 17.1 Components of the T cell receptor/CD3 complex

Subunit	Size (kDa)	Primary structure			Glycosylation	Phosphorylation	Chromosomal location	No. of cM from centromere*
		Ex.	TM	Cyt.				
TCR								
α_H	45–60 (32)	227	27	4	+	–	14q11.2	ND
α_M	45–55 (29)	227	27	4	+	–	14C-D	14(20)
β_H	40–50 (32)	220	23	3	+	–	7q35	ND
β_M	40–55 (32)	220	23	3	+	–	6B	6(19)
γ_H	45–60 (30)	218	25	4	+	–	7p15-14	ND
γ_M	45–60 (30)	218	25	4	+	–	13A2-A3	13(10)
δ_H	40–60 (37)	223	27	5	+	–	14q11.2	ND
δ_M	40–60 (31)	223	27	5	+	–	14C-D	14(20)
CD								
γ_H	25–28 (16)	89	27	44	+	P-Ser/P-Tyr	11q23	
γ_M-cd3d	21 (16)	89	27	44	+	P-Ser/P-Tyr	9	9(26)
δ_H	20 (14)	79	27	44	+	P-Tyr	11q23	
δ_M	28 (16)	79	27	46	+	P/-Tyr	9	9(26)
ε_H	20 (14)	105	26	55	–	P-Tyr	11q23	
ε_M	25 (18)	87	26	55	–	P-Tyr	9	9(26)
ζ_H	16	9	21	112	–	P-Tyr/P-Ser	1q22-q25	
ζ_M	16	9	21	113	–	P-Tyr/P-Ser	1	1(87)
η_M	20	9	21	155	–	P-Tyr?	1	1(87)
ω_H	28	?	?	?	–	?	?	ND
ω_M	28	?	?	?	–	?	?	ND
FcεR1γ_H	9	6	20	42	–	P-Tyr	1q23	ND
FcεR1γ_M	9	6	20	42	–	P-Tyr	1q23	1(92)

*Mouse localizations are from 1994 chromosome committee reports.

antigen are also reflected in the structure of the TCR–CD3 complex itself. Naturally, clonotypically distributed polypeptide chains (TCR αβ or TCR γδ) need to be accommodated by the receptor structure. But, why does the TCR–CD3 complex consist of an ensemble of low molecular weight membrane proteins instead of two transmembrane proteins with variable and constant ectodomains connected with large cytoplasmic tails with multiple functional subdomains? The benefits for the TCR–CD3 complex encompassing this set of polypeptide chains include the necessity to:

1. Permit cell surface expression of alternative receptor protein ensembles, which subserve distinct functions in specialized mature T lymphocytes.
2. Control receptor assembly and surface expression in T lymphocytes.
3. Control surface expression and signal transduction of partial TCR–CD3 complexes during thymocyte development.

The apparent plasticity of the TCR–CD3 complex will be discussed in greater detail in this review.

2 Components of the TCR–CD3 complex

2.1 TCR α, β, γ, and δ

Whereas the TCR α and δ genes are located near each other on the human chromosome 14q11.2 (or murine 14C-D), the TCR β and γ genes are either on opposite ends of the same chromosome (human TCR β, 7q32–35; TCR γ, 7p15–14), or on different chromosomes (murine TCR β, 6B; TCR γ, 13A2–3). A discussion of these complicated rearranging gene systems can be found elsewhere (Chapter 15). Here we will focus on aspects of their protein structure which are relevant for the assembly and structure of the TCR–CD3 complex.

The TCR α, β, γ, and δ polypeptide chains are membrane glycoproteins each consisting of a leader sequence, an externally disposed N-terminal extracellular domain, a single membrane spanning domain, and a cytoplasmic tail (Clevers *et al.* 1988*a*) (Table 17.1). Most T cell receptor αβ heterodimers are covalently linked through disulfide bonds, whilst many TCR γδ receptors associate with one another non-covalently. The αβ and γδ TCR glycoproteins belong to immunoglobulin superfamily and resemble immunoglobulins in that they contain variable and constant domains. The constant regions of the TCR extracellular domains are anchored in the plasma membrane by an α helical transmembrane region; the cytoplasmic domain consisting of only a few amino acids. The charged residues in the transmembrane regions are of importance for the assembly of the receptor as discussed in Sections 3 and 4. In one study utilizing transfection of chimeric proteins into non-T cells, the transmembrane domain of TCR α was shown to be unable to function as an anchor for a heterologous ectodomain in the lipid bilayer of the endoplasmic reticulum (ER). Instead, this CD4-α_{tm} chimeric protein was degraded in the lumen of the ER or was secreted (Shin *et al.* 1993). One interpretation of these data, namely that TCR α does not contain a bona fide transmembrane domain, was favoured by the authors. However, since secretion of TCR α chain by T cells has not been reported, the result could be caused by association with a heterologous ectodomain. Regardless of the behaviour of the independent TCR α polypeptide chain, its incorporation into the different TCR–CD3 complexes is well established as discussed in Sections 3, 4, and 5.

2.2 CD3 γ, δ, and ε

The invariable CD3 γ, δ, and ε polypeptides are encoded by three members of the immunoglobulin supergene family and are found in a cluster on human chromosome 11q23 or murine chromosome 9 (Clevers *et al.* 1988*a*). Since the nucleotide sequences of CD3 γ and δ genes are rather similar and are located only 1.4 kb from one another, it is likely that they arose through gene duplication (Saito *et al.* 1987). A putative gene duplication must have taken place after the divergence of mammalian and avian species, as the single CD3 gene in the chicken (Bernot and Auffray 1991), is equally homologous to CD3 γ and δ (Figures 17.1 and 17.2).

The CD3 γ and δ are glycoproteins which are usually found in single copies in the mature TCR–CD3 complex (Figures 17.1, 17.2, 17.4, and 17.7). In contrast, two copies of the non-glycosylated CD3 ε are detected in the cell surface TCR–CD3 complex (Blumberg *et al.* 1990*a*), and 20–30% of these CD3 ε molecules are found to be disulfide bridged (Sancho *et al.* 1992*a*). In non-T cells, for example in transfected COS or CHO cells or in natural killer cells disulfide bridged CD3 ε homodimers are the predominant species (Figure 17.3 and 17.4). Dimer formation involves solely the CD3 ε ectodomains, as determined by transfection and expression of truncated cDNAs into COS cells (B. Mueller and C. Terhorst, unpublished data). It is likely that in the mature TCR–CD3 complex the dimeric form is sustained by the other protein–protein contacts even in the absence of interchain disulfide bridges. Since some similarities have been found between the CD3 γ, δ, and ε amino acid sequences and members of the immunoglobulin supergene family (Gold *et al.* 1987), intrachain disulfide bridges have been indicated accordingly (Figures 17.3 and 17.4). There is good evidence for intrachain disulfide bonds with CD3 ε based on the difference in mobility in SDS–PAGE with or without reducing agents (Alarcon *et al.* 1988*a*; Ley *et al.* 1989). The transmembrane regions of CD3 γ, δ, and ε contain Asp or Glu, which play a dominant role in interchain contacts with TCR α or β (see Section 4).

In addition to their role in assembly the CD3 γ, δ, and ε proteins are thought to be involved in transduction of the signal initiated by engagement of the TCR variable regions and antigen. In CD3 $\zeta^{-/-}$ cells signal transduction takes place through CD3 γ, δ, and ε (Wegener *et al.* 1992; Liu *et al.* 1993; Ohno *et al.* 1993; Love *et al.* 1993; Malissen *et al.* 1993). All three

```
            Ectodomain
            1                    20                   40
CD3-γH      QSIKGNHLVKVYDYQE*DGSVLLTCDAEAKNITWFKDGKMIGFLTEDKKTLNLGK
CD3-γM      -TN-AKN--Q-DGSRG*--------C---D-T-K-L---SI-SP-NAT-HKW---S
CD3-γR      QQKEEK--F--D-S-QG--------C-FNE-T---L---HR-SPPNAT-S-W---N
CD3-γS      ********---D-N--*-----I-ICVTDE-K---L--M-E-SSGDTN-L-WD--S
CD3γ/δC         -V-*GLSMSVK-VS-K-F-QCQESKDLNTNYLWK-GKEE-GNMRQ*-D-GA

                60                    80
CD3-γH      NAKDPRGMYQCKGSQNKSKPLQVYYRMCQNCIELNAAT
CD3-γM      -------T--CQ-AKET-N--------CE-C----IG-
CD3-γR      G---------CR-AKK--QL------LCE-C----MG-
CD3-γS      ST-----I-EC---S-E--S--I----CQ-C----LA-
CD3γ/δC     IYD----T-TCQRDE-VNST-H-H---C--C--VDAP-

            Transmembrane Domain
                   100
CD3-γH      ISGFLFAEIVSIFVLAVGVYFIA
CD3-γM      ----I--EVI---F--L------
CD3-γR      V---I--E-I---F---------
CD3-γS      VA--I-TE-----L---------
CD3γ/δC     ---IVV-DV-ATVL--IA--C-T

            Cytoplasmic Domain
              120                  140                 160
CD3-γH      GQDGVRQSRASDKQTLLPNDQLYQPLKDREDDQYSHLQGNQLRRN
CD3-γM      ----------------Q-E--Y--L----Y--Y--L------KK
CD3-γR      ----------------Q-E-VY--L----YE-Y-RL----V-KK
CD3-γS      --E-------------N----Y--L-E-----Y--LRKK
CD3γ/δC     ---KGLM-----R-N-IA----Y--LGE-N-G-Y-QLATAKARK
```

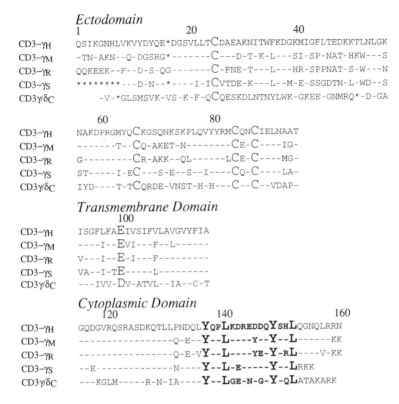

Fig. 17.1 Amino acid sequences of CD3 γ. γH, human; γM, mouse; γR, rat; γS, sheep; γC, chicken. The amino acid sequences of the mature proteins are given and the numbering is based upon the human CD3 γ sequence. The ITAM motif is indicated in bold.

polypeptide chains contain a cannonical sequence in their cytoplasmic tails; the so-called ITAM (Immunoreceptor Tyrosine-based Activation Motif) (Reth 1989; Romeo *et al.* 1992; Cambier *et al.* 1995) (Figures 17.1–17.4). The tyrosine residues in these motifs can be phosphorylated in response to engagements of the T cell receptor (Sancho *et al.* 1993; Weiss 1993; Qian *et al.* 1993*a*,*b*; Perlmutter *et al.* 1993). However, the exact cascade of events following T cell receptor engagement remains to be determined.

2.3 CD3 ζ, η, and FcεR1γ

The CD3 ζ gene is found separately from other TCR and CD3 genes on syntenic regions of the murine chromosome 1 and human chromosome 1q22-q25 (Baniyash *et al.* 1989). Some murine T cells generate a receptor-associated CD3 η chain through alternative splicing of the CD3 ζ mRNA (Jin *et al.* 1990) (Figures 17.4 and 17.5). Thus, some receptors contain ζη or ηη (Clayton *et al.* 1990, 1991). To date, no evidence exists at the protein

```
            Ectodomain
                1                  20                  40
CD3-δH          FKIPIEELEDRVFVNCNTSITWVEGTVGTLLSDITRLDLGKRIL****DPRGIYR*CNGTDIYK
CD3-δM          ---QVT-Y--K---TCN--VMHLD---EGWFAKNK------GV-****------L*CN--EQLA
CD3-δR          ---EVV-Y--K---NCN--IRHLD-S-ERWLTKNKS-I-*-G--****---MGYMCN--EELA
CD3-δS          RALEVL-A--K-ILKCN----LLQ--A-QEVS-NKT-N-----E****---M-Q*CGENAKSF
CD3γ/δC     GVHGLSMSVK-VSGK--LQCQE-KDLNTNYLWKKGKEELGNMRQLDLGAIYD---T-T*CQRDENVN

            60                  80
CD3-δH      KDESTVQVHYRMCQSCVELDPAT
CD3-δM      -VV-S-------C-NC----SG-
CD3-δR      -EV-----Y---C-NC----S--
CD3-δS      ****-L--Y---C-NC----S--
CD3γ/δC     ***--LH-----C-NCI-V-AP-

            Transmembrane Domain
                                100
CD3-δH      VAGIIVTDVIATLLLALGVFCFA
CD3-δM      M--V-FIDL----------YC--
CD3-δR      L--V-I-DL----------YC--
CD3-δS      L--L-I-DI----------YC--
CD3γ/δC     IS--V-AD-V--V---IA-YCIT

            Cytoplasmic Domain
                          120                  140
CD3-δH      GHETGRLSGAADTQALLRNDQVYQPLRDRDDAQYSHLGGNWARNK
CD3-δM      ------P----EV----K-E-LY--L---E-T-Y-RL----P---KS
CD3-δR      ----------V---V--K-E-LY--L-------Y-RL----P---RS
CD3-δS      ------F-R-----V-MG---LY--L-E-N---Y-RL-DK-----
CD3γ/δC     -QDK-LM-R-S-R-N-IA---LY--LGE-N-G-Y-QLATAKA-K
```

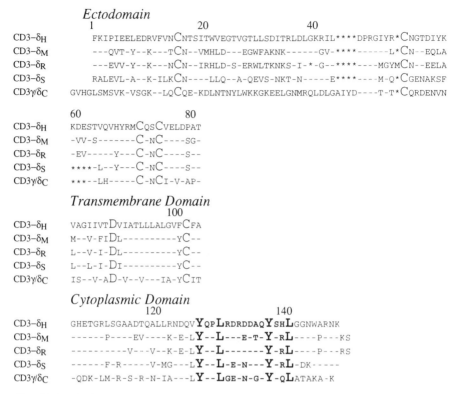

Fig. 17.2 Amino acid sequences of CD3 δ. δH, human; δM, mouse; δR, rat; δS, sheep; δC, chicken. The amino acid sequences of the mature proteins are given and the numbering is based upon the human CD3 δ sequence. The ITAM is indicated in bold.

level for CD3 η chain equivalent alternative splice products in the human or rat (Clayton *et al.* 1990, 1991).

Unlike the other CD3 proteins, CD3 ζ has a very short ectodomain of nine amino acids and a long cytoplasmic tail of 112(H) or 113(M) amino acids (Figures 17.4 and 17.5) (Weissman *et al.* 1988) and does not belong to the Ig supergene family. A truncated form of ζ (deletion Δ66–114, i.e. residues 45–93 in Figure 17.5) can still assemble into the TCR–CD3 complex and arrive at the cell surface, but is functionally inert (Wegener *et al.* 1992). All six tyrosine residues in the three ITAMs, termed ζ_A, ζ_B, and ζ_C, can be phosphorylated upon T cell triggering through the antigen receptor. Phosphorylation of the CD3 ζ tyrosine residues is catalysed by *c-fyn*, CD4–*c-lck*, and/or ZAP-70 (Sancho *et al.* 1992*b*, 1993; Desai *et al.* 1993; Weiss 1993; Wange *et al.* 1993; Perlmutter *et al.* 1993; Exley *et al.* 1994). Chimeric proteins containing ecto- and transmembrane domains from other membrane proteins fused to the cytoplasmic tail of CD3 ζ or to smaller fragments

```
                 Ectodomain
                 1                    20                    40
CD3-εH           DGNEEMGGITQTPYKVSISGTTVILTCPQYP*GSEILWQHNDKNIGGDEDDKNIGSDED
CD3-εM                 DDAENIE---------S-E--C-LDS*DENLK-EK-GQELP********QKHDK
CD3-εS                 DDTE-N--E-----NS-E--C-KDF*ENG-Q-KRNNEQMK********-HN-K
CD3-εD           DEDFKASDDLTSISPEKRF-------E-VV-C-DVFGYDN-K-EK--NLVE********-ASNR

                 60                   80                    100
CD3-εH           HLSLKEFSELEQSGYYVCYPRGSKPEDANFYLYLRARVCENCMEMDVMS
CD3-εM           --V-QD---V-D-----C-TPA-***NK-T----K---C-YCV-L-LTA
CD3-εS           Y-L-DQ---M-S----QCLATEGNT-A-HT*---K---CK-C--VNLLE
CD3-εD           E-SQ-----VDD----AC-ADS***IKEKS--------CA-CI-VNL-A

                 Transmembrane Domain
                              120
CD3-εH           VATIVIVDICITGGLLLLVYYWS
CD3-εM           --I-I--D-C--L---MVI----
CD3-εS           ----IV-D-CV-L----------
CD3-εD           -V--IVAD-CL-L----M-----

                 Cytoplasmic Domain
                          140                 160                   180
CD3-εH           KNRKAKAKPVTRGAGAGGRQRGQNKERPPPVPNPDYEPIRKGQRDLYSGLNQRRI
CD3-εM           ------------T---S-P---------------Y--I-------Y--L---AV
CD3-εS           -S-----T-M---------P----R----------Y--I-------Y--L---GV
CD3-εD           -T---N----M--T---S-P----K-K--------Y--I----Q--Y--L---G-
```

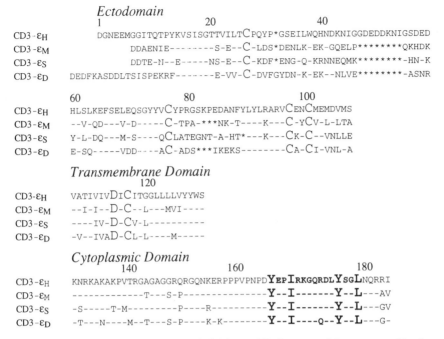

Fig. 17.3 Amino acid sequences of CD3 ε. εH, human; εM, mouse; εD, dog; εS, sheep. The amino acid sequences of the mature proteins are given and the numbering is based upon the human CD3 ε sequence. The ITAM is indicated in bold.

thereof, can be expressed on the cell surface and transduce activation signals when aggregated with monoclonal antibody reagents (Romeo *et al.* 1992; Kolanus *et al.* 1993; Hall *et al.* 1993; Irving *et al.* 1993; Desai *et al.* 1993; Ravichandran *et al.* 1993). Our current hypothesis is that the different signal transducting elements of CD3 ζ (ζ_A, ζ_B, ζ_C) and of CD3 γ, δ, ε play distinct roles during early thymic differentiation, but that co-operation between the different elements is required for a full response during antigen driven T cell activation (Terhorst *et al.* 1995).

A considerable degree of flexibility in T cell signalling is demonstrated by the use of alternative signalling components in different TCR–CD3 complexes. In a subset of αβ and γδ T cells, namely mucosal intestinal epithelial lymphocytes (IEL), the FcεR1γ chain can replace CD3ζ. The FcεR1γ gene is also located at chromosome 1q23. The transmembrane segment of CD3ζ is homologous to the γ subunit of the IgE Fc receptor (FcεR1γ) (Blank *et al.*, 1989; Orloff *et al.* 1990) and permits its expression in the TCR cell surface complex (Figure 17.6). Whereas the ITAM of FcεR1γ can be phosphorylated and FcεR1γ containing receptors functionally active in cell lines, cells carrying such receptors may utilize alternative activation pathways (discussed in Section 5).

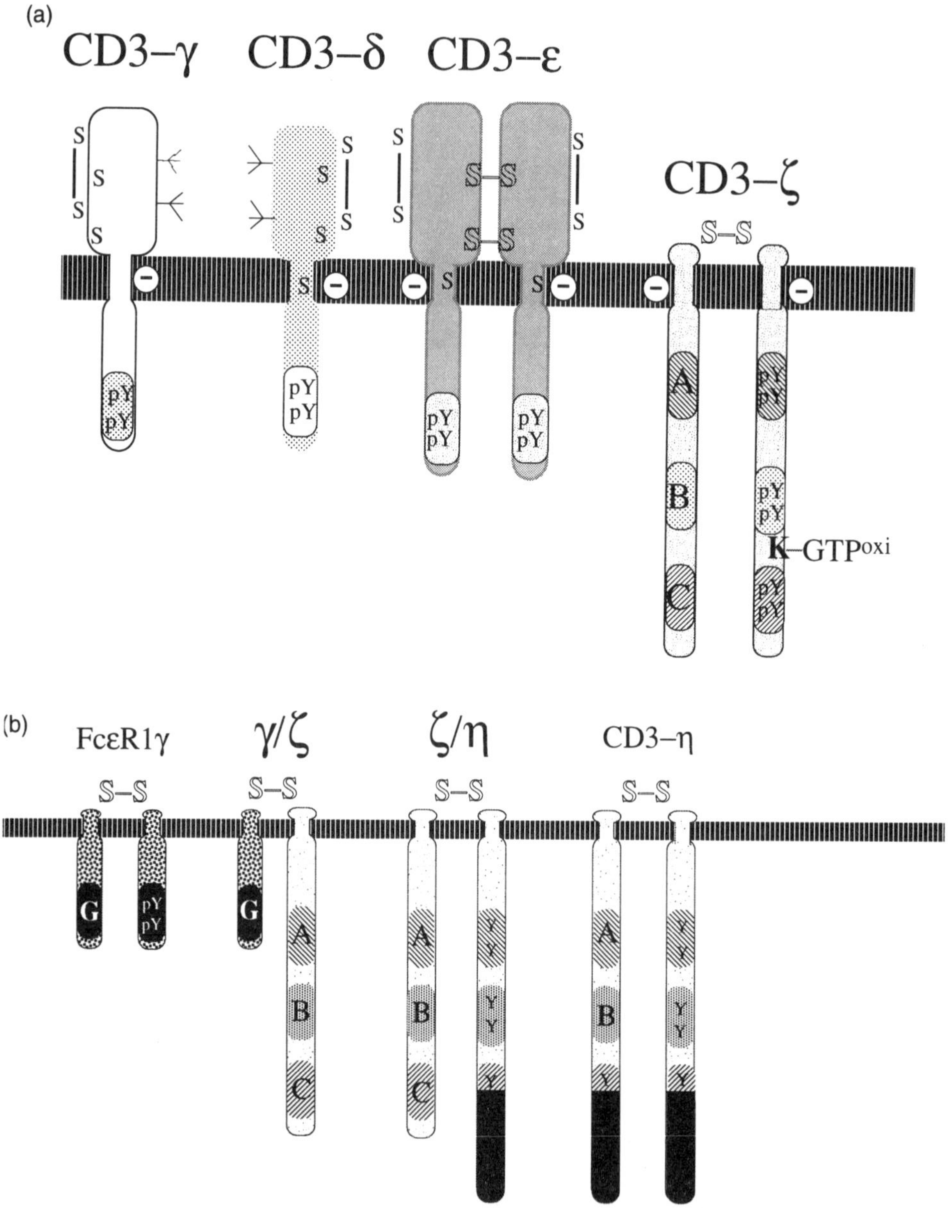

Fig. 17.4 Diagrams of CD3 γ, δ, and ε (A), and CD3 ζ, η, and FcεR1γ (B). S, sulfhydryl group; S-S, disulfide bridge; Y, tyrosine; pY, phosphotyrosine; A, ζA; B, ζB; C, ζC; G, ITAM in FcεR1γ.

```
          Ectodomain Transmembrane      Cytoplasmic Domain
                         Domain
          1          _____________________         40
CD3-ζH    QSFGLLDPKLCYLLDGILFIYGVILTALFLRVKFSRSAEPPAYQQGQ
CD3-ζM    ----------C---D---------I---Y--A-------TA-NL-DP
CD3-ηM    ----------C---D---------I---Y--A-------TA-NL-DP
CD3-ζR    ----------C-M-D---------V---Y--A------DAA--L-DP
CD3-ζS    ----------C---D---------V---F--A------DA----H--
             _______ζA________
                      60                  80
CD3-ζH    NQLYNELNLGR*REEYDVLDKRRGRDPEMGGKP*RRKNPQ
CD3-ζM    ---Y--L----*---Y--LE---A--------QQ------
CD3-ηM    ---Y--L----*---Y--LE---A--------QQ------
CD3-ζR    ---Y--L----*---Y--L--K-P--------QQ--R---
CD3-ζS    -PVY--L-V--*---Y--L-R-G-F--------Q-K---H
             _______ζB________
                       100
CD3-ζH    EGLYNELQKDKMAEAYSEIGMKG**ERRRGKGH
CD3-ζM    --VY-AL--------Y--I-TK-**--------
CD3-ηM    --VY-AL--------Y--I-TK-**--------
CD3-ζR    --VY-AL--------Y--I--K-**--------
CD3-ζS    -VVY--LR-------Y--I--K-DNQ-------
             _______ζC________
            120                140
CD3-ζH    DGLYQGLSTAT*KDTYDALHMQALPPR
CD3-ζM    ---Y--L----*---Y--L---T-A--
CD3-ηM    ---Y-DSHFQAVQFGNRREREGSELTRTLGLRARPKGESTQQSSQSCASVFSIPTLWSPWPPSSSSQL
CD3-ζR    ---Y--L----*---Y--L---T----
CD3-ζS    --VY--L----*---Y--L--------
```

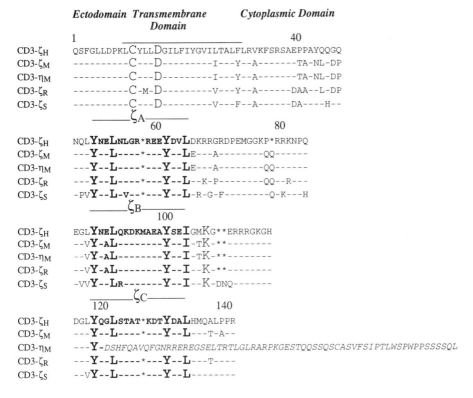

Fig. 17.5 Amino acid sequences of CD3 ζ and η. ζH, human; ζM and ηM, mouse; ζR, rat; ζS, sheep. The amino acid sequences of the mature proteins are given and the numbering is based upon the human CD3 ζ sequence. The ITAMs ζA, ζB, ζC are indicated in bold.

```
              Ectodomain Transmembrane       Cytoplasmic Domain
                              Domain
                  1      ______________________        40
FcεR1γH           LGEPQLCYILDAILFLYGIVLTLLYCRLKIQVRKAAITSYE
FcεR1γM           ------C---D-V------------------------A-R-
CD3-ζH        QSFG-LD-K-C-L-DG---I--VI--A-F--V-FSRSAEPP-YQQ

                    50         60
FcεR1γH       KSDGVYTGLSTRN*QETYETLKHEKPPQ
FcεR1γM       -A-A-Y--LN--S*---Y--L-------
CD3-ζH        GQNQLYNELNLGR*R-EYDVLDKRRGRDPEMGGKPRRKN

CD3-ζH        PQEGLYNELQKDKMAEAYSEIGMKGERRRGK
CD3-ζH        GHDGLYQGLSTAT*KDTYDALHMQALPPR
```

Fig. 17.6 Amino acid sequences of human and murine FcεR1γ chains and a comparison with the human CD3 ζ sequence. FcεR1γH and ζH, human; FcεR1γM, mouse. The amino acid sequences of the mature proteins are given and the numbering is based upon the human FcεR1γ sequence. The ITAM is indicated in bold.

3 A structural model of the T cell receptor for antigen

Antigen recognition by T lymphocytes proceeds in two distinct stages: a non-specific adhesion step without antigen recognition followed by the specific interaction between the TCR and antigen–MHC (Spits *et al.* 1986; Matsui *et al.* 1992). Adhesion permits a T lymphocyte to screen cells throughout the body and decide whether to become functionally engaged. This screening is a necessity because an antigen-presenting cell, which has processed a given protein, will express on its cell surface only a small number of specific MHC–peptide complexes, that can be recognized by a given VαVβ bearing T cell. In view of the low affinities between TCR and MHC–antigen and the small numbers of MHC–antigen complexes in a given area of cell–cell contact (Kanagawa and Ahlem 1989; Davis and Chien 1993), a model has been proposed in which TCR–CD3 complexes and MHC–peptide complexes diffuse into the interface. Thus, a high local density of TCR–CD3 complexes would promote a co-operativity in receptor/antigen binding, involving as few as 100 antigen receptor–ligand interactions per cell (Spits *et al.* 1986; Kanagawa and Ahlem 1989; Christinck *et al.* 1991; Matsui *et al.* 1992; Weber *et al.* 1993; Davis and Chien 1993).

Several recent observations fit into this model of co-operativity through interaction of receptors and ligands in the interface between the two cells. First, as discussed below, TCR–CD3 complexes by themselves have a propensity to form higher order structures on the T cell surface (Exley *et al.* 1995). Secondly, the CD8 co-receptors are found on the surface of human T cells and thymocytes as dimers or multimers (Snow and Terhorst 1983). Thirdly, MHC class II are thought to form dimers on the cell surface (Brown *et al.* 1993).

We favour a divalent model of the TCR–CD3 complex (Figure 17.7) (Exley *et al.* 1991, 1995) which incorporates the following observations:

1. In a T-T hybridoma specific monoclonal antibodies against one TCR αβ pair could co-modulate the second TCR αβ pair (Exley *et al.* 1995).
2. Analyses of sucrose gradients in the presence of mild non-denaturing detergents identified a TCR–CD3 complex of approximately 300 kd (Exley *et al.* 1995). This cell surface receptor, which is more than 100 kD larger than expected from the minimal eight subunit complex ($\alpha\beta\gamma\delta\varepsilon_2\zeta_2$), migrated with a higher sedimentation rate than the B cell receptor (250 kD).
3. An $(\alpha\beta)_2$, γ, δ, ε_2, ζ_2 model accommodates:
   (a) The results of pair formation studies in COS and CHO cells (Table 16.2) (Wileman *et al.* 1993) and in T cells, e.g. TCRβ–CD3γ (Brenner and Strominger 1985).
   (b) Hydrogen bridges between positively and negatively charged residues within the transmembrane regions (Section 4).
   (c) CD3 εε homodimers found in cell surface receptors (Blumberg *et al.* 1990; Hall *et al.* 1991).
   (d) The so-called 'half receptors' (αβ, γ,ε,ζ and αβ, δ,ε,ζ), which have

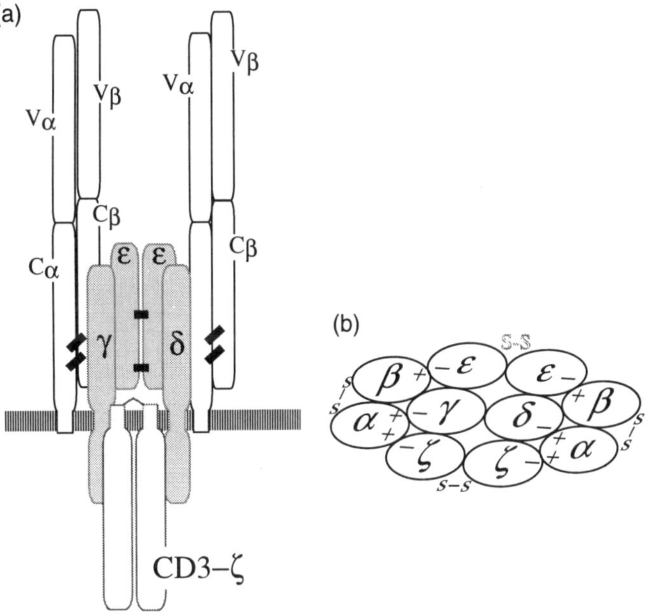

Fig. 17.7 A. Model of the divalent TCR–CD3 complex. B. Interactions between the transmembrane regions. While the disulfide bridge between CD3 ζ molecules is always present, only 10–20% of the CD3 εs are disulfide bridged within the TCR–CD3 complex. This is indicated by different letter types.

been identified as intermediates in the formation of complete TCR–CD3 complexes (Alarcon *et al.* 1991; Kappes and Tonegawa 1991).

(e) A 1:1 ratio of αβ TCR and CD3 ε molecules as detected by quantitative antibody measurements (Meuer *et al.* 1984; Rao *et al.*, in preparation).

It is likely that a divalent model also applies to γδ T cells, since the sedimentation behaviour of γδ TCR in sucrose velocity gradients is identical to that of αβ TCR and since quantitative binding studies with monoclonal antibodies suggests a divalent receptor (Exley *et al.* 1995; Rao *et al.*, in preparation).

4 Assembly in the endoplasmic reticulum

Assembly of a complete TCR–CD3 complex containing all six polypeptide chains (TCR α, β and CD3 γ, δ, ε, ζ) within the endoplasmic reticulum (ER) is an obligate requirement for efficient cell surface expression (Weissman *et al.* 1988; Hall *et al.* 1991; Carson *et al.* 1991). Single subunits and partial complexes are in principle unable to leave the ER. To avoid accumulation of single chains or incomplete TCR–CD3 complexes they are degraded at a rate that is determined by their subunit composition. Whereas the TCR α, β, and

CD3 δ polypeptide chains are degraded rapidly with half-lives of 90 minutes, CD3 γ, ε and ζ, and all subcomplexes that contain them, are relatively long-lived (6–8 hours) (Minami *et al.* 1987; Berkhout *et al.* 1988; Chen *et al.* 1988; Lippincott-Schwartz *et al.* 1988; Bonifacino *et al.* 1989; Wileman *et al.* 1990*a,b*). This selective degradation of TCR proteins implies that proteolysis of newly synthesized proteins is carefully regulated. Survival of the TCR within the ER and consequent cell surface expression thus result from a careful balance between receptor assembly and receptor degradation. This is reflected in the sequence of events followed during the building of the TCR–CD3 complex.

4.1 Hierarchy of assembly in T lymphocytes

During the first minutes of biosynthesis in T cells (Figure 17.8), the first structures that can be detected by immunoprecipitation are associated CD3

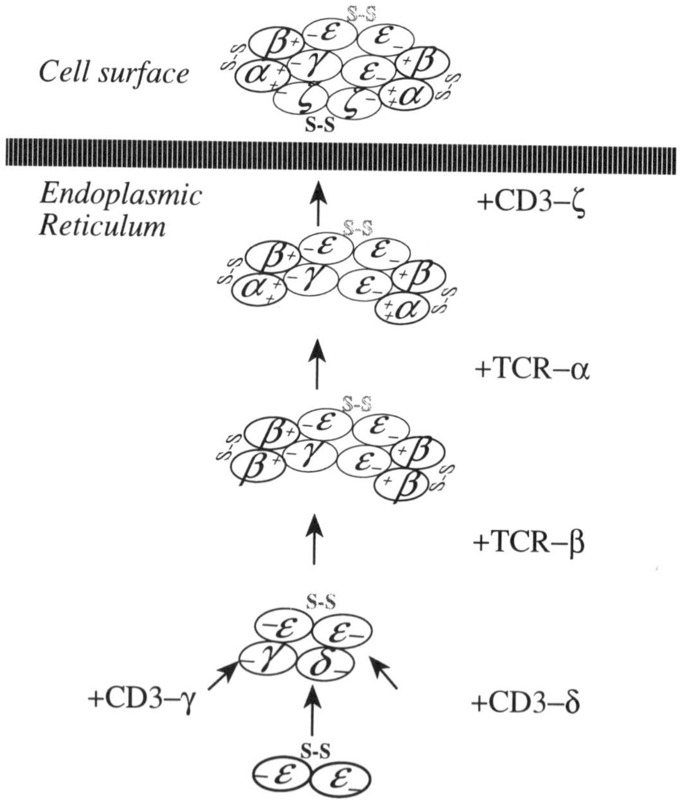

Fig. 17.8 Model for a hierarchy of assembly in T lymphocytes. Sub complexes are indicated by associations between transmembrane regions. Whereas 100% of the free CD3 ε dimers are disulfide linked, this percentage changes during assembly in the ER.

Table 17.2 Detectable associations between TCR α and β and the CD3 chains in the endoplasmic reticulum after cotransfection into COS and CHO cells

Rapidly degraded pairs	Stable pairs
TCR α/CD3 δ[a]	TCR α/CD3 γ[a]
TCR β/CD3 δ[a]	TCR β/CD3 γ
	TCR β/CD3 ε[a]
	CD3 ε/ε[b]
	CD3 δ/ε
	CD3 γ/ε
	CD3 ζ/ζ[b]

[a] Interaction affected by elimination of charged residues in transmembrane region.
[b] Interchain disulfide bridges.

pairs (Table 17.2). Here the core CD3 γεε, CD3 δεε, and/or CD3 δεεγ subcomplexes are assembled (Alarcon *et al.* 1988*a*; de Waal Malefijt *et al.* 1990). T cell receptor chains are then added to the CD3 core structure one at a time in either order, since both TCRα–CD3 and TCRβ–CD3 subcomplexes can be found. Ample evidence exists that in T cells or T cell mutants, the stable CD3 γ and CD3 ε subunits of the CD3 complex prevent rapid degradation of the newly synthesized TCR α, β, and CD3 δ chains (Minami *et al.* 1987; Alarcon *et al.* 1988*a*; de Waal Malefijt *et al.* 1990). Assembly around CD3 ε and γ continues until an αβγδε complex is completed.

The final subunit to assemble with the complex is CD3 ζ (Hall *et al.* 1991; Carson *et al.* 1991; Manolios *et al.* 1991; Rutledge *et al.* 1992). The ζ family members are always isolated as dimers and dimerization occurs prior to assembly with the αβγδε subcomplex. Interestingly, even after addition of the CD3 ζ chain dimer the receptor is not yet ready to leave the endoplasmic reticulum. Although addition of CD3 ζ to the receptor is the last assembly event that we can readily detect, it does not appear that this is the rate limiting step in receptor transport. Addition of CD3 ζ to the pre-assembled αβγδε subcomplex is rapid, yet it takes substantially longer for pulse labelled human T cell receptors to reach the *medial*-Golgi. This is possibly due to conformational maturation of the receptor into a divalent structure which could be rate limiting for subsequent transport to the cell surface.

The charged residues found within the transmembrane anchors of the TCR–CD3 components have long been implicated in receptor assembly. Replacement of the 'forbidden' negatively charged amino acid of the CD3 subunits with a neutral alanine abolished interactions between TCR α and CD3 ε and TCR δ and CD3 ε (Hall *et al.* 1991). Whereas the transmembrane regions of TCR α and TCR δ contain two positively charged amino acids (αH^{Arg239} and αH^{Lys244}; δH^{Arg250} and δH^{Lys255}, respectively), the TCR β and γ transmembrane regions contain only one (βH^{Lys271} and γH^{Lys296}). Removal of charged residues from the membrane anchors of the TCR α and β chains also prevents pairing with CD3 components (Morley *et al.* 1988; Blumberg *et al.* 1990*b*; Bonifacino *et al.* 1990*a*; Cosson *et al.* 1991). The position of the charged residues plays an important role: these interactions are most favoured when the charged residues are located at the same level within the membrane (Letourneur and Klausner 1992*b*). Reconstitution experiments of a TCR α Jurkat cell line showed that both positively charged residues of the transmembrane region of TCR α (i.e. Arg^{239} and Lys^{244}) needed to be modified to prevent incorporation into the complex (Blumberg *et al.* 1990*b*). The negatively charged transmembrane residues in the CD3 subunits do not affect formation of εγ, εε, and εδ dimers for these associations also take place between the charge deleted CD3 mutants (Hall *et al.* 1991; Wileman *et al.* 1993). Surprisingly, the presence of identical (and therefore mutually repulsive) charges in the ζ transmembrane was found to be essential for dimerization and (less surprisingly) subsequent assembly with the TCR–CD3 complex (Rutledge *et al.* 1992).

4.2 ER degradation of the individual TCR–CD3 polypeptide chains

Transfection of TCR α, β, CD3 γ, δ, ε or ζ into COS or CHO cells results in their retention within the ER. Whereas excess amounts of non-assembled TCR α, β, and CD3 δ polypeptide chains are degraded by proteolytic enzymes (half-life 90 minutes), the other polypeptide chains (CD3 γ, ε, and ζ) are relatively stable (half-life 6–8 hours).

Specific amino acid sequences within the membrane spanning domains of the TCR α and β chains have been shown to cause rapid ER degradation (Wileman *et al.* 1990*c*; Bonifacino *et al.* 1990*a,b*; Bonifacino and Lippincott-Schwartz 1991). The TCR α and TCR β ectodomains are by themselves more stable to ER degradation than the complete TCR α or β. TCR α_{tm} or TCR β_{tm} joined to another ectodomain (e.g. IL-2Rα) causes the IL-2Rα ectodomain to be degraded (Bonifacino *et al.* 1990*a,b*; Bonifacino and Lippincott-Schwartz 1991; Wileman *et al.* 1993). In addition to amino acid sequence features, the conformation adopted by the newly synthesized TCR α, β, and CD3 δ proteins will also determine their sensitivity to ER proteases. Moreover, the location of the TCR α, β, and CD3 δ proteins within the ER and further along the exocytic apparatus may contribute to their rapid degradation.

4.2.1 *Necessity for Ca^{2+} and ATP*

In general, the physiological conditions within the ER maintains an oxidizing environment that favours the formation of disulfide bonds, which is in turn linked to correct protein folding (Hwang *et al.* 1992; Braakman *et al.* 1992; De Silva *et al.* 1993). Protein folding requires both oxidizing conditions and ATP (Braakman *et al.* 1992*a,b*), and is facilitated by the high lumenal Ca^{2+} concentration (Shia and Lodish 1989). Significantly, '*in vivo*' studies indicate that depletion of cellular ATP (Stafford and Bonifacino 1991), or the ER calcium store (Wileman *et al.* 1991*b*) markedly accelerate degradation of TCR α and β.

4.2.2 *Importance of the redox potential in the lumen of the ER*

Non-specific thiol modifying reagents such as *N*-ethylmaleimide, iodoacetamide, and 1, 10 phenanthroline complexed with copper (Wileman *et al.* 1991*a,b*; Tsao *et al.* 1992), and the cell permeable oxidizing agent diamide (Stafford and Bonifacino 1991; Inoue *et al.* 1991) are potent inhibitors of ER proteolysis of TCR α, β, and CD3 δ, or chimeric proteins containing their transmembrane regions. This is an important observation because cysteinyl proteases are redox-sensitive enzymes and their activity could, therefore, be regulated by the redox potential within the ER. Indeed, CD3 γ and CD3 ε that are normally long-lived within the ER, become degraded rapidly on addition of DTT (Young *et al.* 1993). Although these results indicate that ER proteolysis may be activated under reducing conditions, there are other possible explanations for this observation. For instance, addition of DTT to cells '*in vivo*' reduces disulfide bonds causing proteins to unfold within the ER thus exposing protease-sensitive sites that are masked under normal oxidizing conditions.

Alternatively, inhibition of ER proteolysis by the alkylating agents mentioned above may also be analogous to the introduction of amines into lysosomes. In the latter case amines act by raising the lumenal pH of the organelle, not by direct chemical inactivation of proteases. It is possible that ER proteases, like their lysosomal counterparts, have multiple specificities, having evolved to act optimally under reducing conditions that enhance activity and promote the exposure of free sulfhydryl groups.

4.2.3 *Proteolysis in ER subdomains*

One possible hypothesis is that the CD3 subunits prevent ER degradation of the TCR αβ polypeptide chains by preventing their movement to '*reducing domains*' within the ER. The presence of dileucine ER retention elements in the CD3 γ, δ, and ε subunits provides a molecular basis for the failure to express individual chains at the cell surface (Letourneur and Klausner 1992*b*; Mallabiabarrena *et al.* 1992). It is thought that these motifs cause integral membrane proteins to recycle, with lumenal KDEL-terminating proteins, through the intermediate and *cis*-Golgi compartments back to the ER (Jackson *et al.* 1993). Assembly with CD3 subunits may, therefore, sequester TCR α and β chains within pre-Golgi membrane compartments that are rich in chaperonins. This would promote folding and disulfide bond formation, and at the same time inhibit ER degradation.

4.2.4 *Lysosomal degradation of CD3 δ containing subcomplexes*

Surprisingly, only 75% of the CD3 δ glycoprotein expressed in COS cells is degraded in the ER. Degradation of the remaining 25% of the CD3 δ polypeptide chain is inhibited by chloroquine, which causes CD3 δ to accumulate in secondary lysosomes. Given that transport from the Golgi to lysosomes is an active, rather than constitutive process (Kornfeld and Mellman 1989), the CD3 δ chain must contain structural information that allows it to be sorted to lysosomes (Wileman *et al.* 1990*b*; Exley *et al.* 1991). Moreover, this observation probably explains the lysosomal degradation of partially assembled CD3 δ containing TCR subcomplexes (Minami *et al.* 1987; Ley *et al.* 1989).

4.2.5 TCR CD3 associated chaperonins

Chaperonins must play an important role in the multitude of individual polypeptide chain and subcomplex folding events. Two candidate chaperonins have been suggested to regulate early and/or late stages of TCR assembly, CD3 ω and IP90/calnexin.

CD3 ω is a less well described member of the TCR–CD3 complex, which has been identified in human and murine T cells (Pettey *et al.* 1987; Alarcon *et al.* 1988*a*; Bonifacino *et al.* 1989). CD3 ω is a 28 kDa intracellular protein which associates with all the components of the TCR–CD3 complex, but only within the ER (Pettey *et al.* 1987; Bonifacino *et al.* 1989). This protein is not part of functional cell surface TCR molecules and it appears to be transiently associated with subcomplexes in the ER, suggesting that CD3 ω could be a chaperonin. CD3 ω is strictly associated with the high mannose ER forms of the intracellular complex (Pettey *et al.* 1987; Bonifacino *et al.* 1989; Neisig *et al.* 1993) and is believed to be lost just prior to the transport of the receptor complex to the *cis*-Golgi. Curiously, CD3 ω can not be detected in association with TCR–CD3 polypeptides expressed in COS or CHO cells (Hall *et al.* 1991; Wileman *et al.* 1990*a,b*). Moreover, transfections with six cDNAs (α,β,γ,δ,ε,ζ) into COS cells results in a low levels of cell surface expression of the intact TCR–CD3 complex. A likely explanation therefore is that the absence of CD3 ω impairs the efficiency of trafficking to the cell surface (Hall *et al.* 1991). Thus, alternatively, CD3 ω may be a key regulator of the TCR–CD3 cell surface expression in T cells

Calnexin was originally found associated with the assembling TCR by intracellular iodination. Subsequently, calnexin (or IP90) was shown to associate with many proteins folded within the ER (including the B cell receptor for antigen and MHC class I) and to be a major ER Ca^{2+} binding protein (Hochstenbach *et al.* 1992; Ahluwalia *et al.* 1992; Ou *et al.* 1993). Calnexin is an integral membrane protein with a large ER luminal domain (461 amino acids), a transmembrane segment (22 amino acids), and a cytoplasmic tail (89 amino acids). The cytoplasmic tail contains a consensus motif RKPRRE implicated in the ER retention of other resident ER proteins. It has now been found that calnexin associates with both monomeric and oligomeric partially folded membrane proteins and may function in promoting correct folding, required for transport of proteins from the ER (Ou *et al.* 1993).

4.3 Association with CD3 γ and ε protects against ER degradation

The stability of CD3 γ, ε, and ζ in the ER plays a dominant role in the first phases of receptor assembly, for the resistance to degradation can be conferred on the unstable polypeptide. Within the ER of doubly transfected COS and CHO cells a large number of polypeptide pairs can be observed (summarized in Table 17.2) (Berkhout *et al.* 1988; Wileman *et al.* 1990*a*, 1993; Hall *et al.* 1991). Specifically, association of the stable CD3 γ and ε polypeptide chains with labile TCR α or β or CD3 δ subunits prevents degradation (Table 17.2). In contrast, dimers composed of intrinsically labile subunits (e.g. TCRα–CD3δ or TCRβ–CD3δ) are degraded rapidly (Berkhout *et al.* 1988; Hall *et al.* 1991; Wileman *et al.* 1993). Thus, incorporation of stable chains into complexes appears to be an obligate requirement for receptor survival.

Studies of polypeptide chain pairing have been refined by site-directed mutagenesis experiments and by the use of chimeric proteins where single domains are coupled to reporter proteins (e.g. the human or murine IL-2Rα ectodomain) (Wileman *et al.* 1990*c*, 1991, 1993; Bonifacino *et al.* 1990*a,b*; Letourneur and Klausner 1992). Interactions between membrane spanning domains is not sufficient for the protection of the TCR β chain from ER proteolysis. When sites of binding between TCR β and CD3 γ, δ, or ε were restricted to the membrane anchor of the TCR β chain, stabilization by CD3 subunits was markedly reduced. The presence of the Cβ domain containing the first 150 amino acids of the TCR ectodomain, greatly increases the stability of all pairs formed in the ER. For assembly with CD3 ε, stability is further enhanced by the Vβ amino acids (Wileman *et al.* 1993). Thus, the efficient neutralization of transmembrane proteolytic targeting information requires associations between membrane spanning domains and the presence of receptor ectodomains. Interactions between receptor ectodomains may slow the dissociation of CD3 subunits from the TCR β chain and prolong the masking of transmembrane targeting information. In similar experiments, after the chimeric TCRα_{tm}–IL-2Rα protein paired with CD3 δ degradation of each chain was reduced (Cosson *et al.* 1991). But, as this was in contrast with the observation that intact TCRα–CD3δ pairs are degraded rapidly, some caution in interpreting chimera experiments may be warranted.

5. Incomplete cell surface TCR–CD3 complexes

5.1 Immunodeficiency patients with dysfunctional T cell receptors

Only two distinct human immunodeficiencies caused by defects in the CD3 genes have been described: the CD3 γ and the CD3 ε mutations (Alarcon *et al.* 1988*b*, 1990; Arnaiz-Villena *et al.* 1992; Timon *et al.* 1993; Soudais *et al.* 1993).

5.1.1 The CD3 γ mutations

A very low level of expression of the TCR–CD3 complex on the surface of most T lymphocytes of two Spanish siblings was detected. Each patient has two mutations in the CD3 γ gene:

1. The paternal mutation was in the translational start codon (ATG to GTG) in exon 1 of the CD3 γ gene.
2. The maternal mutation was on the boundary of intron 2 and exon 3 (cag to cac) resulting in abnormal splicing and in the absence of detectable mRNA (Arnaiz-Villena *et al.* 1992). The CD3 γ mRNA found in both siblings carries the paternal mutation (Alarcon *et al.* 1988*b*; Arnaiz-Villena *et al.* 1992).

Biochemical studies demonstrated that the although the CD3 γ chain was absent, the patient's lymphocytes expressed and αβδεζ TCR–CD3 complex on the surface and contained normal amounts of intracellular TCR α, TCR β, CD3 δ, CD3 ε, and TCR ζ. In this αβδεζ subcomplex the CD3 ζ chain is weakly associated (Alarcon *et al.* 1988*a*; Perez-Aciego *et al.* 1991) (Figure 17.9).

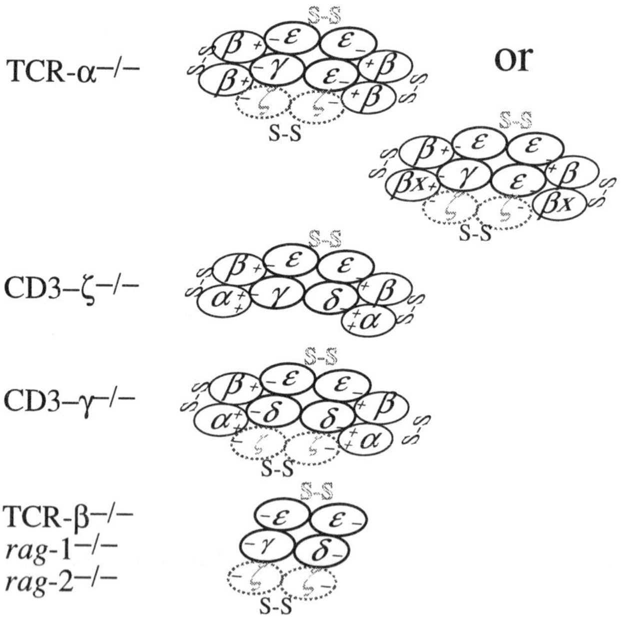

Fig. 17.9 Variant TCR–CD3 structures detected in TCR–CD3 mutants. Subcomplexes are indicated by using transmembrane associations. Since CD3 ζ is sometimes weakly associated its contours are in grey instead of black.

Similar observations were made upon transfection of the five cDNAs into COS or Hela cells (Hall *et al.* 1991; Kappes and Tonegawa 1991).

Although 2% to 12% of the patient's T cells stain with anti-TCR or anti-CD3 reagents, *in vitro* T cell proliferation via the TCR–CD3 pathway is impaired. Anti-CD2 stimulation appears to be normal. Thus, the lack of the CD3 γ chain gives rise to a partially dysfunctional TCR–CD3 complex. This notion was supported by the recent observation showing that a selective proliferation defect of peripheral $CD4^{+}CD45RA^{+}$ and $CD8^{+}$ lymphocytes existed, whilst $CD4^{+}CD45RO^{+}$ T lymphocytes were unaffected (Timon *et al.* 1993).

5.1.2 The CD3 ε mutations

In a second case, a five-year-old French boy had decreased levels of T cell surface expression of the TCR–CD3 complex (Soudais *et al.* 1993). The patient's CD3 ε mRNA was approximately 168 bp shorter than the predicted size of 1040 bp, because of an alternative splicing event in the exon 7 splice site in the maternal chromosome. Whereas the mutant transcript of maternal origin was detected at a level of 40% of normal, a small portion (1–5%) of the mRNA of maternal origin was full-length. It is likely that the the 'full-length' CD3 ε protein found in the low quantities of TCR–CD3 on the cell surface was coded for by this small number of mRNA molecules. Although a mutation inherited from the father resided in codon 59 of exon 6 (TGG to TAG), no transcripts carrying this mutation could be detected by PCR techniques. This mutation must therefore have affected the stability of the resulting mRNA drastically, since the transcription rate was normal as judged by nuclear run-on experiments.

5.2 Variant receptor complexes on the surface of T cells in mutant mice and of murine T cell lines

A systematic study of the relative contribution of the TCR–CD3 proteins has been undertaken in several laboratories by interruption of their respective genes using homologous recombination techniques. Resulting mutant mice in which the genes coding for TCR α, β, δ, CD3 ζ, as well as other mutant mice (e.g. recombinase activating genes RAG-1 or RAG-2) have been disrupted, have become extremely useful in studies of the cell surface expression and/or function of partial TCR–CD3 complexes (Mombaerts *et al.* 1992*a*, *b*; Philpott *et al.* 1992; Shinkai *et al.* 1992; Itohara *et al.* 1993; Liu *et al.* 1993; Ohno *et al.* 1993; Love *et al.* 1993; Malissen *et al.* 1993). Although in several cases the number of available thymus-derived lymphocytes or the level of surface expression of the partial complexes was insufficient for detailed studies, cell lines representing these various mutations have been useful to examine these partial receptors (summarized in Figure 17.9).

5.2.1 TCR $\alpha^{-/-}$ cells

TCR $\alpha^{-/-}$ TCR–CD3 complexes contain all five other polypeptide chains, probably in a configuration which is indicated in Figure 17.9. Although biochemical studies of the TCR $\alpha^{-/-}$ (or TCR β transgenic × RAG-$2^{-/-}$) thymocytes are limited (Mombaerts *et al.* 1992*a*; Shinkai *et al.* 1992), a large amount of data can be derived from TCR α^{-} cell lines. In general, TCR β/β receptors have been detected on the surface of these cell lines (Blumberg *et al.* 1990*b*; Punt *et al.* 1991; Groettrup *et al.* 1992). A propensity to form homodimers can even be recognized upon transfection of the TCR β cDNA into COS cells (Hall *et al.* 1991, 1993). However, on the surface of some TCR β tranfectant cell lines and of thymocytes from TCR β transgenic mice, a PI-linked form of TCR β has been detected on the cell surface (Groettrup *et al.* 1993*a*). This PI-linked protein was not associated with any other polypeptide. The PI-linked TCR β was probably an artefact resulting from over-expression. Recently, another TCR–CD3 complex has been postulated (Groettrup and von Boehmer 1993*b*) in which the TCR β chain associates via disulfide bridges with a potential 'surrogate α chain': a novel smaller polypeptide termed gp33 on the surface of a pre-thymocyte cell line Sci/ET27F/TCR β.

5.2.2 CD3 $\zeta^{-/-}$ cells

Both in CD3 $\zeta^{-/-}$ cell lines and in thymocytes or lymph node cells derived from the CD3 $\zeta^{-/-}$ mouse a subcomplex containing all of the remaining polypeptide chains was detected in the ER (Ashwell 1990; Liu *et al.* 1993; Ohno *et al.* 1993; Love *et al.* 1993; Malissen *et al.* 1993). As addition of CD3 ζ is the last step in building the TCR–CD3 complex, the remaining TCR–CD3 chains form an αβγδε complex that can only be transported out of the endoplasmic reticulum at a low rate (Ashwell 1990; Sancho *et al.* 1989; Exley *et al.* 1994). Only small amounts of these partial receptors could be detected on the thymocyte and T cell surface.

5.2.3 TCR $\alpha^{-/-}\beta^{-/-}$ cells

On the surface of T lymphocytes from TCR $\beta^{-/-}$, RAG-$1^{-/-}$, and RAG-$2^{-/-}$ mice no CD3 components can be detected using biochemical techniques or indirect fluorescence. However, on comparable cell lines this is the case. Earlier, we determined that CD3 γ, δ, ε subcomplexes were on the surface of the $CD4^{-}8^{-}$TCR α^{-}, β^{-} cell line 12.4 (Ley *et al.* 1989). The CD3 γ, δ, ε subcomplex might arrive at the surface because of limited escape from the ER retention mechanism and incomplete degradation of the CD3 δ containing subcomplex in the lysosome. Recent studies with similar cell lines and using different extraction techniques have shown that the cell surface expressed CD3 γ, δ, ε proteins may weakly associate with the CD3 ζ chain (J. Sancho and P. Mombaerts, unpublished data). An alternative model could be that

$gp33_2$/CD3 receptors are expressed on these cells undetected by currently available technology (Groettrup and von Boehmer 1993*b*).

5.3 Incomplete TCR–CD3 complexes during thymocyte development

5.3.1 Prior to TCR β *rearrangement*

It has long been established that the CD3 γ, δ, ε, and ζ genes were expressed
in the earliest identifiable thymocytes before rearrangement of the TCR αβ and possibly before rearrangement of the TCR γδ genes (Figure 17.10) (see chapter by Hayday in this volume). Information about relative timing of expression of the individual CD3 genes has been sketchy, because it is not clear whether the pre-thymocyte cell lines represent developmental stages (Samelson *et al.* 1985; Furley *et al.* 1986*a*; van Dongen *et al.* 1987, 1988). Because of similar reservations the observation that CD3 γ, δ, ε subcomplexes could be detected on the surface of the $CD4^-8^-$ TCR α^-, β^- cell line 12.4 was difficult to interpret (Ley *et al.* 1989). Moreover, signal transduction as

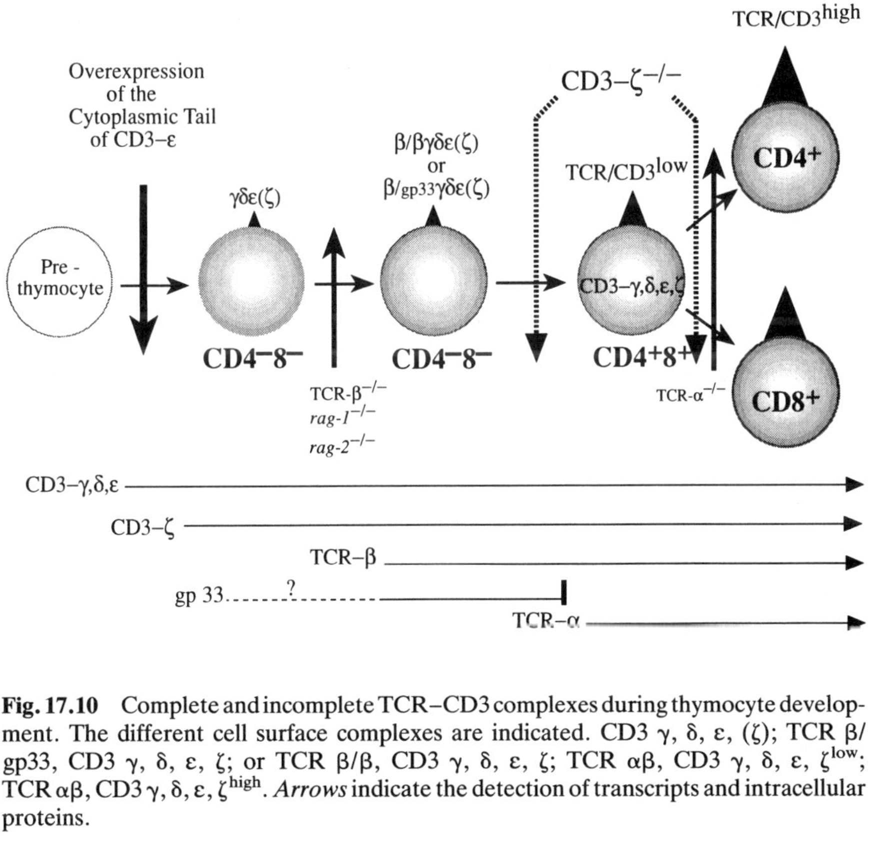

Fig. 17.10 Complete and incomplete TCR–CD3 complexes during thymocyte development. The different cell surface complexes are indicated. CD3 γ, δ, ε, (ζ); TCR β/gp33, CD3 γ, δ, ε, ζ; or TCR β/β, CD3 γ, δ, ε, ζ; TCR αβ, CD3 γ, δ, ε, ζ^{low}; TCR αβ, CD3 γ, δ, ε, ζ^{high}. *Arrows* indicate the detection of transcripts and intracellular proteins.

measured by IL-2 production and Ca^{2+} mobilization could not be measured in these cells unless TCR α and β cDNA were transfected into these cells. Recent experiments by Levelt and Eichmann (1993) (Levelt *et al.* 1993 *a,b,c*) have however demonstrated that the CD3 γ, δ, ε surface complexes are able to transduce functional signals. Their most pertinent experimental reasoning was derived with *in vitro* RAG-$1^{-/-}$ thymocyte cultures and anti-CD3 ε antibody stimulation. They showed that anti-CD3 antibodies induced transcription of the CD4 and CD8 genes in the $CD4^{-}8^{-}$ cells. This result could be reproduced by injecting the same monoclonal antibodies into RAG-1 neonates (Levelt *et al.* 1994). The reasons why this partial CD3 complex might transduce signals is however still obscure, since no ligand for the ecto-domains of these proteins has been described. It is more likely that these polypeptide chains function fully in the presence of the TCR β chain or TCR αβ.

The notion that signal transduction events can be initiated by the CD3 ε cytoplasmic tail in early pre-thymocytes was also supported by a transgenic mouse model. A severe immunodeficiency involving a complete loss of T αβ cells was observed in independent lines of transgenic mice containing more than 30 copies of the human or murine CD3 ε gene. T cell minus mice could also be generated by using a gene fragment coding for the 55 amino acid non-enzymatic CD3 ε cytoplasmic tail (Wang *et al.* 1994). In the pre-thymocyte population, development was abrogated before the Thy-1^{+}Pgp-1^{+} stage and before expression of endogenous CD3 γ, δ, and ε genes or rearrangement of the TCR α and β receptor genes. Transgene coded mRNAs were found in fetal thymuses on day 13 of gestation, whereas on day 14 of gestation transgenic mRNA levels were dramatically higher than those of the endogenous CD3 ε transcripts. Analysis of a series of transgenic lines generated with different genomic CD3 ε constructs revealed that the extent of the immunodeficiency correlated with the number of copies of the transgenes and with the enhancer species used. Taken together, the data demonstrated that high levels of expression of CD3 ε caused the block in early thymocyte development. Since CD3 ε has been found in activated mature human natural killer cells and in human fetal NK cells (Lanier *et al.* 1992; Phillips *et al.* 1992), the influence of the CD3 ε transgene on natural killer cell development was studied. Surprisingly, no NK cells could be found in transgenic mice carrying high copy numbers of CD3 ε gene in their genome. These observations strongly indicated that recruitment of signal transduction molecules by the cytoplasmic tail of CD3 ε played an important role in abrogating development of both cell types, and lend support to the theory that NK cells and T cells stem from a common precursor cell. Since over-expression of CD3 δ (Wang *et al.* 1994) or CD3 η (Hussey *et al.* 1993) did not cause this effect, it is likely that CD3 ε plays a distinct role in early T lymphocyte and NK cell development. Whether CD3 ε recruits unique signal transduction molecules in the different cell types (i.e. pre-thymocytes, mature T cells, and NK cells) will need to be investigated.

5.3.2 *Prior to TCR α rearrangement*

That a subset of the $CD4^-8^-$ thymocytes expresses TCRββ–CD3 complexes is not accepted by those who argue that under physiological conditions only a TCRβ–gp33–CD3 or a $(gp33)_2$/CD3 is expressed (Groettrup and von Boehmer 1993*b*). Whether this provocative observation with one cell line can be extended to $CD4^-8^-$ thymocytes is however not completely certain. The current view of von Boehmer and colleagues (Groettrup and von Boehmer 1993*b*) is that a small population of $CD4^-8^-$ blasts isolated from the TCR $\alpha^{-/-}$ mouse may express the surrogate TCR α chain receptor at high levels. What the ligand for these incomplete receptors might be is uncertain. Levelt and colleagues (1993*c*) reason that this could be a cell surface determinant on the surface of the thymic epithelium. The notion that such a ligand exists is indirectly supported by the studies of Anderson *et al.* (1993) who observed that both MHC class II positive epithelium and mesenchymal cells are needed for maturation from the $CD4^-8^-$ to the $CD4^+8^+$ stage. The function of the TCRββ–CD3 or TCRβ–gp33–CD3 expression is probably to check for functional rearrangement of the TCR-β gene, which then provides the signal which is needed to initiate TCR α rearrangement. Obviously, cell surface expression is not necessarily important for this series of events and the ligand issue may be irrelevant.

Part of the excitement about gp33 also termed 'the surrogate TCR α chain' stems from analogies with early B cell development which utilizes surrogate Ig light chains. The importance of the surrogate light chains became evident only after its gene had been disrupted by homologous recombination techniques resulting in a mutant mouse with a complete block in B cell development (Sakaguchi and Melchers 1986; Pillai and Baltimore 1987; Tsubata and Reth 1990; Kitamura *et al.* 1992; Ehlich *et al.* 1993).

5.3.3 *TCR αβ receptors in the thymus*

Most of the human or murine $CD4^+8^+$ thymocytes express 'low' levels of TCR αβ on their surface. In contrast, single-positive thymocytes and peripheral T lymphocytes express high levels of TCR αβ complexes. Several different hypotheses have been developed to explain the transition from TCR^{low} to TCR^{high}, most of which are connected to models for positive selection on MHC class I– and class II–peptide complexes or to distinct degradation processes in the different cell types. But, to date no unifying hypothesis incorporating the principles of intracellular trafficking, of ER redox potential and TCR αβ pair formation, and accommodating all of the available data can be formulated. It is likely that the level of expression of the complete TCR–CD3 complex (i.e. α,β,γ,δ,ε,ζ) does not affect signal transduction through the complex *per se*. But, signal transduction through TCR–CD3 in $CD4^+8^+$ thymocytes is markedly different from that in single-positive T lymphocytes and differential expression of protein kinases in the major thymocyte subsets (Sancho *et al.* 1992*a*; Perlmutter *et al.* 1993).

Unlike in the case of the TCR β genes, functional rearrangement of the TCR α VDJ segments in one chromosome does not halt their rearrangement on the sister chromosome. Thus, a given mature T lymphocyte could carry two types of receptors on its surface with different TCR α chain associated with one TCR β chain. Indeed, significant populations of mature αβ T cells expressing two antigen receptor pairs utilizing two different TCR αs and one TCR β have recently been identified. Each combination of two of the Vα2, Vα12, and Vα24 chains could be found on the surface of human αβ T cells (Padovan *et al.* 1993). Whether the expression of two TCRα chains per cell is as widespread as suggested by the authors remains to be determined. Examination of this phenomenon of co-expression is complicated by preferences in TCR αβ pair formation (Reno *et al.* 1990; de Waal Malefyt *et al.* 1990.

5.4 The CD3 $\zeta^{-/-}$ mouse

Elimination of the ζ gene from the murine genome by homologous recombination results in an abnormally developed thymus (10–15% of wt size) in which primarily TCR $\alpha\beta^-$, $CD4^-8^-$, and TCR $\alpha\beta^{\text{very low}}$ $CD4^+8^+$ are found (Liu *et al.* 1993; Love *et al.* 1993; Malissen *et al.* 1993; Ohno *et al.* 1993). Our preliminary biochemical data show that, as predicted from studies with cell lines, a minute amount of an αβγδε containing complex is expressed on the surface of the CD3 $\zeta\eta^{-/-}$ thymocytes. Analysis of the D–J joining segments of Vβ4, Vβ8, and Vβ12 in mRNA isolated from $CD4^+CD8^+$ thymocytes and from splenic T cells confirms that cells with productively rearranged Vβ regions are selected. In the CD3 $\zeta^{-/-}$ mouse very few if any 'single-positive' thymocytes are detected (Liu *et al.* 1993; Love *et al.* 1993; Malissen *et al.* 1993; Ohno *et al.* 1993). This could be due to a partial block in the transition from $CD4^+8^+$ to single-positive cells or could be a result of the very small pool of $CD4^+8^+$ cells.

In contrast to the thymus, spleen and lymph nodes contain large numbers single-positive TCR–$CD3^{\text{very low}}CD4^+$ and TCR–$CD3^{\text{very low}}CD8^+$ T lymphocytes. This could be explained by a model in which small numbers of CD3 ζ^- thymocytes are positively selected by interacting with the MHC class I and class II proteins. A second possible interpretation is that the TCR–$CD3^{\text{very low}}$ single-positive cells may have been generated through an alternative pathway not involving positive selection for MHC. Studies with CD3 $\zeta\eta^{-/-}$ × MHC class $I^{-/-}$ mice reveal that $CD8^+$ T lymphocytes are absent in the spleen and lymph nodes, suggesting that positive selection takes place in CD3 $\zeta\eta^{-/-}$ mice. Similarly, $CD4^+$ T cells are not found in CD3 $\zeta\eta^{-/-}$ × MHC class $II^{-/-}$ mice, demonstrating that positive selection occurs even when very low levels of the CD3 ζ^- TCR are present on the cell surface. Surprisingly, the total number of $CD4^+$ T cells is drastically lower in some of the CD3 $\zeta\eta^{-/-}$ × MHC class $I^{-/-}$ mice when compared to the CD3 $\zeta^{-/-}$ animals. Similarly, in some CD3 $\zeta\eta^{-/-}$ × MHC class $II^{-/-}$ animals the number of $CD8^+$ T lymphocytes is reduced compared to CD3 ζ^- mice (Simpson *et al.* 1994). This may reflect the stochastic processes involved in thymic selection. Thus, a model is arising in which the TCR–CD3$^{\text{very low}}$ complexes can still engage in positive selection resulting in the export to the periphery of small numbers of mature thymocytes. How these single-positive T cells expand in the peripheral organs is uncertain. Careful studies of the TCR V region repertoire need to be conducted to examine whether a relatively small number of T cell clones have expanded. Pauciclonality will have to be studied at the mRNA level because V region-specific antibody reagents do not stain the TCR–CD3 ζ^- complex on the surface of spleen cells. Although splenic CD3 $\zeta^{-/-}$ T lymphocytes express several cell surface markers that are connected with T cell activation, the parameters of proliferation are unknown. For example, these cells do not respond to mitogens (Con A, anti-CD3), or in mixed lymphocyte cultures (She *et al.* 1994).

5.4.1 TCR–FcεR1γ$^+$ receptors

Recently, a novel type of T cell was discovered in the murine intestinal epithelium, by studying CD3 $\zeta^{-/-}$ mutant mice (Liu *et al.* 1993; Malissen *et al.* 1993). In these animals high levels of αβ or γδ TCR–CD3 complexes are invariably found on the surface of intraepithelial T cells (IEL) isolated from the small intestine or the colon, but not dendritic epithelial T cells in the skin. This is due to association of TCRαβ–CD3γδε with the FcεR1γ polypeptide chain. Intraepithelial T cells expressing TCR–FcεR1γ$^+$ complexes are functionally indolent, as judged by their poor responses to anti-CD3, anti-TCR, and Con A and in an MLC but they are able to respond to treatment with PMA and ionomycin (Liu *et al.* 1993; Malissen *et al.* 1993; She *et al.* 1994). Redirected CTL assays using fresh populations of IEL from the small intestine, ^{51}Cr labelled P815 cells and anti-CD3 antibodies, do not cause chromium release, in contrast to the *wt* IEL (S. Simpson, unpublished). However, *in vitro* stimulation of these cells may give rise to production of certain cytokines.

The majority of the CD3$^+$ IEL are TCR αβ$^+$, CD8$^{\alpha\alpha+}$ cells, a smaller percentage is TCR αβ$^+$, CD8$^{\alpha\beta+}$ or CD4$^+$. In young animals, a variable fraction of these TCR–FcεR1γ$^+$ IEL are γδ T cells and the percentage γδ T cells increases with age or with an immunological challenge (She *et al.* 1994). The phenotype of TCR–FcεR1γ$^+$ in the IEL of *wt* mice is similar to that in the CD3 $\zeta^{-/-}$ mice. Interestingly, in γδ TCR transgenic mice a high percentage of these γδ T cells in spleen and lymph nodes express TCR–FcεR1γ$^+$ complexes (Qian *et al.* 1993).

Further analyses of the T cells in the intestinal epithelium of CD3 $\zeta\eta^{-/-}$ × MHC class I$^{-/-}$ mice reveal that only the CD8$^{\alpha\beta+}$ IELs and lamina propria lymphocytes of both the small and large intestine, which are predominantly TCR–CD3$^{\text{very low}}$, can be positively selected for class I MHC (i.e. CD8$^+$). Similarly, the CD4$^+$ lymphocytes, which are absent in the intestinal epithelium of CD3 $\zeta\eta^{-/-}$ × MHC class II$^{-/-}$ mice are TCR–CD3$^{\text{very low}}$. By contrast, the majority of TCR–FcεR1γ$^+$ IEL which are CD8$^{\alpha\alpha+}$ and to a lesser extent $CD4^+$ or $CD8^{\alpha\beta+}$, are not selected by either class I or class II MHC antigens. These observations support the concept that the TCR–$Fc\varepsilon R1\gamma^+$ $CD8^{\alpha\alpha+}$ IEL may have developed extrathymically, using alternative restriction elements or without restriction.

Preferential expression of the $Fc\varepsilon R1\gamma$ gene in intestinal epithelial T cells might be caused by induction through environmental factors or be due to an independent programming of precursor lymphocytes involved in extrathymic maturation. At the moment we favour an induction model because of an observation by Mizoguchi *et al.* (1992), who describe an *in vivo* effect of a murine colon carcinoma cell line MC38 which induces an impaired cytotoxic function of splenic $CD8^+$T cells. Interestingly, the splenic T cells completely lack ζ and expressed the TCR–$Fc\varepsilon R1\gamma^+$ complexes on the cell surface. In agreement with the data of the colon cancer model is our observation that the redirected cytotoxic activity of $\zeta^{-/-}$ IELs are completely abrogated when compared to *wt.* However, in contrast to Mizoguchi *et al.*, we have detected a significant level of the *src*-like tyrosine kinases *c-fyn* and *c-lck* in the $\zeta^{-/-}$ IELs. Thus, these colon cancer-induced TCR–$Fc\varepsilon R1\gamma^+$ cells appear similar, but not identical, to the CD3 $\zeta^{-/-}$IELs. None the less, the experiments suggest that a non-responsive TCR–$Fc\varepsilon R1\gamma^+$ cell can be induced by environmental factors, which may in part be provided by intestinal epithelium. Replacement of CD3ζ by the $Fc\varepsilon R1\gamma$ chain in the TCR–CD3 complex resulting in a relatively non-responsive cell type may be a physiological correlate of the observation that $Fc\varepsilon R1\gamma$ chain chimeras function less effectively than CD3ζ chimeras in T cells and optimally in a mast cell line (Letourneur and Klausner 1992).

6 Future directions

Although it is clear that the principal participants of the TCR–CD3 complex can form alternative protein ensembles on the surface of thymocytes and T lymphocytes, the physiological role of the different complexes needs to be determined. Much still needs to be learned about the molecular events during their assembly and signal transduction. As studies of the three-dimensional structures of the TCR and CD3 ectodomains are under way, a major advance in the near future will be a molecular picture of the interacting surfaces between these proteins. Understanding the different co-folding events will however require different experimental approaches. Do different thymocyte and mature lymphocyte subsets use distinct sets of signal transduction molecules and do these relate to different stages of development or functional roles during the immune response? How will the signal transduction events through TCR–CD3 be networked with signal transduction through all of the auxiliary activation structures and adhesion molecules? But a detailed molecular description of the mechanism of action of one of the most complicated cell surface receptor systems will be obtained in the coming years.

Acknowledgement

We should like to thank Karen Moore, Millenium Inc., for chromosomal assignments.

Reference

Ahluwalia, N., Bergeron, J., Wada, I., Degen, E., and Williams, D. (1992). The p88 molecular chaperone is identical to the endoplasmic reticulum membrane protein, calnexin. *J. Biol. Chem.*, **267**, 10914–18.

Alarcon, B., Berkhout, B., Breitmeyer, J., and Terhorst, C. (1988*a*). Assembly of the human T cell receptor-CD3 complex takes place in the endoplasmic reticulum and involves intermediary complexes between the CD3-γ, δ, ε core and single T cell receptor α or β chains. *J. Biol. Chem.*, **263**, 2953–61.

Alarcon, B., Regueiro, J., Arnaiz-Villena, A., and Terhorst, C. (1988*b*). Familial defect in the surface expression of the T cell receptor-CD3 complex. *N. Engl. J. Med.*, **319**, 1203–8.

Alarcon, B., Terhorst, C., Arnaiz-Villena, A., Perez-Aciego, P., and Regueiro, J. (1990). Congenital T cell receptor immunodeficiencies in man. *Immunodef. Rev.*, **2**, 1–16.

Alarcon, B., Ley, S., Sanchez-Madrid, F., Blumberg, R., Ju, S., Fresno, M., *et al.* (1991). The CD3-γ and CD3-δ subunits of the T cell antigen receptor can be expressed within distinct functional TCR/CD3 complexes. *EMBO J.*, **10**, 903–12.

Anderson, G., Jenkinson, E. J., Moore, N. C., and Owen, J. J. T. (1993). MHC class II-positive epithelium and mesenchyme cells are both required for T-cell development in the thymus. *Nature*, **362**, 70–3.

Arnaiz-Villena, A., Timon, M., Corell, A., Perez-Aciego, P., Martin-Villa, J., and Regueiro, J. (1992). Primary immunodeficiency caused by mutations in the gene encoding the CD3-γ subunit of the T lymphocyte receptor. *N. Engl. J. Med.*, **327**, 529–33.

Ashwell, J. A. K. R. (1990). Genetic and mutational analysis of the T cell antigen receptor. *Ann. Rev. Immunol.*, **8**, 139–67.

Azuma, M., Ito, D., Yagita, H., Okumura, K., Phillips, J., Lanier, L., *et al.* (1993). B70 antigen is a second ligand for CTLA-4 and CD28. *Nature*, **366**, 76–8.

Baniyash, M., Hsu, V., Seldin, M. & Klausner, R. (1989). The isolation and characterization of the murine T cell antigen receptor ζ chain gene. *J. Biol. Chem.*, **264**, 13252–7.

Berkhout, B., Alarcon, B., and Terhorst, C. (1988). Transfection of genes encoding the T cell receptor-associated CD3 complex into COS cells results in assembly of the macromolecular structure. *J. Biol. Chem.*, **263**, 8628–36.

Bernot, A. and Auffray, C. (1991). Primary structure and ontogeny of an avian CD3 transcript. *Proc. Natl. Acad. Sci. USA*, **38**, 2550–4

Blank, U., Ra, C., Miller, L., White, K., Metzger, H., and Kinet, J.-P. (1989). Complete structure and expression in transfected cells of high affinity IgE receptor. *Nature*, **337**, 187–9.

Blumberg, R., Ley, S., Sancho, J., Lonberg, N., Lacy, E., McDermott, F., *et al.* (1990*a*). Structure of the T cell antigen receptor: Evidence for two CD3 ε subunits in the T cell receptor/CD3 complex. *Proc. Natl. Acad. Sci. USA*, **87**, 7220–4.

Blumberg, R., Alarcon, B., Sancho, J., McDermott, F., Lopez, P., Breitmeyer, J., and Terhorst, C. (1990*b*). Assembly and function of the T cell antigen receptor: Requirement of either the lysine or arginine residues in the transmembrane region of the α chain. *J. Biol. Chem.*, **265**, 14036–43.

Bonifacino, J., Suzuki, C., Lippincott-Schwartz, J., Wiessman, A., and Klausner, R. (1989). Pre-golgi degradation of newly-synthesized T cell antigen receptor chains: intrinsic sensitivity and role of subunit assembly. *J. Biol. Chem.*, **109**, 73–83.

Bonifacino, J. and Lippincott-Schwartz, J. (1991). Degradation of proteins within the endoplasmic reticulum. *Curr. Opin. Cell Biol.*, **3**, 592–600.

Bonifacino, J., Cosson, P., and Klausner, R. (1990*a*). Co-localized transmembrane determinants for ER degradation and subunit assembly explain the intracellular fate of TCR chains. *Cell*, **63**, 503–13.

Bonifacino, J., Suzuki, C., and Klausner, R. (1990*b*). A peptide sequence confers retention and rapid degradation in the endoplasmic reticulum. *Science*, **247**, 79–82.

Braakman, I., Helenius, J., and Helenius, A. (1992). Role of ATP and disulphide bonds during protein folding in the endoplasmic reticulum. *Nature*, **356**, 260–2.

Brenner, M. T. I. and Strominger, J. (1985). Cross-linking of human T cell receptor proteins: association between the T cell idiotype beta subunit and the T3 glycoprotein heavy sub-unit. *Cell*, **40**, 183–90.

Brown, J., Jardetzky, T., Gorga, J., Stern, L., Urban, R., Strominger, J., *et al.* (1993). Three-dimensional structure of the human class II histocompatibility antigen HLA-DR1. *Nature*, **364**, 33–9.

Cambier *et al.* (1995). *Immunology Today*, **16**, 110.

Carson, G., Kuestner, R., Ahmed, A., Pettey, C., and Concino, M. (1991). Six chains of the human T cell antigen-receptor CD3 complex are necessary and sufficient for processing the receptor heterodimer to the cell surface. *J. Biol. Chem.*, **266**, 7883–7.

Chen, C., Bonifacino, J., Yuan, L., and Klausner, R. (1988). Selective degradation of T cell antigen receptors retained in a pre-Golgi compartment. *J. Cell Biol.*, **107**, 2149–61.

Christinck, E., Luscher, M., Barber, B., and Williams, D. (1991). Peptide binding to class I MHC on living cells and quantitation of complexes required for CTL lysis. *Nature*, **254**, 67–70.

Clayton, L., Bauer, A., Jin, Y.-J., D'Adamio, L., Koyasu, S., and Reinherz, E. (1990). Characterization of thymus-derived lymphocytes expressing Tiα-βCD3γδεζ-ζ, Tiα-βCD3γδδεη-η or Tiα-βCD3γεζ-ζ/η antigen receptor isoforms. *J. Exp. Med.*, **172**, 1243–53.

Clayton, L., D'Adamio, L., Howard, F., Sieh, M., Hussey, R., Koyasu, S. *et al.* (1991). CD3η and CD3ζ are alternatively spliced products of a common genetic locus and are transcriptionally and/or post-transcriptionally regulated during T-cell development. *Proc. Natl. Acad. Sci. USA*, **88**, 5202–6.

Clevers, H., Alarcon, B., Wileman, T., and Terhorst, C. (1988*a*). The T cell receptor/CD3 complex: A dynamic protein ensemble. *Ann. Rev. Immunol.*, **6**, 629–62.

Clevers, H., Dunlap, S., Wileman, T., and Terhorst, C. (1988*b*). Human CD3-ε gene contains three miniexons and is transcribed from a non-TATA promoter. *Proc Natl. Acad. Sci. USA*, **85**, 8156–60.

Cosson, P., Lankford, S., Bonifacino, J., and Klausner, R. (1991). Membrane protein association by potential intramembrane charge pairs. *Nature*, **351**, 414–16.

Davis, M. and Chien, Y. (1993). Topology and affinity of T-cell receptor mediated recognition of peptide-MHC complexes. *Curr. Opin. Immunol.*, **5**, 45–9.

Desai, D., Sap, J., Schlessinger, J., and Weiss, A. (1993). Ligand-mediated negative regulation of a chimeric transmembrane receptor tyrosine phosphatase. *Cell*, **73**, 541–54.

deWaal Malefyt, R., Yssel, H., Spits, H., DeVries, J., Sancho, J., Terhorst, C., *et al.* (1990). Human T cell leukemia virus type 1 prevents cell surface expression of the T cell receptor through down-regulation of the CD3-γ,δ,ε and ζ genes. *J. Immunol.*, **145**, 2297–303.

Ehlich, A., Schaal, S., and Gu, H. (1993). Immunoglobulin heavy and light chain genes rearrange independently at early stages of B cell development. *Cell*, **72**, 695–704.

Exley, M., Terhorst, C., and Wileman, T. (1991). Structure, assembly and intracellular transport of the T cell receptor for antigen. *Semin. Immunol.*, **3**, 283–97.

Exley, M., Varticovski, L., Peter, M., Sancho, J., and Terhorst, C. (1993). Association of phosphatidylinositol 3-kinase associates with a specific sequence in the T cell receptor ζ chain upon T cell activation. *J. Biol. Chem.*, **269**, 15140–6.

Exley, M., Wileman, T., Mueller, B., and Terhorst, C. (1995). A divalent model for the T cell antigen receptor. *Immunol. Cell Biol.*, (submitted for publication).

Furley, A., Mizutani, S., Weilbaecher, K., Dhaliwal, H., Ford, A., Chan, L., *et al.* (1986). Developmentally regulated rearrangement and expression of genes encoding the T cell receptor-T3 complex. *Cell*, **46**, 75–87.

Gold, D., Clevers, H., Alarcon, B., Dunlap, S., Novotny, J., Williams, A., *et al.* (1987). Evolutionary relationship between the T3 chains of the T cell receptor complex and the immunoglobulin supergene family. *Proc. Natl. Acad. Sci. USA*, **84**, 7649–53.

Groettrup, M. and von Boehmer, H. (1993*a*). T cell receptor β chain dimers on immature thymocytes from normal mice. *Eur. J. Immunol.*, **23**, 1393–6.

Groettrup, M. and von Boehmer, H. (1993*b*). A role for a pre-T-cell receptor in T-cell development. *Immunol. Today*, **14**, 610–14.

Groettrup, M., Baron, A., Griffiths, G., Palacios, R., and von Boehmer, H. (1992). T cell receptor (TCR) β chain homodimers on the surface of immature but not mature α, γ, δ, chain deficient T cell lines. *EMBO J.*, **11**, 1735–2746.

Hall, C., Berkhout, B., Alarcon, B., Sancho, J., Wileman, T., and Terhorst, C. (1991). Requirements for cell surface expression of the human TCR/CD3 complex in non-T cells. *Int. Immunol.*, **3**, 359–68.

Hall, C., Sancho, J., and Terhorst, C. (1993). Reconstitution of T cell receptor ζ-mediated calcium mobilization in nonlymphoid cells. *Science*, **261**, 915–18.

Hochstenbach, D., David, V., Watkins, S., and Brenner, M. (1992). Endoplasmic reticulum resident protein of 90 kDa associates with T and B cell antigen receptors and major histocompatibility antigens during their assembly. *Proc. Natl. Acad. Sci. USA*, **89**, 4734–8.

Hogquist, K., Jameson, S., Heath, W., Howard, J., Bevan, M., and Carbone, F. (1994). T cell receptor antagonist peptides induce positive selection. *Cell*, **76**, 17–27.

Hussey, R., Clayton, L., Diener, A., McConkey, D., Howard, F., Rodewald, H.-R., *et al.* (1993). Overexpression of CD3η during thymic development does not alter the negative selection process. *J. Immunol.*, **150**, 1183–94.

Hwang, C., Sinskey, A., and Lodish, H. (1992). Oxidized redox state of glutathione in the endoplasmic reticulum. *Science*, **257**, 1496–502.

Inoue, S., Bar-Nun, S., Roitleman, J., and Simoni, R. (1991). Inhibition of degradation of 3-hydroxy-3-methylglytaryl-coenzyme A reductase *in vivo* by cysteine protease inhibitors. *J. Biol. Chem.*, **266**, 13311–17.

Irving, B., Chan, A., and Weiss, A. (1993). Functional characterization of a signal transducing motif present in the T cell antigen receptor ζ chain. *J. Exp. Med.*, **177**, 1093–103.

Itohara, S., Mombaerts, P., Lafaille, J., Iacomini, J., Nelson, A., Clarke, A., *et al.* (1993). T cell receptor δ gene mutant mice: Independent generation of αβ T cells and programmed rearrangements of γδ TCR genes. *Cell*, **72**, 337–48.

Jackson, M., Nilsson, T., and Peterson, P. (1993). Retrieval of transmembrane proteins to the endoplasmic reticulum. *J. Cell Biol.*, **121**, 317–33.

Jin, Y.-J., Clayton, L., Howard, F., Koyasu, S., Sieh, M., Steinbrich, R., *et al.* (1990). Molecular cloning of the CD3η subunit identifies a CD3ζ-related product in thymus-derived cells. *Proc. Natl. Acad. Sci. USA*, **87**, 3319–23.

Kanagawa, O. and Ahlem, C. (1989). Requirement of the T cell antigen receptor occupancy for the target cell lysis by cytolytic T lymphocytes. *Int. Immunol.*, **1**, 8178–9.

Kappes, D. and Tonegawa, S. (1991). Surface expression of alternative forms of the TCR/CD3 complex. *Proc. Natl. Acad. Sci. USA*, **88**, 337–48.

Kitamura, D., Kudo, A., and Schaal, S. (1992). A critical role of lambda 5 protein in B cell development. *Cell*, **69**, 823–31.

Kolanus, W., Romeo, C., and Seed, B. (1993). T cell activation by clustered tyrosine kinases. *Cell*, **74**, 171–88.

Kornfeld, S. and Mellman, I. (1989). The biogenesis of lysosomes. *Ann. Rev. Cell. Biol.*, **5**, 483–525.

Lanier, L., Chang, C., Spits, H., and Phillips, J. (1992). Expression of cytoplasmic CD3ε proteins in activated human adult natural killer (NK) cells and CD3γ, δ, ε complexes in fetal NK cells. *J. Immunol.*, **149**, 1876–80.

Letourneur, F. and Klausner, R. (1992*a*). Activation of T cells by a tyrosine kinase activation domain in the cytoplasmic tail of CD3-ε. *Science*, **255**, 79–82.

Letourneur, F. and Klausner, R. (1992*b*). A novel di-leucine motif and a tyrosine-based motif independently mediate lysosomal targeting and endocytosis of CD3 chains. *Cell*, **69**, 1143–57.

Levelt, C. and Eichmann, K. (1993). Parallel development of the T cell and its receptor. *Immunologist*, **1**, 151–4.

Levelt, C., Ehrfeld, A., and Eichmann, L. (1993*a*). Regulation of thymocyte development through CD3.1. timepoint of ligation of CD3ε determines clonal deletion or induction of developmental program. *J. Exp. Med.*, **177**, 707–16.

Levelt, C., Carsetti, R., and Eichmann, K. (1993*b*). Regulation of thymocyte development through CD3.II. expression of T cell receptor β CD3ε and maturation to the CD4+8+ stage are highly correlated in individual thymocytes. *J. Exp. Med.*, **178**, ???.

Levelt, C., Mombaerts, P., Iglesias, A., Tonegawa, S., and Eichmann, K. (1993*c*). Restoration of early thymocyte differentiation in T-cell receptor β-chain-deficient mutant mice by transmembrane signaling through CD3ε. *Proc. Natl. Acad. Sci. USA*, **90**, 11401–5.

Levelt, C., Ehrfeld. A., Terhorst. C., and Eichman, K. (1994). Control of transition of TCR-β to TCR-α locus rearrangement by signalling through CD3-ε: Role of CD3-ζ and $p56^{lck}$. *J. Exp. Med.*, (Submitted for publication).

Ley, S., Tan, K., Kubo, R., Sy, M., and Terhorst, C. (1989). Surface expression of CD3 in the absence of T cell receptor (TCR): evidence for sorting of partial TCR/CD3 complexes in a post-endoplasmic reticulum compartment. *Euro. J. Immunol.*, **19**, 2309–17.

Lippincott-Schwartz, J., Bonifacino, J., Yuan, L., and Klausner, R. (1988). Degradation from the endoplasmic reticulum: disposing of newly synthesized proteins. *Cell*, **54**, 209–29.

Liu, C.-P., Euda, R., She, J., Sancho, J., Wang, B., Weddell, G., *et al.* (1993).

Abnormal T cell development in CD3ζ−/− mutant mice and identification of a novel T cell population in the intestine. *EMBO J.*, **12**, 4863–75.

Love, P., Shores, E., Johnson, M., Tremblay, M., Lee, E., Grinberg, A., *et al.* (1993). T cell development in mice that lack the ζ chain of the T cell antigen receptor complex. *Science*, **261**, 918–21.

Malissen, M., Gillet, A., Rocha, B., Trucy, J., Vivier, E., Boyer, C., *et al.* (1993). T cell development in mice lacking the CD3-ζ/η gene. *EMBO J.*, **12**, 4347–55.

Mallabiabarrena, A., Fresno, M., and Alarcon, B. (1992). An endoplasmic reticulum retention signal in the CD3-ε chain of the T cell receptor. *Nature*, **357**, 593–6.

Manolios, N., Letourneur, F., Bonifacino, J., and Klausner, R. (1991). Pairwise, cooperative and inhibitory interactions describe the assembly and probable structure of the T cell antigen receptor. *EMBO J.*, **10**, 1643–51.

Matsui, K., Boniface, J., Reay, P., Schild, H., Fazekas De Groth, B., and Davis, M. (1992). Low affinity interaction of peptide-MHC complexes with T cell receptors. *Science*, **254**, 1788–91.

Minami, Y., Weissman, A., Samelson, L., and Klausner, R. (1987). Building a multichain receptor: synthesis, degradation and assembly of the T cell antigen receptor. *Proc. Natl. Acad. Sci. USA*, **84**, 291–4.

Mizoguchi, H., O'Shea, J., Longo, D., Loeffler, C., MvVicar, D., and Ochoa, A. (1992). Alternations in signal transduction molecules in T lymphocytes from tumor-bearing mice. *Science*, **258**, 1795–8.

Mombaerts, P., Clarke, A., Rudnicki, M., Iacimini, J., Itohara, S., Lefaille, J., *et al.* (1992*a*). Mutations in cell antigen receptor genes α and β block thymocyte development at different stages. *Nature*, **360**, 225–31.

Mombaerts, P., Iacomini, J., Johnson, R., Herrup, K., Tonegawa, S., and Papaioannou, V. (1992*b*). Rag-1 deficient mice have no mature B and T lymphocytes. *Cell*, **68**, 869–77.

Morley, B., Chin, K., Newton, M., and Weiss, A. (1988). The lysine residue in the membrane-spanning domain of the β chain is necessary for cell surface expression of the T cell antigen receptor. *J. Exp. Med.*, **168**, 1971–8.

Neisig, A., Vangsted, A., Zeuthen, J., and Geisler, C. (1993). Assembly of the T-cell antigen receptor. *J. Immunol.*, **151**, 870–9.

Ohno, H., Aoe, T., Taki, S., Kitamura, D., Ishida, Y., Rajewsky, K., *et al.* (1993). Developmental and functional impairment of T cells in mice lacking CD3ζ chains. *EMBO J.*, **12**, 4357–66.

Ou, W.-J., Cameron, P., Thomas, D., and Bergeron, J. (1993). Association of folding intermediates of glycoproteins with calnexin during protein maturation. *Nature*, **364**, 771–5.

Padovan, E., Casorati, G., Dellabona, P., Meyer, S., Brockhaus, M., and Lanzavecchia, A. (1993). Expression of two T cell receptor a chains: Dual receptor T cells. *Science*, **262**, 422–4.

Perez-Aciego, P., Alarcon, B., Arnaiz-Villena, A., Terhorst, C., Timon, M., and Regueiro, J. R. (1991). Expression and function of a variant T cell receptor complex lacking CD3-γ. *J. Exp. Med.* **174**, 319–26.

Perlmutter, R., Levin, S., Appleby, M., Anderson, S., and Alberola-Ila, J. (1993). Regulation of lymphocyte function by protein phosphorylation. *Ann. Rev. Immunol.*, **11**, 451–99.

Pettey, C., Alarcon, B., Malin, R., Weinberg, K., and Terhorst, C. (1987). T3-p28 is a protein associated with the δ and ε chains of the T cell receptor-T3 antigen complex during biosynthesis. *J. Biol. Chem.*, **262**, 4854–9.

Phillips, J., Hori, T., Nagler, A., Bhat, N., Spits, H., and Lanier, L. (1992). Ontogeny of human natural killer (NK) cells: fetal NK cells mediate cytolytic function and express cytoplasmic CD3 εδ proteins. *J. Exp. Med.*, **175**, 1055–66.

Philpott, K., Viney, J., Kay, G., Rastan, S., Gardiner, E., Chae, S., *et al.* (1992). Lymphoid development in mice congenitally lacking T cell receptor αβ-expressing cells. *Science*, **256**, 1448–52.

Pillai, S. and Baltimore, D. (1987). Myristoylation and the post-translational acquisition of hydrophobicity by the membrane immunoglobulin heavy-chain polypeptide in B lymphocytes. *Proc. Natl. Acad. Sci. USA*, **84**, 7654–8.

Punt, J., Kubo, R., Saito, T., Finkel, T., Kathiresan, S., Blank, K., *et al.* (1991). Surface expression of a T cell receptor β (TCR-β) chain in the absence of TCR-α, -δ and -γ proteins. *J. Exp. Med.*, **174**, 775–83.

Qian, D., Sperling, A., Lancki, D., Tatsumi, Y., Barrett, T., Bluestone, J. *et al.* (1993*a*). The γ chain of the high-affinity receptor for IgE is a major functional subunit of the T cell antigen receptor complex in γδ T lymphocytes. *Proc. Natl. Acad. Sci. USA*, **90**, 11875–9.

Qian, D., Griswold-Prenner, I., Rosner, M., and Fitch, F. (1993*b*). Multiple components of the T cell antigen receptor complex become tyrosine-phosphorylated upon activation. *J. Biol. Chem.*, **268**, 4488–93.

Rao, P., Meyer, E., Zivin, R., and Collins, A. (1994). Differential Expression of CD3 epitopes in α/β and γ/δ TCR-bearing T cells: Evidence for a dimeric receptor. *Hum. Immunol.* (manuscript in preparation).

Ravetch. J. (1994). FcReceptors: Ruber Redux. *Cell*, **78**, 553–60.

Ravichandran, K., Lee, K., Songyang, Z., Cantley, L., Burn, P., and Burakoff, S. (1993). Interaction of Shc with the ζ chain of the T cell receptor upon T cell activation. *Science*, **262**, 902–5.

Reno, T., Ley, S., Sugiyama, E., Cantagrel, A., Blumberg, R., Bonventre, J., *et al.* (1990). Defects in signal transduction caused by a T cell receptor β chain substitution. *Eur. J. Immunol.*, **20**, 1417–22.

Reth, M. (1989). Antigen receptor tail clue. *Nature*, **338**, 383.

Romeo, C., Amiot, M., and Seed, B. (1992). Sequence requirements for induction of cytolysis by the T cell antigen/Fc receptor ζ chain. *Cell*, **68**, 889–97.

Rutledge, T., Cosson, P., Manolois, N., Bonifacino, J., and Klausner, R. (1992). Transmembrane helical interactions: zeta chain dimerization and functional association with the T cell antigen receptor. *EMBO J.*, **11**, 3245–54.

Sakaguchi, N. and Melchers, F. (1986). Lambda 5, a new light-chain-related locus selectively expressed in pre-B lymphocytes. *Nature*, **324**, 579–82.

Samelson, L., Lindsten, T., Fowlkes, B., van den Elsen, P., Terhorst, C., David, M., *et al.* (1985). Expression of genes of the T cell antigen receptor complex in precursor thymocytes. *Nature*, **315**, 765–8.

Saito, H., Koyama, T., Georgopoulos, K., Clevers, H., Haser, W. G., LeBien, T., Tonegawa, S., and Terhorst, C. (1987). Close linkage of the mouse and human CD3 γ- and δ-chain genes suggests that their transcription is controlled by common regulatory elements. *Proc. Natl. Acad. Sci. USA*, **84**, 9131–4.

Sancho, J., Chatila, T., Wong, R., Hall, C., Blumberg, R., Alarcon, B., *et al.* (1989). T cell antigen receptor (TCR)-α/β heterodimer formation is a prerequisite for association of CD3-ζ2 into functionally competent TCR-CD3 complexes. *J. Biol. Chem.*, **264**, 20760–9.

Sancho, J., Silverman, L., Castigli, E., Ahern, D., Laudano, A., Terhorst, C., *et al.* (1992*a*). Developmental regulation of transmembrane signalling via the T cell antigen receptor/CD3 complex in human T lymphocytes. *J. Immunol.*, **148**, 1315–321.

Sancho, J., Ledbetter, J., Choi, M., Kanner, S., Deans, J., and Terhorst, C. (1992*b*). CD3-ζ surface expression is required for CD4:p56*lck*-mediated upregulation of TCR/CD3 signaling in T cells. *J. Biol. Chem.*, **267**, 7871–9.

Sancho, J., Franco, R., Chatila, T., Hall, C., and Terhorst, C. (1993). The T cell receptor associated CD3-ε protein is phosphorylated upon T cell activation in the two tyrosine residues of a conserved signal transduction motif. *Euro. J. Immunol.*, **23**, 1636–42.

She, J., Simpson, S., Liu, C. P., Huang, M., Hollander, G., and Terhorst, C. (1994). The -α/β or γ/δ TCR/FceRlg^{+} intestinal epithelium are primarily CD8αα+ and are functionally inert. *J. Immunol.* Submitted for publication.

Shia, M. and Lodish, H. (1989). Two subunits of the asialo glycoprotein receptor have different fates when expressed alone in fibroblasts. *Proc. Natl. Acad. Sci. USA*, **86**, 1158–62.

Shin, J., Lee, S., and Strominger, J. (1993). Translocation of TCRα chains into the lumen of the endoplasmic reticulum and their degradation. *Science*, **259**, 1901–4.

Simpson, S. H., G. She, J., Levelt, C., Huang, M., and Terhorst, C. (1994). Selection of peripheral and intestinal T lymphocytes lacking CD3-ζ. *Int. Immunol.*, **7**, 287–93.

Snow, P., and Terhorst, C. (1983). The T8 antigen is a multimeric complex of two distinct subunits on human thymocytes but consists of homomultimeric forms on peripheral blood T lymphocytes. *J. Biol. Chem.*, **258**, 14675–81.

Soudais, C., de Villartay, J.-P., Le Deist, F., Fisher, A., and Lisowska-Grospierre, B. (1993). Independent mutations of the human CD3-ε gene resulting in a T cell receptor/CD3 complex immunodeficiency. *Nature Genet.*, **3**, 77–81.

Spits, H., van Schooten, W., Keizer, H., van Seventer, G., van de Rijn, M., Terhorst, C., *et al.* (1986). Alloantigen recognition is preceded by nonspecific adhesion of cytotoxic T cells and target cells. *Science*, **232**, 403–5.

Stafford, F. and Bonifacino, J. (1991). A permeabilized cell system identifies the endoplasmic reticulum as a site of protein degradation. *J. Cell Biol.*, **115**, 1225–36.

Terhorst, C., Spits, H., Staal, F., and Exley, M. (1995). *T lymphocyte signal transition* (ed. Hames and Glover). *Molec. Immunol.* In press.

Timon, M., Arnaiz-Villena, A., Rodriguiz-Gallego, C., Perez-Aciego, P., and Regueiro, J. (1993). Selective disbalances of peripheral blood T lymphocyte subsets in human CD3γ deficiency. *Eur. J. Immunol.*, **23**, 1440–4.

Tsao, Y., Ivessa, N., Adesnik, M., Sabatini, D., and Kreibich, G. (1992). Carboxy terminally truncated forms of ribophorin 1 are degraded in a pre-Golgi compartment by a calcium-dependent protease. *J. Cell Biol.*, **116**, 57–67.

Tsubata, T. and Reth, M. (1990). The products of pre-B cell-specific genes (lambda 5 and VpreB) and the immunoglobulin mu chain form a complex that is transported onto the cell surface. *J. Exp. Med.*, **172**, 973–6.

van Dongen, J., Quertermous, T., Bartram, C., Gold, D., Wolvers-Tettero, I., Comans-Bitter, W., *et al.* (1987). The T cell receptor-CD3 complex during early T cell differentiation: Analysis of immature T cell acute lymphoblastic leukemias (T-ALL) at DNA, RNA and cell membrane level. *J. Immunol.*, **138**, 1260–9.

van Dongen, J., Krissansen, G., Wolvers-Tettero, I., Comans-Bitter, W., Adriaansen, H., Hooijkaas, H., *et al.* (1988). Cytoplasmic expression of the CD3 antigen as a diagnostic marker for immature T cell malignancies. *Blood*, **71**, 603–12.

Wang, B., Biron, C., She, J., Sancho, J., Liu, C.-P., Higgins, *et al.* (1994). High level expression of the cytoplasmic tail of CD3-ε in transgenic mice blocks both early T lymphocyte and natural killer cell development. *Proc. Natl. Acad. Sci. USA*, **91**, 9402–116.

Wange, R., Malek, S., Desiderio, S., and Samelson, L. (1993). Tandem SH2 domains of ZAP-70 bind to T cell antigen receptor ζ and CD3 ε from activated Jurkat T cells. *J. Biol. Chem.*, **268**, 19797–801.

Weber, S., Traunecker, A., Oliveri, F., Gerhard, W., and Karjalainen, K. (1993). Specific low-affinity recognition of major histocompatibility complex plus peptide by soluble T-cell receptor. *Nature*, **356**, 793–6.

Wegener, A.-M., Letourneur, F., Hoeveler, A., Brocker, T., Luton, F., and Malissen, B. (1992). The T cell receptor/CD3 complex is composed of at least two autonomous transduction modules. *Cell*, **68**, 83–95.

Weiss, A. (1993). T cell antigen receptor signal transduction: A tale of tails and cytoplasmic protein-tyrosine kinases. *Cell*, **73**, 209–12.

Weissman, A., Hou, D., Orloff, D., Modi, W., Seuanez, H., O'Brien, S. *et al.* (1988). Molecular cloning and chromosomal localization of the human T-cell receptor ζ chain: Distinction from the molecular CD3 complex. *Proc. Natl. Acad. Sci. USA*, **85**, 9709–13.

Wileman, T., Carson, G., Concino, M., Ahmed, A., and Terhorst, C. (1990*a*). The γ and ε subunits of the CD3 complex inhibit pre-Golgi degradation of newly synthesized T cell antigen receptors. *J. Cell Biol.*, **110**, 973–86.

Wileman, T., Pettey, C., and Terhorst, C. (1990*b*). Recognition for degradation in the endoplasmic reticulum and lysosomes prevents the transport of single TCRβ and CD3δ subunits of the T cell antigen receptor to the surface of cells. *Int. Immunol.*, **2**, 743–54.

Wileman, T., Carson, G., Shih, F., Concino, M., & Terhorst, C. (1990*c*). The transmembrane anchor of the T cell antigen receptor β chain contains a structural determinant of pre-Golgi proteolysis. *Cell Reg.*, **1**, 907–19.

Wileman, T., Kane, L., and Terhorst, C. (1991*a*). Degradation of T cell receptor chains in the endoplasmic reticulum is inhibited by inhibitors of cysteine proteases. *Cell Reg.*, **2**, 753–65.

Wileman, T., Kane, L., Carson, G. and Terhorst, C. (1991*b*). Depletion of cellular calcium accelerates protein degradation in the endoplasmic reticulum. *J. Biol. Chem.*, **266**, 4500–7.

Wileman, T., Kane, L., Young, J., Carson, G., and Terhorst, C. (1993). Associations between subunit ectodomains promote T cell antigen receptor assembly and protect against degradation in the ER. *J. Cell Biol.*, **122**, 67–78.

Young, J., Kane, L., Exley, M., and Wileman, T. (1993). Regulation of selective protein degradation in the endoplasmic reticulum. *J. Biol. Chem.*, **268**, 19810–18.

18 The immune recognition unit: the TCR–peptide–MHC complex

KATHERINE L. HILYARD AND
JACK L. STROMINGER

1 Introduction

Triggering of a T cell by engagement with an antigen-presenting cell (APC) involves interactions between numerous different molecules on the cell surfaces for example LFA-1 and ICAM-1, CD2 and LFA-3, and CD28 and B7. The only interaction that is known to confer antigen specificity to this crucial cell–cell interaction is that between the T cell receptor (TCR) and a major histocompatibility complex protein (MHC) presenting a specific peptide antigen. Recent observations suggest that the TCR has a range of signalling capabilities depending on the antigen presented by the MHC. Limited variation in peptide can induce different T cell responses (reviewed in Evavold *et al.* 1993; Sette *et al.* 1994). The molecular mechanisms of this differential signalling are not known but are thought to involve slight perturbations in the TCR–peptide–MHC complex affecting the level of signalling achieved, with a certain threshold of signalling required for full activation. A similar process of differential signalling is thought to be involved in the positive and negative selection of T cells in the thymus (for a review see von Boehmer 1992; Allen 1994). Knowledge of the molecular details of the TCR–peptide–MHC interaction is therefore vital for an understanding of these key processes in the development and function of the immune system.

In addition to interacting with the TCR, non-polymorphic regions of the peptide–MHC complex interact with CD4 or CD8 which are co-receptors on the T cell. CD4 and CD8 are both members of the immunoglobulin (Ig) superfamily, although the former has four Ig domains and the latter only one. They interact with MHC class II and class I molecules respectively and CD4, but not CD8, also interacts with the TCR. These co-receptors appear to have both a cell adhesion function and a signalling function, the latter being through their association with the tyrosine kinase $p56^{lck}$. The details of these functions are reviewed in Vignali *et al.* (1993) and Julius *et al.* (1993) and will not be covered here.

This chapter will review the existing structural information for the TCR based on molecular models generated from protein sequences, and the detailed information provided by the crystal structures of MHC–peptide complexes. A current model for the trimolecular TCR–peptide–MHC complex and the data supporting the model will be discussed. Finally, we shall review the methods available for directly measuring TCR–peptide–MHC interactions and discuss why such experiments are technically challenging.

2 T cell receptor structure

The TCR is composed of two clonotypic transmembrane polypeptide chains, αβ (or γδ in a minor subpopulation of T cells, see Chapter 4) that are associated in the membrane with several non-polymorphic polypeptides, forming the TCR–CD3 complex (reviewed in Chapter 17). Information on the three-dimensional structure of TCR αβ chains is currently limited to molecular models generated from the growing database of TCR sequences. Analysis of the sequences indicate that the α and β chains are members of the Ig superfamily with the extracellular portion predicted to have a structure similar to an antibody Fab, i.e. one variable (V) Ig domain, and one constant (C) Ig domain in each chain (Patten *et al.* 1984; Novotny *et al.* 1986; Davis and Bjorkman 1988; Chothia *et al.* 1988; Claverie *et al.* 1989).

As illustrated in the crystal structures of antibody Fab fragments (reviewed in Davies *et al.* 1990), the Ig domain is composed of two antiparallel β-sheets interacting face-to-face to form a β-sandwich structure, with a disulfide link between the two sheets ('the immunoglobulin fold', Novotny *et al.* 1983). The V domain is a nine stranded β-sandwich and the C domain a seven stranded sandwich. In the Fab fragment, and presumably the TCR, the domains associate into dimers (e.g. VH/VL, CH/CL, Vα/Vβ, Cα/Cβ) which have a β-barrel structure. In the antibody V dimer, the loops connecting the β-strands come together to form a relatively flat platform with a surface of approximately 700 $Å^2$ that is the antigen binding site. The loops, three from each domain, are hypervariable in sequence and are termed complementarity determining regions, or CDRs, 1, 2, and 3 (Figure 18.1). It is proposed that the TCR binding site for the MHC–peptide complex is similarly composed of the loops connecting the β-strands of the Vα and Vβ domains (Davis and Bjorkman 1988; Chothia *et al.* 1988; Claverie *et al.* 1989). The sequence variability of TCR CDRs 1 and 2 is lower than for antibody CDRs 1 and 2, due to a more restricted V gene pool (Davis and Bjorkman 1988). However, there is a higher degree of variability in the CDR3 region of TCR, in part due to the larger number of J gene segments for the TCR compared to antibodies (Davis and Bjorkman 1988). This implies that the CDR3 region in the centre of the binding site, the area proposed to contact the peptide antigen, is more variable in structure than the CDR1/CDR2 region, proposed to contact the MHC molecule. This will be discussed in more detail later. Unlike antibody V domains, the TCR has a fourth hypervariable loop (Jores *et al.* 1990) that has been proposed to be involved in the interaction of a group of molecules called superantigens (Cazenave *et al.* 1990; Pullen *et al.* 1990) (see Chapter 19). These molecules are able to bind to MHC class II molecules

a

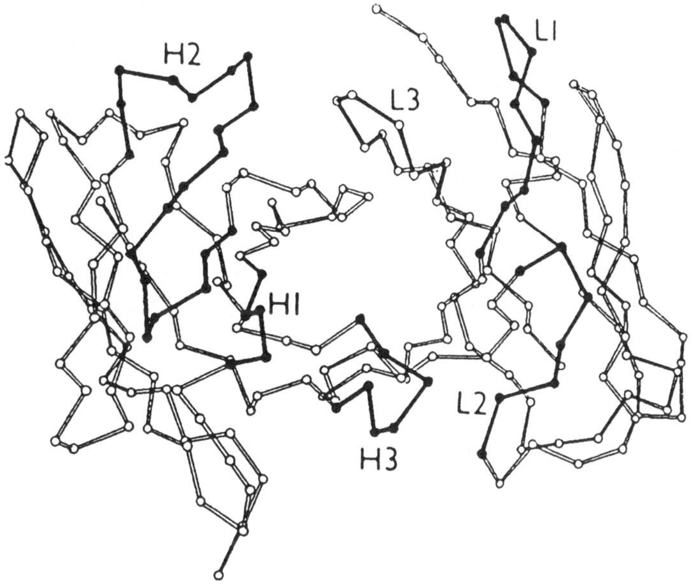

b

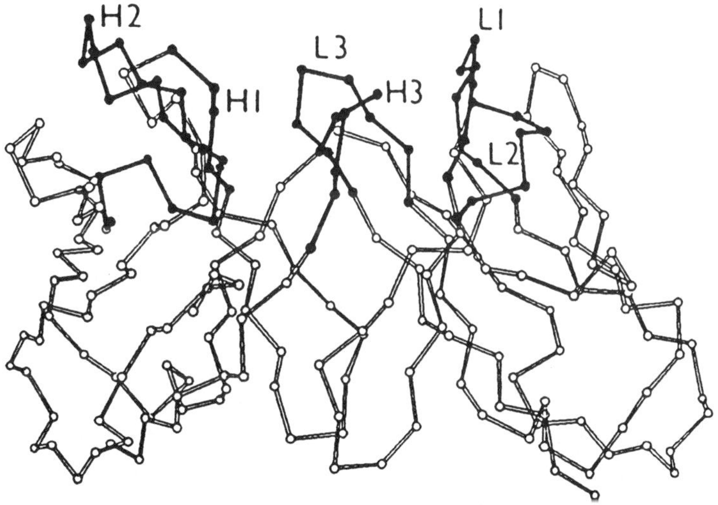

Fig. 18.1 Three-dimensional structure of an antibody V domain dimer (Cα atoms only). (a) View down on to the binding site from the direction of the antigen, and (b) view across the binding site. CDR loops are shown as filled atoms and bonds and are labelled H1, H2, and H3, for CDRs 1, 2, and 3 of the heavy chain, and L1, L2, and L3, for CDRs 1, 2, and 3 of the light chain. Taken from Hilyard (1991), derived from HyHel-5 crystal structure co-ordinates (Sheriff *et al.* 1987).

and interact with TCR in a Vβ-dependent manner, resulting in T cell activation (reviewed in Irwin *et al.* 1993). Recently, complex formation between a superantigen and a soluble TCR in the absence of MHC class II molecules has been demonstrated, with the dissociation constant being in the range of 10^{-5} M (Hilyard *et al.* 1994). Further details of the interactions of superantigens with TCR will not be covered in this chapter.

Comparisons of different antibody structures show the β-barrel structure, or 'framework' region, is highly conserved. The antigen binding specificity is almost completely determined by the CDR conformation. Predicting the structure of an antibody binding site from the primary sequence therefore requires modelling the conformation of the CDR loops on to a known framework structure. Methods that were developed to model antibody V domains have been extended to generate structures for TCR V domains. In one study, a model for a human TCR was produced by generating CDR loop conformations using *ab initio* energy calculations and building these on to an antibody framework structure (Novotny *et al.* 1991). The model was then used to predict exposed hydrophobic residues on the surface of the TCR V domains. Substitution of these positions to hydrophilic residues increased the solubility of the TCR V domain protein when it was produced in *Escherichia coli* (Novotny *et al.* 1991). In another study, CDR loop conformations generated by a combination of *ab initio* calculation and selection of loop conformations from structural databases based on sequence homology and loop length were built on to an antibody framework (Martin *et al.* 1989; Searle and Rees 1995). This method was used to generate models for a panel of human TCRs that were subsequently docked on to the appropriate MHC–peptide structure and the resulting complex used to direct site-specific mutagenesis experiments (Wederburn *et al.* 1995). One problem in using these methods to model TCR V domains is that the framework regions are not as highly conserved as antibody framework regions and so the definition of the CDR is more difficult (Patten *et al.* 1984; Jores *et al.* 1990). Modelling of structures from primary sequences is still extremely difficult and verification of the accuracy of these models requires the experimental determination of TCR structure. Attempts to solve the TCR structure by X-ray crystallography have as yet been unsuccessful. Recently, crystallization of β chain dimers has been reported (Boulot *et al.* 1994) and the determination of this structure is in progress.

Attempts to produce soluble TCR from the TCR–CD3 membrane complex in a form suitable for structural analysis, for example by proteolytic cleavage from the membrane surface, have as yet been unsuccessful. Similarly, attempts to express the α and β chains without CD3 in eukaryotic cells have led to the identification of cellular retention and degradation signals within the transmembrane region that ensure no free α or β chain reach the membrane surface if they are not complexed with CD3 (see Chapter 17). However, expression of the extracellular portion of the α and β chain without these signals, either as a phosphoinositol-linked membrane protein (Lin *et al.* 1990; Slanetz and Bothwell 1991), or as a chimeric protein with the CD3 zeta chain transmembrane and cytoplasmic regions fused to both the α and β chains (Engel *et al.* 1992), has overcome some of these problems. Using these expression systems, αβ heterodimers are detected on the cell surface and can be purified in a soluble form either by phospholipase C hydrolysis or proteolytic cleavage. Other strategies to produce soluble TCR include secretion of various αβ–antibody chimeras (Gascoigne *et al.* 1987; Mariuzza and Winter 1989; Gregoire *et al.* 1991; Weber *et al.* 1992), and secretion of truncated β chain alone (Gascoigne 1990; Pontzer *et al.* 1992). The yield of soluble protein purified from these systems is relatively low and can be very variable (Weber *et al.* 1992). In addition, there are only two examples of αβ heterodimer formation using the antibody chimera approach (Gregoire *et al.* 1991; Weber *et al.* 1992).

Another approach for the production of soluble TCR in a form suitable for structural analysis is to express the V domains of a TCR α and β chain as a single polypeptide chain in *E. coli*. There are now numerous examples of antibody variable fragments (Fv) that have been expressed in *E. coli* as single chain molecules (scFv) with a flexible polypeptide linker joining the C and N termini of the two variable domains (reviewed in Huston *et al.* 1991). It has been shown that soluble TCR scFv fragments produced in *E. coli* react with anti-idiotypic or anti-clonotypic antibodies (Soo Hoo *et al.* 1992; Kurucz *et al.* 1993; Ward 1992, cited as data not shown), and with haptens (Novotny *et al.* 1991; Ganju *et al.* 1992). Recently we have reported a TCR scFv (VαVβ) fragment of a human receptor expressed in *E. coli* that was shown to directly bind to both *Staphylococcus aureus* superantigen B and the relevant MHC–peptide complex, so demonstrating that the V domains are a functional fragment of the TCR molecule (Hilyard *et al.* 1994). The fraction of TCR scFv that was refolded into the native functional form using this construct is unknown and certainly less than 10%. However, additional single chain constructs involving three domains (Vα, Vβ, and Cβ) produced in both eukaryotic cells and *E. coli* appear to make soluble proteins, of which a high proportion is folded in a functional form (Chung *et al.* 1994, 1995).

3 MHC–peptide structure

There are now crystallographically determined structures of class I MHC–peptide complexes for the human alleles HLA-A2 (Bjorkman *et al.* 1987*a*; Saper *et al.* 1991; Madden *et al.* 1993), HLA-Aw68 (Garrett *et al.* 1989; Guo *et al.* 1992; Silver *et al.* 1992), and HLA-B27 (Madden *et al.* 1991, 1992), and the mouse alleles H-2K^{b} (Fremont *et al.* 1992, Matsumura *et al.* 1992*a*; Zhang *et al.* 1992), and H-2D^{b} (Young *et al.* 1994). Recently the structure of the human class II MHC allele, HLA-DR1, has been solved (Brown *et al.* 1993; Stern *et al.* 1994; Jardetzky *et al.* 1994). The structures of class I and class II MHC are similar even though the molecules have different domain organizations (Figure 18.2). Class I MHC has two polypeptide chains, the α heavy chain and β light chain. The heavy chain forms three domains; α1 and α2 are structurally similar and together form an eight stranded antiparallel β-sheet spanned by two long antiparallel helical regions, and α3 that has an Ig fold. The class I β chain, β2-microglobulin, is a single domain and also has an Ig fold similar to the α3 domain. The α3 and β2-microglobulin domains associate together in a manner similar to that found in antibody structures. The class II MHC has two polypeptide chains of similar length that each form two domains. The α1 and β1 domains form an antiparallel β-sheet/α helical structure homologous to the class I α1 and α2 domains, and the α2 and β2 domains have an Ig fold similar to α3 and β2-microglobulin of class I MHC.

The peptide binding site of both class I and class II MHC is in the α1/α2 or α1/β1 domains respectively. In the crystal structures, the peptide lies in an extended conformation along a groove formed by the β-sheet and the two helices (Figure 18.2). The peptide main chain makes extensive hydrogen bond interactions with residues in the walls and the floor of the groove. In addition, the side chains of specific 'anchor residues' in the peptide are found to bind into allele-specific pockets at the base of the groove. It is these pockets that determine the antigen specificity of the MHC molecule and lead to preferences for certain peptide 'motifs' for the different alleles. In class I MHC, the N- and C-termini of the antigen are buried in deep pockets with hydrogen bonds stabilizing the buried charges, with the ends of the pocket 'closed-off' by conserved residues. The binding of the peptide termini into these pockets contributes greatly to the overall binding affinity of the peptide and has the effect of fixing the orientation of the antigen in the groove and restricting the length of the antigen. Elution of naturally processed peptides from class I MHC molecules on the surface of APC have shown that the peptide are generally eight or nine residues in length (Van Bleek and Nathenson 1990; Rotzschke *et al.* 1990; Falk *et al.* 1991; Jardetzky *et al.* 1991; Hunt *et al.* 1992*a*). Longer peptides, up to 11 residues in length, have been eluted from class I (Guo *et al.* 1992). Crystal structures show that variations in length are accommodated by the middle of the peptide bulging out of the groove and protruding into the solvent (Guo *et al.* 1992; Fremont *et al.* 1992; Madden *et al.* 1993). It has recently been demonstrated that very long peptides, up to 33 residues in length, can be eluted from the class I MHC HLA-B27 (Urban *et al.* 1994). These peptides were eluted from a minor subpopulation of the HLA-B27 molecules on the surface and could only be detected when an unusual antibody, MARB4, that binds only 5–20% of the cell surface HLA-B27, was used to affinity purify the MHC molecules (Urban *et al.* 1994). These peptides may bind to the MHC in a class II-like manner, that is with one or both of the peptide termini outside of the cleft.

Comparisons of the antigen binding site of the class II MHC, HLA-DR1, with class I MHC have shown differences in the secondary structure of the class II helices that create a more open groove and therefore may determine why unlike class I, class II binds longer 13–25 residue peptides (Rudensky *et al.* 1991; Chicz *et al.* 1992; Hunt *et al.* 1992*b*). In addition to the secondary

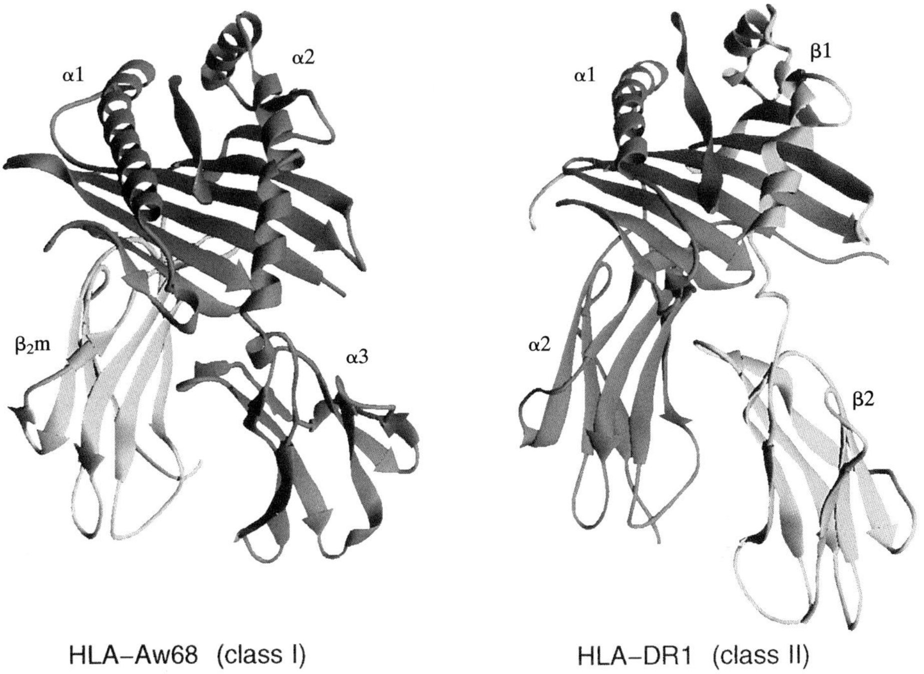

Fig. 18.2 Comparison of MHC class I and class II crystal structures. Ribbon diagrams of class I HLA-Aw68 and class II HLA-DR1 proteins. Green regions are the α domains, yellow regions the β domains. Peptides (blue) bind in the groove formed by the β-sheet and the two α helical regions. Although the domain organization of class I (α1α2α3, β_2m) and class II (α1α2, β1β2) differ, their three-dimensional structures are very similar. β_2m, β2-microglobulin. Produced by Dr Larry Stern. A similar figure appears in Stern and Wiley (1994).

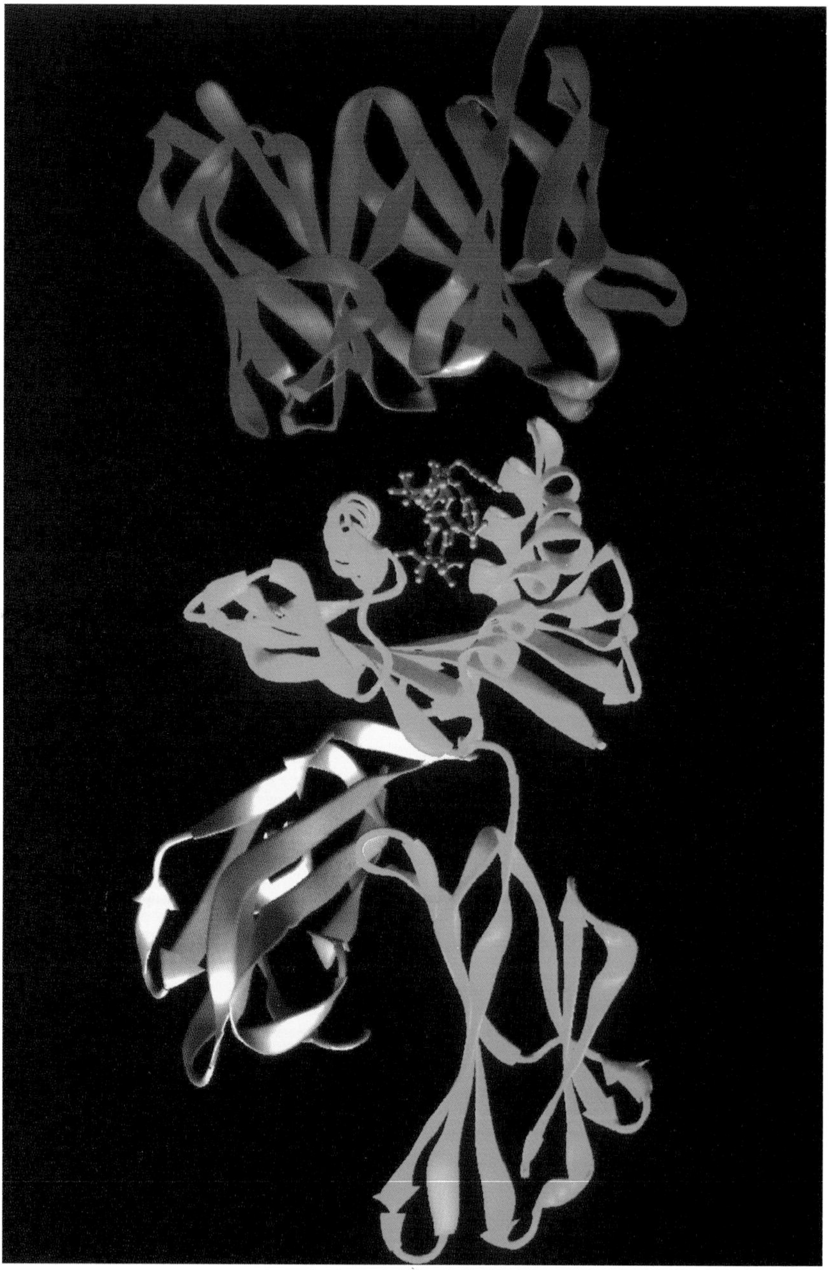

Fig. 18.3 Model of a TCR–peptide–MHC complex. The anti-lysozyme antibody D1.3 VL/VH dimer (PDB:1FDL, red) was docked with the HLA-A2–matrix peptide structure (PDB:1HHI, α chain, green; β2-microglobulin, yellow; peptide, blue). The figure shows the relative disposition of the TCR CDRs (pink) 1 and 2 over the α1/α2 helices and CDR 3 over the peptide. Produced with the help of B. Sherborne (Roche Products Ltd).

structure differences, a number of specific residues that 'close' the ends of the class I groove are not present in the class II site. The conserved residues that bind the peptide N- and C-terminus in the class I structure are also not found in class II sequences. In the structure of HLA-DR1 complexed with a 13 residue influenza virus peptide, the peptide antigen binds as a straight extended chain with a pronounced twist (Stern *et al.* 1994) (Figure 18.2). As with class I, side chains of the peptide are accommodated by polymorphic pockets of the MHC binding groove. Unlike class I, the N- and C-termini extend out of the groove and protrude into the solvent. However, the area of peptide available for direct contact with the TCR (approximately 400–500 $Å^2$) is the same as for several class I–peptide complexes (Stern and Wiley 1994). Other class I–peptide complexes are reported to have an area of 100–300 $Å^2$ of peptide accessible for interaction with the TCR (Fremont *et al.* 1992).

Differences in the peptide binding modes for class I and class II are also evident from *in vitro* peptide binding studies. Loading of peptides on to class I MHC on an APC is thought to normally involve transport of endogenously synthesized protein or peptide into the endoplasmic reticulum where the heavy chain and β2-microglobulin are in the process of folding (see Chapter 20). Attempts to measure MHC class I peptide exchange have had limited success (Chen and Parham 1989). The binding observed (0.3% of the total MHC molecules) was probably due to the low level of 'empty' class I molecules expressed on the cell surface (Schumacher *et al.* 1990). Quantitative binding of peptides to purified MHC has been demonstrated in the presence of excess β2-microglobulin, indicating that peptide exchange is facilitated by β2-microglobulin exchange (Ruppert *et al.* 1993). In general, class I MHC molecules are unstable at 37 °C when they have no peptide antigen bound. However, the number of empty molecules on the cell surface is increased when the cells are incubated at 26 °C (Schumacher *et al.* 1990). Heterologous expression of class I MHC in Drosophila cells which grow at 22 °C produces 'empty' molecules, presumably as the insect cells have no peptide processing machinery (Jackson *et al.* 1992). These molecules have been used to study the kinetics of peptide binding to class I (Matsumura *et al.* 1992*b*) and to produce MHC–peptide complex for crystallographic analysis (Fremont *et al.* 1992). Similar peptide binding assays have been developed using the mouse mutant cell line, RMA-S, which is deficient in peptide processing. This cell line has very low surface expression levels of MHC when cultured at 37 °C. The level of expression can be increased either by culturing in the presence of specific MHC restricted peptide (Townsend *et al.* 1989; Schumacher *et al.* 1990) or by culturing at reduced temperatures to stabilize the empty molecules (Ljunggren *et al.* 1990). MHC class I peptide binding has also been studied with a human cell line that is deficient in antigen processing, T2 (Cerundolo *et al.* 1991). Unlike RMA-S, culturing of T2 at reduced temperatures does not increases the surface expression of MHC class I (Baas *et al.* 1992). This difference may be due to differences in the stability of mouse class I (expressed on RMA-S) compared to human class I (expressed on T2), or to the control mechanisms in the cells that preclude surface expression of empty class I molecules (Baas *et al.* 1992).

An alternative method to obtain purified MHC class I molecules with a single peptide bound is to denature the heavy chain and β2-microglobulin and then refold *in vitro* in the presence of excess peptide (Silver *et al.* 1991; Garboczi *et al.* 1992; Parker *et al.* 1992). This method has been used to produce soluble MHC–peptide complexes for crystallographic structure determination (Garboczi *et al.* 1992; Madden *et al.* 1993). In contrast to the instability of empty class I, class II MHC binds peptide antigen both when on the surface of the APC and when purified (reviewed in Rothbard and Gefter 1991). However, the stability of the class II molecule is increased when peptide is bound (Stern and Wiley 1992).

Another feature of the HLA-DR1 structure that has not been observed for class I MHC is that the molecule was in a dimeric state in the crystal (Brown *et al.* 1993). However, there is as yet no evidence that purified HLA-DR1 is a dimer in solution or that class II dimers can form physiologically in cell membranes. Whether the dimerization was an artefact of the crystallization conditions, or just a particular feature of the HLA-DR1 allele, will require the determination of a number of different class II MHC structures. There is no evidence that MHC class I forms dimers, either in the crystal structures or in solution. However, it is interesting to note that CD8 which is the co-receptor for MHC class I is a dimer, whereas CD4, the co-receptor for class II, is a monomer. It has been proposed that the CD8 dimer could cross-link two class I molecules (Giblin *et al.* 1994). Likewise, if MHC class II were a dimer, binding of CD4 could bring two CD4 molecules into close proximity. If the dimerization of MHC is found to be functionally significant, this will be of great importance in understanding the process of TCR triggering and co-receptor signalling. It has been shown for a number of other receptors (e.g. human growth hormone receptor) that the receptor is stimulated by dimerization induced by ligand binding (De Vos *et al.* 1992).

The fact that the majority of the MHC–peptide binding interactions made are to the peptide main chain atoms, with only a few key anchor residues interacting via side chain atoms, accounts for the very diverse sets of peptides that are able to bind to an MHC allele with high affinity. In peptide elution experiments from class II MHC, over 600 different peptide mass species were eluted from HLA-DR1 (Chicz *et al.* 1992) and 2000 from I-A^d (Hunt *et al.* 1992*b*). For both class I and class II MHC, there is a predominance of peptides derived from endogenously synthesized self-proteins (Jardetzky *et al.* 1991; Rudensky *et al.* 1991; Hunt *et al.* 1992*a*,b; Chicz *et al.* 1992; Guo *et al.* 1992). The TCR therefore has to scan hundreds of different MHC–peptide complexes on the cell surface before engaging in a specific APC–T cell association.

The recent structure determination of HLA-A2 complexed with five different viral peptides (Madden *et al.* 1993) has given insights into the structural features that a TCR has to discern in order to trigger a specific response. In the structures, the HLA-A2 conformation is essentially identical, except for one area of the α2 helix which appears to have two main chain conformations, depending on the peptide present in the groove. Conformational differences in this area of the α2 helix have been observed for the H-2K^b structure with two viral peptides, and were proposed to be important for specific TCR recognition (Fremont *et al.* 1992). However, superimposition of HLA-A2 from peptide complexes that have the same α2 helix conformation indicate that the RMS difference of the binding site structure is less than the co-ordinate error of the structures (Madden *et al.* 1993). There are a few significant conformational adjustments of MHC side chains in direct response to peptide variability. The only potentially accessible side chain is Trp-167, a conserved residue that forms the N-terminal end of the peptide binding groove. This side chain is found in the same position for four of the five structures, but adopts a different position in the fifth complex as a direct consequence of the peptide sequence. Even considering these minor adjustments in MHC structure, it appears for at least some HLA-A2–peptide complexes, the TCR has only differences in the peptide antigen structure to determine specificity.

The five peptides have similar conformations at the N- and C-termini. However, the main and side chain conformations are very different for the four residues at the centre of the peptide. In this region the backbone dramatically adjusts its conformation in response to the sequence of the peptide. In addition, different orientations are seen for the same side chain in identical positions within two peptides. For example, position six in the influenza matrix peptide and the HIV reverse transcriptase peptide is a valine. In the matrix peptide structure the side chain points sideways towards the α1 helix, whereas in the reverse transcriptase peptide it points towards the floor of the peptide binding cleft. Steric interaction between peptide side chains are observed, resulting in one peptide position causing conformational changes at another. Peptide length was also found to affect side chain orientation. Thus, although fixed at the ends, the structure of an MHC bound peptide appears to be dependent on both length and sequence, being sensitive to even small changes in either. Although peptide-specific MHC conformational changes may sometimes contribute to the uniqueness of the complex, the structure of the peptide is the essential determinant of the antigenic identity of the complex.

In addition to making antigen-specific contacts with the peptide, the TCR is also thought to make direct MHC contacts. TCR potentially recognizes the upwardly pointing residues on the MHC α helices. These residues tend to be conserved, with only a few positions showing polymorphism. For example, for class I these positions are 62, 65, and 163 (Bjorkman *et al.* 1987*b*). For class II, DRβ67 may be particularly important as a TCR contact residue (Wucherpfennig *et al.* 1995). It is located in the β1 helix in a similar position to class I α163 on the α2 helix. The MHC specificity of the TCR must therefore be conferred by interaction with these polymorphic residues, with the other conserved residues acting like 'pegs' in the interaction by contributing to the overall affinity. Further details of the trimolecular complex are described in the following section.

4 TCR–peptide–MHC complex model

There are now a wealth of data from cellular immunological studies on the specificity of T cell–APC interactions which have examined the effects of variations in peptide, MHC, or TCR. As yet interpretation of this information at the structural level has had to rely only on three-dimensional models of the TCR–peptide–MHC complex. The models propose that the CDR1 and CDR2 loops of TCR α and β interact with the α helices of the MHC molecule while the CDR3 loops, the most variable region of the TCR, interact directly with the peptide (Figure 18.3). In this model there is a correlation in the degree of variability of the MHC and peptide antigen with the variability of the TCR CDRs. As seen in the MHC crystal structures, the tertiary structure of both α helices of different alleles is relatively conserved. This region is recognized by the CDR1 and CDR2 loops which are themselves relatively conserved due to the size of the V region gene repertoire. In contrast, peptide residues potentially contacting the TCR are very diverse in sequence. In the model this region of the complex is recognized by the CDR3 region of the TCR which is formed by the recombination of the V and J (for the α chain), or V, D, and J (for the β chain) gene segments, with additional variability provided by the random addition of nucleotides by N region diversification, the final result being a highly diverse sequence (Davis and Bjorkman 1988).

In support of the TCR–MHC contacts proposed in the model, amino acid substitutions of a class I molecule have demonstrated that both α helices are simultaneously recognized by a TCR (Ajitkumar *et al.* 1988; Peccoud *et al.* 1990; Jaulin *et al.* 1992). The effects of the mutations varied from one T cell to another, indicating that there are multiple distinct TCR epitopes on the surface of an MHC–peptide complex. Amino acid substitutions in the CDR1 region of a Vα (Nalefski *et al.* 1990) and Vβ chain (White *et al.* 1993; Bellio *et al.* 1994) have identified residues critical for MHC–peptide interaction. Mutagenesis experiments using two class II-specific TCR that differentially recognize polymorphisms in the MHC α chain have shown that MHC specificity can be gained and lost by exchange of the N-terminal portion of the α chain containing the CDR1 and 2 regions (Hong *et al.* 1992).

Indirect evidence that the CDR3 region of the TCR interacts with the peptide antigen has been provided by comparisons of V region sequences. TCR with similar MHC–peptide specificities have been found to have conserved CDR3 regions (Hedrick *et al.* 1988; Acha-Orbea *et al.* 1988; Urban *et al.* 1988; Sorger and Hedrick 1990; Ruberti *et al.* 1993). Likewise, TCR with different fine specificities for peptide have been found to vary only in the

CDR3 region (Winoto *et al.* 1986; Danska *et al.* 1990; Lai *et al.* 1990; Wither *et al.* 1991). Similarly, it has been demonstrated that peptide recognition is affected by mutating the CDR3 region of a TCRβ at a single site (Engel and Hendrick 1988). A more direct approach to studying CDR3–peptide interactions has been taken by Romero *et al.* (1993) using photoaffinity labelling. T cells specific for a peptide antigen–photoreactive label conjugate presented by class I MHC were generated and then labelled. The success of labelling of the receptors correlated with the presence of a tryptophan, a residue known to be reactive with the label, in the CDR3 region of the α chain, thus strongly suggesting that the interaction site for the peptide includes Vα CDR3. As yet there has been no report of mapping of cross-linked reactive groups from MHC restricted peptide antigens on to specific TCR residues, thus providing evidence for direct peptide–TCR interactions.

Experiments using mice transgenic for TCR α or β chains have elegantly shown that changes in the peptide antigen elicit reciprocal changes in the CDR3 regions of the receptor (Jorgensen *et al.* 1992). The mice were immunized with peptides varying at residues that affect T cell recognition but not MHC binding. By using mice transgenic for either TCR α or β, half of the TCR is kept constant in the T cell response. It was observed that peptide changes at one position evoked a better response in the TCR β than the TCR α transgenic mice, with changes at another site a few amino acids away having the reciprocal affect. This suggested that the different parts of the peptide are recognized independently by the Vα and Vβ chains. Sequencing of the V regions of the non-transgenic chain localized the changes in the receptor to be only in the TCR α or TCR β CDR3 region, thus providing strong evidence for direct TCR–peptide contacts. In addition, the data obtained in these experiments suggested an orientation of the TCR on the MHC–peptide complex, that is the TCR α chain interacts with the MHC β chain and the TCR β chain with the MHC α chain. However, other equally compelling experiments have suggested the opposite orientation for the TCR–MHC interaction, namely TCR α interacting with MHC α, and TCR β with MHC β (Hong *et al.* 1992; Wucherpfennig *et al.* 1995). As yet it is not known if there will be a common orientation for the complex, or if it will be different for every TCR–peptide–MHC combination. Furthermore, the TCR–superantigen–MHC complexes may have still different orientations (Hudson *et al.* 1995).

Evidence that all six CDR regions of a TCR are important for the interaction with MHC and peptide has been provided by studying the effects of individual substitutions in each of the CDRs of a class II restricted TCR on antigen recognition (Nalefski *et al.* 1992; Kasibhatla *et al.* 1993). Other data have shown that the CDR regions are not the only determinant required for MHC–peptide recognition. CDR grafting experiments in which the three CDRs of a TCR β chain were expressed on a different Vβ framework did not transfer antigen specificity (Patten *et al.* 1993). In this experiment the two TCR β chains were known to form receptors with a common Vα and recognize similar peptides presented by I-E^K. Analogous CDR grafting experiments with antibody V domains have shown that antigen specificity can be transferred to a different framework, albeit with a reduced affinity (Jones *et al.* 1986). Explanations for why the TCR grafting experiments failed include:

(1) residues outside the CDR regions are crucial for the MHC–peptide interaction;

(2) the CDR loops adopt a different conformation on the two Vβ frameworks;

(3) a reduction in affinity of the TCR–peptide–MHC interaction may result in a lack of signalling.

It has been observed for antibody CDR grafting that framework residues proximal to the CDRs have an effect on CDR conformation. Often these framework residues have to be substituted in addition to the CDR regions in order to reproduce binding affinity and specificity (Morrison 1992). It therefore does not seem unlikely that the same may be true for TCR framework regions. However, a definitive explanation for the results will require the structure determination of the two VαVβ dimers.

Other recent data (Ehrich *et al.* 1993) also suggest that the above model may be over-simplified. Antigen recognition experiments using a group of T cells bearing structurally related TCR and APC expressing mutant MHC molecules and a panel of peptide variants show that the final configuration of the TCR–peptide–MHC complex depends on the antigen. Differences in either the CDR3 region of the TCR or in the peptide affected recognition of MHC variants which had substitutions in the residues on the α helices that point up towards the TCR. It was proposed that TCR–MHC contacts could be made in a variety of ways between the same TCR and MHC, with the final configuration apparently dominated by the antigen. This type of 'induced fit' interaction has been observed for antibody–antigen interactions. The crystal structures of an antibody with and without antigen bound showed that upon antigen binding the CDR3 region of the heavy chain is rearranged resulting in an altered conformation (Rini *et al.* 1992). A certain degree of specificity must be maintained in the TCR–MHC interaction as it has been observed that a single conservative mutation in the MHC α helix can abolish the T cell response (Peccoud *et al.* 1990; Ehrich *et al.* 1993). A similar phenomena has been observed for antibody–antigen binding where a single substitution in the epitope totally abolishes antigen binding (Amit *et al.* 1986).

Even though the data described above are compatible with the model shown in Figure 18.3, there are still results that can not be fully explained. Ultimately, the structures of several different TCR–peptide–MHC complexes will be required in order to determine the general rules of TCR antigen recognition. It must be noted that in all the experiments, the signal measured was T cell triggering and not TCR–MHC binding. The lack of TCR stimulation does not necessarily mean a lack of binding as the relationships between

TCR–MHC binding and signal transduction are still not known (see for example, Vignali and Strominger 1994). It is possible that TCR–MHC engagement could occur without signalling if the kinetics of the interaction or the configuration of the TCR–peptide–MHC complex are such that a signalling complex does not form.

It has been shown for a number of TCR that minor variations in the peptide antigen presented by the MHC can have the effect of being partial agonists or antagonists (reviewed in Evavold *et al.* 1993; Sette *et al.* 1994). As the analogues peptide increases in similarity to the original antigen, the capacity to antagonize the TCR interaction increases, up to a point where the analogues themselves become antigenic (Alexander *et al.* 1993). This can be explained by an affinity-related mechanism whereby a certain affinity is required for signalling through the TCR. Below this level there is sufficient affinity for the MHC-peptide complex to bind the TCR without triggering and thereby prevent interaction with other MHC–peptide molecules, resulting in antagonism. Peptide analogues that interact with the TCR with intermediate affinity may only engage part of the signalling complex and hence result in partial agonism. Another interesting proposal for long peptides bound to class II MHC is the possibility that residues flanking the 'core' region of nine or ten residues (Stern *et al.* 1994), for example positions 13 to 15, may have an important role in recognition of class II MHC–peptide by TCR and in signalling (Vignali and Strominger 1994). In order to further examine these phenomena it will be necessary to measure the direct interaction between TCR and MHC–peptide so that the processes of binding and signalling can be separated, and comparisons of binding affinity and kinetics can be made for the different MHC–analogue peptide complexes. The following section describes the methods that have been developed to measure direct interactions between a TCR and MHC.

5 Methods for measuring TCR–peptide–MHC interactions

There have been three reports demonstrating direct interactions between TCR and MHC. Matsui and co-workers (1991) measured the inhibition of an anti-TCR antibody binding to TCR by soluble class II MHC. Estimates of the dissociation constant for the TCR–MHC class II interaction ranged from 4×10^{-5} M to 6×10^{-5} M. In a different approach, Weber and colleagues (1992) measured inhibition of MHC class II-dependent T cell activation by a soluble TCR–antibody chimeric molecule. The dissociation constant for soluble TCR was estimated to be 5×10^{-6} M. It must be noted that in this case the signal measured was not binding of TCR and MHC, but T cell triggering. We have recently demonstrated direct binding between a soluble TCR and soluble MHC–peptide complex (or superantigen) using a biosensor (Hilyard *et al.* 1994). The V domains of a human TCR were expressed in *E. coli* as a single polypeptide chain and then refolded *in vitro*. After coupling of this material to the biosensor chip it could be demonstrated that there was an interaction with the relevant class I MHC–peptide complex. This binding was only qualitative and no affinity constant could be determined. However, as high concentrations of MHC–peptide complex were required to see the interaction (10^{-5} M), it seems very likely that, in general, the binding constants for TCR binding to MHC class I and class II will be in the order of 10^{-5} M to 10^{-6} M, a figure comparable to that of a naïve antibody.

The principle reasons why direct binding assays for TCR–MHC interactions are difficult are:

1. The interaction has a low affinity and therefore requires high concentrations of ligands to measure binding.
2. The ligands are both membrane proteins and are therefore difficult to obtain at high concentrations in a soluble form.
3. The TCR–MHC complex occurs *in vivo* as a cell–cell interaction with the overall binding affinity having a major contribution from avidity, thus making inhibition experiments more difficult.

The application of biosensor technology to the measurement of biological interactions is likely to provide methods to accurately assay TCR–MHC interactions. Such assays should enable determination of the structural features that contribute to TCR–peptide–MHC specificity and affinity and perhaps begin to explain the phenomena of antigen-dependent differential signalling.

6 Conclusions

This chapter has reviewed our current knowledge of the key recognition element in a T cell interaction with an APC, the TCR–peptide–MHC complex. Unfortunately, structural information on the TCR is restricted to models mainly due to problems in producing protein suitable for crystallographic analysis. However, with the development of additional tools such as anti-TCR antibodies, purification procedures should be more successful.

The structure determination of MHC molecules with single peptides bound has given great insights into the processes of peptide presentation. Especially informative are the structures of HLA-A2 bound to a panel of different peptides which have enabled the identification of features that the TCR has to discriminate between. These features appear to be mainly restricted to the peptide antigen itself, thus making the number of interactions that potentially confer specificity to be relatively limited and may determine the low affinity of the interaction between TCR and MHC–peptide. Considering that a TCR will have at least part of its surface complementary to the MHC α helices, it is feasible that an interaction with a MHC–non-antigenic peptide complex could occur. It is known by experiment that even if such an interaction was to occur, no signal is generated from the TCR. Therefore the crucial factor is whether a TCR signals may not simply be binding or no binding, but may be dependent on the kinetics of the interaction between the TCR and MHC–peptide. The first interactions that occur between an APC and T cell are thought to be between adhesion molecules (Spits *et al.* 1986) with T cell engagement being a secondary event after the TCR has had time to scan the surface MHC–peptide complexes. *In vitro* measurements of the binding kinetics for the adhesion molecules CD48 and CD2 have shown that both the on and off rates of the interaction are fast (van der Merwe *et al.* 1993), although it is possible that the kinetics of the interactions at cell–cell interface may be different due to avidity effects. However, based on the data measured *in vitro*, a model for T cell–APC interactions can be made where the initial adhesion interactions are very fast, allowing for fast exchange of the multiple interactions, but having the effect of holding the cells in proximity and thus allowing the specific low affinity interaction between TCR and MHC–peptide to occur. A slow off rate for the TCR–peptide–MHC interaction should allow for the formation of the signalling complex including CD4 or CD8, $p56^{lck}$ tyrosine kinase, and possibly other kinases such as ZAP-70, which in turn results in an increase in the overall affinity. Thus there may be a certain time threshold of TCR–MHC engagement required in order to achieve signalling. TCR interactions with MHC presenting non-antigenic peptides would not pass this threshold. Likewise, TCR interactions with MHC presenting antagonist peptides may engage for longer than non-antigenic peptide and so cause antagonism, but still not above the threshold required for signalling. Even though models such as this one are easily proposed, the ultimate explanation of these events will require the structure determination of several TCR–peptide–MHC complexes and detailed kinetic analysis of the trimolecular interaction, both of which, unfortunately, will take much longer to achieve.

Acknowledgements

We thank H. Reyburn and A. Lamont for critical reading of the manuscript and L. Stern for the kind gift of the photographs for Figures 18.2.

References

Acha-Orbea, H., Mitchell, D. J., Timmermann, L., Wraith, D. C., Tausch, G. S., Waldor, M. K., *et al.* (1988). Limited heterogeneity of T cell receptors from lymphocytes mediating autoimmune encephalomyelitis allows specific immune intervention. *Cell*, **54**, 263–73.

Ajitkumar, P., Geier, S. S., Kesari, K. V., Borriello, F., Nakagawa, M., Bluestone, J. A., *et al.* (1988). Evidence that multiple residues on both the α-helices of the class I MHC molecule are simultaneously recognized by the T cell receptor. *Cell*, **54**, 47–56.

Alexander, J., Snoke, K., Ruppert, J., Sidney, J., Wall, M., Southwood, S., *et al.* (1993). Functional consequences of engagement of the T cell receptor by low affinity ligands. *J. Immunol.*, **150**, 1–7.

Allen, P. M. (1994). Peptides in positive and negative selection: a delicate balance. *Cell*, **76**, 593–6.

Amit, A. G., Mariuzza, R. A., Phillips, S. E. V., and Poljak, R. J. (1986). Three-dimensional structure of an antigen-antibody complex at 2.8 Å resolution. *Science*, **233**, 747–53.

Baas, E. J., van Santen, H.-M., Kleijmeer, M. J., Geuze, H. J., Peters, P. J., and Ploegh, H. L. (1992). Peptide-induced stabilization and intracellular localisation of empty HLA class I complexes. *J. Exp. Med.*, **176**, 147–56.

Bellio, M., Lone, Y.-C., de la Calle-Martin, O., Malissen, B., Abastado, J-P., and Kourilsky, P. (1994). The Vβ complementarity determining region 1 of a major histocompatibility complex (MHC) class I-restricted T cell receptor is involved in the recognition of peptide/MHC I and superantigen/MHC II complex. *J. Exp. Med.*, **179**, 1087–97.

Bjorkman, P. J., Saper, M. A., Samraoui, B., Bennett, W. S., Strominger, J. L., and Wiley, D. C. (1987*a*). Structure of the human class I histocompatibility antigen, HLA-A2. *Nature*, **329**, 506–12.

Bjorkman, P. J., Saper, M. A., Samraoui, B., Bennett, W. S., Strominger, J. L., and Wiley, D. C. (1987*b*). The foreign antigen binding site and T cell recognition regions of class I histocompatibility antigens. *Nature*, **329**, 512–18.

Boulot, G., Bentley, G. A., Karjalainen, K., and Mariuzza, R. A. (1994). Crystallization and preliminary X-ray diffraction analysis of the β-chain of a T-cell antigen receptor. *J. Mol. Biol.*, **235**, 795–7.

Brown, J. H., Jardetzky, T. S., Gorga, J. C., Stern, L. J., Urban, R. G., Strominger, J. L., *et al.* (1993). Three-dimensional structure of the human histocompatibility antigen HLA-DR1. *Nature*, **364**, 33–9.

Cazenave, P.-A., Marche, P. N., Jouvin-Marche, E., Voegtlé, D., Bonhomme, F., Bandeira, A., *et al.* (1990). Vβ17 gene polymorphism in wild-derived mouse strains: two amino acid substitutions in the Vβ17 region greatly alter T cell receptor specificity. *Cell*, **63**, 717–28.

Cerundolo, V., Elloitt, T., Elvin, J., Bastin, J., Rammensee, H.-G., and Townsend, A. (1991). The binding affinity and dissociation rates of peptides for class I major histocompatibility complex molecules. *Eur. J. Immunol.*, **21**, 2069–75.

Chen, B. P. and Parham, P. (1989). Direct binding of influenza peptides to class I HLA molecules. *Nature*, **337**, 743–5.

Chicz, R. M., Urban, R. G., Lane, W. S., Gorga, J. C., Stern, L. J., Vignali, D. A. A., *et al.* (1992). Predominant naturally processed peptides bound to HLA-DR1 are derived from MHC-related molecules and are heterogeneous in size. *Nature*, **358**, 764–8.

Chothia, C., Boswell, D. R., and Lesk, A. M. (1988). The outline structure of the T-cell αβ receptor. *EMBO J.*, **7**, 3745–55.

Chung, S., Wucherpfennig, K. W., Friedman, S. M., Hafler, D. A., and Strominger, J. L. (1994). Functional three-domain single-chain T-cell receptors. *Proc. Natl. Acad. Sci. USA*, **91**, 12654–8.

Chung, S., Hilyard, K. L., and Strominger, J. L. (1994*b*). Manuscript in preparation.

Claverie, J.-M., Prochnicka-Chalufour, A., and Bougueleret, L. (1989). Implications of a Fab-like structure for the T-cell receptor. *Immunol. Today*, **10**, 10–14.

Danska, J. S., Livingstone, A. M., Paragas, V., Ishihara, T., and Fathman, C. G. (1990). The presumptive CDR3 regions of both T cell receptor α and β chains determine T cell specificity for myoglobin peptides. *J. Exp. Med.*, **172**, 27–33.

Davis, M. M. and Bjorkman, P. J. (1988). T-cell antigen receptor genes and T-cell recognition. *Nature*, **334**, 395–402.

Davies, D. R., Padlan, E. A., and Sheriff, S. (1990). Antibody–antigen complexes. *Annu. Rev. Biochem.*, **59**, 439–73.

de Vos, A. M., Ultsch, M., and Kossiakoff, A. A. (1992). Human growth hormone and extracellular domain of its receptor: crystal structure of the complex. *Science*, **255**, 306–12.

Ehrich, E. W., Devaux, B., Rock, E. P., Jorgenson, J. L., Davis, M. M., and Chien, Y. (1993). T cell receptor interaction with peptide/major histocompatibility complex (MHC) and superantigen/MHC is dominated by antigen. *J. Exp. Med.*, **178**, 713–22.

Engel, I. and Hedrick, S. M. (1988). Site-directed mutations in the VDJ junctional region of a T cell receptor β chain cause changes in antigenic peptide recognition. *Cell*, **54**, 473–84.

Engel, I., Ottenhoff, T. H. M., and Klausner, R. D. (1992). High-efficiency expression and solubilization of functional T cell antigen receptor heterodimers. *Science*, **256**, 1318–20.

Evavold, B. D., Sloan-Lancaster, J., and Allen, P. M. (1993). Tickling the TCR: selective T-cell functions stimulated by altered peptide ligands. *Immunol. Today*, **14**, 602–9.

Falk, K., Rotzschke, O., Stevanovic, S., Jung, G., and Rammensee, H.-G. (1991). Allele-specific motifs revealed by sequencing of self-peptides eluted from MHC molecules. *Nature*, **351**, 290–6.

Fremont, D. H., Matsumura, M., Stura, E. A., Peterson, P. A., and Wilson, I. A. (1992). Crystal structures of two viral peptides in complex with murine MHC class I H-2K^{b}. *Science*, **257**, 919–27.

Ganju, R. K., Smiley, S. T., Bajorath, J., Novotny, J., and Reinherz, E. L. (1992). Similarity between fluorescein-specific T-cell receptor and antibody in chemical details of antigen recognition. *Proc. Natl. Acad. Sci. USA*, **89**, 11552–6.

Garboczi, D. N., Hung, D. T., and Wiley, D. C. (1992). HLA-A2-peptide complexes: refolding and crystallization of molecules expressed in *Escherichia coli* and complexed with single antigenic peptides. *Proc. Natl. Acad. Sci. USA*, **89**, 3429–33.

Garrett, T. P. J., Saper, M. A., Bjorkman, P. J., Strominger, J. L., and Wiley, D. C. (1989). Specificity pockets for the side chains of peptide antigens in HLA-Aw68. *Nature*, **342**, 692–6.

Gascoigne, N. R. J. (1990). Transport and secretion of truncated T cell receptor β-chain occurs in the absence of association with CD3. *J. Biol. Chem.*, **265**, 9296–301.

Gascoigne, N. R. J., Goodnow, C. C., Dudzik, K. I., Oi, V. T., and Davis, M. M. (1987). Secretion of a chimeric T-cell receptor-immunoglobulin protein. *Proc. Natl. Acad. Sci. USA*, **84**, 2936–40.

Giblin, P. A., Leahy, D. J., Mennone, J., and Kavathas, P. B. (1994). The role of charge and multiple faces of the CD8 α/α homodimer in binding to major histocompatibility complex class I molecules: support for a bivalent model. *Proc. Natl. Acad. Sci. USA*, **91**, 1716–20.

Grégoire, C., Rebaï, N., Schweisguth, F., Necker, A., Mazza, G., Auphan, N., *et al.* (1991). Engineered secreted T-cell receptor αβ heterodimers. *Proc. Natl. Acad. Sci. USA*, **88**, 8077–81.

Guo, H.-C., Jardetzky, T. S., Garrett, T. P. J., Lane, W. S., Strominger, J. L., and

Wiley, D. C. (1992). Different length peptides bind to HLA-Aw68 similarly at their ends but bulge out in the middle. *Nature*, **360**, 364–6.

Hedrick, S. M., Engel, I., McElligott, D. L., Fink, P. J., Hsu, M.-L., Hansburg, D., *et al.* (1988). Selection of amino acid sequences in the beta chain of the T cell antigen receptor. *Science*, **239**, 1541–4.

Hilyard, K. L. (1991). Protein engineering of antibody combining sites. D.Phil.Thesis, Oxford University.

Hilyard, K. L., Reyburn, H., Chung, S., Bell, J. I., and Strominger, J. L. (1994). Binding of soluble natural ligands to a soluble human T-cell receptor fragment produced in *Escherichia coli. Proc. Natl. Acad. Sci. USA*, **91**, 9057–61.

Hong, S.-C., Chelouche, A., Lin, R., Shaywitz, D., Braunstein, N. S., Glimcher, L., *et al.* (1992). An MHC interaction site maps to the amino-terminal half of the T cell receptor α chain variable domain. *Cell*, **69**, 999–1009.

Hudson, K. R., Lowe, S. C., Urban, R. G., and Fraser, J. D. (1995). Location of the zinc bridge between Staphylococcal enterotoxin A and major histocompatability class II antigen: a structural model for superantigen T cell activation. Submitted for publication.

Hunt, D. F., Henderson, R. A., Shabanowitz, J., Sakaguchi, K., Michel, H., Sevilir, N., *et al.* (1992*a*). Characterization of peptides bound to the class I MHC molecule HLA-A2.1 by mass spectrometry. *Science*, **255**, 1261–6.

Hunt, D. F., Hanspeter, M., Dickinson, T. A., Shabanowitz, J., Cox, A. L., Sakaguchi, K., *et al.* (1992*b*). Peptides presented to the immune system by the murine class II major histocompatibility complex molecule I-A^d. *Science*, **256**, 1817–20.

Huston, J. S., Mudgett-Hunter, M., Tai, M. S., McCartney, J., Warren, F., Haber, E., *et al.* (1991). Protein engineering of single-chain Fv analogs and fusion proteins. *Methods Enzymol.*, **203**, 46–88.

Irwin, M. J., Hudson, K. R., Ames, K. T., Fraser, J. D., and Gascoigne, N. R. J. (1993). T-cell receptor β-chain binding to enterotoxin superantigens. *Immunol. Rev.*, **131**, 61–78.

Jackson, M. R., Song, E. S., Yang, Y., and Peterson, P. A. (1992). Empty and peptide-containing confomers of class I major histocompatibility complex molecules expressed in *Drosophila melanogaster* cells. *Proc. Natl. Acad. Sci. USA*, **89**, 12117–21.

Jardetzky, T. S., Lane, W. S., Robinson, R. A., Madden, D. R., and Wiley, D. C. (1991). Identification of self peptides bound to purified HLA-B27. *Nature*, **353**, 326–9.

Jardetzky, T. S., Brown, J. H., Gorga, J. C., Stern, L. J., Urban, R. G., Chi, Y., *et al.* (1994). Three-dimensional structure of a human class II histocompatibility molecule complexed with superantigen. *Nature*, **368**, 711–18.

Jaulin, C., Casanova, J.-L., Romero, P., Luescher, I., Cordey, A.-S., Maryanski, J. L., *et al.* (1992). Highly diverse T cell recognition of a single *Plasmodium berghei* peptide presented by a series of mutant H-$2K^d$ molecules. *J. Immunol.*, **149**, 3990–4.

Jones, P. T., Dear, P. H., Foote, J., Neuberger, M. S., and Winter, G. (1986). Replacing the complementarity-determining regions in a human antibody with those from a mouse. *Nature*, **321**, 522–5.

Jores, R., Alzari, P. M., and Meo, T. (1990). Resolution of hypervariable regions in T-cell receptor β chains by a modified Wu-Kabat index of amino acid diversity. *Proc. Natl. Acad. Sci. USA*, **87**, 9138–42.

Jorgensen, J. L., Esser, U., Fazekas de St. Groth, B., Reay, P. A., and Davis, M.

M. (1992). Mapping T-cell receptor-peptide contacts by variant peptide immunization
of single-chain transgenics. *Nature*, **355**, 224–30.

Julius, M., Maroun, C. R., and Haughn, L. (1993). Distinct roles for CD4 and CD8 as co-receptors in antigen receptor signalling. *Immunol. Today*, **14**, 177–83.

Kasibhatla, S., Nalefski, E. A., and Rao, A. (1993). Simultaneous involvement of all six predicted antigen binding loops of the T cell receptor in recognition of the MHC/antigenic peptide complex. *J. Immunol.*, **151**, 3140–51.

Kurucz, I., Jost, C. R., George, A. J. T., Andrew, S. M., and Segal, D. M. (1993). A bacterially expressed single-chain Fv construct from the 2B4 T-cell receptor. *Proc. Natl. Acad. Sci. USA*, **90**, 3830–4.

Lai, M.-Z., Jang, Y.-J., Chen, L.-K., and Gefter, M. L. (1990). Restricted V-(D)-J junctional regions in the T cell response to lambda repressor. Identification of residues critical for antigen recognition. *J. Immunol.*, **144**, 4851–6.

Lin, A. Y., Devaux, B., Green, A., Sagerström, C., Elliot, J. F., and Davis, M. M. (1990). Expression of T cell antigen receptor heterodimers in a lipid-linked form. *Science*, **249**, 677–9.

Ljunggren, H.-G., Stam, N. J., Ohlen, C., Neefjes, J. J., Hoglund, P., Heemels, M.-T., *et al.* (1990). Empty MHC class I molecules come out in the cold. *Nature*, **346**, 476–80.

Madden, D. R., Gorga, J. C., Strominger, J. L., and Wiley, D. C. (1991). The structure of HLA-B27 reveals nonamer self-peptides bound in an extended conformation. *Nature*, **353**, 321–5.

Madden, D. R., Gorga, J. C., Strominger, J. L., and Wiley, D. C. (1992). The three-dimensional structure of HLA-B27 at 2.1Å resolution suggests a general mechanism for tight peptide binding to MHC. *Cell*, **70**, 1035–48.

Madden, D. R., Garboczi, D. N., and Wiley, D. C. (1993). The antigenic identity of peptide-MHC complexes: a comparison of the conformations of five viral peptides presented by HLA-A2. *Cell*, **75**, 693–708.

Mariuzza, R. A. and Winter, G. (1989). Secretion of a homodimeric VαCκ T-cell receptor-immunoglobulin chimeric protein. *J. Biol. Chem.*, **264**, 7310–16.

Martin, A. C. R., Cheetham, J. C., and Rees, A. R. (1989). Modeling antibody variable loops: a combined algorithm. *Proc. Natl. Acad. Sci. USA*, **86**, 9268–72.

Matsui, K., Boniface, J. J., Reay, P. A., Schild, H., Fazekas de St. Groth, B., and Davis, M. M. (1991). Low affinity interaction of peptide-MHC complexes with T cell receptors. *Science*, **254**, 1788–91.

Matsumura, M., Fremont, D. H., Peterson, P. A., and Wilson, I. A. (1992*a*). Emerging principles for the recognition of peptide antigens by MHC class I molecules. *Science*, **257**, 927–34.

Matsumura, M., Saito, Y., Jackson, M. R., Song, E. S., and Peterson, P. A. (1992*b*). *In vitro* peptide binding to soluble empty class I major histocompatibility complex molecules isolated from transfected *Drosophila melanogaster* cells. *J. Biol. Chem.*, **267**, 23589–95.

Morrison, S. L. (1992). *In vitro* antibodies: strategies for production and application. *Annu. Rev. Immunol.*, **10**, 239–65.

Nalefski, E. A., Wong, J. G. P., and Rao, A. (1990). Amino acid substitutions in the first complementarity-determining region of a murine T-cell receptor α chain affect antigen-major histocompatibility complex recognition. *J. Biol. Chem.*, **265**, 8842–6.

Nalefski, E. A., Kasibhatla, S., and Rao, A. (1992). Functional analysis of the antigen binding site on the T cell receptor α chain. *J. Exp. Med.*, **175**, 1553–63.

Novotny, J., Bruccoleri, R. E., Newell, J., Murphy, D., Haber, E., and Karplus, M.

(1983). Molecular anatomy of the antibody binding site. *J. Biol. Chem.*, **258**, 14433–7.

Novotny, J., Tonegawa, S., Saito, H., Kranz, D. M., and Eisen, H. (1986). Secondary, tertiary, and quaternary structure of T-cell-specific immunoglobulin-like polypeptide chains. *Proc. Natl. Acad. Sci. USA*, **83**, 742–6.

Novotny, J., Ganju, R. K., Smiley, S. T., Hussey, R. E., Luther, R. E., Recny, M. A., *et al.* (1991). A soluble, single-chain T-cell receptor fragment endowed with antigen-combining properties. *Proc. Natl. Acad. Sci. USA*, **88**, 8646–50.

Parker, K. C., Carreno, B. M., Sestak, L., Utz, U., Biddison, W. E., and Coligan, J. E. (1992). Peptide binding to HLA-A2 and HLA-B27 isolated from *Escherichia coli*. *J. Biol. Chem.*, **267**, 5451–9.

Patten, P., Yokota, T., Rothbard, J., Chien, Y.-H., Arai, K-I., and Davis, M. M. (1984). Structure, expression and divergence of T-cell receptor β-chain variable regions. *Nature*, **312**, 40–8.

Patten, P. A., Rock, E. P., Sonoda, T., Fazekas de St. Groth, B., Jorgensen, J. L., and Davis, M. M. (1993). Transfer of putative complementarity-determining region loops of T cell receptor V domains confers toxin reactivity but not peptide/MHC specificity. *J. Immunol.*, **150**, 2281–94.

Peccoud, J., Dellabona, P., Allen, P., Benoist, C., and Mathis, D. (1990). Delineation of antigen contact residues on an MHC class II molecule. *EMBO J.*, **9**, 4215–23.

Pontzer, C. H., Irwin, M. J., Gascoigne, N. R. J., and Johnson, H. M. (1992). T-cell antigen receptor binding sites for microbial superantigen staphylococcal enterotoxin. *Proc. Natl. Acad. Sci. USA*, **89**, 7727–31.

Pullen, A. M., Wade, T., Marrack, P., and Kappler, J. W. (1990). Identification of the region of T cell receptor β chain that interacts with the self-superantigen Mls-1^a. *Cell*, **61**, 1365–74.

Rini, J. M., Schulze-Gahmen, U., and Wilson, I. A. (1992). Structural evidence for induced fit as a mechanism for antibody-antigen recognition. *Science*, **255**, 959–65

Romero, P., Casanova, J.-L., Cerottini, J.-C., Maryanski, J. L., and Luescher, I. F. (1993). Differential T cell receptor photoaffinity labeling among H-$2K^d$ restricted cytotoxic T lymphocyte clones specific for a photoreactive peptide derivative. Labeling of the α chain correlates with Jα segment usage. *J. Exp. Med.*, **177**, 1247–56.

Rothbard, J. B. and Gefter, M. L. (1991). Interactions between immunogenic peptides and MHC proteins. *Annu. Rev. Immunol.*, **9**, 527–65.

Rotzschke, O., Falk, K., Deres, K., Schild, H., Norda, M., Metzger, J., *et al.* (1990). Isolation and analysis of naturally processed viral peptides as recognized by cytotoxic T cells. *Nature*, **348**, 252–4.

Ruberti, G., Paragas, V., Kim, D., and Fathman, C. G. (1993). Selection for amino acid sequence and J beta element usage in the beta chain of DBA/2V beta b- and DBA/2V beta a-derived myoglobin-specific T cell clones. *J. Immunol.*, **151**, 6185–94.

Rudensky, A. Y., Preston-Hurlburt, P., Hong, S.-C., Barlow, A., and Janeway, C. A. Jr. (1991). Sequence analysis of peptides bound to MHC class II molecules. *Nature*, **353**, 622–7.

Ruppert, J., Sidney, J., Kubo, R. T., Grey, H. M., and Sette, A. (1993). Predominant role of secondary anchor residues in peptide binding to HLA-A2.1 molecules. *Cell*, **74**, 929–37.

Saper, M. A., Bjorkman, P. J., and Wiley, D. C. (1991). Refined structure of the human histocompatibility antigen HLA-A2 at 2.6 Å resolution. *J. Mol. Biol.*, **219**, 277–319.

Schumacher, T. N. M., Heemels, M.-T., Neefjes, J. J., Kast, W. M., Melief, C. J.

M., and Ploegh, H. L. (1990). Direct binding of peptide to empty MHC class I molecules on intact cells and *in vitro*. *Cell*, **62**, 563–7.

Searle, S. and Rees, A. R. (1994). Manuscript in preparation.

Sette, A., Alexander, J., Ruppert, J., Snoke, D., Franco, A., Ishioka, G., *et al.* (1994). Antigen analogs/MHC complexes as specific T cell receptor antagonists. *Annu. Rev. Immunol.*, **12**, 413–31.

Sheriff, S., Silverton, E. W., Padlan, E. A., Cohen, G. H., Smith-Gill, S. J., Finzel, B. C., *et al.* (1987). Three-dimensional structure of an antibody-antigen complex. *Proc. Natl. Acad. Sci. USA*, **84**, 8075–9.

Silver, M. L., Parker, K. C., and Wiley, D. C. (1991). Reconstitution by MHC-restricted peptides of HLA-A2 heavy chain with β2-microblobulin, *in vitro*. *Nature*, **350**, 619–22.

Silver, M. L., Guo, H.-C., Strominger, J. L., and Wiley, D. C. (1992). Atomic structure of a human MHC molecule presenting an influenza virus peptide. *Nature*, **360**, 367–9.

Slanetz, A. E. and Bothwell, A. L. M. (1991). Heterodimeric, disulfide-linked α/β T cell receptors in solution. *Eur. J. Immunol.*, **21**, 179–83.

Soo Hoo, W. F., Lacy, M. J., Denzin, L. K., Voss, E. W. Jr., Hardman, K. D., and Kranz, D. M. (1992). Characterization of a single-chain T-cell receptor expressed in *Escherichia coli*. *Proc. Natl. Acad. Sci. USA*, **89**, 4759–63.

Sorger, S. B. and Hedrick, S. M. (1990). Highly conserved T-cell receptor junctional regions. Evidence for selection at the protein and the DNA level. *Immunogenetics*, **31**, 118–22.

Spits, H., van Schooten, W., Keizer, H., van Seventer, G., van de Rijn, M., Terhorst, C., *et al.* (1986). Alloantigen recognition is preceded by nonspecific adhesion of cytotoxic T cells and target cells. *Science*, **232**, 403–5.

Stern, L. J. and Wiley, D. C. (1992). The human class II MHC protein HLA-DR1 assembles as empty αβ heterodimers in the absence of antigenic peptide. *Cell*, **68**, 465–77.

Stern, L. J. and Wiley, D. C. (1994). Antigenic peptide binding by class I and class II histocompatibility proteins. *Structure*, **2**, 245–51.

Stern, L. J., Brown, J. H., Jardetzky, T. S., Gorga, J. C., Urban, R. G., Strominger, J. L., *et al.* (1994). Crystal structure of the human class II MHC protein HLA-DR1 complexed with an influenza virus peptide. *Nature*, **368**, 215–21.

Townsend, A., Ohlén, C., Bastin, J., Ljunggren, H.-G., Foster, L., and Kärre, K. (1989). Association of class I major histocompatibility heavy and light chains induced by viral peptides. *Nature*, **340**, 443–8.

Urban, J. L., Kumar, V., Kono, D. H., Gomez, C., Horvath, S. J., Clayton, J., *et al.* (1988). Restricted use of T cell receptor V genes in murine autoimmune encephalomyelitis raises the possibilities for antibody therapy. *Cell*, **54**, 577–92.

Urban, R. G., Chicz, R. M., Lane, W. S., Strominger, J. L., Rehm, A., Kenter, M. J. H., *et al.* (1994). A subset of HLA-B27 molecules contains peptides much longer than nonamers. *Proc. Natl. Acad. Sci. USA*, **91**, 1534–8.

Van Bleek, G. M. and Nathenson, S. G. (1990). Isolation of an endogenously processed immunodominant viral peptide from the class I H-2K^b molecule. *Nature*, **348**, 213–16.

van der Merwe, P. A., Brown, M. H., Davis, S. J., and Barclay, A. N. (1993). Affinity and kinetic analysis of the interaction of the cell adhesion molecules rat CD2 and CD48. *EMBO J.*, **12**, 4945–54.

Vignali, D. A. A. and Strominger, J. L. (1994). Amino acid residues that flank core peptide epitopes and the extracellular domains of CD4 modulate differential signaling through the T cell receptor. *J. Exp. Med.,* **179**, 1945–56.

Vignali, D. A. A., Doyle, C., Kinch, M. S., Shin, J., and Strominger, J. L. (1993). Interactions of CD4 with MHC class II molecules, T cell receptor and p56*lck*. *Phil. Trans. R. Soc. Lond. B*, **342**, 13–24.

von Boehmer, H. (1992). Thymic selection: a matter of life and death. *Immunol. Today*, **13**, 454–8.

Ward, E. S. (1992). Secretion of T cell receptor fragments from recombinant *Escherichia coli* cells. *J. Mol. Biol.*, **224**, 885–90.

Weber, S., Traunecker, A., Oliveri, F., Gerhard, W., and Karjalainen, K. (1992). Specific low-affinity recognition of major histocompatibility complex plus peptide by soluble T-cell receptor. *Nature*, **356**, 793–6.

Wederburn, L. R., Searle, S. J. M., Rees, A. R., Lamb, J. R., and Owen, M. J. (1995). Mapping T cell recognition: the identification of a T-cell receptor residue critical to the specific interaction with an influenza haemagglutinin peptide. *Eur. J. Immunol.* Submitted for publication.

White, J., Pullen, A., Choi, K., Marrack, P., and Kappler, J. W. (1993). Antigen recognition properties of mutant $V\beta3^+$ T cell receptors are consistent with an immunoglobulin-like structure for the receptor. *J. Exp. Med.*, **177**, 119–25.

Winoto, A., Urban, J. L., Lan, N. C., Goverman, J., Hood, L., and Hansburg, D. (1986). Predominant use of a $V\alpha$ gene segment in mouse T-cell receptors for cytochrome *c*. *Nature*, **324**, 679–82.

Wither, J., Pawling, J., Phillips, L., Delovitch, T., and Hozumi, N. (1991). Amino acid residues in the T cell receptor CDR3 determine the antigenic reactivity patterns of insulin-reactive hybridomas. *J. Immunol.*, **146**, 3513–22.

Wucherpfennig, K. W., Hafler, D. A., and Strominger, J. L. (1995). Structure of human T cell receptors specific for an immunodominant myelin basic protein peptide: positioning of T cell receptors on HLA-DR2/peptide complexes. *Proc. Natl. Acad. Sci. USA*. In press.

Young, A. C. M., Zhang, W., Sacchettini, J. C., and Nathenson, S. G. (1994). The three-dimensional structure of H-$2D^b$ at 2.4 Å resolution: implications for antigen-determinant selection. *Cell*, **76**, 39–50.

Zhang, W., Young, A. C., Imarai, M., Nathenson, S. G., and Sacchettini, J. C. (1992). Crystal structure of the major histocompatibility complex class I H-$2K^b$ molecule containing a single viral peptide: implications for peptide binding and T-cell receptor recognition. *Proc. Natl. Acad. Sci. USA*, **89**, 8403–7.

19 A structural model of bacterial superantigen binding to MHC class II and T cell receptors

JOHN D. FRASER AND KEITH R. HUDSON

1 Introduction

The generation of diversity in T cell receptor (TCR) molecules has predominantly been in response to the large number of peptides bound within major histocompatibility complex (MHC) molecules. This has resulted in the concentration of amino acid diversity within the third complementarity determining regions (CDR3) of the antigen binding surface encoded exclusively by the D and J region gene segments. In comparison, the CDR1 and CDR2 regions which are encoded exclusively by TCR V gene segments have developed much less diversity because these regions of the TCR are involved in an essential interaction with the polymorphic region of MHC.

The TCR Vβ domain has another prominent region of hypervariability called HV4 which is thought not to be part of the normal antigen binding surface of the TCR. One of the functions of this region is in the recognition of *superantigens* (SAgs)—proteins which bind simultaneously to MHC class II and TCRs as whole molecules and stimulate large numbers of T cells at femtomolar concentrations (Peavey *et al.* 1970). The best known SAgs are the staphylococcal enterotoxins and structurally unrelated products of the mouse mammary tumour virus (MMTV), commonly known as Mls (for reviews see Janeway *et al.* 1989; Marrack and Kappler 1990; Herman *et al.* 1991*b*). Although MHC class II antigens are an absolute requirement in SAg recognition, the extreme hypervariability which normally governs T cell antigen specificity, does not as a general rule, influence reactivity to SAgs. This is reflected in an almost universal response by any T cell expressing a particular TCR Vβ domain irrespective of either the TCR Vα, Dβ, Jα, and Jβ combination or the MHC class II allele. This contradiction has led to considerable speculation about the need for MHC class II in SAg activation. This chapter details some of the properties of SAgs (listed in Table 19.1) and focuses on staphylococcal enterotoxin A (SEA), the most potent of all the bacterial SAgs. From current data, it is now possible to predict how SEA may be structurally oriented on MHC class II. This in turn, provides some interesting and provocative insights into TCR and MHC class II interactions in general.

Table 19.1 Functional differences between peptide antigens and superantigens

Function	Classical antigen	Bacterial superantigens
Antigen	Small peptide fragments	Intact molecule (22–27 kDa)
MHC binding site	Inside the peptide groove	Conserved region outside the groove
MHC polymorphism	Restricted	Unrestricted
MHC requirement	Class I and class II	Class II only
MHC isotype binding	Restricted	DR > DQ > DP I–E > I–A
T cells responding	1 in 100 000	5–20%
TCR binding site	CDR1, CDR2, CDR3 combined	VβHV4 only

2 Bacterial superantigens

2.1 The staphylococcal and streptococcal enterotoxins

2.1.1 The staphylococcal enterotoxins—prototype superantigens

Much of the seminal work on the identification, purification, and characterization of all the staphylococcal enterotoxins was performed by Merlin Bergdoll and colleagues at the Food Research Institute, Wisconsin and there are several comprehensive reviews by this group on the SEs (Bergdoll 1983; Spero *et al.* 1988; Blomster-Hautamaa and Schlievert 1988).

The enterotoxins produced by *Staphylococcus aureus* consist of eight homologous polypeptides—A, B, C1, C2, C3, D, E, and TSST (toxic shock syndrome toxin). SEA-E cause staphylococcal food poisoning which is characterized by the onset of vomiting and diarrhoea one to six hours after ingestion of food contaminated with *S. aureus* bacteria. Toxic shock syndrome toxin is not enterotoxic but nevertheless causes a life-threatening illness associated with vaginal colonization in menstruating women by *S. aureus* strains producing TSST (Todd *et al.* 1978; Bergdoll *et al.* 1981; Chesney *et al.* 1984). Toxic shock is characterized by the onset of severe fever, shock, and eventual multisystem collapse leading to death in a high proportion of cases, even with antibiotic treatment.

2.1.2 The streptococcal pyrogenic exotoxins

At least three SAgs are also produced by *Streptococcus pyogenes*, called streptococcal pyrogenic exotoxin (SPE) A, B, and C (Wannamaker and Schlievert 1988). They show considerable homology to the staphylococcal enterotoxins suggesting that the genes for these SAgs have been transferred between bacterial species (Figure 19.1). The SPEs were initially characterized by their ability to enhance susceptibility to lethal endotoxin shock—hence their name. SPE-A is known to cause Scarlet fever, another disease with

RESIDUES IN THE MHC CLASS II AND TCR BINDING SITES

```
                             **  oo ooo o                      o  o **
SEA   SEKSEEINEKDLRKKSELQGTALGNLKQIYYYNEKAKTENKESHDQFLQHTILFKGFFTDHSWYNDLLVDFDSKDIVDKY.KGK
SEE   ...SEEINEKDLRKKSELQRNALSNLRQIYYYNEKAITENKESDDQFLENTLLFKGFFTGHPWYNDLLVDLGSKDATNKY.KGK
SED   .....SVKEKELHKKSELSSTALNNMKHSYADKNPIIGENKSTGDQFLENTLLYKKFFTDLINFEDLLINFNSKEMAQHF.KSK

SEB   .ESQPDPKPDELHKSSKFTGL.MENMKVLYDDNHVSAINVK.SIDQFLYFDLIYSIKDTKLGNYDNVRVEFKNKDLADKY.KDK
SEC1  .ESQPDPTPDELHKASKFTGL.MENMKVLYDDHYVSATKVK.SVDKFLAHDLIYNISDKKLKNYDKVKTELLNEGLAKKY.KDE
SEC3  .ESQPDPTPDELHKSSEFTGT.MGNMKYLYDDHYVSATMVK.SVDKFLAHDLIYNISDKKLKNYDKVKTELLNEDLAKKY.KDE
SPE-A ...QQDPDPSQLHRSS.LVKN.LQNIYFLYEGDPVTHENVK.SVDQLLSHDLIYNVSG...PNYDKLKTELKNQEMATLF.KDK
SPE-C ...DSKKDISNV.KS.DLL.YAY.TIT.PYD..YKDCRVNFSTTHT.LNIDTQ.KYRGKD..YY..ISSEM.SYEASQKFKRDD

TST   ...................ST.NDNIKDLLDWYS.SGSDTF.SNSEVLDNSLGSM.RIKN.TDGSISLIIFPSPYYSPAFTKGD
               |         |         |         |         |         |          |         |
                                                      50


            oooooo                   o
SEA   KVDLYGAYYGYQCAGGT..........PNKTACMYGGVTLHDNNRLTEEKKVPINL..WLDGKQNTVPLETVKTNKKNVTVQE
SEE   KVDLYGAYYGYQCAGGT..........PNKTACMYGGVTLHDNNRLTEEKKVPINL..WIDGKQTTVPIDKVKTSKKEVTVQE
SED   NVDVYPIRYSINCYGGE..........IDRTACTYGGVTPHEGNKLKERKKIPINL..WINGVQKEVSLDKVQTDKKNVTVQE

SEB   YVDVFGANYYYQCYFSKKTNDINSHQTDKRKT.CMYGGVTEHNGNQL.DKYR.SITVRVFEDGKN.LLSFD.VQTNKKKVTAQE
SEC1  VVDVYGSNYYVNCYFSSKDN...VGKVTGGKT.CMYGGITKHEGNHFDNGNLQNVLIRVYENKRN.TISFE.VQTDKKSVTAQE
SEC3  VVDVYGSNYYVNCYFSSKDN...VGKVTGGKT.CMYGGITKHEGNHFDNGNLQNVLIRVYENKRN.TISFE.VQTDKKSVTAQE
SPE-A NVDIYGVEYYHLCYLCENA..........ERSACIYGGVTNHEGNHLEIPKK..IVVKVSIDGIQ.SLSFD.IETNKKMVTAQE
SPE-C HVDVFGLFYILNSHTGE................YIYGGITPAQNNKVNHKLLGNLFI..SGESQQN..LNNKIILEKDIVTFQE

TST   KVDLNTKRTKKSNHTSE..........GTYIHFQISGVTNTE..KLPTPIELPLKVKVHG..KDSPLKYG.PKFDKKQLAIST
            |         |         |         |         |         |         |         |
                     100                                               150


                                      Zn                                        Zn
                                      |                  **o  o                |  |
SEA   LDLQARRYLQEKYNLYNSDVFDGKVQRGLIVFHTSTEPSVNYDLFGAQGQYSNT..LLRIYRDNKTINSE.NMHIDIYLYTS
SEE   LDLQARHYLHGKFGLYNSDSFGGKVQRGLIVFHSSEGSTVSYDLFDAQGQYPDT..LLRIYRDNKTINSE.NLHIDLYLYTT
SED   LDAQARRYLQKDLKLYNNDTLGGKIQRGKIEFDSSDGSKVSYDLFDVKGDFPE..KQLRIYSDNKTLSTE.HLHIDIYLYEK

SEB   LDYLTRHYLVKNKKLY..EFNNSPYETGYIKFIENE.NSFWYDMMPAPGDKFDQSKYLMMYNDNKMVDSK.DVKIEVYLTTKKK
SEC1  LDIKARNFLINKKNLY..EFNSSPYETGYIKFIENNGNTFWYDMMPAPGDKFDQSKYLMMYNDNKTVDSK.SVKIEVHLTTKNG
SEC3  LDIKARNFLINKKNLY..EFNSSPYETGYIKFIENNGNTFWYDMMPAPGDKFDQSKYLMMYNDNKTVDSK.SVKIEVHLTTKNG
SPE-A LDYKVRKYLTDNKQLY..TNGPSKYETGYIKFIPKNKESFWFDFFPEP.EFT.QSKYLMIYKDNETLDSN.TSQIEVYLTTK
SPE-C IDFKIRKYLMDNYKIY..DA.TSPYVSGRIEIGTKDGKHEQIDLFDSPNEGT.RSDIFAKYKDNRIINMKNFSHFDIYL.EK

TST   LDFQIRHQLTQIHGLY....RSSDKTGGYWKITMNDGSTYQSDLSKKF.EYN.TEKPPINIDEIKTIEAEIN
       |         |         |         |         |         |         |         |         |
                                    200                                               250
```

Fig. 19.1 Alignment of protein sequences of the staphylococcal and streptococcal toxins. Those residues predicted to contribute to the TCR binding site from the crystal structure of SEB are indicated with open circles above the sequences (Swaminathan *et al.* 1992). Those residues identified by mutational analysis as part of the Vβ binding site are indicated by asterisks (Kappler *et al.* 1992; Hudson *et al.* 1993; Mollick *et al.* 1993). The zinc binding residues which define the MHC class II binding site of the group 1 toxins SEA, SEE, and SED are also identified.

symptoms similar to toxic shock syndrome. Other streptococcal diseases related directly to SAg intoxication include the recently described streptococcal-related toxic shock, a severe, life-threatening form of toxic shock syndrome associated with a new pathogenic strain of *S. pyogenes* that has recently appeared in the United States (Cone *et al.* 1987). Recent data has also linked Kawasaki's disease directly to SAg activation with Vβ2 expansion in acute phase patients (Abe *et al.* 1992). The source of the SAg in this disease has not yet been identified but could well be linked to *S. pyogenes* infection.

2.1.3 Amino acid homology divides the bacterial toxins into three groups

The staphylococcal and streptococcal toxins are single polypeptides between 27–28 kDa (except TSST which is 22 kDa) and all but TSST and SPE-C contain a small disulfide loop in the central part of their sequence. They are very stable to most forms of denaturation and extremely resistant to all known proteases although SEB and SEC have a single hypersensitive tryptic cleavage site within the disulfide loop which does not affect SAg activity (Spero and Morlock 1978). However, complete reduction and alkylation of the single disulfide leads to a loss in T cell stimulation but not MHC class II binding (Grossman *et al.* 1991). Denaturation by chaotrophic agents such as urea or guanidinium is completely reversible. Together, these properties reflect compact, hydrated, globular proteins held together by strong intramolecular forces.

The aligned amino acid sequences of all the known staphylococcal and streptococcal toxins is provided in Figure 19.1. Three clear groups emerge from amino acid sequence comparison and a family tree is provided in Figure 19.2 to indicate the degree of divergence of each member. SEA, SED, and SEE form group 1 with approximately 80% homology while group 2 consists of SEB, SEC1, SEC2, SEC3, SPE-A, and SPE-C. There is about 30% homology between groups 1 and 2. Group 3 consists solely of TSST which shares only very limited homology with the other toxins although there is enough sequence conservation to suggest that TSST represents a very early divergence from a primordial toxin gene. The division of toxins into these groups is not simply an academic exercise. Several important functional differences segregate with these groupings which are discussed in detail below.

2.2 Superantigens from other sources

2.2.1 Mycoplasma arthritidis

A secreted mitogen MAM, has been described from *Mycoplasma arthritidis*, a bacterium which causes infectious arthritis in rats (Atkin *et al.* 1986; Cole *et al.* 1989) and like other SAgs, MAM requires MHC class II for activity and enriches T cells based on the Vβ domain (Table 19.2). Unlike the SEs however, MAM appears quite unstable and this has hampered its character-

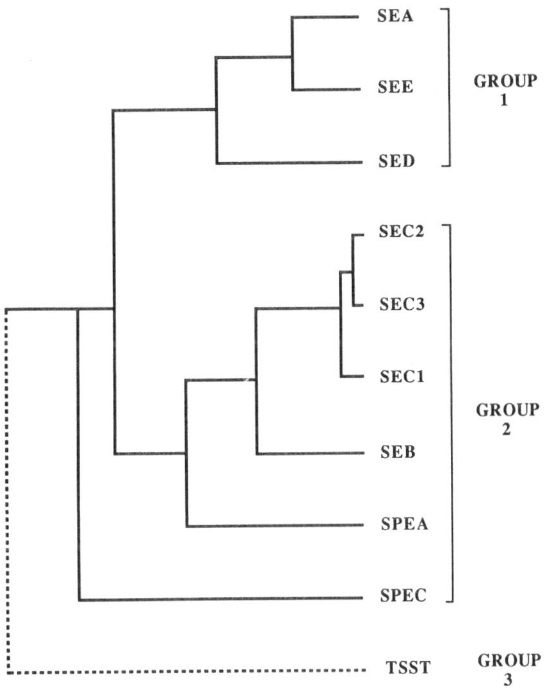

Fig. 19.2 Evolutionary divergence of toxin genes from a primordial superantigen. The enterotoxin family can be divided into three groups based on sequence divergence. Toxic shock syndrome toxin has only limited homology to the group 2 toxins. The homology groupings also reflect differences in function, in particular the region on MHC class II to which they bind and also the general subsets of Vβs stimulated.

ization. Nevertheless, partially purified material is potently mitogenic to unprimed T cells and its activity satisfies all the required criteria of a SAg.

2.2.2 Clostridium perfringens

A SAg has recently been identified from *Clostridium perfringens* another bacteria associated with human food poisoning. Like the SEs, the *C. perfringens* toxin causes vomiting and diarrhoea in humans. Despite the fact that purified toxin stimulated human Vβ6.9 and 22 (Table 19.2), and requires MHC class II (Bowness *et al.* 1992), the *C. perfringens* toxin bears no homology to the staphylococcal enterotoxins (McClane *et al.* 1988). The levels required to stimulate human PBL with this putative SAg is rather high (nanomolar concentrations) compared to the SEs.

2.2.3 Streptococcal M protein

Several reports have suggested that streptococcal M protein behaves as a SAg (Kotb *et al.* 1990; Tomai *et al.* 1992). This cell wall protein is closely linked with streptococcal-related rheumatic diseases. However, preparations of M

Table 19.2 T cell receptor Vβ response to bacterial superantigens

Toxin	Human	Mouse
SEA	1.1, 5.3, 6.3, 6.4, 6.9, 7.3, 7.4, 9.1, 23.1	1, 3, 10, 11, 12, 17
SED	1.1, 5.3, 6.9, 7.4, 8.1	11, 15, 17
SEE	5.1, 6.3, 6.4, 6.9, 8.1	3, 7, 8.3, 11, 17
SEB	3.2, 8.1, 12, 14, 15, 17, 20	7, 8.1–3
SEC1	3.2, 12	8.2, 8.3, 7, 11
SEC2	12, 13, 14, 15, 17, 20	8.2, 10
SEC3	5, 12	7, 8.2
TSST	2.1, 8.1	15, 16
SPE-A	2, 12, 14, 15	ND
SPE-B	8	ND
SPE-C	1, 2, 5.1, 10	ND
MAM	ND	6, 8.1–3
C. perfringens	6.9, 22	ND

Enrichment of murine Vβs have been analysed by a panel of anti-mVβ mAbs (Herman *et al.* 1991*b*). The majority of human specificities have been determined by quantitative PCR of Vβ cDNA (Choi *et al.* 1989; Hudson *et al.* 1993)

protein displaying SAg activity have been derived directly from extracts of *S. pyogenes* and require a relatively high concentration to stimulate T cells. Given that the reported TCR Vβ profiles for M protein are identical to those obtained for the secreted SPEs which are active at extremely low levels (i.e. femtomolar range), there is a possibility that SAg activity of the M protein may in fact be a result of SPE contamination. One report has confirmed this hypothesis (Fleischer *et al.* 1992). Notably a similar situation existed for many years with staphylococcal protein A (SpA). Highly purified preparations of SpA were originally thought to be mitogenic for T cells but this has since been shown to be due to SEA contamination (Smith *et al.* 1983).

2.2.4 Viral superantigens

In 1969, Hillyard Festenstein reported the existence of minor lymphocyte stimulating antigens (Mls) in mice. These endogenous products cause massive T cell stimulation between spleen cells of two mice genotypically identical at all MHC loci. Activity was nevertheless totally dependent on the presence of MHC class II bearing cells (Festenstein 1973). Similarity between the Mls antigens and the SEs was observed 20 years later by Janeway who showed that both products stimulated in a TCR Vβ-specific fashion (Janeway *et al.* 1988, 1989). The Mls products were later shown to be encoded by genes in the 3′ long terminal repeat of the mouse mammary tumour virus (MMTV) (Woodland *et al.* 1991; Dyson *et al.* 1991; Marrack *et al.* 1991; Frankel *et al.* 1991). These endogenous SAgs are described by Acha-Orbea in Chapter 11 of this book and so will not be discussed in detail here.

2.2.5 *Rabies virus nucleoprotein*

Recent reports have implicated the rabies virus nucleoprotein as a SAg behaving in the same way as the prototypic SEs although bearing no sequence homology to either the SEs or Mls products. The rabies nucleoprotein is totally invariant among all known rabies strains and furthermore is not secreted but firmly encapsulated within the virus in conjunction with viral DNA. While structurally unrelated to other SAgs, activity is still characterized by an ability to bind directly to MHC class II and stimulate human T cells via the Vβ8 domain (Lafon *et al.* 1992).

3 Binding of bacterial superantigens

3.1 Regions on MHC class II which bind bacterial toxins

3.1.1 *The staphylococcal toxins bind to most MHC class II molecules*

The SEs bind directly to human class II MHC antigens without prior processing and this feature is an essential requirement for SAg activity (Fleischer and Schrezenmier 1988; Fischer *et al.* 1989; Fraser 1989; Scholl *et al.* 1989*a*; Mollick *et al.* 1989). Human MHC class II molecules are preferred over mouse and HLA-DR isotypes (H-2E in mice) bind all the SEs better than either HLA-DQ (H-2A) or HLA-DP. Nevertheless, all isotypes support activation by all the SEs (Mollick *et al.* 1989; Scholl *et al.* 1990).

The group 1 SEs—A, E, and D, consistently bind with the highest affinity to HLA-DR while group 2 toxins (SEB and SEC1) have affinities about 100-fold and 500-fold lower respectively (Fraser 1989; Purdie *et al.* 1991). The group 3 toxin TSST binds to HLA-DR with an intermediate affinity (Purdie *et al.* 1991).

Despite universal binding of SEs to MHC class II proteins, polymorphisms in the latter do have a strong influence on affinity. For instance SEA exhibits dissociation constants (K_d) for HLA-DR1 and DR6 of 36 nM and 320 nM respectively while SEE/HLA-DR1 has a K_d of 120 nM (Hudson *et al.* 1993). All alleles of DR which have been tested bind SEA and SEE except DRw53 (see below).

Competitive binding experiments among group 1 (SEA), group 2 (SEB), and the group 3 (TSST) toxins, reveal at least three separate sites on MHC class II. SEA which has the highest affinity, competes against both SEB and TSST binding. However, SEB and TSST binding is non-competitive (Fraser 1989; Scholl *et al.* 1989*b*; Purdie *et al.* 1991; Pontzer *et al.* 1991). This suggests that the group 2 and group 3 toxin sites are separate, while the group 1 binding site is located between them and overlaps both. The location of these sites on MHC class II is discussed in more detail below.

SE binding has little direct influence on peptide binding and vice versa. Alanine mutations have been introduced in the H-2A^d molecule along the α helices at positions which abolished binding and presentation of a peptide but none of these mutations affected the ability of SEB to bind and stimulate a T cell hybridoma (Dellbonna *et al.* 1990). In another study, saturating amounts of TSST did not prevent the binding of a biotinylated peptide to an MHC class II molecule (Karp *et al.* 1990). These data point to a binding site for all toxins on the side of the MHC molecule away from the peptide groove. SEs also do not interfere with invariant chain (Ii) binding. A mutant cell line expressing both MHC class II and Ii chain as an undissociated complex on the cell surface binds all the SEs to the same level as the parental cell line expressing only MHC class II on its surface (Karp *et al.* 1992).

3.1.2 SEA and SEE utilize residues in the HLA-DR β1 domain

The DRw53 molecule is the only allele so far examined which does not bind either SEA and SEE. In contrast, DRw53 still binds SEB (Herman *et al.* 1991*a*). While several residues are different between DRw53 and DR1, most of these are at polymorphic positions. The only invariant residue (conserved

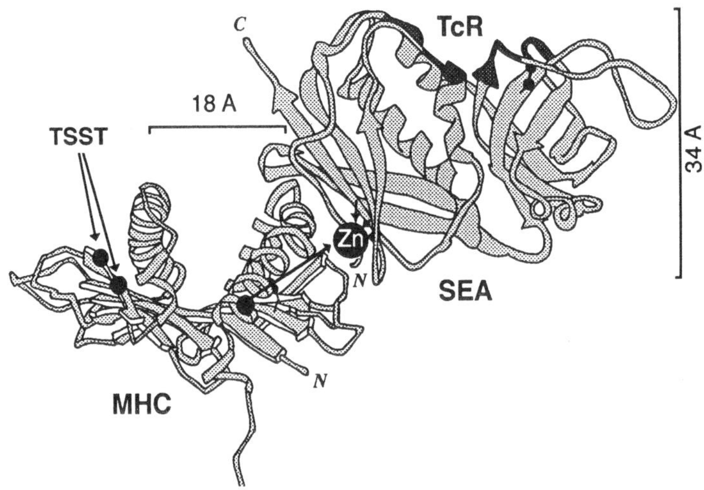

Fig. 19.3 The structures of MHC class II and an enterotoxin together. A picture indicating the relative size of an enterotoxin—in this case SEA (reproduced from the crystal structure by kind permission of Dr S. Swaminathan) in relation to an MHC class II molecule. The crystal structure of HLA-A2 is shown (Bjorkman *et al.* 1987*a,b*) but MHC class I and class II are likely to be very similar in structure (Brown *et al.* 1988). The residues involved in TSST binding (M36, K39) on the α1 domain are shown. The H81 in the β1 region is indicated linked to the zinc binding site on the toxin molecule. The TCR binding groove situated between toxin domains 1 and 2 is indicated by the dark shading. Residues 206 and 207 which determine Vβ specificity between SEA and SEE are located in domain 2 at the bottom of this groove at the beginning of the descending α-helix. When bound to MHC class II, SEA would be rotated 180° so that domain 1 would contact the α1 domain of the MHC molecule. The distance between the zinc atom and the bottom of the TCR binding groove is about 20 Å. Thus the TCR binding site would be about 4–5 Å above the top of the MHC α helices pointing upwards.

in all DRβ sequences) that is naturally altered in DRw53 is H81Y. In two independent studies, a H81Y mutation in DR1 dramatically reduced SEA and SEE binding while a reciprocal Y81H mutation of DRw53 restored SEA and SEE binding although not to normal levels (Herman *et al.* 1991*a*; Karp and Long 1992). H81 is located on the top of the DRβ1 domain adjacent to the disulfide bond between the β1 helix and the N-terminus of the underlying α chain (Figure 19.4). In a model of HLA-DR based on the known crystal structure on HLA-A2, H81 is predicted to point away from the peptide binding groove (Brown *et al.* 1988).

3.1.3 TSST requires residues in both the α1 and β1 domain of MHC class II

TSST utilizes residues in the α1 domain of HLA-DR on the opposite side of the peptide groove to H81 (Figures 19.3 and 19.4). An initial study showed that HLA-DR/DP chimeras expressed in mouse L cells bound high levels of TSST but only when the DRα1 domain was expressed (Karp *et al.* 1990). More recent evidence has shown that residues in both α1 and β1 domains are required for TSST binding (Braunstein *et al.* 1992) particularly residues M36 and K39 in DRα1 which are both essential for TSST binding (Panina-Bourginon *et al.* 1992). These residues are not on the α helix of α1 but are located on a loop just below the helix extending out from the molecule (Figure 18.3). Given that SEA and TSST compete in MHC class II binding studies and all the residues essential for either SEA and TSST binding are located towards one end of the α helices but on either side of the peptide

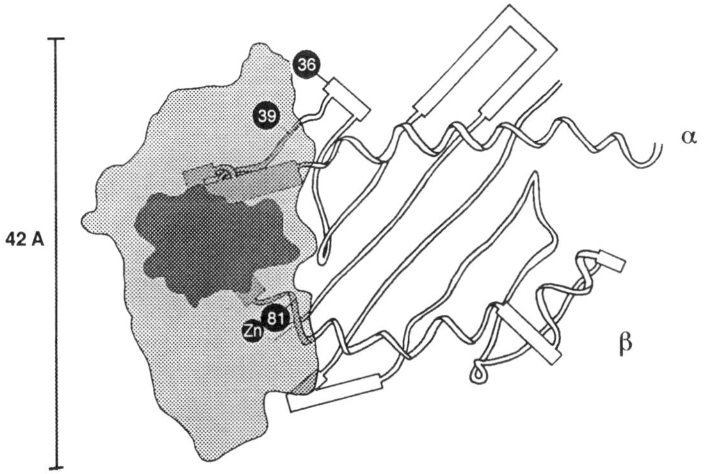

Fig. 19.4 The predicted location of SEA on MHC class II. A view of an SEA–MHC class II structure as seen by a T cell. The MHC class II structure is based on the predicted structure of Brown *et al.* (1988). The dark shaded area indicates the location of the toxin TCR binding groove in SEA containing TCR contact residues. Note that H81 in DRβ is positioned next to the predicted zinc binding site buried beneath domain 2 of SEA.

groove, the conclusion is that SEs bind at the end of the MHC class II molecule. This is discussed in more detail below.

3.2 The region on TCR involved in SAg binding

3.2.1 The fourth hypervariable loop of TCR encoded by Vβ

Several 3-D models of TCR have been constructed based on immunoglobulin Fab crystal structures (Chothia *et al.* 1988; Davis and Bjorkman 1988; Claverie *et al.* 1989; Jores *et al.* 1990). The TCR antigen binding surface is made up of six hypervariable loops or complementarity determining regions (CDRs). Three CDRs are encoded by the β chain and their location along the β chain protein can be determined by using a modified Wu–Kabat analysis of multiple Vβ amino acid sequences (Jores *et al.* 1990). βCDR1 is encoded by residues 24–32, βCDR2 by residues 50–62, and βCDR3 by residues 96–105. Unfortunately the αCDR1, αCDR2 regions are more difficult to identify because hypervariability is evenly spread throughout Vα (Jores *et al.* 1990).

In addition to CDR1, CDR2, and CDR3, a fourth hypervariable loop (HV4) is clearly evident in Vβ between residues 70–74. This region in fact is more hypervariable than βCDR1. The position of the HV4 loop in structural models of the TCR suggest that it lies between the CDR1 and CDR2 loops so that the crest of HV4 is almost flush with the antigen binding surface but out to the side of the binding face (Jores *et al.* 1990) (see Figure 19.5). A large number of studies have now clearly shown that HV4 is the region on TCR that is responsible for SAg recognition and this includes both the SEs and Mls products (White *et al.* 1989; Choi *et al.* 1990; Pullen *et al.* 1990; Pontzer *et al.* 1992). Mutations in this region are sufficient to convert a non-reactive Vβ domain into a reactive one. For the SEs, each toxin stimulates its own unique subset of Vβs although some Vβs react to more than one SE particularly within the toxin groups (Herman *et al.* 1991*b*) (Table 18.2). The patterns of Vβ specificity given in Table 19.2 are obtained quite independently of the MHC phenotype (Choi *et al.* 1989; Herman *et al.* 1991*b*) indicating that MHC class II polymorphisms which normally contribute to the exquisite specificity in peptide recognition, are largely overcome in SAg activation. Thus Vβ specificity is governed solely by diversity within an HV4 binding site(s) located somewhere on the toxin molecule.

3.3 The TCR binding site on sea

3.3.1 The TCR Vβ specificity involves residues Ser206/Asn207

To identify the SE region and residues responsible for Vβ specificity, we constructed a panel of SEA/SEE hybrids and analysed the human TCR Vβ response to them (Hudson *et al.* 1993). SEA and SEE are 83% identical but their Vβ profiles can be distinguished by the fact that SEA stimulates hVβ1.1, 5.3, 7.4, 9.1, and 23.1 while SEE stimulates Vβ5.1 and 8.1 (Table 19.2). In all the SEA/SEE hybrids, specificity mapped to a final C-terminal region

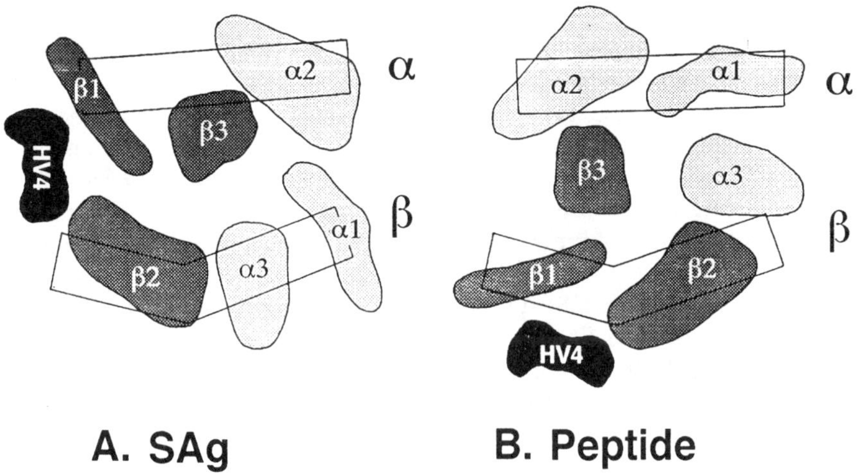

Fig. 19.5 SAgs may induce a rotation of the TCR on MHC class II. A schematic picture of TCR α and β chain CDR1, CDR2, and CDR3 loops positioned over the MHC class II. This view is looking down from the T cell. The α helices of MHC class II are indicated by the boxes beneath the TCR CDR1, CRD2, and CDR3. **A.** The likely TCR orientation based on the predicted location of the SEA Vβ binding site. **B.** The likely orientation of TCR based on the requirement that αCDR3 and βCDR3 must lie directly over the peptide (Chothia *et al.* 1988; Davis and Bjorkman 1988; Claverie *et al.* 1989). As can be seen, a SAg induced orientation requires that the TCR would be rotated 90° so that αCDR3 and βCDR3–peptide contact no longer occurs. Moreover, CDR1 and CDR2 of each TCR chain would contact both MHC class II α helices.

(200–233) containing only five variable positions. Due to alignment gaps, this region is 210–250 in Figure 19.1. All Vβs segregated completely with region 200–233 suggesting that recognition involves a single variable Vβ binding site, rather than multiple toxin sites for each Vβ. The Vβ specificity mapped to two adjacent residues in SEA/SEE-S/P^{206}N/D^{207}. These are residues 220 and 221 in Figure 19.1. Thus the considerable variability in other regions of SEA and SEE contribute very little to the Vβ binding site. These positions in SED are very similar (P206,E207) while in the group 2 toxins SEB and SEC the analogous residues are F208,D209.

In a similar study which also focused on the Vβ difference of SEA and SEE, the same two residues were identified but in addition, the variable residue G/S24 (SEA/SEE) was shown to influence specificity also (Mollick *et al.* 1993).

In a study on SEB, Kappler *et al.* (1992) used random mutagenesis to identify residues involved in TCR binding. In their study, a panel of T cell hybridomas expressing a variety of Vβ elements was used to detect changes in TCR Vβ response. Two regions were located, one involving a highly conserved N25 (in all SEs), the other involving W63,Y64 found in all SEs except SED (F64). Mutation at these positions destroyed the ability of SEB to stimulate T cells but had no effect on MHC class II binding suggesting that the two sites were unrelated.

The results of these two studies at first glance appear contradictory. However, with the publication of the crystal structure of SEB (Swaminathan *et al.* 1992), residues N25 and W63,Y64 identified by Kapper and F220, D221 (corresponding to S206 and N207 in SEA) were found to fold together to form a shallow groove located between the two globular domains of the SEB protein. N25 and W63,Y64 are located in the floor of the groove while F220 and D221 (S206,N207 in SEA) are located on the outer rim of the groove. Positions N25 and W63,Y64 are conserved between SEA and SEE and so their contribution to TCR binding was not seen in the SEA/SEE hybrids. The location of these residues can be seen in Figure 19.3 in the SEB crystal structure. Notably, a groove of this nature would appear to accommodate a loop-like structure consistent with the predicted nature of the TCR HV4 region.

3.3.2 Residues Ser206/Asn207 in SEA are TCR β chain contact residues

Although the natural affinity of TCR to toxin appears to be very low, a simple plate binding assay has been devised by Gascoigne which directly measures binding of soluble TCR β chain to cells expressing a toxin–MHC class II complex. Soluble mouse Vβ3 and Vβ11 protein have been produced by truncating the β chain genes before the transmembrane region. HLA-DR expressing human cells (Raji or LG-2 cells) pre-incubated with toxins such as SEA or SEE bind selectively to plastic plates coated with purified TCR β chains (Gascoigne and Ames 1991). The mVβ3 protein bound DR/SEA while mVβ11 bound DR/SEE. Consistent with the Vβ stimulations, all the SEA/SEE hybrids bound either Vβ3 or Vβ11 and direct binding of TCR β chain was governed solely by residues S/N206 and P/D207 in SEA and SEE (Irwin *et al.* 1992). Thus all evidence suggests that residues 206 and 207 of SEA and SEE, make direct contact with residues in HV4 loop of Vβ.

3.4 The MHC class II binding site on sea

3.4.1 The MHC class II binding site of group 1 toxins contains zinc

Binding of SEA, SED, and SEE to any MHC class II allele is abolished by EDTA. This is due to the removal of an essential zinc atom from the MHC class II binding site (Fraser *et al.* 1992). In contrast, group 2 (SEB, SEC1–3) and group 3 (TSST) toxins are unaffected by EDTA. The binding of ^{65}Zn to SEA has a dissociation constant of about 1 μM. Mapping the residues which co-ordinate zinc has allowed the exact location of the MHC class II binding site at least in the group 1 SEs. All zinc metalloproteins have co-ordination complexes which involve His, Asp, Glu, or Cys side chains and the imidazole nitrogen of His is the most commonly used. Moreover, Cys residues are normally associated with zinc finger-type proteins where zinc is strongly co-ordinated by four Cys residues within the hydrophobic protein core making it inaccessible to EDTA. The co-ordination number of zinc is normally four but can be five or six (Vallee and Auld 1990). In many zinc metalloproteins, where the zinc is part of either a catalytic or co-active site, an activated water molecule is also bound to zinc (Vallee and Falchuk 1993).

Using site-directed mutagenesis, three zinc binding residues have been located in SEA which are also conserved in SEE. These are H187, H225, and D227 (in Figure 19.1, these residues are 201, 242, and 244 and are indicated by the closed circles above the sequence). In SED, the analogous positions are D187, H225, and D227, while in SEB they are I190, K229, and E231 consistent with this toxins inability to bind zinc. Mutation of each of these residues in SEA decrease ^{65}Zn binding, MHC class II binding, and T cell stimulation (Hudson *et al.* manuscript submitted). The most important residues appears to be D227 because while H187A and H225A mutants still retain some activity the D227A mutation completely destroys all detectable MHC binding and T cell proliferation. This suggests that D227 may itself act as a bidentate ligand (i.e. both the hydroxyl and carboxyl group are utilized).

3.4.2 Zinc forms a molecular bridge between MHC class II and SEA

Removal of zinc by EDTA completely destroys SEA binding to all MHC class II molecules. Because zinc is so easily removed by EDTA, it must be readily accessible to water and thus be located on the surface of the toxin molecule. This led to speculation that zinc actively participated in binding through a molecular bridge between SEA and MHC class II. Recent experiments have confirmed this hypothesis (Hudson *et al.* manuscript submitted). There is no detectable difference in ^{65}Zn binding between free SEA and an SEA/DR1 complex (Fraser *et al.* 1992). However, using the SEA–H187A and SEA–H225A mutants, ^{65}Zn binding increases from negligible levels to wild-type levels in the presence of equimolar purified HLA-DR1. Thus these mutants are very useful in highlighting the MHC class II contribution to zinc co-ordination which can not be detected with wild-type SEA. We believe that a water molecule which is bound to zinc is excluded in favour of either a His, Asp, or Glu residue in MHC class II. This exchange does not alter the co-ordination number but exclusion of water from the interface results in a very stable zinc co-ordination complex. Additional experiments have revealed that the zinc atom located within the interface of SEA/DR1, SEA–H187A/DR1, or SEA–H225A/DR1 is no longer accessible to EDTA. Identification of the MHC class II residue which contributes to zinc binding will allow the unequivocal positioning of SEA on MHC class II.

4 Structural models of an SEA–MHC class II complex

4.1 What is the role of MHC class II in SE activation?

There is considerable debate over the involvement of TCR–MHC class II contacts in SAg activation and evidence both for and against has mounted in almost equal proportions. At present the issue is unresolved. Evidence against TCR–MHC class II contacts in SAg activation comes from the following findings. First, the extensive amino acid variability between MHC class II and TCR which normally dictates peptide recognition is irrelevant to an SAg–MHC class II complex with only the TCR Vβ domain determining SAg recognition. Secondly, direct binding of a single TCR β chain to SEA–MHC class II occurs in the absence of TCR α chain (Gascoigne and Ames 1991), a finding which immediately questions the requirement of at least the αCDR1 and αCDR2 in SAg activation. Thirdly, mutations in regions of either MHC class II or TCR which destroy peptide recognition have no affect on the presentation of SEs suggesting that even if TCR–MHC class II contacts are essential, they are not the same as those used in peptide recognition (Dellbonna *et al.* 1990; Jorgensen *et al.* 1992; White *et al.* 1993). However, one could speculate that because the SE–TCR interaction is high affinity and involves multiple contact residues between SE and HV4, mutations of single TCR–MHC contact points are not sufficient by themselves to ablate SE activation and that multiple mutations must be made before any significant affect is seen. Finally, there have been reports of SEs acting on T cells in the absence of any MHC class II (Dohlsten *et al.* 1991; Hewitt *et al.* 1992).

There is also considerable evidence supporting the opposing notion that TCR–MHC class II contacts are essential for SAg activation. For instance, MHC class II polymorphisms do influence the response of individual T–T cell hybridomas in response to a number of SEs (Herman *et al.* 1990). In addition, several recent reports have noted skewed Vα repertoires in SAg activation of peripheral lymphocytes suggesting the Vα–MHC class II contacts have also influenced reactivity (Waanders *et al.* 1993). Finally, some SAgs such as the viral Mls product Mtv-11 are only effective when certain Vα.Vβ combinations are expressed, again suggesting that the Vα–MHC class II contacts have influence on reactivity (Woodland *et al.* 1993). However, these reports do not identify whether the Vα influence is indirect, affecting the conformation of HV4 or direct, involving αCDR1 and αCDR2 contacts with MHC class II. Furthermore, it could be argued that Vα is somehow involved in the signalling mechanism and that some Vαs are more effective than others in this regard. We feel that the most compelling argument in favour of essential TCR/MHC class II contacts is the fact that in the overwhelming majority of cases, SAgs have an essential requirement for MHC class II. Those few situations where SEs have been shown to operate in the absence of any classical MHC class II molecules are exceptions that merely prove the rule.

There are currently two possible structural models which attempt to explain the role of MHC class II in SAg activation. The first is that a combination of both SAg–TCR Vβ and TCR–MHC class II binding energies is required for signalling. The second model proposes that no TCR–MHC class II contacts are made and that MHC class II is required to induce a conformational change in the SAg structure which increases its affinity for TCR. However, in this second model, it is hard to envisage similar conformational changes occurring in the structurally unrelated bacterial and viral SAgs, induced by the same MHC class II molecule.

4.2 The TCR site and MHC class II site are on opposite sides of the toxin

The crystal structure of SEB reveals a globular protein consisting of two β-barrel domains. The larger domain 2 is bounded by residues 127–239 while domain 1 is bounded by residues 33–126. The first 32 amino acids are part of domain 2 (Swaminathan *et al.* 1992). The TCR binding site is clearly visible as a shallow groove between the two domains. The equivalent amino acids to the zinc binding residues H187, H225, and D227 of SEA are located on SEB at the very bottom of domain 1 approximately 20 Å from the TCR site (see Figure 19.3). H187 is located on the inner side of the bottom loop of domain 2 while H225 and D227 are located on the inner β-strand facing this loop. All three zinc binding side chains point to the left and would be in sufficient proximity to chelate a zinc atom (S. Swaminathan, personal communication). The zinc site lies directly underneath the very N-terminus of the toxin and this is consistent with reports that peptides to the N-terminus of SEA inhibit binding (Pontzer *et al.* 1989). One study has shown that a truncated SEA molecule consisting only of domain 2 (107–233) can still bind to MHC class II at fivefold lower affinity than wild-type SEA but no longer stimulates T cells (Hedlund *et al.* 1991). This is consistent with the exclusive location of the zinc in domain 2.

Perhaps surprising is the distance between the TCR contact residues 206 and 207 and the zinc binding site (approx. 20 Å). This distance must determine the proximity of the TCR HV4 loop and the residues on MHC class II which co-ordinates the zinc atom. A very good candidate for this residue is position 81 in DRβ1—first because it is a zinc reactive histidine which points out from the α-helix. Secondly, mutation of H81 selectively destroys SEA and SEE binding but not SEB binding (Herman *et al.* 1991*a*; Karp and Long 1992), and thirdly, synthetic peptides spanning H81 inhibit SEA binding but not SEB or TSST (Russell *et al.* 1991).

4.3 A predicted structure of the SEA–MHC class II complex

Using H81 as the zinc co-ordinating residue in MHC class II juxtaposed to the zinc binding site of SEA, a model can be constructed which allows SEA to bind across the end of the peptide groove so that domain 1 of SEA also contacts regions of the DR α-chain (Figure 19.4). In this position, the TCR binding site points upwards towards TCR. The TCR contact residues would lie about 4–5 Å above the top of the MHC class II α-helices in a prime position to bind directly to the bottom of the HV4 loop of TCR. Moreover, SEA would also interfere with TSST binding which utilizes residues M36 and K39 in the DRα1 domain (Panina-Borginon *et al.* 1992).

Importantly, this model predicts continued contact between TCR CDR1, 2, and 3 loops and the α helices of MHC class II.

4.4 How is TCR oriented on MHC class II in superantigen binding?

Both TCR and MHC class II display diad symmetry but it is not yet known whether TCRs bind in one, two, or even multiple orientations on MHC class II. It is most likely however, given the fact that TCRs bind to syngeneic, allogeneic, and even xenogeneic MHC that this fundamental reactivity results from a single constrained orientation.

We make the assumption that SAgs require TCR and MHC class II contacts to contribute to the total binding energy. In order to accommodate the very high numbers of TCRs which react to a single SE, all TCRs must therefore be oriented in the same way on MHC class II to allow the HV4 loop to bind to the fixed SAg site which we have positioned at the end of the peptide groove (Figure 19.4).

Chothia *et al.* (1988) have provided a model which gives the predicted spatial arrangement of CDR1, CDR2, and CDR3 regions of TCR (see Figure 19.5B). The relative distances between CDRs fit well with the known dimensions of the MHC α helices (Bjorkman *et al.* 1987*a,b*; Davis and Bjorkman 1988). In the most favoured model (Figure 19.5B) the βCDR1 and βCDR2 loops are predicted to contact one α helix while αCDR1 and αCDR2 contact the other α helix. This allows both αCDR3 and βCDR3 to be located directly over peptide residues. This orientation clearly fits very well with the MHC and peptide molecule but does not fit quite so well with the orientation we predict for SAg binding (Figure 19.5A). Rotating the TCR slightly so that HV4 is positioned over the predicted toxin binding site, would mean that each chain of the TCR would contact both MHC helices. However, the CDR3 regions are no longer located directly over the peptide (Figure 19.5A).

These two models conflict but only if it is assumed that the TCR–MHC class II orientation does not change between peptide and SAg recognition. However, TCRs may have two distinct orientations on MHC class II, the first may be dictated by peptide specificity and would thus require CDR3–peptide interactions (Figure 19.5B). The second more stable orientation (Figure 19.5A) may exclude CDR3–peptide interactions and instead be stabilized by the additional contribution of accessory molecules such as CD4 or SAgs. Without the assistance of SAgs, the second state would obviously be contingent on the first. Moreover, the transition betweeen these states may be the physical event which induces TCR signalling. While this idea is purely speculative, the fact that there are several mutations of residues in MHC class II which abolish TCR contact during peptide recognition but do not affect SAg activation, may suggest that the relevant CDR loop which normally would make contact with these residues is in a different position during SAg activation.

5 Future directions

There is much to learn from the way in which SAgs bypass the normal MHC restriction process of peptide recognition and the reasons why such proteins have evolved at all. Do they benefit the host or the bacteria and how has the TCR Vβ repertoire been shaped by SAgs? What is the involvement of HV4 in other recognition processes?

Clearly HV4 recognizes a diverse structure(s) but to date SAgs seem to be the only proven candidates. It would seem unlikely that HV4 diversity has been driven solely by a handful of microbial products, given the degree of diversity in HV4 and the apparently limited distribution of SAgs in nature. One possibility is that SAg recognition is a remnant of an original primordial TCR which may have been a ββ homodimer. Prior to the development of the TCR α chain and junctional region diversity, recognition by the ββ homodimer may have been predominantly SAg-like, since at this stage, MHC class II may not have developed the ability to bind and present peptides. There is both phenotypic and genotypic evidence to support the existence of such a β chain homodimeric receptor during ontogeny (Toyonaga and Mak 1987) and recent reports indicate that a ββ TCR generated by transgenic expression of a single β chain cDNA in a RAG-$2^{-/-}$ mouse has important functional activity in T cell development (Shinkai *et al.* 1993).

In summary, SAgs are powerful tools which define a novel and powerful means of overcoming the restrictions imposed on peripheral TCRs. Whether SAgs mimic the normal process of peptide recognition or define a novel TCR–MHC class II orientation remains to be seen. However, they have already proved extremely useful in determining the nature of MHC–TCR interactions at least in SAg activation. The continued study of these unique proteins will no doubt provide invaluable insight into other aspects of the T cell recognition process including the events that lead to signalling through the antigen receptor complex.

Note added in proof

Since the writing of this manuscript, the crystal structures of SEB/HLA-DR1 (Jardetzky *et al.* (1994) *Nature*, **368**, 711–18) and TSST/DR1 (Kim *et al.* (1994) *Science*, **266**, 1870–4) have been resolved. The structures reveal that both toxins bind to a conserved hydrophobic groove in the α-chain of MHC class II in similar but not identical fashions. Notably, TSST appears to mask a major portion of the polymorphic α-helices of class II suggesting that there is likely to be minimal contact between TcR and MHC class II during TSST activation.

Acknowledgements

We wish to thank Drs Carol Horgan, John Marbrook, and Jim Watson for critical reviews of this manuscript. We wish to acknowledge the generous support of the Wellcome Trust (UK), The Health Research Council of New Zealand, and The Auckland Medical Research Foundation.

References

Abe, J., Kotzin, B. L., Jujo, K., Melish, M. E., Glode, M. P., Kohsaka, T., *et al.* (1992). Selective expansion of T cells expressing T-cell receptor variable regions V beta 2 and V beta 8 in Kawasaki disease. *Proc. Natl. Acad. Sci. USA*, **89**, 4066–70.

Atkin, C. L., Cole, B. C., Sullivan, G. J., Washburn, L. R., and Wiley, B. B. (1986). Stimulation of mouse lymphocytes by a mitogen derived from *Mycoplasma arthritidis*. *J. Immunol.*, **137**, 1581–9.

Bergdoll, M. S. (1983). The enterotoxins. In *Staphylococcal and streptococcal infections* (ed. C. S. F. Easom and C. Aslam), p. 559. Academic Press, New York.

Bergdoll, M. S., Crass, B. A., Reiser, R. F., Robbins, R. N., and Davis, J. P. (1981). A new staphylococcal enterotoxin F, associated with toxic-shock syndrome *Staphylococcal aureus* isolates. *Lancet*, **1**, 1017–21.

Bjorkman, P. J., Saper, M. A., Samraoui, B., Bennet, W. S., Strominger, J. L., and Wiley, D. C. (1987*a*). Structure of the human class I histocompatibility antigen HLA-A2. *Nature*, **329**, 506–12.

Bjorkman, P. J., Saper, M. A., Samraoui, B., Bennet, W. S., Strominger, J. L., and Wiley, D. C. (1987*b*). The foreign antigen binding site and T cell recognition regions of class I histocompatibility antigens. *Nature*, **329**, 512–18.

Blomster-Hautamaa, D. A. and Schlievert, P. M. (1988). Non-enterotoxic staphylococcal toxins. In *Handbook of bacterial toxins*, Vol. 4 (ed. C. M. Hardegree and A. T. Tu), pp. 297–330. Marcel Dekker, New York.

Bowness, P., Moss, P. A. H., Tranter, H., Bell J. A., and McMichael, A. J. (1992). Clostridium perfringens enterotoxin is a superantigen reactive with Vβ6.9 and Vβ22. *J. Exp. Med.*, **176**, 893–6.

Braunstein, N. S., Weber, D. A., Wang, X. C., Long, E. O., and Karp, D. (1992). Sequences in both class-II major histocompatibility complex alpha and beta chains contribute to the binding of the superantigen toxic shock syndrome toxin-1. *J. Exp. Med.*, **175**, 1301–5.

Brown, J. H., Jardetzky, T., Saper, M. A., Samraoui, B., Bjorkman, P. J., and Wiley, D. C. (1988). A hypothetical model of the foreign antigen binding site of class II histocompatibility molecules. *Nature*, **332**, 845–50.

Chesney, P. J., Bergdoll, M. S., Davis, J. P., and Vergeront, J. M. (1984). The disease spectrum, epidemiology, and etiology of toxic shock syndrome. *Annu. Rev. Microbiol.*, **38**, 315–38.

Choi, Y., Kotzin, B., Herron, L., Callahan, J., Marrack, P., and Kappler, J. (1989). Interaction of *Staphylococcus aureus* 'superantigens' with human T cells. *Proc. Natl. Acad. Sci. USA*, **86**, 8941–5.

Choi, Y. W., Herman, A., DiGiusto, D., Wade, T., Marrack, P., and Kappler, J.

(1990). Residues of the variable region of the T-cell-receptor beta-chain that interact with *S. aureus* toxin superantigens. *Nature*, **346**, 471–3.

Chothia, C., Boswell, D. R., and Lesk, A. M. (1988). The outline structure of the T-cell αβ receptor. *EMBO J.*, **7**, 3745–55.

Claverie, J.-M., Prochnicka-Chalufour, A., and Bourgeleret, L. (1989). Implications of a Fab-like structure for the T-cell receptor. *Immunol. Today*, **10**, 10–14.

Cole, B. C., Kartcher, D. R., and Wells, D. J. (1989). Stimulation of mouse lymphocytes by a mitogen derived from *Mycoplasma arthritidis*. *J. Immunol.*, **142**, 4131–7.

Cone, L., Woodward, D. R., Schlievert, P. M., and Tomory, G. S. (1987). Clinical and bacteriologic observations of a toxic shock like syndrome due to *Streptococcal pyogenes*. *N. Eng. J. Med.*, **317**, 146–8.

Davis, M. M. and Biorkman, P. J. (1988). T-cell antigen receptor genes and T-cell recognition. *Nature*, **334**, 395–402.

Dellbona, P., Peccoud, J., Kappler, J., Marrack, P., Benoist, C., and Mathis, D. (1990). Superantigens interact with MHC class II molecules outside the antigen binding groove. *Cell*, **62**, 1115–21.

Dohlstein, M., Hedlund, G., Segren, S., Lando, P. A., Herrmann, T., Kelly, A. P., *et al.* (1991). Human major histocompatibility complex class-II negative colon carcinoma cells present staphylococcal superantigens to cytotoxic T lymphocytes: evidence for a novel enterotoxin receptor. *Eur. J. Immunol.*, **21**, 1229–34.

Dyson, P. J., Knight, A. M., Fairchild, S., Simpson, E., and Tomanari, K. (1991). Genes encoding ligands for deletion of Vβ11 T cells cosegregate with mammary tumour virus genomes. *Nature*, **349**, 531–2.

Festenstein, H. (1973). Immunogenetic and biological aspects of *in vitro* lymphocyte allo-transformation (MLR) in the mouse. *Transplant. Rev.*, **15**, 62–88.

Fischer, H., Dohlsten, M., Lindvall, M., Sjogren, H., and Carlsson, R. (1989). Binding of staphylococcal enterotoxin A to HLA-DR on B cell lines. *J. Immunol.*, **142**, 3151–7.

Fleischer, B. and Schrezenmeier, H. (1988). T cell stimulation by staphylococcal enterotoxins: clonally variable response and requirement for major histocompatibility complex class II molecules on accessory or target cells. *J. Exp. Med.*, **167**, 1697–707.

Fleischer, B., Smiidt, K.-L., Gerlach, D., and Kohler, W. (1992). Separation of T cell stimulating activity from streptococcal M protein. *Infect. Immunol.*, **60**, 1767–70.

Frankel, W. N., Rudy, C., Coffin, J. M., and Huber, B. (1991). Linkage of Mls genes to endogenous mammary tumour viruses of inbred mice. *Nature*, **349**, 526–8.

Fraser, J. D. (1989). High-affinity binding of staphylococcal enterotoxins A and B to HLA-DR. *Nature*, **339**, 221–3.

Fraser, J. D., Urban, R. G., Strominger, J. L., and Robinson, H. (1992). Zinc regulates the function of two superantigens. *Proc. Natl. Acad. Sci. USA*, **89**, 5507–11.

Gasgoigne, N. R. J. and Ames, C. T. (1991). Direct binding of secreted T-cell receptor β chain to superantigen associated with class II major histocompatibility complex protein. *Proc. Natl. Acad. Sci. USA*, **88**, 613–16.

Grossman, D., Van, M., Mollick, J. A., Highlander, S. K., and Rich, R. R. (1991). Mutation of the disulphide loop in staphylococcal enterotoxin A consequences for T cell recognition. *J. Immunol.*, **147**, 3274–81.

Hedlund, G., Dohlsten, M., Herrmann, T., Buell, G., Lando, P. A., Segren, S., *et al.* (1991). A recombinant C-terminal fragment of staphylococcal enterotoxin A binds to human MHC class II products but does not activate T cells. *J. Immunol.*, **147**, 4082–5.

Herman, A., Croteau, G., Sekaly, R.-P., Kappler, J., and Marrack, P. (1990). HLA-DR alleles differ in their ability to present staphylococcal enterotoxins to T cells. *J. Exp. Med.*, **172**, 709–17.

Herman, A., Labrecque, N., Thibodeau, J., Marrack, P., Kappler, J. W., and Sekaly, R. P. (1991*a*). Identification of the staphylococcal enterotoxin A superantigen binding site in the beta 1 domain of the human histocompatibility antigen HLA-DR. *Proc. Natl. Acad. Sci. USA*, **88**, 9954–8.

Herman, A., Kappler, J. W., Marrack, P., and Pullen, A. M. (1991*b*). Superantigens: mechanism of T cell stimulation and role in immune responses. *Annu. Rev. Immunol.*, **9**, 745–72.

Hewitt, C. R. A., Lamb, J. R., Hayball, J., Hill, M., Owen, M. J., and Ohehir, R. E. (1992). Major histocompatibility complex independent clonal anergy by direct interaction of *Staphylococcus aureus* enterotoxin B with the T cell antigen receptor. *J. Exp. Med.*, **175**, 1493–9.

Hudson, K. R., Robinson, H., and Fraser, J. D. (1993). Two adjacent residues in staphylococcal enterotoxins A and E determine T cell receptor Vβ specificity. *J. Exp. Med.*, **177**, 175–84.

Irwin, M. J., Hudson, K. R., Fraser, J. D., and Gascoigne, N. R. J. (1992). Enterotoxin residues determining T-cell receptor Vbeta binding specificity. *Nature*, **359**, 841–3.

Janeway, C. A., Chalupny, J., Conrad, P. J., and Buxser, S. (1988). An external stimulus that mimics Mls locus responses. *J. Immunogenet.*, **15**, 161–8.

Janeway, C. J., Yagi, J., Conrad, P. J., Katz, M. E., Jones, B., Vroegop, S. *et al.* (1989). T-cell responses to Mls and to bacterial proteins that mimic its behavior. *Immunol. Rev.*, **107**, 61–88.

Jores, R., Alzari, P. M., and Meo, T. (1990). Resolution of hypervariable regions in T-cell receptor β-chains by modified Wu-Kabat index of amino acid diversity. *Proc. Natl. Acad. Sci. USA*, **87**, 138–42.

Jorgensen, J. L., Reay, P. A., Ehrlich, E. W., and Davis, M. M. (1992). Molecular components of T-cell recognition. *Annu. Rev. Immunol.*, **10**, 835–73.

Kappler, J. W., Herman, A., Clements, J., and Marrack, P. (1992). Mutations defining functional regions of the superantigen staphylococcal enterotoxin B. *J. Exp. Med.*, **175**, 387–96.

Karp, D. R., Teletski, C. L., Scholl, P., Geha, R., and Long, E. O. (1990). The alpha 1 domain of the HLA-DR molecule is essential for high-affinity binding of the toxic shock syndrome toxin-1. *Nature*, **346**, 474–6.

Karp, D. R. and Long, E. O. (1992). Identification of HLA-DR1 beta chain residues critical for binding staphylococcal enterotoxins A and E. *J. Exp. Med.*, **175**, 415–24.

Karp *et al.* 1990.

Karp, D. R., Jenkins, R., and Long, E. O. (1992). Distinct binding sites on HLA-DR for invariant chain and staphylococcal enterotoxins. *Proc. Natl. Acad. Sci. USA*, **89**, 9657–61.

Kotb, M., Majumdar, G., Tomai, M., and Beachey, E. H. (1990). Accessory cell-independent stimulation of human T cells by streptococcal M protein superantigen. *J. Immunol*, **145**, 1332–6.

Lafon, M., Lafage, M., Martinez-Arends, A., Ramirez, R., Vuillier, F., Charron, D., *et al.* (1992). Evidence for a viral superantigen in humans. *Nature*, **358**, 507–10.

Marrack, P. and Kappler, J. (1990). The staphylococcal enterotoxins and their relatives. *Science*, **248**, 705–11.

Marrack, P., Kushnir, E., and Kappler, J. (1991). A maternally inherited superantigen encoded by a mammary tumour virus. *Nature*, **341**, 524–6.

McClane, B. A., Hanna, P. C., and Wnek, A. P. (1988). Clostridium perfringens exotoxin. *Microb. Pathol.*, **4**, 317–21.

Mollick, J. A., Cook, R. G., and Rich, R. R. (1989). Class II MHC molecules are specific receptors for staphylococcus enterotoxin A. *Science*, **244**, 817–20.

Mollick, J. A., McMasters, M. L., Grossman, D., and Rich, R. R. (1993). Localisation of a site on bacterial superantigens that determines T cell receptor β chain specificity. *J. Exp. Med.*, **177**, 283–93.

Panina-Bordignon, P., Fu, X. T., Lanzavecchia, A., and Karr, R. W. (1992). Identification of HLA-DR alpha chain residues critical for binding of the toxic shock syndrome toxin superantigen. *J. Exp. Med.*, **176**, 1779–84.

Peavy, D. L., Adler, W. H., and Smith, R. T. (1970). The mitogenic effects of endotoxin and staphylococcal enterotoxin B on mouse spleen cells and human peripheral lymphocytes. *J. Immunol.*, **105**, 1453–8.

Pontzer, C. H., Russell, J. K., and Johnson, H. M. (1989). Localization of an immune functional site on staphylococcal enterotoxin A using the synthetic peptide approach. *J. Immunol.*, **43**, 280–4.

Pontzer, C. H., Russell, J. K., and Johnson, H. M. (1991). Structural basis for differential binding of staphylococcal enterotoxin A and toxic shock syndrome toxin 1 to class II major histocompatibility molecules. *Proc. Natl. Acad. Sci. USA*, **88**, 125–8.

Pontzer, C. H., Irwin, M. J., Gascoigne, N. R. J., and Johnson, H. M. (1992). T-cell antigen receptor binding sites for the microbial superantigen staphylococcal enterotoxin A. *Proc. Natl. Acad. Sci. USA*, **89**, 7727–31.

Pullen, A. M., Wade, T., Marrack, P., and Kappler, J. W. (1990). Identification of the region of T cell receptor β chain that interacts with the self-superantigen Mls-1a. *Cell*, **61**, 1365–74.

Purdie, K., Hudson, K. R., and Fraser, J. D. (1991). Bacterial superantigens. In *Antigen processing and presentation* (ed. J. MacCluskey), pp. 193–214. CRC Press, Boca Raton, Fl.

Russell, J. K., Pontzer, C. H., and Johnson, H. M. (1990). The I-A beta b region (65–85) is a binding site for the superantigen, staphylococcal enterotoxin A. *Biochem. Biophys. Res. Commun.*, **168**, 696–701.

Russell, J. K., Pontzer, C. H., and Johnson, H. M. (1991). Both alpha-helices along the major histocompatibility complex binding cleft are required for staphylococcal enterotoxin A function. *Proc. Natl. Acad. Sci. USA*, **88**, 7228–32.

Scholl, P., Diez, A., Mourad, W., Parsonnet, J., Geha, R. S., and Chatila, T. (1989*a*). Toxic shock syndrome toxin 1 binds to major histocompatibility complex class II molecules [published erratum appears in *Proc. Natl. Acad. Sci. USA*, 1989 Sep; **86**(18): 7138]. *Proc. Natl. Acad. Sci. USA*, **86**, 4210–14.

Scholl, P. R. Diez, A., and Geha, R. S. (1989*b*). Staphylococcal enterotoxin B and toxic shock syndrome toxin-1 bind to distinct sites on HLA-DR and HLA-DQ molecules. *J. Immunol.*, **143**, 2583–8.

Scholl, P. R., Diez, A., Karr, R., Sekaly, R. P., Trowsdale, J., and Geha, R. (1990). Effects of isotypes and allelic polymorphisms on the binding of staphylococcal exotoxins to MHC class II molecules. *J. Immunol.*, **144**, 226–30.

Shinkai, Y., Koyashu, S., Nakayama, K., Murphy, K. M., Loh, D. Y., Reinherz, E. L., *et al.* (1993). Restoration of T cell development in RAG-2 defficient mice by functional TCR transgenes. *Science*, **259**, 822–5.

Smith, E. M., Johnson, H. M., and Blalock, J. E. (1983). *Staphylococcus aureus* protein A induces the production of interferon-α in human lymphocytes and interferon α/β in mouse spleen cells. *J. Immunol.*, **130**, 773–6.

Spero, L. and Morlock, B. A. (1978). Biological activities of the peptides of staphylococcal enterotoxin C formed by limited tryptic hydrolysis. *J. Biol. Chem.*, **253**, 8787–92.

Spero, L., Johnson-Winegar, A., and Schmidt, J. J. (1988). Enterotoxins of staphylococci. *In Handbook of bacterial toxins*, Vol. 4 (ed. C. M. Hardegree and A. T. Tu), pp. 131–63. Marcel Dekker, New York.

Swaminathan, S., Furey, W., Pletcher, J., and Sax, M. (1992). Crystal structure of staphylococcal enterotoxin-B, a superantigen. *Nature*, **359**, 801–6.

Todd, J., Fishaut, M., Kapral, F., and Welch, T. (1978). Toxic shock syndrome associated with phage-group-1 staphylococci. *Lancet*, **ii**, 1116–18.

Tomai, M. A., Schlievert, P. M., and Kotb, M. (1992). Distinct T-cell receptor Vβ gene usage by human T lymphocytes stimulated with the streptococcal pyrogenic exotoxins and pep M5 protein. *Infect. Immunol.*, **60**, 701–4.

Toyonaga, B. and Mak, T. (1987). Genes of the T-cell antigen receptor in normal and malignant cells. *Annu. Rev. Immunol.*, **5**, 585–620.

Vallee, B. L. and Auld, D. S. (1990). Active-site zinc ligands and activated H_2O of zinc enzymes. *Proc. Natl. Acad. Sci. USA*, **87**, 220–4.

Vallee, B. L. and Falchuck, K. H. (1993). The biochemical basis of zinc physiology. *Physiol. Rev.*, **73**, 79–118.

Waanders, G. A., Lussow, A. R., and MacDonald, H. R. (1993). Skewed T cell receptor Vα repertoire among superantigen reactive murine T cells. *Int. Immunol.*, **5**, 55.

Wannamaker, L. W. and Schlievert, P. M. (1988). Exotoxins of group A streptococci, In *Handbook of bacterial toxins*, Vol. 4 (ed. C. M. Hardegree and A. T. Tu), pp. 267–95. Marcel Dekker, NewYork.

White, J., Herman, A., Pullen, A. M., Kubo, R., Kappler, J. W., and Marrack, P. (1989). The V beta-specific superantigen staphylococcal enterotoxin B: stimulation of mature T cells and clonal deletion in neonatal mice. *Cell*, **56**, 27–35.

White, J. A., Pullen, A., Choi, K., Marrack, P., and Kappler, J. W. (1993). Antigen recognition properties of mutant Vβ3+ T cell receptors are consistent with an immunoglobulin-like structure for the T cell receptor. *J. Exp. Med.*, **177**, 119–25.

Woodland, D. L., Happ, M. P., Gollob, K. J., and Palmer, E. (1991). An endogenous retrovirus mediating deletion of αβ T cells? *Nature*, **349**, 529–30.

Woodland, D. L., Smith, H. P., Surman, S., Le, P., Wen, R., and Blackman, M. (1993). Major histocompatibility complex-specific recognition of Mls-1 mediated by multiple elements of the T cell receptor. *J. Exp. Med.*, **177**, 433–42.

20 Biosynthesis of MHC products and its relevance for antigen presentation

HIDDE PLOEGH

1 Introduction

When T cells engage their target cells, they do so via a variety of surface receptors, amongst which are those that are specific for antigen. It has been known since the mid-seventies that this type of recognition is restricted by the products of the major histocompatibility complex (MHC). In the face of extensive polymorphism of the products encoded by the MHC, the T cell, as a rule, will engage in antigen-specific recognition only if the target presents the same (set) of MHC products that the T cell grew up with. T cells from one individual will therefore recognize the appropriate target cells from an unrelated individual only if they are matched with respect to the particular MHC products that the T cell has a preference for. The structure of the antigen-specific receptors on T cells, and what they recognize, is dealt with elsewhere in this volume. The present chapter aims to give an overview of the events that precede recognition of target cells by T cells. These events, which result in the generation of MHC molecules complexed with peptides, are referred to as antigen processing and presentation.

On the one hand, the solution at atomic resolution of the MHC class I structure has at once clarified the phenomenon of MHC restriction in molecular detail (Bjorkman *et al.* 1987*a,b*), on the other hand, progress in the area of cell biology has done much to promote our understanding of the relationship between the biosynthesis of MHC products and the generation of MHC peptide complexes (Neefjes 1991; Unanue 1992; Neefjes and Ploegh 1992).

2 Structure of MHC products

To appreciate the intricacies of the biosynthesis of MHC products, some insight into the structure of these products is required. For more complete accounts, the reader is referred to Bjorkman *et al.* (1987*a,b*) and Brown *et al.* (1993). Relevant for this chapter is the notion that peptide should be considered an essential structural subunit for MHC class I products. Given the strong structural similarity between MHC class I and MHC class II products (Brown *et al.* 1988, 1993), a similar role for peptide in maintaining the structure of MHC class II molecules may be proposed (Dornmair 1990; Germain and Hendrix 1991).

MHC class I proteins (Figure 17.2) consist of two subunits, a membrane embedded heavy chain, in tight but non-covalent association with the light chain, β2-microglobulin. The extracellular portion of the molecule can be divided into roughly two regions. The β2M and α3 domains are both immunoglobulin-like domains (even though their contacts are not strictly equivalent to those found in the immunoglobulins proper). These two Ig-like domains are membrane proximal, and support an eight stranded β-pleated sheet, topped by two α-helices. It is this portion of the molecule that binds peptide. The size of the pocket, formed by the α helices and the β-sheet floor (Bjorkman *et al.* 1987*a,b*) accommodates peptides of an average size of eight to ten residues. The cleft is closed at either end by bulky aromatic side chains, and a conserved salt bridge, respectively. This arrangement prevents peptides from protruding out of the cleft with their termini. When longer peptides are accommodated, they do so by bulging out in the middle (Guo *et al.* 1992). In terms of lodging the peptide into the peptide binding groove, there are important interactions with both the αNH_2 and αCOOH groups (obviously conserved amongst peptides, barring exceptional cases such as *N*-formylated peptides accommodated by non-classical class I gene products). Furthermore, the polymorphism of class I products is concentrated largely in the α1α2 domains which comprise the peptide binding groove. Thus, each allelic form of a class I molecule possesses a rather unique set of residues that line the peptide binding groove. In addition to 'pockets' that accommodate the NH_2 and COOH-termini, there are additional pockets that interact with specific side chains of the bound peptide. These interactions can be quite specific: some pockets preferentially accept an Arg residue (as is the case for the P2 position in HLA-B27) (Jardetzky *et al.* 1991) whereas other pockets have a preference for an aromatic side chain (as is the case for the pocket that accommodates Tyr or Phe at P5 in K^b binding peptides) (Rötzschke *et al.* 1990). The presence of such specific pockets manifests itself in the occurrence of so-called anchor residues: positions in the peptide sequence that are most frequently occupied by a particular amino acid side chain (Rötzschke and Falk 1991). In addition to these major pockets, the peptide binding groove is lined with depressions and protrusions (minor pockets) that likewise contribute to specific peptide–MHC interactions (Madden *et al.* 1991). A considerable part of the peptide is actually buried and is not solvent accessible, when assembled with a class I molecule (Fremont *et al.* 1992). The surface of the class I peptide complex that presents itself for recognition will thus be treated as a single contiguous surface by the T cell receptor. Not all polymorphic positions are confined to the peptide binding groove proper. Some polymorphic residues project their side chains away from the peptide, and are clearly available for interaction with T cell receptors. By imposing specificity on the peptide interaction, and by presenting unique features of the class I molecule to the T cell receptor, MHC polymorphism precludes interaction of T cells with inappropriately filled ('wrong antigen') or allogeneic ('wrong MHC allele') MHC products. The phenomenon of MHC restriction can thus be understood in molecular terms.

MHC class II proteins (HLA-DR1) have also been crystallized, and their structure has been solved (Brown *et al.* 1988, 1993). The class II structure is remarkably similar to that of a class I molecule (Figure 17.2), the main difference being the different arrangement of the functional domains over the two constituent subunits. Class II molecules consist of an α and a β chain, where each subunit contributes an Ig-like domain, and roughly half of the residues needed to construct the peptide binding groove. The architecture of this groove differs from that of class I, in that it lacks the occlusions found in the class I peptide binding site. Consequently, peptides can protrude from both ends of the groove, and bind to the class II protein via a central core. There are no 'pockets' that accommodate the free αNH_2 and αCOOH-termini. Consequently, longer peptides bind to class II molecules: their average size has been reported to be 14–16 residues. Pockets similar to those seen in class I molecules are present in HLA-DR1. It appears that many of the MHC side chains hydrogen bond to main chain atoms in the peptide and thus stabilize the interaction.

A most remarkable feature of the class II structure is the occurrence of a dimer of dimers. In the crystal structures solved, pairs of αβ dimers occur in an orientation that could easily be envisioned to exist at the cell surface as well. So far, no functional role for this dimerization has been established, but given the paradigm of ligand-induced receptor dimerization exemplified by the receptor for epidermal growth factor, the possibility of functional significance has received attention (Brown *et al.* 1993).

To appreciate the fundamental difference in presentation of peptides via class I and class II molecules, notwithstanding their similar overall structure, a few immunological facts must be presented here, which will be elaborated on later. First, presentation of antigen synthesized by the antigen-presenting cell itself can occur via class I molecules, and if it occurs, it is usually sensitive to agents that interfere with egress of proteins from the endoplasmic reticulum. Only under very special circumstances will presentation of exogenously delivered proteins via class I molecules be observed (Heemels and Ploegh 1993). In contrast, proteins administered via the endocytic pathway will be presented by class II molecules in a fashion that is usually sensitive to lysosomotropic drugs. This global distinction holds fairly well, and immediately suggests that the intracellular site where MHC class I and class II molecules pick up their cargo must differ. Much of the arguments presented below will show that this is indeed the case: the route travelled by MHC products determines the compartments these molecules can sample for peptides.

For the sake of the present chapter, we shall therefore consider the phenomena of biosynthesis of MHC products and antigen processing as two intimately connected series of events.

3 Biosynthesis of MHC class I products: the protein subunits

The mRNAs that specify the MHC class I subunits encode typical type I membrane proteins (for review, see Germain 1994). For class I heavy chains, there is a cleavable signal sequence, and a hydrophobic stop transfer signal that anchors the primary translation product in the membrane. The mRNA for β2M is also typical for a secretory protein: it too specifies a cleavable N-terminal signal sequence. Following translocation of the nascent chains into the lumen of the ER, co-translational glycosylation takes place for the heavy chain. By analogy with other type I membrane proteins (Braakman 1992), it may be assumed that folding on the lumenal side is dependent on ATP, commences prior to completion of the polypeptide chain, and includes the rapid formation of the intrachain disulfide bonds (Braakman 1992). The early stages of assembly of heavy and light chains have been explored by *in vitro* translation systems, supplemented with either dog pancreas microsomes (for the mouse class I products) (Ribaudo 1992; Bijlmakers *et al.* 1993), and microsomes prepared from human lymphoid cells (for human class I proteins) (Kvist and Hamann 1990). Particularly relevant here is the notion that peptide is the third structural subunit, a statement based largely on the analysis of cell lines defective in peptide transport (see below). At what time does this requirement for peptide first manifest itself?

Interestingly, for successful *in vitro* translation of mouse class I molecules, an absolute requirement for the inclusion of oxidized glutathione has been reported (Ribaudo 1992; Bijlmakers *et al.* 1993), as most translation systems are prepared and run under reducing conditions. However, the lumenal side of the ER maintains a far more oxidized redox environment owing to the selective import of oxidized GSSG (Hwang *et al.* 1992). No assembly of class I heavy chain and β2M was observed in the absence of oxidized GSSG *in vitro*. Heterodimers were detectable even in the complete absence of added peptide. In contrast, the human HLA-B27 molecule has been reported to assemble well in the absence of oxidized GSSG, but requires the co-translational presence of peptide to do so (Kvist and Hamann 1990). In living cells, newly synthesized class I molecules are found associated with calnexin, a protein that binds Ca^{2+} and has been proposed to act as a chaperonin (Degen and Williams 1991). The ER is a known storage site of Ca^{2+}, and several of the proteins that reside there may have been selected to function in the ER environment and may thus be strictly dependent on high local Ca^{2+} concentrations. In addition, Ca^{2+} is known to be important to allow egress of newly synthesized proteins from the ER and their delivery to the subsequent stations in the intercellular pathway (Beckers and Balch 1989). The association of newly synthesized class I molecules with calnexin is an early event and can be detected by chemical cross-linking in pulse chase experiments (mouse) (Degen and Williams 1991; Degen *et al.* 1992) or by co-immuno-precipitation as such (human) (Hochstenbach *et al.* 1992). The association with calnexin is not unique to MHC class I products: incompletely assembled TCR as well as class II molecules and a host of other proteins (Hochstenbach *et al.* 1992) associate with this chaperonin. While its function is not completely understood, it may act to slow down the folding and assembly process and thus allow each of the subunits sufficient time to search for its proper conformation (entry into the proper folding pathway) and subunit partner. The calnexin–class I interaction is destroyed at a time that coincides with the acquisition of complex-type oligosaccharides on the *N*-linked glycans borne by class I molecules. Recently it was suggested that calnexin may monitor protein folding by interacting with incompletely trimmed *N*-linked oligosaccharides (Hammond *et al.* 1994). The possible association of nascent and newly synthesized class I proteins should also be considered. While hard evidence is lacking, there have been suggestions of other proteins forming a complex with class I molecules in the ER. Prominent amongst them is Grp94, an ER resident chaperonin (Li and Srivastava 1993). By exploring peptide import into the ER (see below), it was observed that peptides interact with class I molecules in the ER in a manner that is reversible by inclusion of ATP (Schumacher *et al.* 1994). The protein(s) responsible for this ATP dependency remain to be identified, but these observations underscore the need for further study of the earliest stages of assembly of both MHC class I and class II molecules (Cresswell 1994).

The *N*-linked glycan transferred co-translationally to class I molecules may serve to guide their folding pathway: if *N*-linked glycosylation is suppressed by pharmacological agents such as tunicamycin, some class I molecules no longer associate properly with β2M, a finding most easily interpreted as a defectively folded class I heavy chain that is no longer competent to associate (Neefjes and Ploegh 1988). In the presence of agents that inhibit *N*-linked processing glycosidases and thus prevent the remodelling of *N*-linked glycans once transferred, no effects on intracellular transport or stability of class I molecules have been observed. This is perhaps not surprising, as all proteins that carry *N*-linked glycans essentially start out with the same glycan structure. Because folding of proteins is likely to be largely completed prior to exit from the ER, clearly the folding pathway of most proteins must be compatible with this type of high mannose oligosaccharides, and should not be dependent on subsequent modification of the oligosaccharide side chain. Following dissociation of calnexin and removal of the peripheral glucose and mannose residues, the protein has now entered the Golgi apparatus. In this organelle, glycan modifications continue and are complete by the time class I molecules exit from the *trans*-Golgi or *trans*-Golgi network. From the TGN, class I proteins are transferred to the cell surface, in a manner presumed to be characteristic of the constitutive secretory pathway. There is no evidence that specialized structures might be uniquely involved in ferrying class I molecules to the cell surface. The half-life of class I molecules at the cell surface has been studied in cell lines and shows considerable variation: it ranges from 3–16 hours. Breakdown of class I molecules has not been studied in detail, but is proposed to involve the loss of a subunit, either peptide or β2M, followed by internalization of the now unfolded heavy chain (Neefjes *et al.* 1993).

3.1 Where does peptide binding take place?

Based on the analysis of properties of class I molecules in mutant cells, it has been inferred that loading of class I molecules with peptides must be complete prior to acquisition of EndoH resistance. This statement is based on the observation that in such mutant cells, unable to engage in class I restricted antigen presentation by a defect in the machinery that delivers peptides to the site of assembly of class I molecules, the class I proteins are retained intracellularly, and largely in an endoglycosidase H sensitive form (Ljunggren *et al.* 1990). This enzyme will cleave high mannose-type oligosaccharides and sensitivity to the enzyme is essentially lost subsequent to the action of *N*-acetylglucosaminyltransferase, an enzyme that was initially positioned in the medial Golgi, but which may in fact show far less specialization in terms of its intracellular location, and could even be present in the *trans*-Golgi. Hence, the use of EndoH sensitivity as a tool to obtain topological information must be used with caution. It has been observed that class I molecules synthesized in cells exposed to the fungal antibiotic brefeldin A, which blocks egress of newly synthesized proteins from the ER, have the physicochemical and antigenic properties of peptide loaded class I molecules. In mouse cells infected with CMV, newly synthesized class I molecules no longer reach the cell surface, even though they are still peptide loaded (Del Val *et al.* 1992). Again, intracellular arrest takes place at the endoglycosidase H sensitive stage. Finally, there is a preliminary report concerning the intracellular localization of (a subunit of) the proposed peptide transporter (Kleijmeer *et al.* 1992). It is found in the ER, but a location at other intracellular sites can not be excluded at present. Along a different tack, the intracellular location of empty MHC class I molecules has been explored using immunocytochemistry (Hsu *et al.* 1991; Baas *et al.* 1992). The results suggest that empty class I molecules may penetrate into the Golgi, but probably not beyond. Combined, the data clearly suggest that peptide loading takes place prior to exit of class I molecules from the Golgi apparatus, and in all likelihood the process is largely completed in the ER.

3.2 What do we know about the peptide transporter?

The discovery of components involved in peptide loading of MHC molecules was made possible by the availability of mutants carrying large deletions in the MHC class II region. These mutants were unable to load their class I molecules with endogenous peptides. Taking inventory of the genes that map to this deletion, genes encoding proteins with remarkable similarity to previously identified ATP-dependent transporter molecules were discovered.

Independently, other workers examining the murine and rat MHC discovered the homologous genes independently. The two genes identified in the MHC are called TAP1 and TAP2 (DeMars and Spies 1992). They encode homologous membrane proteins that are members of the ATP binding cassette family of transporter molecules, to which also belong the multidrug resistance glycoprotein, the cystic fibrosis Cl^- channel, and the yeast Ste6 gene. Antibodies have been raised against synthetic peptides corresponding to the COOH-terminal portion of the molecule, as well as to the ATP binding domain expressed as a fusion product in bacteria (Cromme *et al.* 1994). For the human gene products, a physical association of the TAP1 and TAP2 product has been shown (Powis *et al.* 1991). The existence of higher order structures (e.g. $(TAP1)_2(TAP2)_2$) can not be excluded at present. The topology of the TAP1 and TAP2 subunits has not been addressed. Much remains to be learned about the mode of membrane insertion, and the possible association of the TAP1–TAP2 complex with other proteins.

By immunocytochemistry, the TAP1 gene product has been localized to the ER, but the actual levels of staining observed were such that the presence of at least TAP1 at other locations can not be excluded at present (Kleijmeer *et al.* 1992). The issue of involvement of ATP in the translocation of peptides, or even of TAP1 altogether has been raised also based on experiments in which peptide translocation into microsomal membranes was investigated (Heemels *et al.* 1993; Neefjes *et al.* 1993; Shepherd *et al.* 1993). Both human and canine microsomes apparently allow access of peptides to the lumenal side in the absence of ATP (Levy *et al.* 1991; Bijlmakers *et al.* 1993). Moreover, for human microsomes, exposure of microsomes to proteolytic digestion in no way impairs peptide translocation. Most surprisingly, there was no difference in the ability of microsomes from normal or TAP deficient cells to translocate peptides (Levy *et al.* 1991).

More recently, the involvement of TAP in peptide translocation was shown directly. In streptolysin-O permeabilized T2 cells, no delivery of a peptide containing an *N*-linked glycosylation site was observed. In T2 cells transfected with TAP cDNAs, such translocation was readily observed and was ATP dependent (Neefjes *et al.* 1993). By taking advantage of the TAP1 knock-out mouse as a negative control, a translocation assay for peptides was established using liver microsomes (Shepherd *et al.* 1993). These studies revealed the involvement of TAP, as well as the requirement for nucleoside triphosphates. The translocation assays all rely on the presence of a retention device that keeps peptide on the lumenal side of the ER, both in the semi-intact cell system and in the microsome-based system. This retention device can either be the bulky *N*-linked glycan, transferred to peptide on the lumenal side of the ER, or a class I binding site capable of interacting with the reporter translocation substrate. Most interestingly, the data suggest the existence not only of an ATP-dependent peptide import system, but ATP- and temperature-dependent efflux of peptide out of microsomes has also been observed (Schumacher *et al.* 1994). The latter phenomenon suggests the possibility that in living cells peptide may be refluxed continuously through the ER at low steady state concentration. This would ensure the preferential binding of high affinity peptides. Note that if TAP were to pump peptides into the ER against a concentration gradient, then abundant peptides of low affinity for class I could compete effectively for binding of less abundant, high affinity peptides.

Using the *in vitro* system, data have been obtained that suggest the COOH-terminal residue of a peptide to be a prime determinant in translocation (Momburg *et al.* 1994; Schumacher *et al.* 1994). In the mouse, only peptides terminating in I, L, M, V, Y, F, and W are translocated efficiently. It has been known for some time that all peptides extracted from murine class I molecules have an aliphatic C-terminal residue. Thus the specificity of the peptide transporter fits the specificity of the class I molecules remarkably well. While no extensive data have been published yet on the specificity of the human transporter, it would be predicted that it would at least encompass the C-termini already identified in class I bound peptides. These COOH-termini are far more diverse than their counterparts in the mouse (including L, I, K, R, Y, V) and suggest that the human TAP complex may have a broader specificity in this regard than the mouse. In terms of peptide length, the TAP complex appears selective, in that peptides shorter than eight residues do not compete efficiently for uptake of the appropriate reporter substrates (Schumacher *et al.* 1994).

While limited polymorphism has been observed for the TAP genes in mouse and man, the situation appears dramatically different in the rat, where a number of allelic forms, differing by as many as 25 positions, have been identified. Most interestingly, the rat transporters can be classified into two types: those that serve the rat class I molecule RT1Aa, and those that are less efficient at doing so. The less efficient transporters can not provide an adequate source of peptides to the RT1Aa molecule, and this deficiency ultimately leads to the intracellular arrest of the class I molecules along with their inability to present certain antigens to T cells of appropriate specificity. This phenomenon has been called the class I modifier, or cim phenomenon, for which two allelic forms, *cima* and *cimb* have been described (Livingstone 1989; Powis *et al.* 1992). Using microsomes from rats of *cima* and *cimb* type, it could be shown that the major difference between the two types of transporters again concerns the choice of C-terminal peptide residues (Heemels *et al.* 1993). The *cima* transporter is capable of translocating peptides with a C-terminal His, Lys, or Arg, while the *cimb* transporter can not. The available evidence is compatible with a role for other positions in the peptide sequence that contribute to the ability to serve as a TAP substrate, but the nature of these features has not yet been established.

3.3 Sources of peptides for MHC class I molecules

The peptides that are found associated with class I molecules endogenously are usually derived from proteins with a cytosolic disposition. The source proteins include both relatively abundant and minor proteins. In addition, mitochondrial proteins (encoded by mtDNA) have been identified as source proteins. The experiments of Bevan *et al.* have shown that any protein that can gain access to the cytosol is in principle a target for class I restricted antigen presentation (Moore *et al.* 1988). Cytoplasmic loading of ovalbumin can sensitize for lysis by $CD8^+$ cells. Generally, the presentation of any protein antigen by class I molecules is eliminated by disruptions in the TAP peptide loading machinery. Of late, a number of exceptions to this rule have emerged. RMA-S cells carry a mutation in the TAP2 gene and consequently are unable to load their mouse class I molecules (or human molecules produced following transfection) with peptide. Presentation of most viral antigens, consequently, is defective. However, given sufficiently high doses of virus, the presentation defect can be bypassed as has been shown for VSV (Hosken and Bevan 1992). In addition the T2 cell line transfected with $H\text{-}2K^b$ is perfectly capable of presenting Sendai virus, and does so in a BFA-insensitive manner (Zhou *et al.* 1993). For these cell lines, one should bear in mind that the patterns of membrane traffic and/or import of materials into the ER that they use may not be representative of normal cells. In particular the T2 cell line, arrived at by fusing a B lymphoblastoid cell and a T leukaemic cell line, may display peculiarities in this regard. The T2 cell line is also capable of presenting peptide epitopes that arise as the consequence of transfecting minigenes encoding signal sequence-less peptide epitopes (Anderson *et al.* 1991; Heemels and Ploegh 1993). In view of the complete deficiency of TAP products in this cell line, it is a remarkable result that implies either the existence of other transport pathways into the ER, or might depend on unusual pathways travelled by class I molecules (Zweerink *et al.* 1993). A further exploration of TAP deficient mice should prove illuminating, where the *in vivo* relevance of the various pathways is concerned. A very interesting finding concerns the presentation to $CD8^+$ T cells of antigens produced by bacteria. By feeding bacteria expressing ovalbumin to phagocytic cells, presentation to a $CD8^+$ T cell of the appropriate specificity could be obtained. This result suggested the interesting possibility of 'regurgitation' of peptides by phagocytic cells, to deposit them on neighbouring cells.

Combined, the current picture for class I restricted presentation would still be that the vast preponderance of peptides presented to $CD8^+$ cells is derived from cytosolic proteins. Importantly, the dependency on TAP is rather stringent, but may not be absolute. Any protein that can gain access to the lumen of the ER, or to the cytosol, is conceivably presentable as a peptide by class I molecules. Our thinking should even include specialized cases where phagocytic cells may participate in the generation of peptides. For the latter possibility, we note in passing that regurgitation of peptides and sensitization of innocent bystanders for lysis has often been looked at with very sensitive methods, and usually with negative results. It is a process that may well be confined to cellular interactions at close quarters, as they occur in lymph nodes.

Our knowledge about the mechanisms that are responsible for the generation of peptides is still limited. Protein breakdown must take place in the cytosol, and stratagems that enhance cytoplasmic breakdown—such as the attachment of ubiquitin—apparently enhance antigen presentation (Goldberg and Rock 1992). The ubiquitin-dependent degradation pathway makes use of proteasomes, but proteasomes are not strictly dependent on ubiquitination for them to function. Much interest was aroused by the discovery of genes in the MHC that encode polypeptides not only homologous to proteasome subunits, but even retrieved, by immunoprecipitation, in a complex with proteasomes (Monaco and McDevitt 1986). These genes are inducible by IFNγ, a treatment that enhances antigen presentation, possibly due to the combination of increased expression of the MHC subunits themselves, as well as increased representation of LMP-2 and LMP-7 containing proteasomes (Brown *et al.* 1991; Glynne *et al.* 1991). Proteasomes are capable of degrading a large variety of proteins in an ATP-dependent fashion, without pronounced substrate specificity (multicatalytic protease). Until it is known with certainty what types of substrates are handled by the TAP1–TAP2 complex and at what concentrations, we must reserve judgement on the relevant intermediates generated in the course of proteolysis. It is difficult to even guess what the intracellular concentrations of peptides suitable for delivery to class I molecules could be.

3.4 Complexity of peptides

The complexity of the sets of peptides generated must be extremely high: on the assumption that a given cell may express anywhere from 5000–20 000 distinct proteins, each of which could conceivably give rise to tens of different peptides, the enormity of this problem is readily apparent. There is considerable variation in the relative abundance of each of these proteins, and surely not all of them are broken down with similar kinetics. Of the most abundant peptides generated, many may not be capable of being delivered via the TAP1–TAP2 machinery, and even if they did, their ability to bind to class I molecules would depend on additional factors, such as sequence and length. From surface class I molecules, the complexity of the peptide mixture as it can be extracted suggests the presence of hundreds of peptides for a given class I allele, and with great variation in relative abundance. The ability of a $CD8^+$ T cell to be triggered by a small number of appropriately occupied class I molecules (the estimates vary from 10–200 molecules/cell), and the number of class 1 molecules on a typical target cell (100 000) indicate that for a productive interaction between CTL and target, no more than 0.2% of all class I molecules need to carry the relevant peptide epitope.

3.5 The act of peptide binding

Convenient binding assays with which to measure class I–peptide interactions have been developed based on the use of TAP deficient cell lines. Townsend and co-workers observed that the TAP2 deficient cell line RMA-S could be rendered strongly class I positive by incubation with peptides restricted by the K^b and D^b molecules expressed by this cell line. It is now thought that this peptide feeding effect is due largely to the capture of peptides by empty molecules as they emerge at the cell surface (Schumacher *et al.* 1990). The empty molecules can be stabilized not only by peptide, but also by a reduction in the temperature at which the cells are maintained (Ljunggren *et al.* 1990). Using 'empty' class I molecules, initially from TAP mutant cells but more recently produced in insect cells or bacteria, kinetic parameters of peptide binding have been explored. In principle, both free class I chains and empty molecules can bind peptide, but free chains do so with much lower affinity (due to fast off-rates). The act of binding peptide facilitates recruitment of β2M to generate a class I complex. There has been a long-standing discussion of the role of β2M in peptide binding. Most of these studies concern binding experiments in intact cells, and in any case using a starting population of class I molecules likely to contain both free chains and assembled 'empties'. The beneficial effect of β2M seen on peptide binding is most readily explained by the generation of a surplus of 'empties' that now engage in peptide binding. This issue has been discussed in detail by Eliott.

4 MHC class II molecules

MHC class II molecules consist of two membrane embedded, MHC encoded subunits. One of the more gratifying observations of the recent past has been the remarkable structural similarity between class I and class II molecules (Bjorkman *et al.* 1987*a,b*; Brown *et al.* 1988, 1993). The α1, α2 and β1 domains of class II are homologous to α1, β2M, and α2 of class I, and the β2 domain of class II is homologous to α3 in class I. Notwithstanding the striking resemblance in folding pattern, the architecture of the peptide binding groove is different, and so is the mode of peptide binding. Class II molecules contain a peptide binding groove that is open at both ends, and thus permits the binding of peptides that show wide variation in size. In addition, many of the interactions that stabilize the peptide–class II complex involve MHC side chains making contacts with the main chain atoms of the bound peptide. For the class II structure solved—HLA-DR1—there is a clear pocket that corresponds to the type of pocket seen accommodating the so-called peptide anchor residues in class I molecules. From the immunological point of view, perhaps the most striking feature is the strict functional dichotomy in cells that utilize class II molecules as restriction elements and that require the CD4 molecule as a co-receptor. The source of peptides presented by class II molecules almost invariably concerns proteins that can gain access to the endocytic pathway. Both proteins delivered exogenously, and proteins produced by the antigen-presenting cell itself and that make it into the endocytic pathway are potential targets for presentation. How do class II molecules gain access to the endocytic pathway?

4.1 Biosynthesis of class II molecules

MHC class II molecules are synthesized in the ER in a manner similar to class I molecules, but in the ER, they associate with a third polypeptide, a type II membrane protein referred to as the invariant chain (Ii) or γ chain (Kvist *et al.* 1982; Machamer and Cresswell 1982; Cresswell *et al.* 1987). The Ii chain fulfils at least three known functions.

1. It prevents peptide binding, as long as it is associated with the AB heterodimer (Roche and Cresswell 1990; Teyton *et al.* 1990).
2. It serves as a chaperone that somehow guides the folding of class II molecules and promotes their egress from the ER (Layet and Germain 1991; Anderson and Miller 1992; Schaiff *et al.* 1992).
3. It carries, in its cytoplasmic NH_2-terminus, the signal(s) necessary and sufficient to direct newly synthesized class II molecules to endocytic compartments (Bakke and Dobberstein 1990; Lotteau *et al.* 1990).

In vitro translation experiments have shown that the $\alpha\beta$ heterodimer can form only if proper disulfide bonding can be attained, and that the $\alpha\beta$ heterodimer can associate with Ii homotrimers (Bijlmakers *et al.* 1994). This association can involve newly synthesized $\alpha\beta$ heterodimers that latch on to pre-existing Ii homotrimers, or conversely, newly synthesized homotrimers can associate with pre-existing $\alpha\beta$ heterodimers. Peptide added in the course of translation can interact with newly synthesized $\alpha\beta$ heterodimers in the ER environment, and once a stable interaction between peptide and the $\alpha\beta$ heterodimers has formed, the resulting complex appears less able to recruit an invariant chain homotrimer. The latter attribute would presumably allow escape of ER loaded class II $\alpha\beta$ heterodimers from the endosomal targeting signals and could result in direct deposition of these complexes into the secretory pathway for transfer to the cell surface, bypassing endocytic compartments. This type of peptide loading may be relevant for class II restricted antigen presentation via the so-called endogenous pathway.

While the role of Ii in targeting newly synthesized class II molecules to endocytic compartments is rather well established, the identity of the proteins that interact with the cytoplasmic tail of Ii, or with the complex of Ii and $\alpha\beta$ cytoplasmic tails, remains elusive. Deletion mutants in which portions of the Ii cytoplasmic tail were progressively removed, provide clear evidence not only for an endosomal targeting signal, but also for an ER retention signal borne by a form of Ii that arises from an alternative translational start site (Bakke and Dobberstein 1990; Lotteau *et al.* 1990). In addition, the presence or absence of an alternatively spliced exon in the lumenal domain has been observed (Strubin *et al.* 1986; O'Sullivan *et al.* 1991), and its presence may have consequences for antigen presentation (Peterson and Miller 1992).

The manner by which Ii prevents peptide binding to the αβ heterodimer is still a matter of debate. The two most attractive possibilities are either direct occlusion of the peptide binding pocket by some portion of Ii, or an allosteric modification of the αβ heterodimer by interaction with Ii, without any direct contact of Ii and the peptide binding pocket proper. The data presently available do not allow a choice between these possibilities.

Peptide loading of class II molecules takes place after exit of the αβ–Ii complex from the Golgi. Much discussion has centred around the identity of the compartment in which peptide loading takes place. There is consensus on only one point; that this peptide loading compartment is considered part of the endocytic pathway (Cresswell 1985; Guagliardi *et al.* 1990; Neefjes *et al.* 1990; Lamb *et al.* 1991; Peters *et al.* 1991; Pieters *et al.* 1991). Class II molecules have been located by immunocytochemical methods in virtually all aspects of the endocytic pathway, from early endosomes to late endosomes to lysosomes. Perhaps most provocative have been the suggestions in favour of a novel intracellular compartment, first observed by immunoelectron microscopy in human B lymphoblastoid cells, and referred to as the MHC class II compartment, or MIIC (Peters *et al.* 1991). This compartment bears a number of characteristics usually found associated with the later aspects of the endocytic pathway, and appears distinct from lysosomes only in its unusually high MHC class II content. There is now evidence from subcellular fractionation studies that this compartment can in fact be physically separated either by density gradient centrifugation, or by free flow electrophoresis, from early and late endosomes as well as lysosomes. The further characterization of this compartment is an important future goal that should shed more light on the physiology of MHC class II restricted antigen presentation.

Peptide binding in the endocytic compartments can take place only after removal of Ii (Roche and Cresswell 1991). How much of Ii needs to be removed to allow peptide binding is not clear. There may be important cell type-specific differences in this regard, and it is not known if and how the rate of Ii breakdown could affect targeting of class II molecules to distinct subcompartments of the endocytic pathway, and hence determine the sets of peptides available at that location for binding to class II molecules. Once peptide binding, thought to be favoured by the locally acidic conditions (Jensen 1990), has taken place, class II molecules can reach the cell surface. Whether they do so constitutively, or whether specialized signals involving other endosomal proteins or even surface receptors play a role, is not yet known. The peptide bound state is conveniently scored for by the remarkable stability of peptide liganded class II molecules to high concentrations of SDS at ambient temperature (Springer *et al.* 1977; Germain and Hendrix 1991; Neefjes and Ploegh 1992; Stern and Wiley 1992). The structural features of class II molecules responsible for SDS stability have not been identified, but it is clear that not all peptide occupied class II molecules are necessarily SDS stable (Riberdy *et al.* 1992).

The proteases involved in generating antigenic peptides are almost certainly the major lysosomal proteases, cathepsins B and D (Buus and Werdelin 1986; Puri and Factorovich 1988; Takahashi *et al.* 1989; Diment 1990; van Noort *et al.* 1991; Michalek *et al.* 1992). They play a role not only in degradation of Ii, but also in the generation of the class II peptide ligands. A role for cathepsin E has been invoked, based on the results obtained with a selective cathepsin E inhibitor (Bennett *et al.* 1992). Proteolysis may impinge on class II peptide loading at several levels. If the rate of Ii breakdown and its consequent release from class II αβ heterodimers controls the rate at which the αβ molecules reach their destination, or perhaps even their final destination itself, then the rate and mode of proteolysis (relative amount of the enzymes involved) could be of considerable importance. It is likely that not all APC possesses the same relative content of these proteolytic enzymes (some of them, such as cathepsin E, have a fairly restricted tissue distribution). The application of gene disruption technology will surely allow the generation of animals with defined deficiencies in lysosomal protease content, and will shed light on the identity and roles of the proteases involved in Ii breakdown and antigen processing.

An unresolved issue is the route travelled by class II molecules to the processing compartment, and thence to the cell surface. Biosynthetically labelled class II molecules can be addressed by biosynthetic tracers, and it has been claimed that this can occur prior to surface deposition of class II molecules (Cresswell 1985; Neefjes *et al.* 1990). At the same time, there are reports that document the presence of class II-associated Ii at the cell surface (Wraight *et al.* 1990; Koch *et al.* 1991; Roche *et al.* 1993). These complexes still possess the targeting signal(s) borne by Ii, and consequently could still be internalized and delivered to the appropriate endocytic compartments. Based on the published evidence, it can not even be excluded that the cell surface is an obligate way-station in the routing of class II molecules to their final destination. A transient appearance at the cell surface, followed by rapid internalization, would be difficult to detect with the methods brought to bear on the issue to date. This is an obvious target for further study, and has relevance not only for targeting of class II molecules *per se*, but also for the more general issues concerning origin and maintenance of endocytic compartments.

5 What are some of the main questions outstanding?

Clearly one of the main items on the agenda would be to obtain more structural data, first and foremost for the T cell receptor. The efforts now underway in a number of laboratories may reasonably be expected to bear fruit within the foreseeable future. The availability of a TCR structure, preferably with its attendant co-receptors, should allow the design of further experiments that probe the role of TCR–MHC interactions, and perhaps examine the role of MHC–peptide complexes not only during effector phase recognition, but also during T cell development. In addition, the structure of MHC class I molecules devoid of peptide, and the structure of MHC class II molecules in a complex with Ii are obvious targets.

For the biogenesis of class I molecules, key questions concern the exact mode of peptide loading in the ER environment. Are there chaperonin-type proteins, in addition to calnexin, that play a role? Furthermore, the available assays for TAP-dependent peptide translocation should put to rest the controversies concerning the possible involvement of polymorphisms in TAP as factors that contribute to autoimmune diseases. Very little is known about the mechanistic aspects of TAP-dependent peptide transport, and quantitative data on variables such as cytosolic and ER lumenal peptide concentrations, and their effects on class I assembly are altogether lacking. While there is an emerging consensus that the multicatalytic protease or proteasomes are likely involved in the generation of peptides presented by class I molecules (and perhaps in exceptional circumstances by class II molecules), this point remains to be more firmly supported by hard experimental evidence.

The role for the Ii chain in class II restricted antigen presentation and in trafficking of class II molecules remains a very actively studied issue. Thus far, little attention has been paid to the professional APC in this regard, most of the work having been done with established cell lines, mostly of B cell lineage, and with transfectants. Nothing is known about the proteins that interact with Ii and that must be responsible for some of the sorting events. The availability of semi-intact cell systems, or even complete *in vitro* trafficking assays, has not yet had any impact on these particular questions.

Finally, it is clear that not all the components necessary for antigen processing and presentation have been identified. The usefulness of mutant cell lines, such as those contributed by DeMars and co-workers, or Karre *et al.* can not be overstated. Other genes in the class II region of the MHC have already been implicated in class II restricted antigen presentation. Establishing further mutant cell lines, or mutant strains of mice with defects in antigen processing and presentation, will surely provide a resource for biochemists, cell biologists, and immunologists alike.

Note added in proof

For recent, more extensive reviews the reader is referred to Heemels and Ploegh (1995) and Wolf and Ploegh (1995).

References

Anderson, M. S. and Miller, J. (1992). Invariant chain can function as a chaperone protein for class II major histocompatibility complex molecules. *Proc. Natl. Acad. Sci. USA*, **89**, 2282–6.

Anderson, K., Cresswell, P., Gammon, M., Hermes, J., Williamson, A., and Zweerink, H. (1991). Endogenously synthesized peptide with an endoplasmic reticulum signal sequence sensitizes antigen processing mutant cells to class I restricted cell mediated lysis. *J. Exp. Med.*, **174**, 489–92.

Baas, E. J., van Santen, H. M., Kleijmeer, M. J., Geuze, H. J., Peters, P. J., and Ploegh, H. L. (1992). Peptide-induced stabilization and intracellular localization of empty HLA class I complexes. *J. Exp. Med.*, **176**, 147–56.

Bakke, O. and Dobberstein, B. (1990). MHC class II-associated invariant chain contains a sorting signal for endosomal compartments. *Cell*, **63**, 707–16.

Beckers, C. J. M. and Balch, W. E. (1989). Calcium and GTP: essential components in vesicular trafficking between the endoplasmic reticulum and golgi apparatus. *J. Cell Biol.*, **108**, 1245–56.

Bennett, K., Levine, T., Ellis, J. S., Peanasky, R. J., Samloff, I. M., Kay, J., *et al.* (1992). Antigen processing for presentation by class II major histocompatibility complex requires cleavage by cathepsin E. *Eur. J. Immunol.*, **22**, 1519.

Bijlmakers, M. J. E., Neefjes, J. J., Wojcik-Jacobs, E. H. M., and Ploegh, H. L. (1993). The assembly of H2-K^b class I molecules translated *in vitro* requires oxidized glutathione and peptide. *Eur. J. Immunol.*, **23**, 1305–13.

Bjorkman, P. J., Saper, M. A., Samraoui, B., Bennett, W. S., Strominger, J. L., and Wiley, D. C. (1987*a*). The foreign antigen binding site and T cell recognition regions of class I histocompatibility antigen, HLA2. *Nature*, **329**, 512–18.

Bjorkman, P. J., Saper, M. A., Samraoui, B., Bennett, W. S., Strominger, J. L., and Wiley, D. C. (1987*b*). Structure of the human class I histocompatibility antigen HLA-A2. *Nature*, **329**, 506–12.

Braakman, I., Helenius, J., and Helenius, A. (1992*a*). Role of ATP and disulphide bonds during protein folding in the endoplasmic reticulum. *Nature*, **356**, 260–2.

Braakman, I., Helenius, J., and Helenius, A. (1992*b*). Manipulating disulfide formation and protein folding in the endoplasmic reticulum. *EMBO J.*, **11**, 1717–22.

Brown, J. H., Jardetzky, T., Saper, M. A., Samraoui, B., Bjorkman, P. J., and Wiley, D. C. (1988). A hypothetical model of the foreign antigen binding site of class II histocompatibility molecules. *Nature*, **332**, 845–50.

Brown, M. G., Driscoll, J., and Monaco, J. J. (1991). Structural and serological similarity of MHC linked LMP and proteasome (multicatalytic proteinase) complexes. *Nature*, **353**, 355–7.

Brown, J. H., Jardetzky, T. S., Gorga, J. C., Stern, L. J., Urban, R. G., Strominger, J. L., *et al.* (1993). Three-dimensional structure of the human class II histocompatibility antigen HLA-DR1. *Nature*, **364**, 33–9.

Buus, S., Sette, A., Colon, S. M., Jenis, D. M., and Grey, H. M. (1986). Isolation and characterization of antigen-la complexes involved in T cell recognition. *Cell*, **47**, 1071–7.

Cresswell, P. (1985). Intracellular class II HLA antigens are accessible to transferrin-neuraminidase conjugates internalized by receptor-mediated endocytosis. *Proc. Natl. Acad. Sci. USA*, **82**, 8188–92.

Cresswell, P. (1994). Assembly, transport and function of MHC class II molecules. *Annu. Rev. Immunol.*, **12**, 259–93.

Cresswell, P., Blum, J. S., Kelner, D. N., and Marks, M. S. (1987). Biosynthesis and processing of class II histocompatibility antigens. *CRC Crit. Rev. Immunol.*, **7**, 31–53.

Cromme, F. V., Airey, J., Heemels, M.-T., Ploegh, H. L., Meijer, C. J., and Walboomers, J. M. (1994). Loss of transporter protein, encoded by the TAP-1 gene is highly correlated with loss of HLA expression in malignant cells. *J. Exp. Med.*, **179**, 335–40.

Degen, E. and Williams, D. B. (1991). Participation of a novel 88-kD protein in the biogenesis of murine class I histocompatibility molecules. *J. Cell. Biol.*, **112**, 1099–115.

Degen, E., Cohen-Doyle, M., and Williams, D. (1992). Efficient dissociation of the p88 chaperone from major histocompatibility complex class I molecules requires both β2-microglobulin and peptide. *J. Exp. Med.*, **175**, 1653–61.

Del Val, M., Hengel, H., Hacker, H., Hartlaub, U., Ruppert, T., Lucin, P., *et al.* (1992). Cytomegalovirus prevents antigen presentation by blocking transport of peptide-loaded major histocompatibility complex class I molecules into the medial golgi compartment. *J. Exp. Med.*, **176**, 729.

DeMars, R. and Spies, T. (1992). New genes in the MHC that encode proteins for antigen processing. *Trends Cell. Biol.*, **2**, 81–6.

Diment, S. (1990). Different roles for thiol and aspartyl proteases in antigen presentation of ovalbumin. *J. Immunol.*, **145**, 417.

Dornmair, K. and McConnell, H. M. (1990). Refolding and reassembly of separate α and β chains of class II molecules of the major histocompatibility complex leads to increased peptide-binding capacity. *Proc. Natl. Acad. Sci. USA*, **87**, 4134–8.

Fremont, D. H., Matsumura, M., Stura, E. A., Peterson, P. A., and Wilson, I. A. (1992). Crystal structures of two viral peptides in complex with murine MHC class I H-2K^b. *Science*, **257**, 919–27.

Germain, R. N. (1994). MHC-dependent antigen processing and peptide presentation: providing ligands for T lymphocyte activation. *Cell*, **76**, 287–99.

Germain, R. N. and Hendrix, L. R. (1991). MHC class II structure, occupancy and surface expression determined by post-endoplasmic reticulum antigen binding. *Nature*, **353**, 134–9.

Glynne, R., Powis, S. H., Beck, S., Kelly, A., Kerr, L.-A., and Trowsdale, J. (1991). A proteasome-related gene between the two ABC transporter loci in the class II region of the human MHC. *Nature*, **353**, 357–60.

Goldberg, A. L. and Rock, K. L. (1992). Proteolysis, proteasomes and antigen presentation. *Nature*, **357**, 375–9.

Guagliardi, L. E., Koppelman, B., Blum, J. S., Marks, M. S., Cresswell, P., and Brodsky, F. M. (1990). Co-localization of molecules involved in antigen processing and presentation in an early endocytic compartment. *Nature*, **343**, 133–9.

Guo, H., Jardetzky, T. S., Garrett, T. P. J., Lane, W. S., Strominger, J. L., and Wiley, D. C. (1992). Different length peptides bind to HLA-Aw68 similarly at their ends but bulge out in the middle. *Nature*, **360**, 364–6.

Hammond, C., Braakman, I., and Helenius, A. (1994). Role of N-linked oligosaccharide recognition, glucose trimming, and calnexin in glycoprotein folding and quality control. *Proc. Natl. Acad. Sci. USA*, **91**, 913–17.

Heemels, M.-T. and Ploegh, H. L. (1993). Untapped peptides. *Curr. Biol.*, **3**, 380–3.

Heemels, M. T. and Ploegh, H. L. (1995). Generation, translocation and presentation of MHC Class I-restricted peptides. *Ann Rev. Biochem.*, **64**, 463–91.

Heemels, M.-T., Schumacher, T. N. M., Wonigeit, K., and Ploegh, H. L. (1993). Different peptides are translocated by allelic variants of the transporter associated with antigen presentation (TAP). *Science*, **262**, 2059–63.

Hochstenbach, F., David, V., Watkins, C., and Brenner, M. B. (1992). *Proc. Natl. Acad. Sci. USA*, **89**, 4734.

Hosken, N. A. and Bevan, M. J. (1992). An endogenous antigenic peptide bypasses the class I antigen presentation pathway of RMA-S. *J. Exp. Med.*, **175**, 719–29.

Hsu, V. W., Yuan, L. C., Nuchtern, J. G., Lippincott-Schwartz, J., Hammerling, G. J., and Klausner, R. D. (1991). A recycling pathway between the endoplasmic reticulum and the Golgi apparatus for retention of unassembled MHC class I molecules. *Nature*, **352**, 441–4.

Hwang, C., Sinskey, A., and Lodish, H. F. (1992). Oxidized redox state of glutathione in the endoplasmic reticulum. *Science*, **257**, 1496–502.

Jackson, M. R., Song, E. S., Yang, Y., and Peterson, P. A. (1992). Empty and peptide containing conformers of class I major histocompatibility complex molecules expressed in *Drosophila melanogaster* cells. *Proc. Natl. Acad. Sci. USA*, **89**, 12117–21.

Jardetzky, T. S., Lane, W. S., Robinson, R. A., Madden, D. R., and Wiley, D. C. (1991). Identification of self peptides bound to purified HLA-B27. *Nature*, **353**, 326.

Jensen, P. E. (1990). Regulation of antigen presentation by acidic pH. *J. Exp. Med.*, **171**, 1779–84.

Kleijmeer, M., Kelly, A., Geuze, H. J., Slot, J. W., Townsend, A., and Trowsdale, J. (1992). Location of MHC encoded transporters in the endoplasmic reticulum and cis-golgi. *Nature*, **357**, 342–4.

Koch, N., Moldenhauer, G., Hofmann, W. J., and Moller, P. (1991). Rapid intracellular pathway gives rise to cell surface expression of the MHC II-associated invariant chain (CD74). *J. Immunol.*, **147**, 2643.

Kvist, S. and Hamann, U. (1990). A nucleoprotein peptide of influenza A virus stimulates assembly of HLA-B27 class I heavy chains and β2-microglobulin translated *in vitro*. *Nature*, **348**, 446–8.

Kvist, S., Wiman, K., Claesson, L., Peterson, P. A., and Dobberstein, B. (1982). Membrane insertion and oligomeric assembly of HLA-DR histocompatibility antigens. *Cell*, **29**, 61–9.

Lamb, C. A., Yewdell, J. W., Bennink, J. R., and Cresswell, P. (1991). Invariant chain targets HLA class II molecules to acidic endosomes containing internalized influenza virus. *Proc. Natl. Acad. Sci. USA*, **88**, 5998–6002.

Layet, C. and Germain, R. N. (1991). Invariant chain promotes egress of poorly expressed, haplotype-mismatched class II major histocompatibility complex AαAβ dimers from the endoplasmic reticulum/cis-Golgi compartment. *Proc. Natl. Acad. Sci. USA*, **88**, 2346–50.

Levy, F., Gabathuler, R., Larsson, R., and Kvist, S. (1991). ATP is required for *in vitro* assembly of MHC class I antigens but not for transfer of peptides across the ER membrane. *Cell*, **67**, 265–74.

Li, Z. and Srivastava, P. K. (1993). Tumor rejection antigen gp96/grp94 is an ATPase: implications for protein folding and antigen presentation. *EMBO J.*, **12**, 3143–51.

Livingstone, A. M. (1989). A trans-acting major histocompatibility complex-linked gene whose alleles determine gain and loss changes in the antigenic structure of a classical class I molecule. *J. Exp. Med.*, **170**(3), 777–95.

Ljunggren, H. G., Stam, N. S., Ohlen, C., Neefjes, J. J., Hoglund, P., Heemels, M. T., *et al.* (1990). Empty MHC class I molecules come out in the cold. *Nature*, **346**, 476–80.

Lotteau, V., Teyton, L., Peleraux, A., Nilsson, T., Karlsson, L., Schmid, S. L., *et al.* (1990). Intracellular transport of class II MHC molecules directed by invariant chain. *Nature*, **348**, 600–5.

Machamer, C. E. and Cresswell, P. (1982). Biosynthesis and glycosylation of the invariant chain associated with HLA-DR antigens. *J. Immunol.*, **129**, 2564–9.

Madden, D. R., Gorga, J. C., Strominger, J. L., and Wiley, D. C. (1991). The structure of HLA-B27 reveals nonamer self-peptides bound in an extended conformation. *Nature*, **353**, 321–5.

Michalek, M., Benacerraf, B., and Rock, K. (1992). The class II MHC-restricted presentation of endogenously synthesized ovalbumin displays clonal variation, requires endosomal/lysomal processing, and is up-regulated by heat shock. *J. Immunol.*, **148**(4), 1016–24.

Momburg, F., Roelse, J., Howard, J. C., Butcher, G. W., Hammerling, G., and Neefjes, J. J. (1994). Selectivity of MHC encoded peptide transporters from human, mouse and rat. *Nature*, **367**, 648–51.

Monaco, J. J. and McDevitt, H. O. (1986). The LMP antigens: a stable MHC controlled subunit protein complex. *Hum. Immunol.*, **75**, 416–26.

Moore, M. W., Carbone, F. R., and Bevan, M. J. (1988). Introduction of soluble protein into the class I pathway of antigen processing and presentation. *Cell*, **54**, 777–85.

Neefjes, J. J. and Ploegh, H. L. (1988). Allele and locus specific differences in cell surface expression and the association of HLA class I heavy chain with β2-microglobulin: differential effects of inhibition of glycosylation on class I subunit association. *Eur. J. Immunol.*, **18**, 801–10.

Neefjes, J. J., Schumacher, T. N. M., and Ploegh, H. L. (1991). Assembly and intracellular transport of major histocompatibility complex molecules. *Curr. Opinion. Cell. Biol.*, **3**, 601–9.

Neefjes, J. J., and Ploegh, H. L. (1992). Intracellular transport of MHC class II molecules. *Immunol. Today*, **13**(5), 179–84.

Neefjes, J. J. and Ploegh, H. L. (1992). Inhibition of endosomal proteolytic activity by leupeptin blocks surface expression of MHC class II molecules and their conversion to SDS resistant αβ heterodimers. *EMBO J.*, **11**, 411–16.

Neefjes, J. J., Stollorz, V., Peters, P. J., Geuze, H. J., and Ploegh, H. L. (1990). The biosynthetic pathway of MHC class II but not class I molecules intersects the endocytic route. *Cell*, **61**, 171–83.

Neefjes, J. J., Momburg, F., and Hammerling, G. (1993). Selective and ATP-dependent translocation of peptides by the MHC-encoded transporter. *Science*, **261**, 769–71.

O'Sullivan, D., Arrhenius, T., Sidney, J., Del Guercio, M.-F., Albertson, M., Wall, M., *et al.* (1991). On the interaction of promiscuous antigenic peptides with different DR alleles. Identification of common structural motifs. *J. Immunol.*, **147**, 2663.

Peters, P. J., Neefjes, J. J., Oorschot, V., Ploegh, H. L., and Geuze, H. J. (1991). Segregation of MHC class II molecules from MHC class I molecules in the golgi complex for transport to lysosomal compartments. *Nature*, **349**, 669–76.

Peterson, M. and Miller, J. (1992). Antigen presentation enhanced by the alternatively spliced invariant chain gene product p41. *Nature*, **357**, 596–8.

Pieters, J., Horstmann, H., Bakke, O., Griffiths, G., and Lipp, J. (1991). Intracellular transport and localization of Major Histocompatibility Class II molecules and associated invariant chain. *J. Cell Biol.*, **115**, 1213–23.

Powis, S. J., Townsend, A., Deverson, E. V., Bastin, J., Butcher, G. W., and Howard, J. C. (1991). Restoration of antigen presentation to the mutant cell line RMA-S by an MHC-linked transporter. *Nature*, **354**, 528–31.

Powis, S. J., Deverson, E. V., Coadwell, W. J., Ciruela, A., Huskisson, N. S., Smith, H., *et al.* (1992). Effect of polymorphism of an MHC-linked transporter on the peptides assembled in a class I molecule. *Nature*, **357**, 211–15.

Puri, J. and Factorovich, Y. (1988). Selective inhibition of antigen presentation to cloned cells by protease inhibitors. *J. Immunol.*, **141**, 3313–17. [Published erratum appeared in *Immunol.*, **142**(5), 1781.]

Ribaudo, R. K. and Margulies, D. H. (1992). Independent and synergistic effects of disulfide bond formation, β2-microglobulin, and peptides on MHC folding and assembly in an *in vitro* translation system. *J. Immunol.*, **149**, 2935–44.

Riberdy, J. M., Newcomb, J. R., Surman, M. J., Barbosa, J. A., and Cresswell, P. (1992). HLA-DR molecules from an antigen-processing mutant cell line are associated with invariant chain peptides. *Nature*, **360**, 474–6.

Roche, P. A. and Cresswell, P. (1990). Invariant chain association with HLA-DR molecules inhibits immunogenic peptide binding. *Nature*, **345**, 615–18.

Roche, P. A. and Cresswell, P. (1991). Proteolysis of the class II-associated invariant chain generates a peptide binding site in intracellular HLA-DR molecules. *Proc. Natl. Acad. Sci. USA*, **88**, 3150–4.

Roche, P. A., Teletski, C. L., Stang, E., Bakke, O., and Long, E. O. (1993). Cell surface HLA-DR invariant chain complexes are targeted to endosomes by rapid internalization. *Proc. Natl. Acad. Sci. USA*, **90**, 8581–5.

Rotzschke, O. and Falk, K. (1991). Naturally occurring peptide antigens derived from the MHC class-I-restricted processing pathway. *Immunol. Today*, **12**, 447–55.

Rotzschke, O., Falk, K., Wallny, H. J., Faath, S., and Rammensee, H. G. (1990). Characterization of naturally occurring minor histocompatibility peptides including H-4 and H-Y. *Science*, **249**, 283.

Schaiff, W. T., Hruska, K. A., McCourt, D. W., Green, M., and Schwartz, B. D. (1992). HLA-DR associates with specific stress proteins and is retained in the endoplasmic reticulum in invariant chain negative cells. *J. Exp. Med.*, **176**, 657–66.

Schumacher, T. N. M., Heemels, M.-T., Neefjes, J. J., Kast, W. M., Melief, C. J. M., and Ploegh, H. L. (1990). Direct binding of peptide to empty MHC class I molecules on intact cells and *in vitro*. *Cell*, **62**, 563–7.

Schumacher, T. N. M., Kantesaria, D. V., Heemels, M.-T., Ashton-Rickardt, P. G., Shepherd, J. C., Fruh, K., *et al.* (1994). Peptide length and sequence specificity of the mouse TAP1/TAP2 translocator. *J. Exp. Med.* **179**, 533–40.

Shepherd, J. C., Schumacher, T. N. M., Ashton-Rickardt, P. G., Imaeda, S., Ploegh, H. L., Janeway, C., *et al.* (1993). TAP1-dependent peptide translocation *in vitro* is ATP dependent and peptide selective. *Cell*, **74**, 577–84.

Springer, T. A., Kaufman, J. F., Siddoway, L. A., Mann, D. L., and Strominger, J. L. (1977). *J. Biol. Chem.*, **252**, 6201–7.

Stern, L. J. and Wiley, D. C. (1992). The human class II MHC protein HLA-DR1 assembles as empty αβ heterodimers in the absence of antigenic peptide. *Cell*, **68**, 465–77.

Strubin, M., Berte, C., and Mach, B. (1986). Alternative splicing and alternative initiation of translation explain the four forms of the Ia antigen-associated invariant chain. *EMBO J.*, **5**, 3483–8.

Takahashi, H., Cease, K., and Berzofsky, J. (1989). *J. Immunol.*, **142**, 2221.

Teyton, L., O'Sullivan, D., Dickson, P. W., Lotteau, V., Sette, A., Fink, P., *et al.* (1990). Invariant chain distinguishes between the exogenous and endogenous antigen presentation pathways. *Nature*, **348**, 39–44.

Unanue, E. R. (1992). Cellular studies on antigen presentation by class II MHC molecules. *Curr. Opin. Immunol.*, **4**, 63.

van Noort, J. M., Boon, J., van der Drift, A. C. M., Wagenaar, J. P. A., Boots, A. M. H., and Boog, C. J. P. (1991). Antigen processing by endosomal proteases determines which sites of sperm-whale myoglobin are eventually recognized by T cells. *Eur. J. Immunol.* **21**, 1989.

Wolf, P. and Pleogh, H. L. (1995). Class II restricted antigen presentation. *Ann. Rev. Cell Biol.* (in press).

Wraight, C. J., Van Endert, P., Möller, P., Lipp, J., Ling, N. R., MacLennan, I. C. M., *et al.* (1990). Human major histocompatibility complex class II invariant chain is expressed on the cell surface. *J. Biol. Chem.*, **265**, 5787–92.

Zhou, X., Glas, R., Ljunggren, H. G., and Jondal, M. (1993). Antigen processing mutant T2 cells present viral antigen restricted through H2-K^b. *Eur. J. Immunol.* **23**, 1796–1801.

Zweerink, H. J., Gammon, M. C., Utz, U., Sauma, S. Y., Harrer, T., Hawkins, J. C., *et al.* (1993). Presentation of endogenous peptides to MHC class I restricted cytotoxic T lymphocytes in transport deletion mutant T2 cells. *J. Immunol.*, **150**, 1763–71.

Index

Page numbers in *italics* refer to figures and tables